Courtaney

W9-BHK-350

Atomic Masses and Numbers

Name	Symbol	Atomic number	Atomic mass	Name	Symbol	Atomic number	Atomic mass
Actinium	Ac	89	227.0278	Molybdenum	Mo	42	95.94
Aluminium	Al	13	26.98154	Neodymium	Nd	60	144.24
Americium	Am	95	(243)	Neon	Ne	10	20.179
Antimony	Sb	51	121.75	Neptunium	Np	93	237.0482
Argon	Ar	18	39.948	Nickel	Ni	28	58.70
Arsenic	As	33	74.9216	Niobium	Nb	41	92.9064
Astatine	At	85	(210)	Nitrogen	N	7	14.0067
Barium	Ba	56	137.33	Nobelium	No	102	(259)
Berkelium	Bk	97	(247)	Osmium	Os	76	190.2
Beryllium	Be	4	9.01218	Oxygen	O	8	15.9994
Bismuth	Bi	83	208.9804	Palladium	Pd	46	106.4
Boron	B	5	10.81	Phosphorus	P	15	30.97376
Bromine	Br	35	79.904	Platinum	Pt	78	195.09
Cadmium	Cd	48	112.41	Plutonium	Pu	94	(244)
Caesium	Cs	55	132.9054	Polonium	Po	84	(209)
Calcium	Ca	20	40.08	Potassium	K	19	39.0983
Californium	Cf	98	(251)	Praseodymium	Pr	59	140.9077
Carbon	C	6	12.011	Promethium	Pm	61	(145)
Cerium	Ce	58	140.12	Protactinium	Pa	91	231.0359
Chlorine	Cl	17	35.453	Radium	Ra	88	226.0254
Chromium	Cr	24	51.996	Radon	Rn	86	(222)
Cobalt	Co	27	58.9332	Rhenium	Re	75	186.207
Copper	Cu	29	63.546	Rhodium	Rh	45	102.9055
Curium	Cm	96	(247)	Rubidium	Rb	37	85.4678
Dysprosium	Dy	66	162.50	Ruthenium	Ru	44	101.07
Einsteinium	Es	99	(252)	Samarium	Sm	62	150.4
Erbium	Er	68	167.26	Scandium	Sc	21	44.9559
Europium	Eu	63	151.96	Selenium	Se	34	78.96
Fermium	Fm	100	(257)	Silicon	Si	14	28.0855
Fluorine	F	9	18.998403	Silver	Ag	47	107.868
Francium	Fr	87	(223)	Sodium	Na	11	22.98977
Gadolinium	Gd	64	157.25	Strontium	Sr	38	87.62
Gallium	Ga	31	69.72	Sulfur	S	16	32.06
Germanium	Ge	32	72.59	Tantalum	Ta	73	180.9479
Gold	Au	79	196.9665	Technetium	Tc	43	(98)
Hafnium	Hf	72	178.49	Tellurium	Te	52	127.60
Helium	He	2	4.00260	Terbium	Tb	65	158.9254
Holmium	Ho	67	164.9304	Thallium	Tl	81	204.37
Hydrogen	H	1	1.0079	Thorium	Th	90	232.0381
Indium	In	49	114.82	Thulium	Tm	69	168.9342
Iodine	I	53	126.9045	Tin	Sn	50	118.69
Iridium	Ir	77	192.22	Titanium	Ti	22	47.90
Iron	Fe	26	55.847	Tungsten (Wolfram)	W	74	183.85
Krypton	Kr	36	83.80	(Unnilhexium)	(Unh)	106	(263)
Lanthanum	La	57	138.9055	(Unnilpentium)	(Unp)	105	(262)
Lawrencium	Lr	103	(260)	(Unnilquadium)	(Unq)	104	(261)
Lead	Pb	82	207.2	Uranium	U	92	238.029
Lithium	Li	3	6.941	Vanadium	V	23	50.9415
Lutetium	Lu	71	174.967	Xenon	Xe	54	131.30
Magnesium	Mg	12	24.305	Ytterbium	Yb	70	173.04
Manganese	Mn	25	54.9380	Yttrium	Y	39	88.9059
Mendelevium	Md	101	(258)	Zinc	Zn	30	65.38
Mercury	Hg	80	200.59	Zirconium	Zr	40	91.22

SOURCE: Adapted from *Pure Appl. Chem.* **51**, 405 (1979). Values in parentheses are for nonnaturally occurring elements and are the mass numbers of the longest lived isotope of the element.

JOHN W. LEHMAN

Lake Superior State College

OPERATIONAL ORGANIC CHEMISTRY

A Laboratory Course

Allyn and Bacon, Inc.   Boston   London   Sydney   Toronto

To My Parents

Library of Congress Cataloging in Publication Data

Lehman, John W
 Operational organic chemistry.

 Bibliography: p.
 Includes index.
 1. Chemistry, Organic—Experiments. I. Title.
QD261.L39   547'.0028   80-28118
ISBN 0-205-07146-5

Printed in the United States of America.

10 9 8 7   85

Production editors: David Dahlbacka and Greg Giblin

Contents

PART II   Alternate and Supplementary Experiments

Systematic Organic Qualitative Analysis   PART III

Advanced Projects   PART IV

PART V   The Operations

Elementary Operations

Operations for Conducting Chemical Reactions

Separation Operations

Appendixes and Bibliography

Preface

This book began more than seven years ago as a modest collection of supplemental experiments for the organic chemistry laboratory course at Lake Superior State College. It has been expanded and improved considerably in the interim, and a year's respite from teaching in 1978–79 gave me the time and opportunity to complete the present manuscript. Throughout the development of the text I have been guided by certain personal convictions: (1) that an "operational" approach to the organic chemistry laboratory is superior to the more traditional "cookbook" approach; (2) that motivation is as important as ability in assuring a student's success in the laboratory; (3) that a solid and reasonably rigorous laboratory course can also be "fun" and can be taught without subjecting the student to too much unpleasantness; (4) that exploding costs necessitate the use of comparatively inexpensive chemicals and basic apparatus in the undergraduate laboratory; (5) that students should be made safety-conscious and aware of the potential impact of chemicals on the environment.

I have tried to emphasize throughout the book that a synthetic experiment, for example, is not a unique process requiring detailed step-by-step directions, but is rather a series of easily mastered experimental operations performed in sequence and adapted to the requirements of the preparation. This approach is exemplified by the detailed operation descriptives in Part V, and by extensive Methodology sections in each experiment which explain the experimental approach. Reactant quantities are given in moles or millimoles to promote the student's understanding and appreciation of reaction stoichiometry. A motivational component is furnished by the use of imaginary "situations" and considerable background information stressing the relevance of each experiment. This kind of material is particularly important for students in the life sciences and pre-professional fields, who are often turned off by the abstract nature of chemical reactions and mechanisms. When feasible, natural products are used because they often cost less, smell nicer, and are more pleasant to work with than their purely synthetic analogs. Although numerous experiments are designed to use gas chromatography and spectrometric methods, nearly all can be performed satisfactorily without the instruments by furnishing students with authentic spectra or chromatographic data. Magnetic stirring is suggested in a number of procedures, but is required for only one experiment utilizing phase-transfer catalysis; and all the basic operations can be performed using the glassware found in a typical 19/22 organic labkit. Safety is emphasized throughout the text, with special symbols to indicate chemical hazards and a summary of hazards and precautions in each experiment. Proper methods for handling chemicals are discussed, and the student is frequently reminded of the need to dispose of harmful chemicals safely.

Operational Organic Chemistry incorporates a wide variety of laboratory experiences to help maintain interest and avoid unnecessary repetition. This makes the book suitable for many different kinds of students and courses—it should be as useful for biology, pre-medical, pre-pharmacy, medical technology and other majors as it is for chemistry students. The main-sequence experiments in Part I are suitable for a short course in organic chemistry. A more rigorous general or majors course can be built around a core of main-sequence experiments with judicious selections from the supplementary experiments in Part II, along with some qualitative analysis experiments based on Part III and perhaps a few advanced projects from Part IV. Additional flexibility is provided by the Minilabs and the Experimental Variations section found in each experiment.

Many people have been involved in the preparation of this book and their contributions should be acknowledged. The manuscript in various stages of development was reviewed by Newton D. Werner (Cerritos College), Kurt C. Schreiber (Duquesne University), Bruce Jarvis (University of Maryland), Robert H. Feiertag (Ohio State University), James B. Ellern (University of Southern California), Fred M. Dewey (Metropolitan State College), H. Leroy Nyquist (California State University at Northridge), Gerald F. Koser (University of Akron), Constance Suffredini (University of California at Irvine), Hance H. Hamilton (Eastfield College), and Robert E. Kohrman (Central Michigan University). The reviewers provided many valuable suggestions which resulted in a number of changes and improvements. Thanks are due to my laboratory students at Lake Superior State College who have class-tested most of the procedures and suffered through a few of my less successful efforts during the long process of development. I would like to express my special appreciation to Professor David Todd of Worcester Polytechnic Institute who, with his graduate students, tested a dozen of the experiments in Part II; to George Sypniewski (Michigan Technological University), who supervised the class-testing of several experiments at L.S.S.C. during my absence; and to the following graduate and undergraduate students who lab-tested the remainder of the experiments: Mike Lingo (Central Michigan University), Edwin Tewes (Tufts University), Greg Bosch (Massachusetts Institute of Technology), Masayuki Nakajima (University of Massachusetts—Arlington) and Kunio Kano (University of Massachusetts—Boston). Acknowledgements are in order for the assistance provided by Professors Gerald D. Weatherby and Purna Chandra of Lake Superior State College, who read several experiments and offered helpful comments; to Professor Dagmar Ponzi, who provided laboratory space at M.I.T. for some of the lab testing; and to Linda Dicks, Deborah Morley, and other laboratory assistants at L.S.S.C., who checked some of the experimental procedures before they were used in the laboratory course.

Special thanks go to the administration and staff at Massachusetts Institute of Technology for granting me temporary faculty privileges, which greatly facilitated the library research that went into the preparation of the final manuscript; to Professor C. G. Swain, who sponsored my visit to M.I.T. and reviewed the kinetics experiments; and to the administration and staff of Lake Superior State College, which approved my leave of absence and generally supported my efforts. Finally I would like to express my appreciation to Kathy McCaskey, Barbara Vilenski, Deborah Brooks, and Deborah Hannon, who typed the various drafts of the manuscript; to James M. Smith, Greg Giblin, David Dahlbacka, Judy Fiske, Lorraine Perrotta, and other Allyn and Bacon personnel who furnished invaluable assistance during the manuscript preparation and other stages of publication; and to Larry Largray, who illustrated the book. Many other individuals provided help and moral support throughout the development and publication of this book, and their contributions are appreciated.

J. W. L.

Introduction and
Advice for the Student

Purpose and Organization of the Textbook

As implied by the title, *Operational Organic Chemistry* stresses an operational approach to the laboratory practice of organic chemistry. Rather than viewing a given analysis or synthetic preparation as a unique process unrelated to any other, I have treated each experiment as a series of operations performed in a logical sequence to fulfill certain objectives. In turn, I have described each operation in sufficient depth to enable the student not only to master the basic manipulative techniques but also to improvise when necessary.

Perhaps even more important than providing a student with the *means* to complete an experiment successfully is to supply the *motivation* which will make him or her approach the experiment with anticipation (or at least without outright abhorrence) and thus learn from the experience. The motivational component is supplied by including enough background material to demonstrate the everyday relevance (and some of the fascination) of organic chemistry, and by providing "real-life" situations with each experiment.

Following the introduction and laboratory safety section, the textbook is divided into five parts. Part I contains enough experiments for a short course in organic chemistry and can give each student experience with nearly all of the operations described in Part V. The very fundamental operations, which will be practiced repeatedly throughout the course, are introduced in the first five experiments. These experiments are meant to perform the same function as the "techniques exercises" found in some laboratory textbooks, but without the concentrated, repetitive exposure to isolated techniques which many students find so tedious. Each operation is reinforced by later applications, often at a somewhat higher level of difficulty, so that mastery can be accomplished over a reasonable period of time. Experiment 6, which introduces some common functional groups, might be considered a "breather" in that it teaches no complicated operations and requires no special manipulative skills. The remaining experiments of Part I are keyed to topics covered in most organic chemistry lecture textbooks; they include more advanced operations as well as advanced applications of the basic ones.

Part II provides a large selection of experiments that can be incorporated into a full-year laboratory course in organic chemistry and/or substituted for some of the Part I experiments following Experiment 5. Although the general level of difficulty (with some exceptions) is higher than in Part I, most can be performed satisfactorily by average sophomore-level students who have been provided with adequate instruction and equipment.

To allow additional flexibility and provide motivated students with a variety of special projects, each experiment in Parts I and II is accompanied by a selection of "experimental variations" and suggestions for further work. This section may also include a "minilab" requiring from half an hour to an hour for completion. Although each minilab appears in conjunction with an experiment on the same theoretical topic, it can be combined with a different experiment if desired, or done separately.

Part III provides a comprehensive introduction to the practice of organic qualitative analysis using chemical and spectral methods. Part IV describes some open ended, research-type projects that can be assigned to advanced or highly motivated students. Part V contains the operation descriptions, which are referred to by number (preceded by OP) throughout the text.

The Appendix provides useful information, tables, and spectra along with a guide to the chemical literature, keyed to the Bibliography. Reference to sources in the Bibliography is made throughout the text in the form (Bib–D2), where the notation in the example refers to bibliographic source number 2 under category D, General Laboratory Techniques.

Organization of the Experiments

Operations. Experimental operations to be used in performing the experiment are listed by number and name. Boldface OP numbers indicate operations that are being used in this textbook for the first time. You are expected to read each operation description thoroughly before the laboratory period in which it will first be used. Sometimes it will not be necessary to read the entire description the first time, in which case the page numbers of the appropriate section are indicated in the Prelab Exercises. Previously used operations should be reviewed as needed before the lab period. If you have performed an operation once or twice before, it may only be necessary to read the operation summary (if there is one) to refresh your memory.

Objectives. All the experiments are designed to fulfill a number of "experimental" and "learning" objectives. *Experimental objectives* are those that will presumably be accomplished by successfully completing the experiment, such as identifying an unknown or preparing a substance. *Learning objectives* relate to information, theoretical concepts, and experimental methods that will be learned or reinforced while reading and performing the experiment. (In most experiments, it will be understood that one of the objectives is to gain additional experience with the experimental methods learned previously.)

Situation.   Organic chemistry is not always regarded as exciting or relevant, and many students do poorly in it because they lack interest in the subject and thus the desire to learn. Throughout Parts I and II you will find a number of more-or-less imaginary "situations" that are meant to make the experiments more interesting and show how the laboratory work might be applied to the solution of real-life problems. Some are purely imaginary and may be quite fantastic or absurd; others realistically portray the type of research problem that might be encountered by a professional chemist. The use you choose to make of the situations is limited only by your imagination and your willingness to accept a challenge.

Background.   A laboratory course is not meant to teach techniques alone, but to put in practice, and thus reinforce, the facts and concepts dealt with in lecture. If the laboratory experience is to have any meaning to you, it is important that you see it in perspective and appreciate its relationship to the "real world" of organic chemistry. The background material included with each experiment is meant to demonstrate its relevance and tie it in with the lecture course by discussing some of the underlying theoretical concepts, describing the applications of analogous systems and methods, or relating historical sidelights and interesting facts.

Methodology.   Students often carry out a chemistry experiment as if they were preparing a soufflé—by reading the directions and following them mechanically. Typically they will perform each operation without knowing why, or what to do if something unexpected happens, or how to improvise when the situation demands it. The Methodology section of each experiment outlines the experimental approach, explains the rationale for certain procedures, and often discusses alternatives. It may include enough general information about the applications and limitations of a given reaction type or experimental method to provide a basis for improvisation or more advanced preparations.

Prelab Exercises.   For efficient use of laboratory time, it is essential that some preliminary work be done before the start of each laboratory period. The experiment should be read through and the Methodology and Procedure sections studied in detail. New operations must be read and the old ones reviewed. A checklist should be prepared for most experiments, and preliminary calculations often must be carried out before starting work. The Prelab Exercises section lists various preliminary tasks to be performed, and generally requires that you submit a prelab writeup to the instructor at the beginning of each laboratory period. Although writing a checklist is not explicitly required in the Part II experiments, it is strongly recommended that you do so; your instructor will indicate whether or not the checklist must be submitted for approval at the beginning of the laboratory period. Some of the Topics for Report (see below) may also be assigned as Prelab Exercises, and your instructor may require additional preparatory work as described in Appendix III.

Reactions and Properties.   In this section—a subsection under Prelab Exercises—equations for major chemical reactions and data on the physical properties of reactants, products, and solvents are given. Dangerous properties of chemicals (Hazards) and precautions to follow in their use are also given.

Procedure.   Most of the experimental procedures are not written in a detailed, "cookbook" style; it will not be possible to carry out an experiment successfully by simply "following directions." You must have previously read the experiment, carried out the Prelab Exercises, and prepared a laboratory checklist. At that point you should be able to perform the experiment with the procedure as your guide.

Experimental Variations.   This section may include alternative procedures for an experiment, or additional experiments that can be performed at the option of the instructor. You are encouraged to do as many of the additional assignments (including the short minilabs) as you have time for; however, you must obtain advance permission from the instructor before using an alternative procedure or starting an unassigned experiment.

Calculations.   Most experiments require routine stoichiometric calculations (see Appendix IV) which should be submitted with the Prelab Exercises or with the final report. When additional calculations are required, they will be specified under this heading.

Topics for Report.   You will be expected to submit a report or turn in a laboratory notebook after each experiment, following the instructions in Appendix II or III. The Topics for Report may specify additional information to include with the report, ask questions related to the experiment or background material, assign problems to be solved, or describe library projects requiring consultation of the chemical literature (see Appendix VIII). You are not expected to report on all the topics included with an experiment, but only on those assigned by your instructor. It is recommended that topics 1 and 2 be included with each report, along with one or more of the remaining topics.

Laboratory Organization

Because of wide variations in individual working rates, it is usually not possible to schedule experiments so that everyone can be finished in the allotted time; if all laboratories were geared to the slowest student, the objectives of the course could not be accomplished in the limited time available. As a result, some students will invariably get behind during a lab period and find it necessary to put in extra hours outside their scheduled laboratory section in order to complete the course. The students who fall into this group do not necessarily lack ability—some of the brightest students may also be among the slowest—but they are usually not well organized and thus fail to make the best use of their time. The following suggestions should help you work more efficiently in the laboratory:

1. *Be prepared to start the current experiment the moment you reach your work area.* Don't waste the precious minutes at the start of a laboratory period doing calculations, reading the experiment, washing glassware, or carrying out other activities that should have been performed at the end of the previous period or during the intervening time. The first half hour of any lab period is the most important—if you can collect your reagents, set up the apparatus, and get the initial operation (reflux, distillation, etc.) underway in that time, you should be able to complete the experiment in the designated time period.

2. *Organize your time effectively.* Read the experiment and operation descriptives before coming to the laboratory, and plan ahead so that you know approximately what you will be doing at each stage of the experiment. A laboratory checklist (Appendix V) can be invaluable in helping you to break down the experiment into a logical sequence of events and to schedule activities for the "dead" periods during such operations as reflux and drying.

3. *Organize your work area.* Before performing any operation, all of the equipment and supplies you will need to use during the operation should be set out neatly on your bench top in approximately the order in which they will be used. Small objects like spatulas and any item that might be contaminated by contact with the bench top should be placed on a paper towel, laboratory tissue, or mat. After each item is used it should be removed to an out-of-the-way location (e.g., dirty glassware to a washing trough in the sink) from which it can be cleaned and returned to its proper location when time permits. It is important to keep your locker well organized, with each item placed in a specific location after use, so that you can immediately find the equipment you need. This will also enable you to discover quickly that a piece of equipment has been misplaced or stolen, so that it can be hunted down or reported to the instructor without delay.

Laboratory Etiquette

This section should not bring to mind "Ettie Kett's Guide to What's Proper in the Chemistry Laboratory." Instead it provides a set of common sense rules that, if followed, should increase efficiency and maintain peace in the laboratory.

1.   Return all chemicals and supplies to the proper location after use. It has been said that 99.47% of all laboratory motion follows the question. "Who took the #★&@! silver bromide!?" This expresses the frustration experienced when, after searching in every obvious location for an essential reagent, one finally comes across it at another student's station in a far corner of the laboratory. Chemicals that are being weighed out may be left at the balance if there are students waiting to use them; otherwise, they should be returned to the reagent shelves. Containers should be taken to the reagents and filled there; reagent bottles should not be taken to your lab station.

2.   Measure out only what you need. Liquids and solutions should be measured into graduated containers so that you will take no more than you expect to use for a given operation. Solids can usually be weighed directly from their containers or measured from a special solids dispenser.

3.   Prevent contamination of reagents. Do not use pipets or droppers to remove liquids from reagent bottles, and do not return unused reagent to a stock bottle. Be sure to close all bottles tightly after use—particularly those containing anhydrous chemicals and drying agents.

4.   Tell those nearby that you are about to light a burner, unless they are already using burners. This will enable them to cover any containers of flammable solvents and to limit or modify their use of such solvents during your operation. In certain circumstances, as when ether extractions are being performed, you should be prepared to use another heat source, move your operation to a safe location (for instance, under a fume hood) or find something else to do while flammable solvents are in use.

5.   Leave all community property where you found it. Some items, such as ringstands, steam baths, labkits, clamps, and condenser tubing, may not be supplied in student lockers. Such items may be part of the standard equipment found at each laboratory station, or may be obtained from the stockroom at the beginning of the laboratory period. Since such items may be needed by students in other lab sections, they should always be returned to the proper storage space at the end of the period.

6.   Clean up for the next person. There are few experiences more annoying than finding that the labkit you just checked out is full of dirty glassware, or that your lab station is cluttered with paper towels, broken glass, and spilled chemicals. The last 15 minutes or so of every laboratory period should be set aside for cleaning up your lab station and the glassware used during the experiment. Put things away so that your station looks uncluttered, clean off the bench top with a towel or wet sponge, remove debris (including condenser tubing and other community supplies) from the sink, and thoroughly wash dirty glassware that is to be returned to the stockroom, as well as that from your locker. Any spills and broken glassware should be cleaned up immediately, unless the chemical spilled is very toxic by inhalation or lachrymatory (like thionyl chloride), in which case the instructor should be informed and the area evacuated if necessary.

7. *Heed Gumperson's Second Law,* which states: "In the **laboratory** attention should be directed toward increasing that commodity indicated by the first five letters, and toward diminishing that commodity indicated by the last seven." This does not mean that all conversation in the laboratory must come to a halt; but it suggests that discussion of extraneous subject matter, such as the quality of food in the college cafeteria or the mayhem committed at the last football game, should be kept within reasonable proportions. Quiet conversation during a lull in the experimental activity is okay if it doesn't annoy your neighbors; a constant stream of chatter directed at another student during a delicate operation is distracting and may lead to an accident. For the same reason radios and tape players should not be used in the laboratory, since for every person who finds country-western or rock music soothing, there is another who may be driven to distraction by the sounds of Conway Twitty or Led Zeppelin. So if you don't want your radio or tapes confiscated, leave them at home!

Laboratory Safety

Preventing Laboratory Accidents

In any laboratory course, particularly one that involves the use of hazardous chemicals, there is always the chance of a serious accident. Organic compounds are often toxic and flammable, glassware can break and inflict severe cuts, and chemicals can cause burns or inflammation. Anyone who enters a chemistry laboratory without being acquainted with the fundamentals of laboratory safety is gambling, with his or her health and well-being at stake. Following these safety rules will help minimize accidents:

1. Wear appropriate glasses or safety goggles in the laboratory at all times. Prescription glasses should have safety lenses, if possible. Contact lenses should not be worn in the laboratory since corrosive fumes or chemicals may get underneath them. The location and operation of the eyewash fountain should be learned at the first laboratory session. Remember —you don't get a second chance to grow an eye!

2. Never smoke in the laboratory or use open flames in operations involving low-boiling flammable solvents. Ethyl ether and petroleum ether are particularly hazardous, but other common solvents such as acetone and ethanol can be dangerous as well. The location and operation of the fire extinguishers, fire blanket, and safety shower must be learned at the first laboratory session. Smoking is prohibited inside chemical laboratories because of the fire hazard.

3. Consider all organic chemicals poisonous unless there is unequivocal information to the contrary. Avoid breathing the vapors of volatile solvents, particularly halogenated hydrocarbons like chloroform and other solvents with a high inhalation toxicity index (see the Chemical Hazards section, page 12). Never taste chemicals unless you are specifically requested to do so, and wash thoroughly after handling them. Food and drink are not allowed in the laboratory because of the danger of accidentally ingesting toxic chemicals. For the same reason, mouth suction should not be used while pipetting liquids. Contact of chemicals with the skin, eyes, and clothing should be avoided, since some chemicals cause burns or severe irritation, while others can be absorbed through the skin and cause systemic poisoning.

4. Handle strong acids or bases, bromine, and other corrosive chemicals with great care and never allow them in contact with skin or eyes. Spills of such chemicals must be cleaned up immediately, using an appropriate neutralizing agent and plenty of water.

5. *Use safe techniques for inserting and removing glass tubing.* Proper procedures for working with glass should be learned by reference to operation ⟨OP–4⟩. Great care must be exercised when inserting or removing thermometers and glass tubing from stoppers and thermometer adapters, since severe lacerations can result from accidental breakage.

6. *Wear appropriate clothing.* Servicable shoes (not sandals) should always be worn in the laboratory to provide adequate protection against spilled chemicals and broken glass. Hairnets for long hair may be advisable when flames are in use (human hair is very flammable!). Clothing should be adequate to offer protection against accidental spills of corrosive chemicals, but the laboratory is obviously not the place to wear your best clothes.

7. *Dispose of chemicals in the proper containers and in the manner specified by your instructor.* Organic chemicals, for reasons of safety and environmental protection, should not ordinarily be washed down the sinks. Whenever possible, they should be disposed of in waste crocks (for solids) or in labeled waste-solvent cans. In most cases, dilute aqueous solutions can be safely washed down the drain, but the instructor should be consulted if there is any question about the best method for disposing of particular chemicals or solutions.

8. *Never work alone in the laboratory or perform unauthorized experiments.* If you find it necessary to work in the laboratory when no formal lab period is scheduled, you must obtain permission from the instructor and be certain that others will be present while you are working.

Reacting to Accidents: First Aid

Any serious accident involving poisoning or bodily injury should be treated by a competent physician, but to minimize the damage from such an accident all students and instructors should be familiar with some basic first-aid procedures. If you have an accident that requires quick action to prevent permanent injury (flushing chemical burns with water, etc.), take the appropriate action as described below and then, as soon as possible, see that the instructor is informed of the accident and leave any further first aid to him or her. If you *witness* an accident you should summon the instructor immediately and leave the first aid to him *unless:* (1) no instructor or assistant is in the laboratory area; (2) the victim requires immediate attention because of stopped breathing, heavy bleeding, etc.; (3) you have had formal training in emergency first-aid procedures and the instructor consents.

If an accident victim has difficulty breathing or goes into shock as a result of any kind of accident, standard procedures for artificial respiration and treating shock should be applied. Detailed first-aid procedures can be found in the *Handbook of Laboratory Safety,* 2nd ed. (Bibliography–C4), pages 23–48, and in various first-aid manuals.

Eye Injuries. If any chemical enters your eyes, they should *immediately* be flooded with water from an eyewash fountain, with the eyelids kept open. If contact lenses are being worn they must be removed before irrigation. Irrigation should continue for at least 15 minutes, and the eyes should then be examined by a physician. The use of boric acid or other neutralizing solutions is not recommended for eye injuries, since this sometimes causes more damage than no irrigation at all.

If foreign bodies such as glass particles are propelled into an eye, the injured person should get immediate medical attention. Removal of such particles is a job for a specialist.

Chemical Burns. The affected area should *immediately* be flushed with water, using a safety shower if the area of injury is extensive or if it is inaccessible to washing from a tap. Speed in washing is the most important factor in reducing the extent of injury. Water should be applied continuously for 10–15 minutes and the burned area covered with sterile gauze or other clean cloth. The burn should be examined by a physician unless the skin is only reddened in a small area. *The Merck Index* (Bib–A3), p. MISC–22 ff., or another reference, should be consulted for specific procedures to be used for burns caused by bromine, hydrofluoric acid, and certain other substances. The use of neutralizing solutions, ointments, greases, etc., on chemical burns is not recommended unless specifically indicated; in any case, flooding with water should precede the use of such agents.

If the burn is extensive or severe, the victim should be made to lie down with the head and chest a little lower than the rest of the body. If the victim is conscious and able to swallow, he or she should be provided with plenty of liquid (nonalcoholic) to drink until a physician arrives.

Thermal Burns. For small thermal burns such as those inflicted by handling hot glass, a sterile gauze pad should be soaked in a baking soda solution (2 tablespoons sodium bicarbonate to a quart of lukewarm water, about $2\frac{1}{2}\%$ $NaHCO_3$), placed over the burn, then bandaged loosely in place. If the skin is not broken, the burned part can be immersed in clean, cold water, or ice can be applied to reduce the pain. Blisters should not be opened. Unless the skin is only reddened over a small area, the burn should be seen by a nurse or physician.

In case of an extensive thermal burn, the burned area should be covered with the cleanest available cloth material and the victim should be made to lie down, with the head and chest lower than the rest of the body, until a physician arrives. If the injured person is conscious and able to swallow, he or she should be provided with plenty of nonalcoholic liquid to drink (water, tea, coffee, etc.).

Bleeding, Cuts, and Abrasions. In case of a minor cut or abrasion, the wound and surrounding skin should be cleansed with soap and lukewarm water, wiping away from the wound. A sterile pad should be held over the wound until bleeding stops and then be replaced by a clean pad, which should be secured loosely with a triangular or rolled bandage. The pad and bandage should be replaced as necessary to keep them clean and dry. Contact of the wound with mouth, fingers, handkerchiefs, or other unsterile material should be avoided. Antiseptics should not be used on an open wound.

In case of a major wound that involves heavy bleeding, do not waste time cleansing the wound but *apply pressure immediately* directly over the wound. Use a pad (clean handkerchief or other cloth) and press firmly with one or both hands to reduce the bleeding as much as possible. The victim should be made to lie down with the bleeding part higher than the rest of the body, and the pad held in place with a strong bandage (a necktie may be adequate, or cloth strips torn from a shirt). A physician should be called as soon as possible, and the victim kept warm with a blanket or coat in the meantime. If the victim is conscious and able to swallow, he or she should be provided with plenty of nonalcoholic liquid to drink.

Poisoning. Chemical poisoning may be suspected whenever, after handling or accidentally ingesting chemicals, there is a pain or burning sensation in the throat, discoloration of lips or mouth, stomach cramps, nausea and vomiting, confusion, or loss of consciousness.

If poison is swallowed, a physician or ambulance should be called immediately, and the victim should be given 2–4 glasses of water or milk to drink. If it is safe to do so, vomiting should be induced by placing a finger at the back of the victim's throat or by having the victim drink 2 teaspoons of salt in a glass of warm water (10 g in 200 ml), or one ounce (30 ml) of ipecac syrup. Vomiting should not be induced if the patient is unconscious, in convulsions, or has severe pain and burning sensations in the mouth or throat, nor if the poison is a petroleum product or a strong acid or alkali. When vomiting begins the victim should be placed face down with head lower than hips, and vomiting should continue until the vomitus is clear. When there is time, the poison should be identified (if possible) and an appropriate antidote given (see *The Merck Index,* 9th Edition, pages MISC–22 ff., or call a poison control center). If the poison cannot be identified and a physician is not present, a heaping tablespoon (15 g) of the following Universal Antidote can be given in a half glass of warm water: 2 parts activated charcoal, 1 part magnesium oxide (milk of magnesia), 1 part tannic acid. A sample of the poison should be saved for the physician if possible; if the poison is unknown, the vomitus should be saved for examination.

If a poison has been inhaled, the victim must be taken to fresh air and a physician called immediately. If the victim shows evidence of difficulty in breathing, mouth-to-mouth resuscitation should be administered. The victim should be kept warm and as quiet as possible until professional help is available. If the poison is a highly toxic gas such as hydrogen cyanide, hydrogen sulfide, or phosgene, the persons attempting to rescue the victim should wear gas masks (if available) while they are in contact with the vapors.

In case of skin contamination by toxic substances, the same general procedure as for chemical burns should be followed. *The Merck Index* or other references should be consulted for specific procedures to be used for certain substances.

Reacting to Accidents: Fires

In case of a serious fire, one's first reaction should be to *get away from it* as quickly as possible. If the fire is small and confined to a container such as a flask or beaker, it can often be extinguished by quickly placing an asbestos pad or watch glass over the mouth of the container. Otherwise, you should get out of the area and let the instructor or an assistant extinguish it. If no instructor or assistant is in the vicinity, you may obtain a fire extinguisher of the appropriate type (dry chemical extinguishers are best for most flammable liquid fires) and attempt to put out the fire by aiming the extinguisher at the base of the fire (maintain a safe distance). Be prepared to call the fire department and arrange to evacuate the area if there is a chance that the fire cannot be controlled. Even after a fire has been extinguished, it is a good procedure to report it to the fire department so that they can check the area for smoldering embers.

If your hair or clothing should catch on fire, go directly to the nearest fire blanket or safety shower and attempt to extinguish the fire by wrapping yourself in the blanket or drenching yourself with water. If another person has caught on fire, try to prevent panic and wrap the person in a coat or fire blanket or lead him or her to a safety shower.

More detailed information on fire and rescue procedures can be found in the *CRC Handbook of Laboratory Safety,* 2nd ed., pages 15–22 and 187–91.

Chemical Hazards

No chemical substance can be considered entirely hazard-free, though some are considerably safer than others. It is a good practice to regard all chemicals as flammable and toxic unless there is definite information to the contrary. Even then, such information may be erroneous since the long-term effect of many chemicals is unknown; some commonly used chemicals regarded as relatively safe years ago are now known to cause cancer on prolonged exposure. The best practice is to *exercise caution in handling* all chemicals and *minimize your exposure* to them.

In many of the following experiments and operation descriptions, the hazards posed by various chemicals will be summarized by means of octagonal hazard signs containing the following hazard indexes:

Hazard Index Numbers and Symbols

Health (including contact and vapor hazards)
    4 Can cause death or major injury despite medical treatment
    3 Can cause serious injury despite medical treatment
    2 Can cause injury. Requires prompt treatment
    1 Can cause irritation if not treated
    0 No known hazard
    CA Classified as a carcinogen (cancer-causing agent) on the basis of tests with laboratory animals, or known to cause cancer in humans

Fire
    4 Very flammable gases or very volatile flammable liquids
    3 Can be ignited at all normal temperatures
    2 Ignites if moderately heated
    1 Ignites after considerable preheating
    0 Will not burn

Reactivity
    4 Readily detonates or explodes
    3 Can detonate or explode, but requires strong initiating force or heating under confinement
    2 Normally unstable, but will not detonate
    1 Normally stable, but unstable at high temperature and pressure
    0 Normally stable
    OX Strong oxidant; may react explosively with combustible material
    P Polymerizes readily
    W Reacts violently with water

Hazard Signs Used in This Text

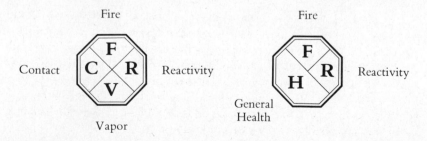

In the four-part hazard signs the contact index (**C**) in the left-hand quadrant rates the hazard due to eye contact, skin contact, or ingestion, whichever is highest. For carbon tetrachloride, the skin contact and ingestion numbers (adapted from the *CRC Handbook of Laboratory Safety*) are both 1 while the eye contact index is 2, so the latter number is used. The Vapor Toxicity (**V**) index in the lower quadrant rates the danger of inhaling the vapors of a chemical. When specific contact and vapor hazard data is unavailable, a three-part hazard sign containing only a single health hazard (**H**) number (from the National Fire Protection Agency system) is used. (The 1–5 scale of numbers used in the CRC source has been converted to a 0–4 scale to make the two systems more compatible.) The NFPA's fire (**F**) index number rates the fire danger posed by a chemical, and their reactivity (**R**) number assesses the danger of a violent reaction or explosion. A blank in any section of the hazard sign means that no hazard index number was reported in the sources available to the author; it does not necessarily mean "no hazard."

Examples:

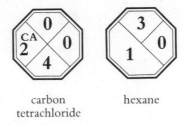

carbon   hexane
tetrachloride

For additional information on safety procedures and hazards, refer to sources described under Category C, Laboratory Safety and Chemical Hazards, in Appendix VIII and to the corresponding section in the Bibliography.

PART I

Main-sequence Experiments

PRELIMINARIES

Checking In

During the first laboratory period you should obtain a locker and a list of the equipment in it and check the items on your list to see if anything is missing or damaged. (Illustrations of typical items of glassware and other supplies can be found in Appendix I.) Any mysterious items that you cannot identify should be referred to your instructor. Pieces of glassware with chips, cracks, or star fractures should be replaced; they may cause cuts, break on heating, or shatter under stress. You should also use this time to clean any dirty glassware in your locker or labkit after reading operation ⟨OP–1⟩ on page 440.

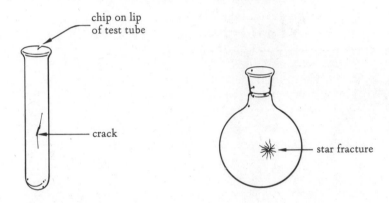

Laboratory Orientation

In order to work efficiently in the laboratory, you should know where instruments and supplies are located and where to find the chemicals you will need. It is essential that you learn where the safety equipment is so that you can get to it immediately in case of an accident. During the first period, you should walk around the laboratory (your instructor may conduct a guided tour) and locate all of the items listed below that are provided in your laboratory. (Additional items may be specified by your instructor.) You should understand the function of each item of safety equipment and learn how it is operated.

     Fire extinguishers

     Safety showers

     Fire blankets

     Eyewash fountains

     First-aid supplies

     Spill cleanup supplies

     Telephone and emergency numbers

Liquid disposal containers

Solid disposal containers (waste crocks)

Balances

Hoods

Drying ovens

Reagent shelves

Shelves for consumable supplies

At your instructor's request, you may also learn how to operate the balances and practice some weighings after reading operation ⟨OP–5⟩.

Making Useful Laboratory Items

The following items of equipment will be used frequently in future experiments and can be prepared either during the first laboratory period or during the experiment in which they will first be used. You should read ⟨OP–3⟩, Using Corks and Rubber Stoppers, and ⟨OP–4⟩, Basic Glass Working, before beginning. Have the instructor inspect and approve your items when you are finished.

Flat-bottomed Stirring Rod. Prepare the stirring rod by cutting a 20–25 cm length of 5–6 mm (diameter) soft glass rod, flattening one end so that it flares out to a diameter of 10 mm or more, and rounding the other end. A flat-bottomed stirring rod can be used to pulverize solids for melting point determinations, wash solids on a Buchner filter, break up a filter cake, dissolve crystals in a recrystallization solvent, and for many other purposes besides stirring.

Vacuum Filtration Apparatus. Select a rubber stopper that will fit your filter flask and, using a cork borer having about the same diameter as the *midpoint* of your Buchner funnel stem, bore a hole through the middle of the stopper and assemble the filtering apparatus illustrated in Figure 1 on page 471. Construct a filter trap using a heavy walled bottle (250 ml or larger), a two-hole rubber stopper of appropriate diameter, and two lengths of 8-mm soft glass tubing. It is best to wrap the bottle with plastic tape (or some other strong adhesive tape) to prevent possible injury from implosion. Alternatively, a large filter flask or a quart milk bottle can be used [see the *Journal of Chemical Education,* Vol. 54, page 611 (1977)]. One tube should be long enough to extend 5 cm or so into the bottle after it is bent; the second need only be long enough to extend all the way through the stopper. Be sure to use a suitable lubricant (such as glycerine) when inserting the glass tubes through the stopper. The trap, as constructed, can be used for solvent evaporation ⟨OP–14⟩ as well as vacuum filtration. If the trap is to be used only for vacuum filtration, the longer tube can extend nearly to the bottom of the bottle.

Boiling Point Tube.   Obtain a piece of 5-mm O.D. soft glass tubing and carefully seal it at one end, then cut it to a length of 8–10 cm and fire-polish the open end. Test the tube according to instructions in ⟨OP–4⟩ to make sure it is sealed. The tube will be used for obtaining micro boiling points, as described in ⟨OP–29*a*⟩.

Cleanup Routine

You should never leave the laboratory until the following tasks have been completed:

1.   Clean up all dirty glassware.

2.   Clean up your work area—wiping the bench with a sponge or wet paper towel—and remove any refuse or equipment from the sink.

3.   Turn in any items that you checked out from the stockroom; return other items to their proper location.

4.   Be certain that all of the items on your locker list are safely inside, and lock your locker.

The Analysis of "Panacetin," Part 1

Separation Methods

Operations

⟨**OP–5**⟩ Weighing

⟨**OP–11**⟩ Gravity Filtration

⟨**OP–12**⟩ Vacuum Filtration

⟨**OP–13**⟩ Extraction

⟨**OP–14**⟩ Solvent Evaporation

⟨**OP–21**⟩ Drying Solids

Experimental Objective

To separate the components of "Panacetin" and determine the composition of the mixture.

Learning Objectives

To learn and apply some of the operations used for separating pure organic compounds from mixtures.

To learn about the history and uses of some analgesic/antipyretic drugs.

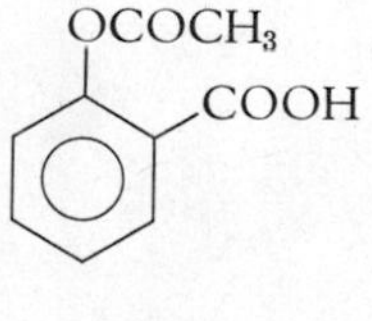

SITUATION

As an analytical chemist working for the Food and Drug Administration (FDA), you are investigating an allegation that an over-the-counter drug manufactured by the Mithridates Pharmaceutical Company of Pontus, West Jersey, has been adulterated with a potentially toxic substance. The drug, Panacetin, is sold as an **analgesic/antipyretic** drug and lists aspirin and acetaminophen as its active ingredients. It is suspected that, because of the rising cost of acetaminophen, the chemically similar but dangerously toxic substance acetanilide has been used in its place. Your assignment is to confirm the presence of acetanilide as an adulterant in Panacetin and determine the composition of the drug.

aspirin   acetaminophen

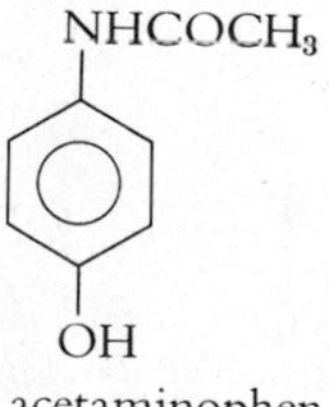

analgesic: pain reducing
antipyretic: fever reducing

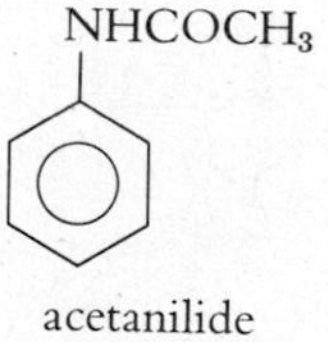

acetanilide

BACKGROUND

Discovering Painkillers by Accident and by Design

Constituents of over-the-counter drugs

APC: aspirin, phenacetin, and caffeine

Anacin: aspirin and caffeine

Bufferin: aspirin and buffers

Excedrin: aspirin, acetaminophen, caffeine, and salicylamide

Tylenol: acetaminophen

Most over-the-counter analgesic and antipyretic drugs contain either aspirin or acetaminophen (p-hydroxyacetanilide) as the major ingredient. Some include additional ingredients such as buffers (magnesium carbonate, sodium bicarbonate) to reduce acidity and prevent stomach upset, stimulants (caffeine), and other analgesic or antipyretic substances (phenacetin, salicylamide, etc.). A small amount of starch is generally used as a filler to hold the tablets together. The constituents of several well-known analgesic drugs are shown in the margin.

At one time acetanilide, dispensed under the name "antifebrin," was considered an important pain killing and fever reducing drug. Excessive use of acetanilide, however, can cause a serious type of anemia called methemoglobinemia, which involves alteration of the body's hemoglobin so that it can no longer carry sufficient oxygen through the bloodstream. For that reason acetanilide has been largely replaced by its less toxic cousins, acetaminophen and phenacetin, and by aspirin.

The story of antifebrin is an excellent example of the kind of "lucky accident" that occasionally leads to a major scientific discovery—although the haphazard methods used by Cahn and Hepp would hardly be condoned today! It began in 1886, when the director of a German medical clinic assigned a new job to a pair of assistants named Arnold Cahn and Paul Hepp. They were to develop a drug for use against a worm that had the nasty habit of burrowing into the intestinal walls of humans. The trick was to find a chemical that would kill the worm but not the patient. Their initial tests, using an ancient bottle of naphthalene that someone brought up from the stockroom, were not encouraging. Then they decided to try the "naphthalene" on a patient who had every intestinal malady in the book, including worms. It didn't faze the worms, but it reduced the patient's fever markedly. Cahn and Hepp were excitedly preparing to announce the discovery of a new antipyretic drug when they re-examined the substance and discovered that it didn't smell at all like naphthalene, and that the label on the bottle was completely illegible. Fortunately, Hepp had a cousin who was a chemist in a nearby dye factory, and they sent a sample of the compound to him for analysis. The mysteri-

naphthalene

Naphthalene is the strong smelling component of old-fashioned mothballs.

ous white powder turned out to be acetanilide, which soon became a popular antipyretic drug.

Scarcely six months after the discovery of antifebrin, another important drug was developed because of a storage problem. Carl Duisberg, director of research at the Friedrich Bayer Company's dye factory, had about fifty tons of a yellow powder on his hands that was taking up entirely too much space. The powder was *para*-aminophenol, then considered a useless by-product of dye manufacture, and Duisberg was faced with two alternatives—either pay a teamster to haul the stuff away, or change it into something Bayer could sell. Being a practical businessman as well as a research chemist, Duisberg preferred the second alternative; he was seeking possible uses for *para*-aminophenol when the report of Cahn and Hepp caught his attention. He knew it would be a simple matter to form a "modified" acetanilide molecule by **acetylating** the amino group of *para*-aminophenol, and he reasoned that such a substance might be as effective as acetanilide against fevers, or even more so. It was known that an OH group attached directly to a benzene ring could be quite toxic, however, so he chose to "mask" this group by substituting an ethyl radical for the hydrogen atom, forming phenacetin by the synthetic route shown in the margin. This new compound proved to be a remarkably safe, effective, and inexpensive antipyretic/analgesic drug. Ironically, the substance Duisberg chose to bypass by masking the hydroxy group, *para*-hydroxyacetanilide, is also a safe and effective drug that is widely used under the generic name acetaminophen. Both acetanilide and phenacetin are thought to be converted in the body to acetaminophen, which may be the active form of all three drugs.

METHODOLOGY

Organic compounds seldom occur in a pure form, either in nature, in commercial preparations, or in chemical reaction mixtures. In order to obtain a pure compound from such mixtures, one must separate it from all other substances present in the mixture. Separation procedures are based on differences in the physical and chemical properties of the components in a mixture. For example, two solids having very different solubilities in a given liquid can be separated by filtration; liquids having different boiling points can often be separated by distillation; and acidic or basic com-

acetylate: to replace a hydrogen atom by an acetyl (CH_3CO-) group.

Synthesis of phenacetin

NH_2

p-aminophenol

OH

NH_2

O$\boxed{CH_2CH_3}$ ← ethyl "masking" group

$NHCOCH_3$

acetyl group

OCH_2CH_3
phenacetin

pounds can be separated from neutral substances by extraction into a basic or acidic medium.

Of the three components likely to be present in your sample of Panacetin (aspirin, the adulterant, and starch), only starch is insoluble in the organic solvent methylene chloride (CH_2Cl_2). If a sample of Panacetin is dissolved as completely as possible in methylene chloride, the insoluble starch can be filtered out, leaving adulterant and aspirin in solution.

Although the adulterant and aspirin are both quite insoluble in water at room temperature, the sodium salt of aspirin is very soluble in water but insoluble in methylene chloride. Because aspirin is a reasonably strong acid, it can be converted to this salt, sodium acetylsalicylate, by reaction with the base sodium hydroxide. Adding a dilute aqueous solution of sodium hydroxide to the methylene chloride solution results in two sharply separated liquid layers. The methylene chloride layer, being denser than water and insoluble in it, will be on the bottom. If the two layers are thoroughly mixed, the aspirin in the bottom layer will react with the sodium hydroxide in the top layer and be converted to sodium acetylsalicylate, which then migrates to the aqueous layer and can be easily separated in a separatory funnel. Adding some dilute hydrochloric acid to the aqueous solution regenerates free aspirin as an insoluble white solid; evaporating the solvent from the bottom layer leaves the adulterant behind.

Interconversion of aspirin and its sodium salt

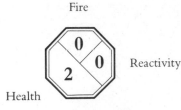

aspirin   sodium acetylsalicylate

PRELAB EXERCISES

1. Read the experiment and understand its objectives.

2. Read the sections in Part V describing operations ⟨OP–5⟩, ⟨OP–11⟩, ⟨OP–12⟩, ⟨OP–13⟩, ⟨OP–14⟩, and ⟨OP–21⟩.

3. Read Appendix II, Writing Laboratory Reports, or Appendix III, Keeping a Laboratory Notebook, according to your instructor's directions.

PROCEDURE

Weigh ⟨OP–5⟩ approximately 3.0 g of Panacetin to the nearest 0.01 gram and transfer it to an Erlenmeyer flask.

Hazard

Fire

0

2   **0**   Reactivity

Health

methylene chloride

(See Chemical Hazards, p. 13 for explanation of symbols.)

Measure 50 ml of methylene chloride into a graduated cylinder, add it to the solid, and stir the mixture vigorously, using a flat bottomed stirring rod to break up any lumps.

Do not breathe the vapors of methylene chloride; avoid contact of the liquid with your skin and eyes.

WARNING

When it appears that no more of the solid will dissolve, filter the mixture by gravity ⟨OP–11⟩, collect the undissolved solid on the filter paper, and set it aside to dry.

Transfer the filtrate to a separatory funnel and extract it ⟨OP–13⟩ with two 25 ml portions of aqueous 1M sodium hydroxide. Drain the methylene chloride layer and save it, then transfer the aqueous (top) layer to an Erlenmeyer flask. Acidify the aqueous layer by slow addition, with stirring, of 20 ml of 3M hydrochloric acid. Cool the mixture to room temperature or below by swirling the flask under a cold water tap, then collect the aspirin by vacuum filtration ⟨OP–12⟩. Wash the aspirin on the filter with cold distilled water and air dry it on the filter, then dry it thoroughly ⟨OP–21⟩ for weighing.

Evaporate the solvent ⟨OP–14⟩ from the methylene chloride solution under aspirator vacuum to obtain the adulterant and dry it if necessary ⟨OP–21⟩.

When the starch, aspirin, and adulterant are completely dry, transfer them to tared (preweighed) vials and weigh them ⟨OP–5⟩ to the nearest 0.01 gram. Save the adulterant for use in Experiment 2, and turn in the other products to the instructor in labeled vials. The labels should include the information indicated in the example and, if requested, the tare weight as well.

It is advisable to weigh the filter paper beforehand, so that the weight of the starch can be obtained without transfer.

Do not leave the aspirin in the basic solution too long, or it may partially hydrolyze to form salicylic acid.

Make sure the solution is strongly acidic by testing it with litmus paper or adding a few extra drops of HCl to test for complete precipitation.

The adulterant may remain liquid after all of the solvent is removed, but it will solidify on cooling.

Aspirin
Exp. 1
2.15 g (net)
John Smith
9/15/82

A typical label

Calculations

1. Calculate the percent recovery of the Panacetin constituents by adding their weights, dividing the sum by the initial weight of Panacetin, and multiplying the quotient by 100.

2. Calculate the approximate percent composition of the drug, based on the amount recovered (the percentages should add up to 100).

Experimental Variation

The methylene chloride layer containing the adulterant can be dried ⟨OP–20⟩, using calcium sulfate or another suitable drying agent, just before the solvent is removed.

Topics for Report

1. Why do you think it is important to cool the acidified aspirin mixture before filtering the aspirin? Why was the solution warm in the first place?

2. Describe what might happen if, during the extraction, the two liquid layers were not thoroughly mixed. Indicate what substances would be isolated when the organic layer was evaporated and how your experimental results would be affected.

Acetaminophen is one of the phenols, which are weak acids. Most phenols are more acidic than water but less acidic than carbonic acid, H_2CO_3. Aspirin is more acidic than carbonic acid.

3. Suppose that, contrary to your expectations, the Panacetin were not adulterated with acetanilide but contained only acetaminophen with the aspirin and starch. *(a)* At what stage would the acetaminophen be isolated and which other component would be mixed with it? *(b)* Suggest a way of separating the acetaminophen from this component.

4. Diagram the separation process using a flow chart format similar to that illustrated in Figure 1, Experiment 5.

Library Topic

Refer to section L of the Bibliography for possible sources of information on this topic.

Look up some of the uses and side effects of aspirin and acetaminophen, and discuss their relative advantages and disadvantages. Indicate under what circumstances a patient might take acetaminophen (Tylenol) to reduce pain, but not aspirin.

The Analysis of Panacetin, Part 2

Purification and Analysis Techniques

Operations

⟨OP–5⟩ Weighing
⟨OP–21⟩ Drying Solids
⟨**OP–23**⟩ Recrystallization
⟨**OP–28**⟩ Melting Point

Experimental Objective

To purify the "drug adulterant" isolated in Experiment 1, verify its identity, and assess its purity.

Learning Objectives

To learn how to purify solids by recrystallization, how to dry them, and how to obtain a melting point.

To learn and understand some basic principles relating to the purification and analysis of organic compounds.

To learn how to use some basic chemistry references to obtain information about the properties of organic compounds.

METHODOLOGY

A compound that has just been separated from a mixture is seldom ready for immediate analysis or for applications requiring a pure substance. Contamination may result from a variety of causes:

1. The separation procedure may be imperfect, leaving small quantities of other substances in the compound after separation.

2. Additional contaminants may be introduced during the separation, either accidentally or as an inevitable consequence of the separation procedure used.

3. Chemical reactions may occur prior to or during the separation, adding new impurities.

Various methods have been devised for purifying organic compounds. Solids can be purified by such methods as recrystallization, chromatography, and sublimation; liquids can be purified by distillation and chromatography, among other methods. The recrystallization technique will be applied in this experiment.

After a compound is purified, it is usually analyzed to determine if it is, in fact, the expected compound and to obtain a measure of its purity. The analytical procedure can be very complex, requiring a series of operations on special instruments such as gas chromatographs, infrared spectrophotometers, and nuclear magnetic resonance spectrometers; or it may require only the determination of a physical constant, such as a melting point.

Since different compounds seldom have exactly the same melting point, and since the melting point of a compound is usually lowered by impurities, a melting point determination can be used both to help identify a compound and to estimate its purity. However, for a melting point to be of much use, one must have a fairly good idea of the identity of the compound to begin with, and the melting point of the pure compound must be recorded somewhere in the chemical literature.

Acetanilide is a well-known substance whose physical properties have been fully reported in the literature. One good source of physical constants is the *CRC Handbook of Chemistry and Physics* (Bib–A1), which lists approximately 15,000 organic compounds. Figure 1 reproduces the *Handbook* entry for acetanilide, which is listed under the name "acetic acid, N-phenylamide."

The entry reveals that acetanilide has a molecular weight of 135.17; that it crystallizes from water in the form of rhombic crystals or plates; that it has a melting point of 114.3°C (although another source reports a slightly higher *melting range*); that it boils at 304°C at a pressure of 760 **torr;** and that it has a density of 1.2190 g/ml at 15°C. It is sparingly soluble (δ) in water at room temperature; soluble (s) in ether, benzene, toluene and hot water; and very soluble (v) in ethyl alcohol, acetone, chloroform, carbon tetrachloride, methanol, and hot toluene. A solution of acetanilide dissolved in alcohol shows an optical absorption band at 242 **nanometers** with an absorptivity of 1.4×10^4 (antilog 4.16).

760 **torr** = 1 atmosphere

1 **nanometer** (nm) = 10^{-9} meter

Figure 1. *CRC Handbook* Entry for acetanilide. (Reprinted with permission from the *CRC Handbook of Chemistry and Physics*, 59th Edition, pp. C–83 and C–85. Copyright— The Chemical Rubber Company, CRC Press, Inc.)

No.	Name	Synonyms and Formula	Mol. wt.	Color, crystalline form, specific rotation and λ_{max} (log ϵ)	m.p. °C	b.p. °C	Density	n_D	Solubility						Ref.
									w	al	eth	ace	bz	other solvents	
	Acetic acid														
Ω a60	—, amide..........	Acetamide. Ethanamide.* CH_3CONH_2	59.07	trg mcl (al-eth) λ^{MeOH}	82.3	221.2^{760} 120^{20}	0.9986^{65} 1.159^{20}_{20}	1.4278^{78}	s	v	i	.....	δ	chl s Py s	$B2^2,177$
$\rightarrow \Omega$ a164	—, —, N-phenyl....	Acetanilide. Antifebrin. $CH_3CONHC_6H_5$	135.17	rh or pl(w) λ^{al} 242(4.16)	114.3 (115–6)	304^{760}	1.2190^{15}	....	δ s^A	v	s	v	s	chl, CCl$_4$, MeOH v to s, v^A	$B12^2,137$
a165	—, —, N-phenyl-N-propyl-	$CH_3CON(C_6H_5)CH_2CH_2CH_3$	177.25	mcl lf (eth, lig)	49 (56)	266^{12}	....	....	i	v	v	.....	.....	.....	$B12,246$

This may be more than you really wanted to know about acetanilide, but it is helpful to have such information at your disposal when needed.

Since you will be recrystallizing an "adulterant" believed to be acetanilide, the solubility behavior of acetanilide is of interest. The facts that crystals of acetanilide have been obtained using water as a solvent, and that the compound is only sparingly soluble in cold water but soluble in hot, suggest that water may be a suitable recrystallization solvent.

From the *CRC Handbook* tables, pure acetanilide should have a melting point of about 114°C. Unfortunately, over fifty other compounds are listed with the same melting point, and when one allows for experimental error and possible melting point variations due to small amounts of impurities, it can be seen that determining a melting point alone cannot establish the identity of a compound beyond reasonable doubt. Most uncertainty can be removed, however, by measuring the melting point of a mixture containing approximately equal quantities of the purified compound and an authentic sample of acetanilide. If the melting point of the mixture is considerably lower than that of the acetanilide, the compounds in the mixture must be different; if the melting points are the same, or nearly so, then the two compounds are identical.

See ⟨OP–28⟩ for a discussion of mixture melting points.

PRELAB EXERCISES

1. Read the experiment and understand its objectives.

2. Read all of ⟨OP–23⟩ and all of ⟨OP–28⟩. Read or review the other operations as needed.

3. The solubility of acetanilide in hot water (80°C) is reported to be 3.5 g/100 ml. Calculate the volume of hot water that should just dissolve all of the acetanilide you recovered from Experiment 1.

Be prepared to show this calculation to your instructor, and include it with your report.

PROCEDURE

Purify the crude solid saved from Experiment 1 by recrystallizing it ⟨OP–23⟩ from boiling distilled water. Begin with the amount of water calculated in the Prelab Exercises and add more if necessary. Dry the purified crystals thoroughly using one of the methods described in ⟨OP–21⟩.

Obtain melting points ⟨OP–28⟩ of the dry crystals, of an authentic sample of acetanilide, and of a 1:1 mixture of

Deionized water is also suitable.

More acetanilide will dissolve at the boiling point of water than at 80°C; but using too little water will cause the excess acetanilide to separate as an oil on cooling.

the dry crystals with the authentic sample. The melting point of the authentic sample can be obtained while the crystals are drying. Each melting point should be measured on at least two samples—more than two if additional practice seems desirable or if the melting points are imprecise or inaccurate.

If the purified sample is not completely dry, its melting point will be in error.

Weigh the purified crystals ⟨OP–5⟩ in a tared (preweighed) vial and turn in the labeled vial to your instructor.

Experimental Variations

1. If time permits, the aspirin from Experiment 1 can be purified by recrystallizing it from an ethanol/water mixture as described in Experiment 4 so that its melting point can be measured. Because aspirin partially decomposes on heating, its melting point may not be as sharp or as accurate as that of acetanilide.

2. The instructor may provide a sample of impure acetanilide containing small amounts of sucrose, sand, or other impurities in place of the product from Experiment 1. It may then be necessary to filter the hot recrystallization solution to remove insoluble impurities.

Topics for Report

1. **(a)** Calculate the percent recovery of the purified solid and try to account for any losses. **(b)** Give your conclusion regarding the identity of the "adulterant" in Panacetin, with supporting evidence.

2. The solubility of acetanilide at 25°C is reported to be 0.56 gram per 100 ml of water. Using your experimental data, calculate the *maximum* quantity of acetanilide that could have been recovered after your sample was recrystallized, assuming that the temperature of the cooled recrystallization mixture was about 25°C. Compare the calculated amount with the quantity actually recovered.

3. The melting point of an unknown organic compound is measured as about 121°C. A mixture of the unknown with benzoic acid melts at 102°C, a mixture with *m*-aminophenol melts at 122°C, and a mixture with phenyl succinate melts well below 100°C. Give the probable identity of the unknown and explain your reasoning.

Table 1. List of possible unknowns for topic 3.

Compound	m.p.
o-toluic acid	102°C
benzoic acid	121°C
phenyl succinate	121°C
m-aminophenol	122°C

4. **(a)** List the kinds of intermolecular forces which are responsible for the solubility of a substance in water or organic solvents, and discuss the meaning of the rule-of-thumb "like dissolves like." **(b)** Which kind of intermolecular interaction is most responsible for the solubility of acetanilide in benzene? Of sodium acetylsalicylate in water? Of aspirin in ethyl alcohol (C_2H_5OH)?

Library Topics

1. Look up the properties of aspirin (listed as "benzoic acid, 2-hydroxy, acetate") in the *CRC Handbook of Chemistry and Physics* and, referring to the list of abbreviations preceding the tables, write a short paragraph reporting on those properties (without using abbreviations), as was done for acetanilide in the Methodology.

Example of journal citation

J. Org. Chem., *9*, 148 (1944).

journal   page

volume   year

2. Locate citations in the *Dictionary of Organic Compounds* (Bib–A7) for one or more journal articles that give procedures for preparing aspirin. Give the full name of the journal in which the procedure appears, the volume number, page number, and year. Full names of journals can be found in the *Chemical Abstracts Service Source Index* (1907–1974 Cumulative). Journal citations in North American literature generally give the name of the journal, volume number, page, and year in that order; European publications may give the year first, then the volume number (if any) and page number.

Alternatively, you may use the Utermark and Schicke (Bib–A11) tables.

3. Give the names of at least ten compounds in the latest edition of the *CRC Handbook* that have a listed melting point of 114°C. Use the index in the back of the *Handbook* to locate the appropriate tables.

The Synthesis of Salicylic Acid from "Wintergreen Oil"

Preparation and Purification of Organic Solids I

Operations

⟨**OP–2**⟩ Using Standard Taper Glassware

⟨OP–5⟩ Weighing

⟨**OP–7**⟩ Heating

⟨**OP–7a**⟩ Refluxing

⟨OP–12⟩ Vacuum Filtration

⟨OP–21⟩ Drying Solids

⟨OP–23⟩ Recrystallization

⟨OP–28⟩ Melting Point

Experimental Objective

To prepare salicylic acid from synthetic "wintergreen oil" and purify it for use in preparing aspirin.

Learning Objectives

To learn how to conduct an organic reaction under reflux and to gain additional experience with the separation and purification techniques learned previously.

To learn how to perform calculations for organic synthesis.

To learn how to use a laboratory checklist to organize laboratory time efficiently.

To learn and understand some basic principles relating to organic synthesis and its application in the laboratory.

To learn some characteristics of natural and synthetic compounds.

SITUATION

You are a professor of organic chemistry at the Universite d'Artichaut located in the independent island nation of Chaud-Froid. In the past, most of the nation's petroleum was imported from the Arab bloc; but your government offended the oil sheiks by limiting foreign ownership of property, and they cut off your oil in retaliation. Anticipating a severe petroleum shortage, your government has now prohibited the use of petroleum for all but essential purposes.

Proposed route to aspirin

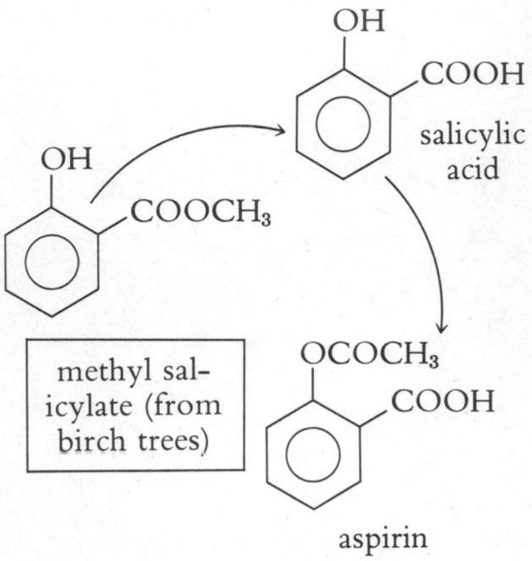

Commercial route to aspirin

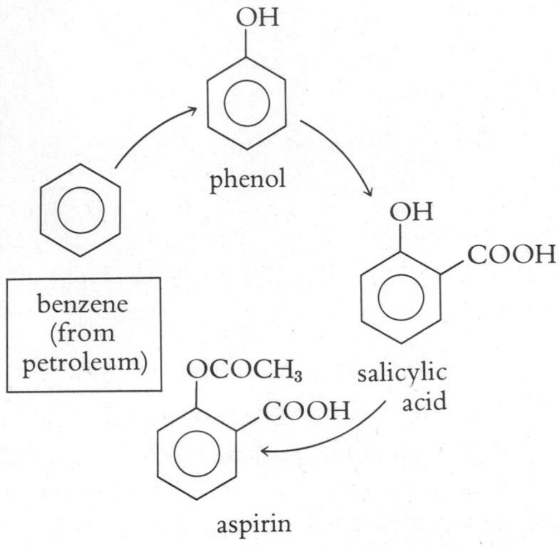

Due to the bad news of an impending energy crisis, Chaud-Froidians are suffering from more frequent headaches, which has caused a rising demand for aspirin. Aspirin is manufactured commercially from salicylic acid, which requires the use of benzene as a raw material. As it happens, benzene is a product of petroleum refining. Consequently, there is such a shortage of salicylic acid that the island's only manufacturer of aspirin, Compagnie de la Maldetete, cannot keep up with the demand. Its board of directors has initiated a desperate search for alternative sources of salicylic acid, and the company's research department is now investigating the possible use of the island's vast stands of sweet birch trees. A process has already been developed for distilling methyl salicylate, commonly known as "oil of wintergreen," from the twigs, roots, and bark of the tree.

You once served as a consultant for the Compagnie de la Maldetete, and they have just asked you to develop an economically feasible process for converting methyl salicylate to salicylic acid. If you should succeed in this endeavor, you will then (in Experiment 4) try to convert your salicylic acid to aspirin.

BACKGROUND

Wintergreen Oil—Natural or Synthetic?

Methyl salicylate was first isolated as a constituent of the oil obtained from the leaves of the wintergreen plant, *Gaultheria procumbens*. This "oil of wintergreen" can be obtained even more readily from other sources such as sweet birch (*Betula lenta*) and yellow birch (*Betula alleghanensis*)—most "natural" methyl salicylate is distilled from the bark of *Betula lenta*. When cheap raw materials became available from petroleum, a more economical commercial process for synthesizing methyl salicylate was devised that involved the **esterification** of salicylic acid with methyl alcohol. A worldwide petroleum shortage could alter the economics of production so that the natural sources of methyl salicylate again become the important ones. Thus, the possibility that aspirin might someday be manufactured from birch trees is not as unlikely as one might think.

Regardless of its source, methyl salicylate is methyl salicylate; there is no difference whatever between the **natural** and **synthetic** compounds provided both are of equal purity. Both forms are composed of the same kind of molecules and must therefore have the same properties. While it is true that there are minor differences between oil of wintergreen obtained from *Gaultheria* or *Betula* species and synthetic methyl salicylate, these differences are due to the 1% or so of impurities in the natural preparations.

Methyl salicylate, both synthetic and natural, has been used for many years as a flavoring agent because of its very pleasant, penetrating odor and flavor. A good wintergreen tea can be made from either the leaves of the wintergreen plant or the twigs and inner bark of sweet and yellow birch; the distinctive wintergreen flavor is found in everything from root beer to mouthwash. Medicinally, wintergreen oil has some of the pain-killing properties of the other salicylates, including aspirin. It can be absorbed through the skin, producing an astringent but soothing sensation, and is frequently used in preparations for muscular aches and arthritis.

Commercial route to synthetic "wintergreen oil"

esterification: combination of an organic acid with an alcohol to form an ester.

natural: obtained directly (or nearly so) from plant or animal matter.

synthetic: prepared by chemically altering other compounds, which may themselves be either natural or synthetic.

Ben-Gay contains methyl salicylate and menthol.

METHODOLOGY

Organic synthesis can be described as the art of preparing pure organic compounds by chemically modifying other compounds. To prepare salicylic acid, for instance, a methyl (CH_3) group must be removed from each molecule of methyl salicylate and replaced by a hydrogen atom. Since we are talking about replacing some 22 sextillion methyl groups by an equal number of hydrogen atoms, the process is going to take some time; so it is not surprising that an extended reaction period is required.

The time required for organic reactions is usually much longer than that for typical inorganic reactions. When silver nitrate and sodium chloride are mixed, their reaction is almost instantaneous, taking place as quickly as each pair of ions can come together to form silver chloride. In a given organic reaction, the reacting molecules may come together about as frequently as the ions in an inorganic reaction, but they may not collide with enough energy or at the right orientation to react. Ions need only "stick together" to form a product; organic molecules must undergo a complicated process of bond making and bond breaking before being converted to product molecules.

The reaction, then, is the crucial step in an organic synthesis—it may be over in a few minutes or require a few weeks, but enough time should be allowed to bring the reaction as nearly as possible to completion before proceeding to the next steps. These steps have already been described in Experiments 1 and 2: the desired product must be *separated* from the reaction mixture, *purified* to remove residual contaminants, and *analyzed* to verify its identity and purity.

In this experiment, the reaction will be carried out by heating methyl salicylate (an oily liquid) and sodium hydroxide under reflux, which will convert methyl salicylate to the disodium salt of salicylic acid. Since methyl salicylate is the **limiting reactant** in this synthesis, it should be carefully weighed so that the theoretical yield of salicylic acid can be accurately calculated. This can be conveniently done by measuring the approximate volume of methyl salicylate (calculated from the Prelab Exercises) into a preweighed, graduated container, then weighing the liquid by difference. A large excess of NaOH is used to insure a rapid, complete reaction. In a fast subsequent reaction, dilute sulfuric acid is added to the salt to generate salicylic acid. The separation of salicylic acid from the reaction mixture will be accomplished by vacuum filtration and the crude product will then be

5 g salicylic acid = 0.036 moles = 22×10^{21} molecules

"Completion" implies that as many reactant molecules as possible have been converted to product molecules under the conditions of the reaction.

See ⟨OP–7a⟩ for a discussion of refluxing.

The **limiting reactant** is the one that determines how much product can (in theory) be formed from a reaction —i.e., the reactant that is not present in excess.

purified by recrystallization from water. Since there may be some insoluble impurities in the crude salicylic acid, it is advisable to filter the hot recrystallization solution.

PRELAB EXERCISES

In this and all subsequent experiments, "read the experiment and understand its objectives" will be an implicit prelab assignment, although it will not be listed as such.

1. Read ⟨OP–2⟩, ⟨OP–7⟩, and ⟨OP–7a⟩. Review ⟨OP–23⟩ and read or review the other operations as needed.

2. Read Appendix IV and calculate the weight and volume of methyl salicylate that will, if completely converted to product, yield 5.00 g of salicylic acid.

3. Read the laboratory checklist in Appendix V and be prepared to use it in the laboratory.

Reactions and Properties

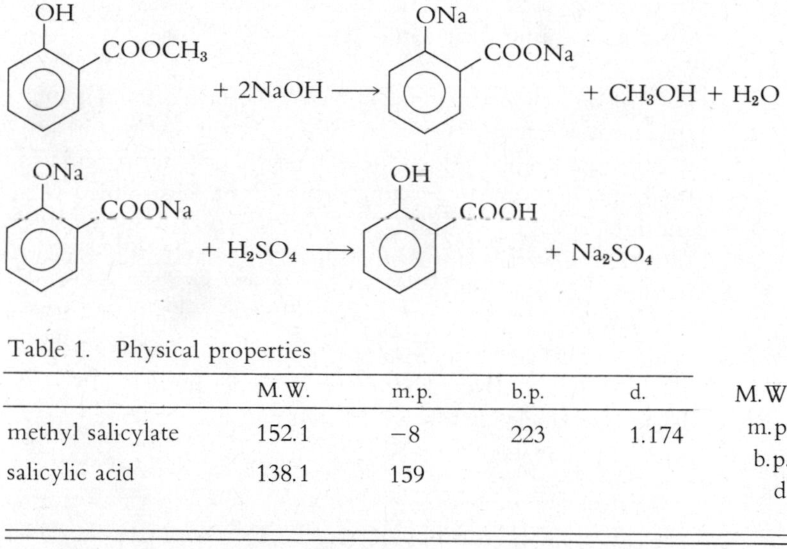

OH
COOCH$_3$

+ 2NaOH ⟶

ONa
COONa

+ CH$_3$OH + H$_2$O

ONa
COONa

+ H$_2$SO$_4$ ⟶

OH
COOH

+ Na$_2$SO$_4$

Table 1.   Physical properties

	M.W.	m.p.	b.p.	d.
methyl salicylate	152.1	−8	223	1.174
salicylic acid	138.1	159		

M.W. = molecular weight (g/mole)
m.p. = melting point (°C)
b.p. = boiling point (°C)
d. = density (g/ml)

Hazards

Aqueous sodium hydroxide is very corrosive. Avoid contact with skin, eyes, or clothing, and clean up spills immediately with large amounts of water. Both methyl salicylate and salicylic acid can irritate the skin and damage eyes, so avoid

sodium hydroxide

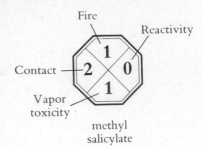

methyl
salicylate

Remember to use boiling chips.

The white solid that forms will dissolve on heating.

The solubility of salicylic acid is approximately 5.3 g/100 ml at 98°C, so it should not be necessary to use more than 100 ml of water.

Salicylic
Acid
Exp. 3
———— g
mp ————°C
Your Name
Current Date

contact. Do not inhale the vapors of methyl salicylate over a long period of time.

PROCEDURE

While carrying out this experiment, refer to the laboratory checklist in Appendix V to help you organize your time efficiently.

Reaction. Assemble ⟨OP–2⟩ the apparatus for reflux ⟨OP–7⟩, ⟨OP–7a⟩ using standard taper glassware if available. Combine 50 ml of 5M sodium hydroxide with the calculated amount of methyl salicylate (from the Prelab Exercises) in the boiling flask and reflux the mixture ⟨OP–7a⟩ for 30 minutes. Let the solution cool to room temperature and transfer it to a beaker large enough to accommodate the solution and the sulfuric acid to be added. Begin stirring the solution and slowly add 70 ml of 2M sulfuric acid. Test the solution with pH paper and, if necessary, add more sulfuric acid to bring it to a pH of 1–2.

Separation. Cool the mixture to room temperature or below (use of an ice bath is recommended) and separate the solid by vacuum filtration ⟨OP–12⟩.

Purification. Recrystallize ⟨OP–23⟩ the crude salicylic acid from water, using 70 ml of hot water to begin with and adding more if needed. When little or no solid material remains, filter the hot solution. After the solution has cooled long enough to crystallize fully, collect the crystals by vacuum filtration ⟨OP–12⟩, dry them ⟨OP–21⟩, and weigh the product ⟨OP–5⟩ in a tared vial.

Analysis. Measure the melting point ⟨OP–28⟩ of the pure salicylic acid and save it for use in Experiment 4. The vial should be labeled as shown in the margin.

Experimental Variations

1. An equivalent volume of dilute hydrochloric acid can be used in place of the sulfuric acid to neutralize the reaction mixture.

2. Methyl or ethyl benzoate can be hydrolyzed to benzoic acid by the procedure described above for methyl salicylate. The quantities of sodium hydroxide and sulfuric acid given are suitable for preparing about 5 grams of benzoic acid.

3. The melting point of an authentic sample of salicyclic acid and a mixture melting point of the authentic sample with the purified product can be obtained for additional practice in determining melting points.

Topics for Report

1. The solubility of salicylic acid in water is approximately 0.25 g/100 ml at 25°C and 0.10 g/100 ml at 0°C. Calculate the maximum amount of salicylic acid that could be recovered after crystallizing 5.00 g of the acid from 90 ml of water if the water were cooled (**a**) to 25°C; (**b**) to 0°C. Calculate the percent recovery in each case, and comment on whether the additional cooling is justified.

2. Calculate the number of moles and equivalents of sodium hydroxide and sulfuric acid used in this experiment, and show that an excess of the sulfuric acid was used to precipitate salicylic acid.

3. The value of K_2 for salicylic acid is 3.6×10^{-14}. If the hydroxide ion concentration of your reaction mixture after refluxing is 3.5M, calculate the approximate percentage of the salicylic acid that will be in the form of the monoprotic salt *A* in that solution.

4. What is the white solid that forms when methyl salicylate is mixed with aqueous sodium hydroxide? Draw its structure.

Library Topics

1. (**a**) Find out how salicylic acid was first prepared and give the literature reference. (**b**) Describe a current method for manufacturing salicylic acid and summarize the steps in a typical manufacturing process.

2. Report on the natural and synthetic sources of Vitamin C (ascorbic acid).

3. Comment on the statement, "Marijuana is a natural drug, so it must be harmless; but aspirin is dangerous because it's man-made." Give examples to support your point of view.

Second ionization step for salicylic acid

A   *B*

$$K_2 = \frac{[B][H^+]}{[A]} = 3.6 \times 10^{-14}$$

ascorbic acid

The Synthesis of Aspirin

Preparation and Purification of Organic Solids II

Operations

⟨OP–5⟩ Weighing

⟨OP–7⟩ Heating

⟨OP–11⟩ Gravity Filtration

⟨OP–12⟩ Vacuum Filtration

⟨OP–21⟩ Drying Solids

⟨**OP–23b**⟩ Recrystallization from Mixed Solvents

Experimental Objective

To synthesize aspirin from the salicylic acid prepared in Experiment 3 and assess its purity.

Learning Objectives

To learn how to recrystallize aspirin from a solvent mixture and to gain additional experience with basic techniques used in the synthesis of organic solids.

To learn how to write a laboratory checklist.

To learn about the history and uses of aspirin.

BACKGROUND

Hermann Kolbe's Blunder

Even today the reasons for aspirin's effectiveness remain somewhat of a mystery, although many theories have been proposed to account for its action.

Although aspirin is the most important pain reliever of modern times, it was considered so unremarkable by its discoverer that it was forgotten for forty years before being resurrected by chemists working for the Bayer Company. The "aspirin story" began even before its discovery when, in 1763, the clergyman Edward Stone read a paper on the uses of willow bark to the Royal Society of London. Stone had found the bark of the white willow, *Salix alba,* quite effective in reducing the intermittent fever (ague) of malaria.

Because of its derivation from *Salix* species, the active principle of willow bark was named salicin. Salicin is a fairly complex substance that can be broken down in the presence of water and an oxidant to form the simpler molecules of glucose (simple sugar) and a sweet-smelling liquid named salicylaldehyde. In 1838 it was discovered that salicylaldehyde obtained from another source, the meadowsweet plant (*Spiraea* species), reacted with strong alkali to yield a white solid on neutralization. The new compound was named spirsäure (*Spiraea* + *säure,* the German name for an acid) by its German discoverers, but called salicylic acid by the English, who traced its lineage back to the white willow tree.

Salicylic acid and its derivatives, known collectively as the salicylates, comprise an important series of medicinally useful compounds. The properties of salicylic acid were discovered as the result of a "brilliant mistake" by one of the most eminent chemists of his day, Hermann Kolbe. Around 1870 Kolbe was approached by a physician who was looking for a safe substitute for carbolic acid (phenol). This acid had long been known to be an effective germicide for external use, but it was much too caustic for internal application against such diseases as typhoid and pneumonia. Twenty years earlier, Kolbe had invented a synthesis of salicylic acid from phenol and carbon dioxide. Perhaps, he reasoned, salicylic acid would break down in the human body to regenerate phenol and thus act as a safe, *internal* source of this germicide. He carried out a number of experiments that "proved" to his satisfaction that salicylic acid was indeed an effective germ-killer, and soon recommended its use on patients suffering from a variety of bacterial diseases and infections. The first reports seemed promising—patients were still dying after salicylic acid treatments, but they felt much better while doing so! Pretty soon doctors began to suspect that the salicylic acid "cured" only those patients who would have survived without any medicine. The blunder, of course, was Kolbe's—salicylic acid does not break down into phenol in the body, and his test results were later discarded as invalid.

Still, it was apparent that the acid provided some benefits: it made the patients feel more comfortable and reduced fever, and its curative effect on rheumatism was remarkable. It was probably about this time that the Rev. Stone's willow-bark treatment was recalled and the chemical relationship between salicin and salicylic acid was recognized. Salicin was known to be mildly effective against both fevers and rheumatism, so why not salicylic acid as well? In fact, the

The aspirin family tree

salicin

salicylaldehyde

$C_6H_{12}O_6$
glucose

salicylic acid

aspirin
(acetylsalicylic acid)

other salicylates

phenol

acid was soon discovered to relieve the pain of neuralgia, sciatica, neuritis, and headaches, as well as rheumatism. Salicylic acid might not be worth much as a germicide, but it was unsurpassed as a fever-fighter and painkiller.

Aspirin Rediscovered

Salicylic acid has one serious drawback as a pain remedy—it irritates the mucous membranes lining the mouth, esophagus, and stomach. The irritation can be so severe that some patients treated with it almost preferred to live with their rheumatism rather than take the cure. One of these disgruntled patients was the father of Felix Hoffman, a staff chemist working for the Bayer Company. By a happy coincidence, Hoffman's company was interested in finding a substitute for salicylic acid at about the same time his father was suffering from its side effects, and this convergence of Hoffmann's personal and research interests provided the motivation which led to aspirin's rediscovery in 1893.

Hoffmann knew that **phenolic compounds** were corrosive because of the presence of a free OH group, and he reasoned that "masking" the OH with some easily removed substituent might provide the benefits of salicylic acid without the irritation. While studying some of the known derivatives of salicylic acid, Hoffmann came across one in which the OH hydrogen was replaced by an acetyl group; this was acetylsalicylic acid, which had been prepared some forty years earlier by Charles Gerhardt. Tests proved that acetylsalicylic acid was superior to all known pain-killers, both in its effectiveness against pain and fever and its freedom from serious side effects, and it was soon being marketed under the Bayer trade name, "aspirin."

Hoffmann's intuition about the function of the acetyl group in aspirin turned out to be quite accurate. Aspirin apparently passes through the acidic environment of the stomach intact, but loses its acetyl group in the alkaline medium of the small intestine to regenerate salicylic acid. The salicylic acid is then absorbed into the bloodstream and carried to the sites where it is needed.

METHODOLOGY

Acetic anhydride is an effective acetylating agent which reacts rapidly with alcohols or phenols in the presence of a

A **phenolic compound** contains an OH group on a benzene ring. Salicylaldehyde, salicylic acid, and phenol are all phenolic, but aspirin is not.

aspirin

"Aspirin" is derived from *acetylspirs*äure, the German name for acetylsalicylic acid.

acetic anhydride

catalyst to yield acetate esters. In this experiment, acetic anhydride will be used (in approximately 100% excess) to convert salicylic acid to aspirin and to serve as a solvent for the
reaction. Since acetic anhydride is so reactive, it will not be
necessary to heat the reactants to reflux temperatures;
merely warming them in a 50–60° water bath will be sufficient. When the reaction is complete, water is added to destroy the excess acetic anhydride (converting it to water-soluble acetic acid) and to precipitate the crude aspirin.

The most likely impurities in the crude product are salicylic acid itself and polymers formed by various combinations of salicylic acid with aspirin molecules. The polymers
can be removed by dissolving the product in aqueous sodium bicarbonate solution and filtering off any insoluble
material, since aspirin reacts with sodium bicarbonate to
form a water-soluble sodium salt whereas the polymeric by-
product is largely insoluble. Aspirin is reprecipitated from
the solution of its salt by acidifying the filtrate with hydrochloric acid. The salicylic acid impurity remaining in the aspirin is removed by recrystallizing it from an ethanol-water
mixture. Because aspirin can partially hydrolyze back to salicylic acid and acetic acid when dissolved in boiling water, a
special recrystallization procedure will be used in which the
aspirin is first dissolved in hot ethanol and the solution is
brought to the saturation point by adding warm water.

Since salicylic acid is the most likely impurity in aspirin,
it will be tested for in both the crude solid and in the recrystallized aspirin. Like most other phenols, salicylic acid reacts
with ferric chloride to form a highly colored complex ion;
aspirin, in which the phenolic OH group is acetylated, does
not. The absence of color in the test solutions thus suggests a
high degree of purity.

*The interconversion of aspirin with its so-
dium salt is described in Experiment 1.*

*The melting point of aspirin is not a reli-
able indicator of its purity, because it
partly decomposes on heating.*

PRELAB EXERCISES

1.   Read ⟨OP–23*b*⟩ and read or review the other operations as needed.

2.   Read Appendix V, Writing a Laboratory Checklist,
and prepare a checklist for this experiment following the
format illustrated or one recommended by your instructor.
The checklist should be submitted to your instructor for approval at the beginning of the laboratory period.

Reaction and Properties

$$\underset{\text{salicylic acid}}{\begin{array}{c}\text{OH}\\\text{COOH}\end{array}} \; + \; \underset{\text{acetic anhydride}}{(CH_3CO)_2O} \; \xrightarrow{H_2SO_4} \; \underset{\text{aspirin}}{\begin{array}{c}\text{OCOCH}_3\\\text{COOH}\end{array}} \; + \; \underset{\text{acetic acid}}{CH_3COOH}$$

Table 1.   Physical properties

"d" by a melting point means that the substance decomposes while melting.

	M.W.	m.p.	b.p.	d.
salicylic acid	138.1	159		
acetic anhydride	102.1	−73	140	1.082
aspirin	180.2	135 d		

acetic anhydride

sulfuric acid

If your yield was less than three grams, obtain additional salicylic acid from your instructor or scale down the quantities given.

If crystals remain in the reaction flask, use a little filtrate to wash them into the funnel.

Hazards

Acetic anhydride causes severe burns to skin and eyes, its vapors irritate the respiratory system, and it reacts violently with water. Prevent contact, do not breathe vapors; use of rubber gloves and a fume hood are advisable. Sulfuric acid is very corrosive and reacts violently with water. Handle with great care, avoid contact with skin or eyes, keep away from water.

Review the introductory section, Reacting to Accidents: First Aid, to learn what to do in case of accidental contact with these chemicals.

PROCEDURE

Reaction.   Mix your weighed, *dry* salicylic acid from Experiment 3 with 10 ml of pure acetic anhydride (*Caution!* See Hazard notes above) in a 125-ml Erlenmeyer flask and add ten drops of concentrated sulfuric acid (*Caution!* See Hazard notes above). Swirl or stir the mixture in a 50–60° warm water bath ⟨OP–7⟩ until the solid dissolves; then leave the solution in the bath for 10 minutes with occasional swirling. Allow the solution to cool to room temperature, or until crystallization begins, then add 75 ml of water and cool the mixture in an ice bath to complete the crystallization.

Separation.   Separate the aspirin from the reaction mixture by vacuum filtration ⟨OP–12⟩ (washing the crystals with cold water) and allow them to air dry briefly on the filter.

Purification.   Save a few crystals (about 10 mg) of the crude product for analysis and transfer the rest to a 250-ml beaker. Slowly add 60 ml of a 5% aqueous solution of sodium bicarbonate and stir the mixture until all signs of reaction (fizzing noise) have stopped. An extra 5 ml or so of the bicarbonate solution can be added to make certain the reaction is complete. Filter the solution by gravity ⟨OP–11⟩ through a coarse, fluted filter paper and pour it, while stirring, into 20 ml of 3M hydrochloric acid. The aspirin should precipitate at this point; if precipitation is not complete, another 5 ml of 3M HCl can be added. Cooling the mixture in an ice bath will help insure maximum recovery of the crystals. Collect the aspirin by vacuum filtration ⟨OP–12⟩ and air dry it thoroughly on the filter.

Be careful that the solution does not foam over the top of the beaker.

Test for complete precipitation by allowing the precipitate to settle and adding a few drops of HCl to the clear liquid.

Recrystallize the aspirin from ethanol/water by the following modification of the mixed solvent method ⟨OP–23b⟩. Dissolve the aspirin in a minimum (measured) volume of boiling ethanol and add another 1–2 ml of hot ethanol (total volume 15 ml or less). Stir in twice that total volume of warm (50–60°) water. If crystals form at this point, dissolve them by gentle heating. Allow the solution to cool until crystallization is complete, collect the pure crystals by vacuum filtration, ⟨OP–12⟩ and wash them with cold water. Dry ⟨OP–21⟩ and weigh ⟨OP–5⟩ the aspirin, set aside a few crystals for analysis, and turn in the remainder to your instructor.

Heat the ethanol on a steam bath or hot plate.

Usually about 20–30 ml of hot water is required.

If crystals do not form on cooling, evaporate some of the solvent on a steam bath.

Analysis.   Place approximately equal quantities (about 10 mg) of your crude aspirin, the recrystallized aspirin, and pure salicylic acid in three small test tubes, dissolve each sample in 1 ml of ethanol, and add a drop of 1% ferric chloride to each test tube. Results of the test should be recorded and interpreted in your report.

The crystals need not be completely dry for analysis.

Experimental Variations

1.   Ordinary reagent-grade salicylic acid can be used in place of the Experiment 3 product.

2.   If time is limited and the instructor permits, the first purification step can be omitted.

3.   Aspirin can also be recrystallized from an ethyl ether/petroleum ether mixture (see ⟨OP–23b⟩).

Aspirin may decompose on standing, so it will be instructive to test both old and freshly purchased bottles of aspirin.

4.   Commercial aspirin tablets can be tested for the presence of salicylic acid by the method described above. The presence of starch as a binder in aspirin tablets can also be detected by boiling 10 mg of commercial aspirin in 2 ml of water and adding a drop of iodine/potassium iodide solution. Starch forms a deep blue-violet complex with iodine.

Topics for Report

A 5% (by weight) solution contains 5 g of solute per 100 g of solution.

1.   Calculate the volume of 3M HCl needed to just neutralize the 60 ml of 5% sodium bicarbonate used in the purification step. The density of 5% $NaHCO_3$ is 1.034 g/ml.

2.   Write balanced equations for the reaction of aspirin with sodium bicarbonate and the neutralization of the resulting solution with hydrochloric acid.

Market prices, July 1979
acetic anhydride: $0.27/lb. (r.r. tank cars)
salicylic acid: $0.90/lb. (U.S.P. crystal, drums)

3.   A small bottle of 5-grain aspirin tablets holds 100 tablets, each containing about 0.325 g of aspirin. Calculate the cost of the acetic anhydride and salicylic acid required to prepare the aspirin in such a bottle, using the market prices given in the margin and assuming equimolar quantities of the reactants. At your instructor's request, you can use the current market prices listed in *Chemical Marketing Reporter*.

Library Topics

1.   Acetic anhydride is one of the top fifty chemicals in the U.S. in terms of production volume, being manufactured at the rate of about two billion pounds per year. Report on some commercial uses for acetic anhydride other than producing aspirin.

2.   It has been speculated that aspirin exerts its analgesic effect by inhibiting the action of bradykinin, a "pain molecule." Find out what you can about the role of bradykinin in causing pain and the apparent effect of aspirin on this process.

3.   Look up the structures and report on the drug uses of some other salicylates such as salicylamide, sodium salicylate, and phenyl salicylate.

The Preparation of Synthetic "Banana Oil"

Preparation and Purification of Organic Liquids. Modifying Literature Procedures.

Operations

⟨OP–5⟩ Weighing

⟨OP–7*a* ⟩ Refluxing

⟨OP–11⟩ Gravity Filtration

⟨**OP–19**⟩ Washing Liquids

⟨**OP–20**⟩ Drying Liquids

⟨**OP–25**⟩ Simple Distillation

Experimental Objective

To prepare isoamyl acetate from acetic acid and isoamyl alcohol by modifying a procedure used for preparing butyl acetate.

Learning Objectives

To learn and apply some of the basic operations used in preparing and purifying organic liquids.

To learn how to revise a literature procedure so that it can be applied to the synthesis of a different (but chemically related) product.

To learn about some of the properties and uses of esters, and about the formulation of artificial flavoring agents.

SITUATION

You are an agricultural chemist employed by the Destileria Bullir y Burbujear in the tropical republic of El Platano Grande. The most important agricultural product of El Platano Grande is bananas, and one of its most popular exports is your company's potent banana liqueur, Ojo Rojo

The banana plant (*Musa cavendishii* and other species) is actually a gigantic herb and not a real tree; each year the foliage-bearing part withers away and is replaced by new growth from an underground stem.

$$CH_3\overset{\displaystyle O}{\overset{\|}{C}}OCH_2CH_2CH-CH_3$$
$$|$$
$$CH_3$$

isoamyl acetate

fusel oil: a by-product of carbohydrate fermentation that contains mostly isoamyl alcohol.

$$CH_3\overset{\displaystyle O}{\overset{\|}{C}}OCH_2CH_2CH_2CH_3$$

butyl acetate

$$R-\overset{\displaystyle O}{\overset{\|}{C}}-O-R'$$

acid   alcohol
portion   portion

an ester

Antiguo. This liqueur is prepared from a very secret recipe that has been passed down from generation to generation since the time of Axolotl I, legendary king of ancient Platano. Its ingredients include a specially prepared banana extract and a distillate obtained from fermented banana mash.

A catastrophic hurricane has just destroyed most of the banana trees in El Platano Grande and, as a result, the Destileria Bullir y Burbujear can no longer obtain its normal supply of banana extract. The only solution to its dilemma appears to be the rapid development of a synthetic formulation that can be mixed with genuine banana extract without causing too obvious an alteration in its qualities.

You have been told to research the problem and to come up with a synthetic "banana oil" that will do the job. While hunting for leads, you went over some old lecture notes from your organic chemistry course at the Calendula Agricultural Institute and discovered a marginal notation stating that isoamyl acetate has a strong odor resembling that of ripe bananas! Unfortunately, your company's technical library is sadly deficient, and your search for a synthetic procedure led only to a 1878 article in an obscure foreign publication called (in rough translation) *The Journal of the Society for the Practice of Chemistry, Organic Gardening, and Gourmet Cooking.* Your translation of the procedure reads as follows:

*In our quest to obtain a superior essence of ripe bananas we mixed purified **fusel oil** with essence of dehydrated vinegar in an iron pot and cooked the mixture over an open hearth. By an unfortunate accident, the liquid boiled over, and for weeks and weeks thereafter it reeked of bananas all over the place! We thereupon gave up our banana-pursuit and went back to growing celery.*

This is not much help. Luckily, while reviewing your old laboratory text from Calendula A.I., you came across a procedure for preparing butyl acetate, a very similar ester. Now you have only to make some minor revisions in the procedure in order to prepare synthetic "banana oil," isoamyl acetate.

BACKGROUND

Esters and Artificial Flavorings

Esters are organic compounds that consist of an acid portion and an alcohol (or phenol) portion. That is, they can in theory be prepared by eliminating water from the molecules of a carboxylic acid and an alcohol or phenol and combining the two fragments. Several of the compounds you worked with in previous experiments were esters; methyl salicylate is the ester of salicylic acid and methanol, whereas aspirin is the ester of acetic acid and salicylic acid.

Most of the volatile esters, including methyl salicylate, have strong, pleasant odors often described as "fruity." The ester you will be preparing in this experiment, isoamyl acetate (also named isopentyl acetate or 3-methylbutyl acetate) has a strong odor of ripe bananas when pure; in dilute solution, its odor resembles pears. Some other esters with odors and flavors resembling natural ones are illustrated in Table 1.

Table 1.   Esters as flavor ingredients

Name	Structure	Odor/Flavor
Propyl acetate	$CH_3\overset{\displaystyle O}{\overset{\|}{C}}-OCH_2CH_2CH_3$	pears
Octyl acetate	$CH_3\overset{\displaystyle O}{\overset{\|}{C}}-O(CH_2)_7CH_3$	oranges
Benzyl acetate	$CH_3\overset{\displaystyle O}{\overset{\|}{C}}-OCH_2-$⟨◯⟩	peaches, strawberries
Isopentenyl acetate	$CH_3\overset{\displaystyle O}{\overset{\|}{C}}-OCH_2CH=\overset{\displaystyle CH_3}{\overset{\|}{C}}-CH_3$	"Juicy-Fruit"
Isobutyl propionate	$CH_3CH_2\overset{\displaystyle O}{\overset{\|}{C}}-OCH_2\overset{\displaystyle CH_3}{\overset{\|}{C}H}-CH_3$	rum
Ethyl butyrate	$CH_3CH_2CH_2\overset{\displaystyle O}{\overset{\|}{C}}-OCH_2CH_3$	pineapples

Some esters that do not exist in nature have been used to construct new flavors; isopentenyl acetate, for example, is the main flavor ingredient of Juicy-Fruit gum. More often, esters are used to manufacture artificial flavorings that resemble the genuine article. They may be used alone or in combination with other esters and with additional flavoring ingredients such as those listed in Table 2.

The ingredients in one formulation for an artificial raspberry flavoring include the seven esters isobutyl acetate, phenylethyl isobutyrate, phenylethyl anthranilate, hexyl butyrate, isoamyl acetate, ethyl butyrate, and ethyl methyl

Table 2.   Other flavor ingredients

Name	Structure	Odor/Flavor
Anisaldehyde	 $CH_3O-\bigcirc-CHO$	hawthorne flowers
Vanillin	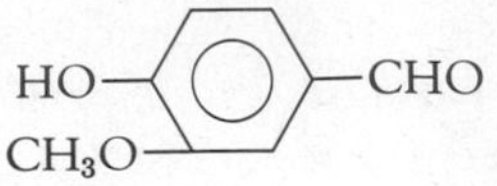 $HO-\bigcirc-CHO$ CH_3O	vanilla
Phenylethyl alcohol	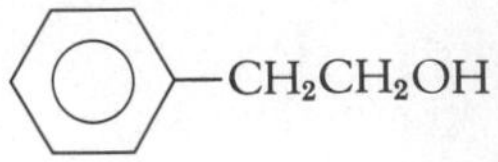 $\bigcirc-CH_2CH_2OH$	roses
β-Ionone	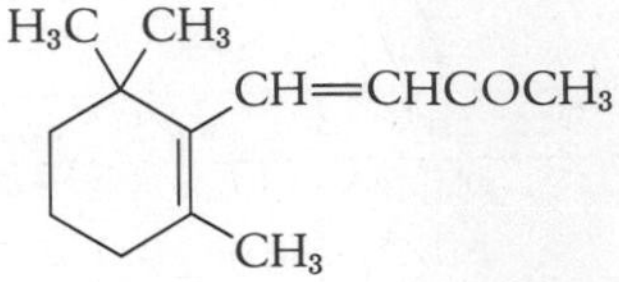 $\underset{CH_3}{\overset{H_3C\ \ CH_3}{\bigcirc}}CH=CHCOCH_3$	violets

Fixatives

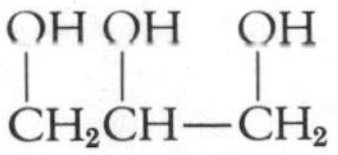

$$\underset{\text{glycerine}}{\overset{OH\ OH\ \ \ OH}{CH_2CH-CH_2}}$$

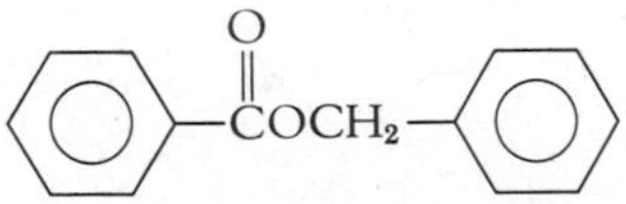

benzyl benzoate

Vehicle

CH_3CH_2OH
ethyl alcohol
(ethanol)

p-tolyl glycidate, along with some anisaldehyde, vanillin, phenylethyl alcohol, β-ionone, diallyl sulfide, and rose oil.

In addition to artificial flavor ingredients, some natural flavors such as the essential oils of plants or extracts from plant and animal sources may be included in a particular formulation. High boiling compounds like benzyl benzoate or glycerine are added to retard vaporization of volatile components; such an ingredient is called a fixative. Finally, all of the ingredients are dissolved in a solvent called the vehicle, which helps to blend and hold the components. The most commonly used vehicle is ethyl alcohol. Because of the extreme complexity of natural flavors (more than 100 compounds are said to make up the flavor of ripe strawberries), such artificial preparations may only be a crude approximation of the real thing. Superior flavor essences are most often composed of a large proportion of the natural fruit extract, fortified with a few synthetics to replace flavor elements lost in the extraction process and to support and enhance the natural flavor.

METHODOLOGY

As stated in the Situation, the procedure that follows is for preparing *n*-butyl acetate and must be revised to make it suitable for isoamyl acetate. Since the two esters are very similar in structure and properties and only one of the reactants (the alcohol) is different, only a few modifications will be necessary. Isoamyl alcohol has a higher molecular weight than *n*-butyl alcohol (1-butanol), so more of it will be needed to provide the same number of moles. Isoamyl acetate boils at a higher temperature than *n*-butyl acetate, so the distillation range must be changed.

In general, a procedure utilizing a reactant **A** can be revised for use with a similar reactant **B** through the following steps:

1.   Calculate the number of moles of **A** in the original procedure;

2.   Calculate the amount (mass or volume) of **B** needed to provide the same number of moles;

3.   Change the values for any physical properties (distillation boiling ranges, etc.) of the original product so that they are applicable to the new product.

In preparing aspirin, acetic anhydride served as the acetylating agent; in this experiment, the less reactive compound acetic acid is used instead. Consequently a longer reaction time and a higher temperature will be required to bring the reaction to equilibrium. In contrast to the aspirin synthesis, the reaction forming isoamyl acetate is reversible. By using a large excess of the acetic acid, the equilibrium can (by Le Chatelier's principle) be shifted to favor the products, but even then it will be impossible to obtain a quantitative (99% +) yield of the ester under the conditions used.

The operations used for isolating and purifying organic liquids differ from those used for solids in several respects. A liquid cannot be separated from a reaction mixture by filtration or purified by recrystallization, and its purity cannot be assessed by a melting point determination. In this experiment, the isoamyl acetate will be separated from other components in the reaction mixture by washing it with water and sodium bicarbonate. This removes the excess acetic acid, the sulfuric acid catalyst, and other water-soluble impurities. Any residual water in the ester is taken up by a drying agent, and the crude product is purified by distillation to remove unreacted isoamyl alcohol and any remaining impurities. The distillation boiling range gives a rough indication of purity.

These processes are summarized by the flow diagram in Figure 1.

$$CH_3\overset{\displaystyle O}{\overset{\|}{C}}\!-\!OH$$
acetic acid

Figure 1.   Flow diagram for the
synthesis of isoamyl acetate

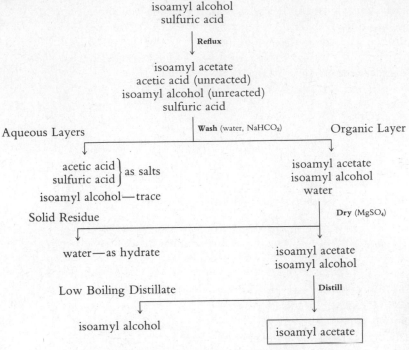

PRELAB EXERCISES

1.   Read ⟨OP–19⟩, ⟨OP–20⟩ and ⟨OP–25⟩, and read or review the other operations as needed.

2.   Revise the literature procedure given for preparing butyl acetate so that it can be used for isoamyl acetate, following the general approach outlined in the Methodology.

3.   Make up a checklist for the experiment, including the quantities of all reagents, catalysts, wash liquids, and drying agents.

Reaction and Properties

Table 3.   Physical properties

	M.W.	m.p.	b.p.	d.
acetic acid	60.1	17	118	1.049
isoamyl alcohol	88.1		129	0.809
isoamyl acetate	130.2	−79	142	0.867
n-butyl alcohol	74.1	−90	117	0.810
n-butyl acetate	116.2	−78	125	0.883

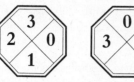

$$CH_3\overset{O}{\overset{\|}{C}}-OH + HOCH_2CH_2\overset{CH_3}{\overset{|}{C}}HCH_3 \rightleftharpoons CH_3\overset{O}{\overset{\|}{C}}-OCH_2CH_2\overset{CH_3}{\overset{|}{C}}HCH_3 + H_2O$$

acetic acid   isoamyl alcohol   isoamyl acetate

Hazards

alcohol	sulfuric acid	acetic acid	ester
3 / 2 / 0 / 1	0 / 3 / W̶ / 2	2 / 4 / 1 / 2	3 / 1 / 0 / 2

Hazard numbers are for isobutyl acetate and isobutyl alcohol, but the isoamyl compounds should be similar.

Acetic acid is very corrosive and a strong irritant. Do not breathe vapors; avoid contact with eyes and skin. Isoamyl acetate and isoamyl alcohol are both quite flammable and are respiratory irritants. The alcohol is a strong eye irritant. Keep them away from open flames; avoid breathing vapor and prolonged contact with skin. Avoid all contact with sulfuric acid; see Experiment 4 for hazard data.

PROCEDURE FOR BUTYL ACETATE

Italicized material must be changed as described above.

Reaction.   Mix together *11.1 g (13.7 ml) of* n-*butyl alcohol* and 17.2 ml of glacial acetic acid in a 100-ml round-bottom flask, and carefully add 1 ml of concentrated sulfuric acid. Attach a water-cooled reflux condenser, and reflux⟨OP–7a⟩ the mixture gently for one hour.

Use acid-resistant boiling chips.

Separation.   Wash the reaction mixture⟨OP–19⟩ in a separatory funnel with 50 ml of water, followed by two portions of saturated sodium bicarbonate solution (see ⟨OP–13⟩ regarding the volume of wash solvent required). Dry the ester ⟨OP–20⟩ with anhydrous magnesium sulfate, and filter by gravity ⟨OP–11⟩ into a dry 50-ml round-bottom boiling flask (for distillation) or into another container.

Important: *Stir layers until CO_2 evolution subsides before stoppering flask. Be certain to save the right layer.*

Purification.   Distill the product ⟨OP–25⟩, collecting the pure *butyl acetate at 120–126°.* Weigh ⟨OP–5⟩ and turn in the product.

All glassware must be dry and thermometer placement correct.

Experimental Variations

1. The purity of the product can be checked by measuring its refractive index ⟨OP–30⟩.

2. The ester can also be separated from the reaction mixture by distilling it until about three ml of liquid remain in the flask, then washing the distillate with water and sodium bicarbonate solution. It can then be dried and purified as described above.

3. Many other esters of primary alcohols can be prepared by the general procedure described here. Students might use different alcohols of three carbons or more to synthesize such esters as propyl acetate, octyl acetate, or benzyl acetate, and compare the yields and properties of their products.

4. A small amount of the isoamyl acetate can be dissolved in 95% ethanol and changes in its odor quality on dilution observed.

5. If samples of a number of different esters are available, their odors may be observed and described during the reflux period.

Topics for Report

1. The equilibrium constant (K) for the formation of isoamyl acetate is about 4.2. **(a)** Calculate the maximum possible amount of isoamyl acetate that could be recovered if you started with 0.15 moles each of isoamyl alcohol and acetic acid. **(b)** Repeat the calculation with the quantities of starting materials used in this experiment. **(c)** Calculate the percentage of the theoretical yield (for 100% completion) this represents in each case, and compare the results with your experimental yield.

2. **(a)** Write equations for the reactions of sulfuric acid and acetic acid with the aqueous sodium bicarbonate solution. **(b)** Explain why these reactions result in the removal of both acids from the organic layer containing isoamyl acetate.

3. How would the procedure have to be revised to prepare isobutyl propionate? Give the names, structures, and quantities of all starting materials required.

Isoamyl acetate equilibrium
$i\text{-AmOH} + \text{HOAc} \rightleftharpoons i\text{-AmOAc} + \text{H}_2\text{O}$

$$K = \frac{[i\text{-AmOAc}][\text{H}_2\text{O}]}{[i\text{-AmOH}][\text{HOAc}]}$$

Ac = acetyl, $\text{CH}_3\text{CO}-$
i-Am = isoamyl, $\underset{\underset{\text{CH}_3}{|}}{\text{CH}_3\text{CHCH}_2\text{CH}_2-}$

$$\underset{\text{isobutyl propionate}}{\text{CH}_3\text{CH}_2\overset{\overset{\text{O}}{\|}}{\text{C}}-\text{OCH}_2\overset{\overset{\text{CH}_3}{|}}{\text{CH}}\text{CH}_3}$$

Library Topics

1.   Look up some industrial processes for synthesizing such esters as butyl and isoamyl acetate and compare them with the laboratory synthesis you carried out.

2.   Describe two or more methods (other than using an excess of one reactant) that are used in esterification reactions to shift the equilibrium toward the products.

3.   Many acetate esters, including isoamyl acetate, act as pheromones for various species of insects. These "chemical messengers" may function as sex attractants, alarm signals, path markers, etc. **(a)** Look up the structures of gyplure and propylure and find out which insects secrete them and what function they perform. **(b)** Describe how the common honeybee makes use of such pheromones as isoamyl acetate.

Breaking the Chemical Code

Physical and Chemical Properties. Functional Group Chemistry.

Operation

⟨OP–6⟩ Measuring Volume

Experimental Objective

To classify each of eight unknown organic compounds in the correct chemical family and to compare the odors and physical properties of the compounds representing each family.

Learning Objectives

To learn how to use the chemical and physical properties of organic compounds in their identification and characterization.

To learn how to recognize some important functional groups.

To learn how the physical and chemical properties of organic compounds are affected or determined by their structures, and in particular by their functional groups.

To gain experience in using chemistry reference books to locate the physical properties of organic compounds.

SITUATION

You are a graduate student at a prestigious eastern university, the Miskatonic Institute of Lucubration, where you have recently been granted a teaching assistantship in organic chemistry. Your duties include setting up and supervising some undergraduate laboratory sections.

For tomorrow's experiment, which deals with organic qualitative analysis, you have transferred eight different organic liquids into eight numbered bottles according to a code prepared by the MIL professor who supervises your work. Upon returning from an extended coffee break you discover, to your dismay, that an uncoordinated work-study student has used your code sheet to wipe up a sulfuric acid spill. After the initial hysteria has passed, you pick up the coffee-cup fragments and begin thinking of ways to salvage the situation, since if the lab is not set up on time you will end up scrubbing beakers for a month.

You know the *names* of the eight compounds, because they are on the (now empty) reagent bottles you poured them from, but you don't know which name corresponds to which number. You cannot make up a new code and prepare another set of liquids because the stockroom has just closed and you have no way of obtaining more chemicals. Your only options are to: (**1**) pretend to have a nervous breakdown and spend the next month in a rest home; or (**2**) run a series of tests on the compounds to find out which one is in each bottle. Being a highly motivated and conscientious graduate student, you choose option 2.

BACKGROUND

Chemical "Taxonomy"

A botanist who wishes to identify an unknown flowering plant will generally examine the flowering parts first, to detect features that indicate to which family of plants it belongs. Then by a more detailed examination of the whole plant, he or she can determine the genus and species. For example, a plant having four symmetrical flower petals and six stamens (four long and two short) might be classified with the *Cruciferae,* or mustard, family. White flower petals and the presence of lyre shaped leaves and a red, globular root classify the plant as a specimen of *Raphanus sativus.* Just as an experienced gardener might immediately identify that plant as an over-mature radish, an experienced chemist might recognize some familiar organic compounds by their odors and general appearance; but for most purposes a systematic approach to identification is needed.

The identification, or qualitative analysis, of organic compounds is analogous in some ways to **plant taxonomy.** To classify an organic compound into a given family requires detecting a specific **functional group** on the molecules of the compound. Obviously one cannot detect such groups visually, but because the kind of functional group present determines a compound's chemical properties (and, to a large extent, its physical properties) one can classify a compound into a specific family by finding out what kinds of

Raphanus sativus = garden radish

plant taxonomy: the systematic classification of plants according to their structural features and presumed natural relationships.

functional group: the atom or group of atoms that is characteristic of a family of organic compounds and that determines the chemical properties of the family.

A list of the functional groups encountered in this experiment is given in Table 1.

For a more comprehensive discussion of qualitative analysis, see Part III.

chemical reactions it undergoes. Once its chemical family has been determined, a compound can be identified by various methods, including measurement of physical constants, preparation of derivatives, and spectrometric analysis. In this experiment, all of the unknown compounds belong to different families, and since the names of the eight compounds are given it will only be necessary to determine the family of each compound in order to learn its identity.

Table 1.　Common functional groups and their families

Functional group	Functional group name	Family
$-C=C-$	carbon-carbon double bond	alkene
$-Cl$	chlorine atom	alkyl chloride
$-OH$	hydroxyl group	alcohol (also phenol)
$-\overset{\overset{O}{\|\|}}{C}-H$ $(-CHO)$	carbonyl group, with H on carbonyl carbon	aldehyde
$-\overset{\overset{O}{\|\|}}{C}-$ $(-CO-)$	carbonyl group, no H on carbonyl carbon	ketone
$-\overset{\overset{O}{\|\|}}{C}-OH(-COOH)$	carboxyl group (*carb*onyl + hyd*roxyl*)	carboxylic acid
$-NH_2$	amino group	amine (primary)

Condensed representations of some functional groups are shown in parentheses.

There are far fewer chemical than botanical families; but an organic compound can belong to more than one family. A specimen of *Raphanus sativus* could not belong to both the *Cruciferae* and the *Violaceae,* even if a radish plant were to unaccountably bloom violets, because taxonomic rules do not allow it. But as simple a compound as vinyl chloride is both an alkene and an alkyl chloride because it contains the functional groups of both families; and vanillin, the principal constituent of vanilla extract, is at the same time an aldehyde, phenol, and ether.

To simplify matters, each compound used in this experiment contains only one functional group and thus belongs to only one chemical family. The names and structures of the eight compounds, along with the families they belong to, are given in Table 2. The compounds, all clear, more-or-

Vinyl chloride, $CH_2{=}CHCl$, is a gas used to manufacture polyvinyl chloride (PVC) plastics.

aldehyde f.g.　CHO
ether f.g.　OCH₃
OH　alcohol & phenol f.g.
vanillin

less colorless liquids, have been carefully selected for molecular weights in the range 70–80, so that the properties they exhibit will be due almost entirely to their functional groups rather than to molecular weight variations. This will allow you to compare their odors, boiling points, densities, and water solubilities and draw some conclusions about the relationship of structure to properties.

Table 2.   Compounds to be identified

Compound	Structure	Family
butanal	$CH_3CH_2CH_2\underline{CHO}$	aldehyde
1-butanol	$CH_3CH_2CH_2CH_2\underline{OH}$	alcohol
2-butanone	$CH_3CH_2\underline{CO}CH_3$	ketone
butylamine	$CH_3CH_2CH_2CH_2\underline{NH_2}$	amine
1-chloropropane	$CH_3CH_2CH_2\underline{Cl}$	alkyl chloride
pentane	$CH_3CH_2CH_2CH_2CH_3$	alkane
2-pentene	$CH_3CH_2\underline{CH=CH}CH_3$	alkene
propanoic acid	$CH_3CH_2\underline{COOH}$	carboxylic acid

Note: The functional groups are underlined.

Molecular Structure and Odor

Compounds in a given chemical family often have similar odors, or odors with some characteristics in common, although alterations in molecular size and shape can introduce marked differences. Within a given family, odor intensity will often reach a maximum at some intermediate molecular weight, then drop off as the compounds become less volatile; odor quality may vary considerably as well. The odors you observe in this experiment will not necessarily be characteristic of all members of the corresponding families, but they should be characteristic at least of the lower-molecular-weight compounds of similar structure.

It is difficult to describe odors with any accuracy, since their perception is a very subjective experience. One attempt to classify odors proposed seven primary odor types, corresponding to seven different kinds of olfactory receptor sites. Each receptor would be activated when a molecule that had the right size and shape entered and fit into it. The seven basic types postulated were camphoraceous, musky, floral, pepperminty, ethereal, pungent, and putrid.

If it is true that there are primary odors, just as there are primary colors, it should be theoretically possible to reconstruct any given odor by mixing primary odor standards in specific proportions.

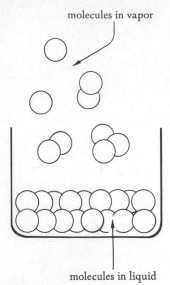

molecules in vapor

molecules in liquid

Figure 1.   Boiling of a liquid

London dispersion forces are sometimes referred to as "Van der Waals forces."

$$\underset{\delta+ \quad \delta-}{H_2C=O}$$
$$\underset{\delta- \quad \delta+}{O=CH_2}$$

dipole-dipole
interaction
(formaldehyde)

$$\underset{\delta+ \quad \delta-}{H-O}$$
$$H \; \delta+$$
$$\underset{\delta- \quad \delta+}{O-H}$$
$$\underset{\delta+}{H}$$

hydrogen
bonding
(water)

More detailed information on the various kinds of intermolecular forces can be found in most general chemistry textbooks.

Approximate atomic radii and nuclear masses

   C: 77 pm, 12 D
   N: 75 pm, 14 D
   O: 74 pm, 16 D
   Cl: 100 pm, 35.5 D

A dalton (D) is one atomic mass unit (amu); 1 pm = 10^{-12} m.

Molecular Structure and Boiling Points

Boiling occurs in a liquid when the kinetic energy of its component molecules becomes high enough to overcome the forces between them, allowing them to leave the surface of the liquid and enter the gaseous state. Since the kinetic energy of molecules increases with temperature and the energy required to separate molecules depends on the strength of the forces between them, we can expect liquids having strong intermolecular forces also to have high boiling points.

The three important kinds of intermolecular forces occurring between organic molecules are, in order of increasing strength: (1) London dispersion forces; (2) dipole-dipole interactions; and (3) hydrogen bonding. *Dispersion forces* are caused by alternating transient charge separations on the surfaces of molecules. Although they occur between all organic molecules, their effect is negligible in many molecules containing oxygen or nitrogen atoms, for which the polar attractive forces predominate. *Dipole-dipole interactions* result when molecules having permanent bond dipole moments line up with the negative end of one molecule's dipole opposite the positive end of another's, and vice versa. *Hydrogen bonding* is a special kind of dipole-dipole interaction involving the attraction of a highly polarized hydrogen atom for an electron donating atom (such as O or N) on another molecule. In organic chemistry, hydrogen bonding makes important contributions to boiling points only in compounds containing OH and NH bonds, with the OH compounds forming the strongest hydrogen bonds.

Molecular Structure and Density

The density of an organic molecule depends, to an extent, on the mass/volume ratio of its constituent atoms. Atoms having a high nuclear mass confined within a small atomic volume, such as those in the top right-hand corner of the periodic table, have a high atomic density and thus increase the density of molecules containing them. Among the atoms encountered in this experiment (other than hydrogen) oxygen has the highest mass/volume ratio, followed by chlorine, nitrogen, and carbon. The density of an organic liquid will thus depend on the kind of "heavy" atoms (other than C and

H) it contains and on the fraction of its molecular mass they contribute. For example, chlorobenzene (C_6H_5Cl) has a higher density than phenol (C_6H_5OH) even though Cl has a lower atomic density than O, because chlorine contributes 32% of the molecular mass of chlorobenzene while oxygen makes up only 17% of the molecular mass of phenol.

Molecular Structure and Solubility

A compound will generally be soluble in a given solvent if the forces between its own molecules are much the same as those between the molecules of the solvent, or if it can form hydrogen bonds with the solvent. Hexane easily dissolves in benzene because both are hydrocarbons whose molecules are held together by dispersion forces. Ethyl alcohol dissolves in water because both compounds form strong intermolecular hydrogen bonds. Some polar compounds like formaldehyde, which cannot form hydrogen bonds between their own molecules, can hydrogen-bond to hydroxylic solvents such as water, and are therefore soluble in those solvents.

$$CH_3CH_2CH_2CH_2CH_2CH_3$$
hexane

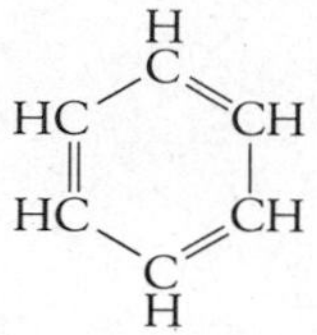

benzene

Hydrogen bonding interactions

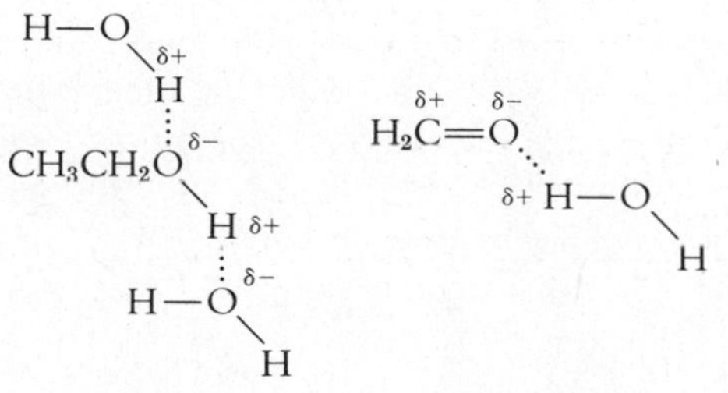

ethanol
and water

formaldehyde
and water

As a general rule it can be said that polar compounds tend to dissolve in polar solvents and nonpolar compounds in nonpolar solvents; or, put more succinctly, "Like dissolves like." It must be emphasized that "solubility" is a relative term, and that different degrees of solubility are observed. Terms commonly used to indicate the extent to which one compound dissolves in another are, in order of decreasing solubility, *miscible* (∞), *very soluble* (v), *soluble* (s), *sparingly soluble* (δ), and *insoluble* (i).

METHODOLOGY

Some odor types and examples:

acrid (harsh & irritating, like smoke)
almond (maraschino cherry)
anise (licorice)
ammoniacal (ammonia)
burnt (burnt sugar)
camphoraceous (camphor, mothballs)
ethereal (chlorinated dry cleaning solvents)
floral (roses, violets)
fruity (bananas, "Juicy-Fruit")
garlicky (garlic)
grassy (grass)
musky (musk perfume)
pepperminty (peppermint)
pungent (vinegar)
putrid (rotten eggs)
spicy (cloves, cinnamon, nutmeg)
spiritous (ethyl alcohol)
sweaty (dirty socks)
sweet (vanilla)

The eight organic liquids may be distributed in numbered vials or obtained from reagent bottles. The odor and solubility of each unknown will be observed first, then tests will be performed to determine the identity of each compound. The density and boiling point of each compound should be recorded from a *CRC Handbook of Chemistry and Physics* (or another appropriate reference book) at some time during the laboratory period, unless your instructor directs otherwise. Organic compounds are often listed under the name of the parent carbon chain in the *CRC Handbook*. For example, 1-chloropropane is listed as propane, 1-chloro and butylamine as butane, 1-amino.

Descriptive terms that may be helpful to you in characterizing odors are listed in the margin. Some odors may be characterized by comparing them with something familiar, like the odor of ripe bananas in the case of isoamyl acetate. Others may best be described as a combination of several different odor types.

The compounds will be classified into families using simple chemical tests. Some of these tests are specific for a

Figure 2.   Flow chart for classifying unknown compounds.

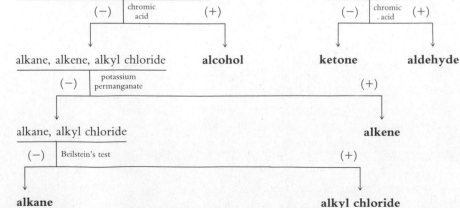

given functional group; others give a positive result with more than one functional group, and their results must be interpreted with care. The procedure will follow the flow sheet in Figure 2, each compound being set aside when it is identified and the remaining compounds being tested until all have been identified. For example, the 2,4-dinitrophenyl-hydrazine test will be carried out on only those six compounds that give negative results in the litmus test, etc.

Unless your instructor directs otherwise, it is not important for the purposes of this experiment that you become familiar with the chemistry involved in each test; but only that you recognize that the differences in chemical behavior are due to the variations in functional groups, and that these differences can be used to classify each unknown compound as a member of a given chemical family.

The functional group chemistry is, where appropriate, described in the classification test references in Part III.

Water-soluble carboxylic acids are acidic enough to turn blue litmus paper red in solution. Water soluble amines in solution are quite basic and generally turn red litmus paper blue. Aldehydes and ketones both possess a carbonyl group (C=O) and will form a yellow precipitate with 2,4-dinitro-phenylhydrazine (DNPH) solutions. Aldehydes (but not ketones) are oxidized by chromic acid solutions to yield an opaque blue-green suspension within half a minute. Most alcohols (which do not react with DNPH) also form a blue-green suspension with chromic acid solutions, reacting within two or three seconds. Alkenes (and some other oxidizable compounds) will decolorize a purple solution of potassium permanganate and form a brown precipitate. Alkyl chlorides burn with a green flame when heated on a copper wire (Beilstein's test). Alkanes give negative results with all of these tests and are recognized by a process of elimination. Procedures for the tests will be found in Part III, Systematic Organic Qualitative Analysis, on the pages listed in Table 3.

A slight pink color with blue litmus should not be interpreted as a positive test —some of the unknowns may contain acidic impurities.

Potassium permanganate solution gives a positive test with aldehydes also.

PRELAB EXERCISES

1. Read ⟨OP–6⟩ on measuring liquids by pipetting and read the procedures for all tests listed in Table 3.

2. Review the material from Experiment 2, page 26, on use of the *CRC Handbook of Chemistry and Physics*.

PROCEDURE

Obtain a 1-ml sample of each unknown using eight numbered, stoppered test tubes or vials.

Table 3.   Location of procedures for classification tests

No.	Test	Page
C–11	dinitrophenyl-hydrazine	401
C–9	chromic acid	400
C–19	potassium permanganate	406
C–5	Beilstein's test	398

Most of the unknown liquids are flammable and their vapors may be harmful; some can cause burns on contact with eyes or skin. Keep them away from open flames, avoid unnecessary contact or inhalation of vapors.

Observation of Odors

If your instructor prefers, this part may be done in small groups and the odor descriptions arrived at by consensus.

Smell each of the eight compounds by holding the test tube or vial about six inches from your face and "fanning" the vapors toward your nose with your hand. Never hold a container with a volatile liquid directly under your nose, as the vapors may be irritating, acrid, or corrosive. It may be necessary to move to an area free of interfering odors, and to wait after each test to allow your olfactory receptors time to recover from the previous odor. The intensity (strength) as well as the character of each odor should be recorded.

Determination of Solubility

Do not pipet organic liquids using mouth suction, as many are toxic or corrosive. Use a pipetting bulb such as the one illustrated in ⟨OP–6⟩.

Use a graduated cylinder to measure the water for this step.

Determine the approximate water solubility of each compound by the following procedure. Using a pipet, accurately measure ⟨OP–6⟩ 0.20 ml of the organic liquid into a 10-cm test tube. Then add 0.20 ml of water, stopper the tube, and shake it vigorously for 15 seconds. After allowing the mixute to stand undisturbed for a minute or more, observe whether two layers separate, whether the mixture is cloudy, or whether it contains droplets of one liquid in the other. If not, and the mixture appears entirely homogeneous, the compound is miscible in water and need not be tested further. If the liquid is not miscible, add another 6 ml of water to the test tube and repeat the process. If the compound dissolves at this point, designate it as soluble; if not, as insoluble. Save the solutions of all miscible and soluble liquids for the litmus tests in the next part.

Classification of Unknowns

If any unknown gives an ambiguous test, repeat the test or carry it through the subsequent tests until its identity can be established.

Following the flow sheet and the appropriate test procedures from the Methodology, determine the chemical family to which each unknown belongs. Use the solutions from the previous part for carrying out the litmus tests.

Obtaining Physical Constants from the Literature

Using the *CRC Handbook of Chemistry and Physics* (or another appropriate reference book), look up the density at 20°C and the boiling point at 760 torr of each compound listed in Table 2. Densities should be reported to at least three significant figures, boiling points to the nearest degree.

Experimental Variations

1. The density and boiling point of one or more of the unknowns may be measured in the laboratory. Densities can be determined by accurately pipetting 1 ml of liquid into a tared vial, which is then stoppered and weighed to the nearest milligram. Boiling points are determined by one of the methods in ⟨OP–29⟩.

2. Water solubility can be approximated more closely by using, in succession, total water volumes of approximately 0.2, 1, 6, 100 ml. Substances dissolving in these amounts are designated as miscible, very soluble, soluble, and sparingly soluble, respectively.

3. It has been reported that about 2% of the population cannot smell the organic compound isobutyric acid except in extremely high concentrations. An inability to detect odors of one particular type (while having a normal perception of other odors) is called specific anosmia. Students can test themselves for this kind of "odor-blindness" with dilute aqueous solutions of isobutyric acid. Most normal individuals should be able to detect its odor at concentrations of 60 parts per million or higher.

J. E. Amoore et al., "Measurement of specific anosmia," *Percept. Motor Skills* *26,* 143 (1968).

$$
\begin{array}{c}
\quad\quad\text{O} \\
\quad\quad\parallel \\
CH_3CHC\!-\!OH \\
\mid \\
CH_3
\end{array}
$$
isobutyric acid

Isobutyric acid is said to smell like "unclean goat" or dirty socks.

4. Additional tests can be used, if necessary, to identify the unknowns with certainty. Examples are Tollen's test (C–23) for aldehydes, sodium iodide in acetone (C–22) for alkyl halides, and bromine in carbon tetrachloride (C–7) for alkenes.

See Part III for procedures.

Topics for Report

1. Tabulate your results, giving the name and chemical class of each numbered unknown along with the experimental data and literature values you obtained. Justify your conclusions regarding the identity of each compound.

2. (a) Classify your eight organic compounds into five groups according to boiling point, by grouping together those compounds having boiling points within 5°C of each other. Explain the variations in boiling point among these groups (not necessarily between the compounds in each group) based on the discussion of boiling points in the Background. Be as specific as possible, indicating the kinds of intermolecular forces operating in each case. (b) Classify the compounds into five different groups according to density, combining those having densities within 0.050 g/ml of each other. Explain the variations in density among the groups based on the Background discussion. (c) Explain the variations in solubility among the eight compounds, based on the Background discussion. Specify and illustrate the intermolecular forces involved in each case where solubility or miscibility is observed.

3. Compound X, having the molecular formula $C_4H_6O_2$, is believed to be one of the following:

$$
\begin{array}{ll}
\textit{1.} \quad
\underset{\displaystyle CH_3CH_2\overset{\displaystyle \|}{C}-\overset{\displaystyle \|}{CH}}{\overset{\displaystyle O \quad\ \ O}{}}
&
\textit{3.} \quad
\underset{\displaystyle CH_2{=}CH\overset{\displaystyle \|}{C}H-\overset{\displaystyle \|}{CH}}{\overset{\displaystyle OH \quad\ O}{}}
\\[3em]
\textit{2.} \quad
\underset{\displaystyle CH_3\overset{\displaystyle \|}{C}-\overset{\displaystyle \|}{C}CH_3}{\overset{\displaystyle O \quad\ \ O}{}}
&
\textit{4.} \quad
\underset{\displaystyle CH_3CH{=}CH\overset{\displaystyle \|}{C}-OH}{\overset{\displaystyle O}{}}
\end{array}
$$

(a) State the chemical family or families to which each compound belongs. (b) Write a flow chart diagramming a procedure that could be used to determine the identity of compound X, using chemical tests described in Part III.

4. Referring to the general reaction equations in Part III, write balanced chemical equations for the positive chemical tests that you observed in this experiment. Use the chemical structures given in Table 2 in writing your equations.

Library Topics

1. Using literature sources cited in the Bibliography, discuss the effect of chain branching and molecular weight on odor and also the effect of osmophores such as benzene rings and carbon-carbon double bonds. Give specific examples where possible.

2.   Write a short paper describing the stereochemical theory of odors proposed by J. E. Amoore. Illustrate also the different shapes proposed for the olfactory receptor sites. (Parts of Amoore's theory have since been withdrawn, but his speculations are thought provoking.)

Sci. Amer., Feb. 1964, p. 42.

Identification of a High-octane Hydrocarbon

Alkanes. Measurement of Physical Constants.

Operations

⟨OP–5⟩ Weighing

⟨OP–6⟩ Measuring Volume

⟨**OP–25***a*⟩ Semimicro Distillation

⟨**OP–29**⟩ Boiling Point

⟨**OP–29***a*⟩ Micro Boiling Point

⟨**OP–30**⟩ Refractive Index

Experimental Objective

To determine the identity of an unknown alkane by measuring its physical constants and comparing them with known values.

Learning Objectives

To learn how to determine the boiling point, refractive index, and density of a liquid.

To learn about some methods used in the identification of specific organic compounds.

To learn about the production and composition of gasoline.

SITUATION

You are a chemical technologist working for the Rococo Oil Company, which is the leading oil producer and distributor on the Ergonic Gulf. Your company is facing a serious challenge from a competitor, Renaissance Oil, which has mounted a massive advertising campaign to promote its so-called "miracle additive," SRTC-491. Renaissance Oil claims that its "Super Premium with SRTC-491" burns more cleanly than ordinary gasoline, gives increased power and mileage, and reduces engine noise.

Your analytical division analyzed Renaissance's product by gas chromatography and found no evidence of unusual non-hydrocarbon additives—but the alkane–cycloalkane make-up of the gasoline differs from the customary formulation. The director of your analytical division suspects that their alleged "additive" is nothing more than an ordinary high-octane hydrocarbon for which Renaissance Oil has developed an improved refining process. Based on the analytical results, he has concluded that the additive can be any one of nine possible alkanes or cycloalkanes, but he cannot narrow down that list without more information.

By lucky coincidence, the sister of a Rococo Oil vice-president is married to Dr. Benedict Hepp-Taine, a research chemist on the staff of Renaissance Oil. Somehow she managed to persuade Dr. Hepp-Taine to pilfer a vial of SRTC-491 from his company's product research laboratory, and then passed it on to her brother at Rococo Oil. Now it is up to you to determine the true identity of the elusive hydrocarbon contained in that vial.

BACKGROUND

Concocting a Chemical Soup

Gasoline is a kind of "chemical soup" containing an incredibly large number of ingredients, all carefully selected and blended to produce a fuel with the desired properties. A typical high-octane gasoline might contain a mixture of straight and branched chain alkanes, some cycloalkanes (also called naphthenes), a few alkenes, and several different aromatic hydrocarbons as the main fuel components. Then a dash of tetraethyllead might be added to increase the octane number; along with a lead scavenger like ethylene bromide, a quick-start additive like butane to facilitate cold-weather starting, some antioxidants and metal deactivators for stability, antifreeze to prevent carburetor icing, and dyes for identification and eye appeal.

Typical gasoline ingredients

$CH_3CH_2CH_2CH_2CH_2CH_3$
hexane
(straight-chain alkane)

$CH_3CH_2CH_2CH{=}CH{-}CH_3$
2-hexene
(alkene)

$$CH_3\underset{\underset{\displaystyle CH_3}{|}}{\overset{\overset{\displaystyle CH_3}{|}}{C}}{-}CH_2\underset{\underset{\displaystyle CH_3}{|}}{CH}{-}CH_3$$

2,2,4-trimethylpentane
(branched alkane)

cyclohexane
(cycloalkane)

benzene
(aromatic hydrocarbon)

OH
|
$CH_3CH—CH_3$
isopropyl alcohol
(de-icer)

disalicyl-1,2-propanediimine
(metal deactivator)

2,6-t-butyl-4-methylphenol (BHT)
(antioxidant)

$(CH_3CH_2)_4Pb$
tetraethyllead (antiknock additive)

$BrCH_2CH_2Br$
ethylene bromide (lead scavenger)

$CH_3CH_2CH_2CH_3$
butane (quick-start additive)

The chemistry involved in these reactions is described in many textbooks of organic chemistry.

Nearly all of the major components of gasoline are derived either directly or indirectly from petroleum, which must be refined before a useable fuel is obtained. The word "refine" suggests a simple separation and purification process, but the refining of petroleum is a much more complex operation involving many different chemical reactions.

The first major process to which petroleum is subjected is fractional distillation (fractionation), which separates the components according to their boiling ranges. The fraction boiling between about 50°C and 150°C is often called "straight-run" gasoline. Straight-run gasoline is not a good motor fuel by itself, because it contains a large proportion of unbranched hydrocarbons like heptane and hexane in addition to branched alkanes and napthenes. Straight-chain alkanes burn very rapidly, generating a kind of shock wave in the combustion chamber that reduces power and can damage the engine. This "knocking" does not occur with many branched-chain alkanes, which burn slowly and evenly enough to deliver the optimum amount of power to the cylinder head.

The octane number of a fuel is a measure of its antiknock qualities. The highly branched alkane 2,2,4-trimethylpentane (sometimes called "isooctane") is a very good fuel and has arbitrarily been assigned an octane number of 100; n-heptane, with no branching, has an octane number of zero. The performance of all grades of gasoline is measured in relation to these two alkanes; for example, a fuel performing as well as a mixture containing 70% 2,2,4-trimethylpentane and 30% heptane is assigned an octane number of 70.

A major objective of petroleum refining is to convert low-octane components of petroleum into high-octane compounds having the proper boiling range. This can be accomplished by chemical reactions such as isomerization, which converts straight-chain alkanes to branched alkanes; cracking, which breaks down large molecules into smaller fragments; alkylation, which combines short-chain alkane and alkene molecules to form longer, branched molecules; and reformation (also called aromatization), which converts aliphatic hydrocarbons to aromatics. Aromatic hydrocarbons like toluene have especially high octane numbers, and are used to increase the octane rating of unleaded fuels.

Gasoline for use in automobile engines is prepared by combining varying amounts of straight-run gasoline, cracked gasoline, alkylated gasoline, reformate, and other hydrocarbon mixtures in the right proportions to give the desired boiling range and octane number. The properties of the fuel may be further adjusted by adding lead compounds to raise the octane number and low boiling alkanes to insure quick starts in cold climates or for winter driving. Leaded gasoline contains either tetramethyl- or tetraethyllead. Both of these are unstable compounds that decompose in the combustion chamber to form lead and lead oxide, which interrupt the fast free-radical chain reactions that lead to knocking. A lead scavenger such as ethylene bromide (1,2-dibromoethane) must be added to convert the excess lead and its oxide to lead bromide, which is volatile enough to pass out of the combustion chamber with the exhaust gases. Antioxidants are used to prevent or retard **gum** formation, particularly when a high percentage of alkenes is present. Certain metals, such as copper and iron, can catalyze gum forming reactions, so a chelating compound like disalicyl-1, 2-propanediimine is usually added to gasoline to combine with and deactivate these metals.

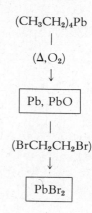

Reactions of tetraethyllead

$$(CH_3CH_2)_4Pb$$

$$|$$

$$(\Delta, O_2)$$

$$\downarrow$$

$$\boxed{Pb, PbO}$$

$$|$$

$$(BrCH_2CH_2Br)$$

$$\downarrow$$

$$\boxed{PbBr_2}$$

Gums are high-molecular-weight substances that form as a result of oxygen-initiated free-radical reactions. They can cause carburetor malfunctions.

METHODOLOGY

An organic compound is often identified by the preparation of a derivative, a solid compound whose melting point can give a clue to its identity. Alkanes and cycloalkanes are comparatively unreactive and it is difficult to convert them to suitable derivatives, so they are usually identified by their spectra and physical properties. Among the properties measured are boiling point, refractive index, and density.

A compound's boiling point can be measured either by distilling the compound and recording the observed temperature range during the distillation, or by using a special boiling-point apparatus. It is best, in any case, to distill an unknown liquid in order to purify it before measuring its other properties. Then a more precise value for the boiling

A small-scale (semimicro) apparatus should be used for distilling small quantities of a liquid.

point can be obtained by a subsequent distillation or, as in
this experiment, by a micro boiling point determination.
The refractive index of a liquid can be measured with great
accuracy and is therefore valuable for characterizing organic
compounds. Its density can be obtained by accurately weighing a measured volume of liquid. The volume is measured
with a pipet and the weight determined to the nearest milligram on an accurate balance.

The physical constants of the hydrocarbons listed in
Table 1 are different enough that an accurate determination
of all three constants should allow the certain identification
of an unknown as one of the nine.

PRELAB EXERCISES

1. Read the new operations ⟨OP–25a⟩, ⟨OP–29⟩,
⟨OP–29a⟩, and ⟨OP–30⟩ and read or review the others as
needed.

2. Review safety rules 1–3 (p. 8) and the section
Reacting to Accidents: Fires.

Structures and Properties

Table 1. List of possible hydrocarbons in SRTC-491

Name	Structure	b.p.	n_D^{20}	d^{20}
cyclopentane		49	1.4065	0.746
2,2-dimethylbutane		50	1.3688	0.649
2,3-dimethylbutane		58	1.3750	0.662
methylcyclopentane		72	1.4097	0.749
2,4-dimethylpentane		80	1.3815	0.673

Table 1.   List of possible hydrocarbons in SRTC-491 (*continued*)

Name	Structure	b.p.	n_D^{20}	d^{20}
cyclohexane		81	1.4266	0.779
2,3-dimethylpentane	$CH_3\overset{\overset{\displaystyle CH_3}{\mid}}{CH}-\overset{\overset{\displaystyle CH_3}{\mid}}{CH}-CH_2CH_3$	90	1.3919	0.695
2,2,4-trimethylpentane	$CH_3\overset{\overset{\displaystyle CH_3}{\mid}}{\underset{\underset{\displaystyle CH_3}{\mid}}{C}}-CH_2\overset{\overset{\displaystyle CH_3}{\mid}}{CH}-CH_3$	99	1.3915	0.692
methylcyclohexane		101	1.4231	0.769

Note: n_D^{20} = refractive index at 20° using sodium D line; d^{20} = density at 20°

Hazards

All of the unknowns are very flammable and should not be handled when there are flames in the vicinity. Their health hazards are generally low, but prolonged inhalation should be avoided. Some (like 2,2,4-trimethylpentane) may cause skin irritation.

C_5-C_8 alkanes and cycloalkanes

PROCEDURE

Obtain a 5–10-ml sample of the unknown hydrocarbon from your instructor and purify it by semimicro distillation ⟨OP–25a⟩. Record the boiling range ⟨OP–29⟩ of the main fraction and the distillation temperature when approximately half the liquid has distilled (median boiling point). Obtain the micro boiling point ⟨OP–29a⟩ of the purified liquid. Record the barometric pressure in the laboratory so that a boiling point correction can be made.

The micro boiling point should correspond closely to the median distillation boiling point.

The use of open flames should be avoided in the distillation if possible.   **WARNING**

2.00 ml pipet

Determine the density of the unknown by measuring exactly 1 ml of the purified liquid ⟨OP–6⟩ into a tared vial, stoppering the vial immediately, and weighing it ⟨OP–5⟩ to the nearest milligram. Measure the refractive index of the pure liquid ⟨OP–30⟩ to at least three decimal places (four if possible) and record the temperature at which the measurement was made.

Correct the boiling point and refractive index to 20°C and 760 torr and record them, along with the density, in your laboratory notebook. Deduce the identity of the unknown from your results.

Experimental Variations

1.   A combustion test can be performed on your unknown by placing one drop of the liquid in an evaporating dish and cautiously igniting it with a burning wood splint. A similar test with toluene should be carried out and the observations reported.

2.   Unknowns other than those listed in Table 1 may be used if a list of physical properties is supplied.

Future references to this journal will be written in the form:

J. Chem. Educ. 53, 306 (1976)

3.   Experiments using the "butane" in lighter fluid cannisters are outlined in the *Journal of Chemical Education,* Volume 53, page 306 (1976). For example, the average molecular weight of the gas (which is mostly isobutane with some *n*-butane and propane) can be estimated using the ideal-gas law.

4.   Experiments utilizing the gas chromatographic analysis of gasoline are described in *J. Chem. Educ. 49,* 764 (1972) and *J. Chem. Educ. 53,* 51 (1976).

Topics for Report

CH_3CCl_3
1,1,1-trichloroethane

1.   The dry cleaning solvent 1,1,1-trichloroethane is a colorless, water-insoluble liquid with a boiling point of 74°C, a density of 1.339 g/ml, and a refractive index of 1.4379 at 20°C. Suppose you have two cans of dry-cleaning solvents at home, one containing a hydrocarbon solvent and the other trichloroethane, both of which are missing their labels. How could you tell which is which (other than by odor) using only what is available in an ordinary kitchen?

2.   The following properties are measured for an unknown hydrocarbon in a laboratory with an ambient temperature of 28°C and a barometric pressure of 28.9 inches of mercury:

boiling point: 78.2°C

refractive index: 1.3780

mass of 5 ml: 3.346 g

Correct the refractive index and boiling point to 20°C, 1 atmosphere, and calculate the density of the unknown. If the unknown is one of the hydrocarbons listed on Table 1, what is its probable identity?

3.   Give names and chemical structures for all additional isomers of 2,3- and 2,4-dimethylpentane.

4.   Show how *n*-butane can be converted to 2,2,4-trimethylpentane using reactions mentioned in the Background.

Library Topics

1.   (*a*) Find the octane numbers of as many of the Table 1 compounds as you can. (*b*) Giving examples, discuss the apparent effect of chain branching, chain length, unsaturation, and cyclic structural units on octane number.

2.   Find out how ethylene bromide can be synthesized from sea water and petroleum and give equations for the chemical reactions involved. (*Hint:* Look up the Dow process for preparing bromine.)

The Dehydration of 4-Methyl-2-Pentanol

Preparation of Alkenes. Carbonium-ion Elimination Reactions. Dehydration of Alcohols. Analysis of Reaction Mixtures.

Operations

⟨OP–5⟩ Weighing

⟨OP–19⟩ Washing Liquids

⟨OP–20⟩ Drying Liquids

⟨OP–25*a*⟩ Semimicro Distillation

⟨OP–27⟩ Fractional Distillation

⟨OP–32⟩ Gas Chromatography

Experimental Objective

To carry out the dehydration of 4-methyl-2-pentanol, determine the composition of the product mixture, and provide a mechanistic explanation for the results.

Learning Objectives

To learn how to carry out a distillation using a fractionating column and to learn how to analyze a liquid by gas chromatography.

To learn how Le Chatelier's principle can be applied to drive a chemical reaction to completion.

To learn how to predict products of elimination reactions and how to write carbonium-ion mechanisms.

To learn about the discovery and history of ethylene.

SITUATION

Journal of the American Chemical Society *66*, 1649 (1944).

In 1944, A. L. Henne and A. H. Matuszak reported the results of their experiments with a series of alcohols, each of which was dehydrated to a mixture of alkenes. The purpose of their study was to test several different theories of elimination reactions—including the theory of Frank C. Whitmore, who postulated the formation of a carbonium-ion intermediate. Henne and Matuszak worked up their product mixtures by careful fractional distillation to separate the pure alkenes, which they then identified by ozonolysis, yielding known aldehydes and ketones. Such a method would be considered time consuming and imprecise today, when organic liquids can be rapidly separated by gas chromatography and accurately identified by any of a multitude of spectral methods. Nevertheless, their results have gone unchallenged for more than 35 years.

Your assignment is to carry out the dehydration of one of the compounds they studied, 4-methyl-2-pentanol, and analyze the product mixture by gas chromatography to find out if their results and conclusions are still valid.

$$\underset{\text{4-methyl-2-pentanol}}{CH_3CH-CH_2CH-CH_3} $$

$$\overset{OH}{CH_3CH}-CH_2\overset{}{CH}-CH_3$$
$$\underset{CH_3}{|}$$

4-methyl-2-pentanol

BACKGROUND

Ethylene: The Unfinished Molecule

Great scientific discoveries are seldom made by amateurs; but in 1794 a doctor, a merchant, a druggist, and a botany professor came together to "discuss chemistry and conduct chemical experiments"—and accidentally discovered the first and most important member of a whole new family of chemical compounds. These four, the so-called "Dutch chemists" Deiman, Paets van Troostwyck, Lauwerenburgh, and Bondt, were trying to repeat an experiment that had already been performed successfully, the preparation of ethyl ether from the reaction of ethyl alcohol and sulfuric acid. By varying the proportions of alcohol and sulfuric acid, they found that a mysterious gas was produced along with the ether—mysterious because of its unusual reaction with chlorine. When the gas was combined with chlorine over water, the volume of the mixture decreased markedly and a pearl-gray oil appeared on the surface of the water. Because of this behavior, they gave their discovery the unwieldy name *gaz hydrogene carbone huileux,* which was later shortened to olefiant gas (oil-forming gas) and eventually changed to *ethylene.*

The discovery of ethylene led chemists to pose questions about its structure that required most of the next century for satisfactory answers. It also resulted in a feverish search for similar substances—one intrepid chemist even hunted for ethylene in the stomach of an elephant that died

Synthesis of ethylene by the Dutch chemists

$$CH_3CH_2OH \xrightarrow{H_2SO_4} CH_2{=}CH_2 + H_2O$$

Reaction of ethylene with chlorine

$$CH_2{=}CH_2 + Cl_2 \longrightarrow \overset{Cl\ \ Cl}{\underset{}{CH_2CH_2}}$$

ethylene chloride
("Dutch oil")

L. N. Vauquelin, "Analyse du gaz trouve dans l'abdomen de l'elephant, mort au Museum," *Journal de Pharmacie et de Chimie, 3,* 205 (1817).

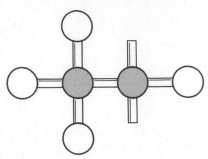

Figure 1.   Von Hofmann's model of ethylene

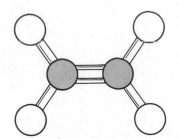

Figure 2.   Double bond model of ethylene

Refer to your organic chemistry textbook for a discussion of carbonium-ion reaction mechanisms.

at the Paris museum! Thus propylene and butylene (1-butene) were found in the oil gas resulting from thermal decomposition of petroleum oil; 1-hexene and 1-nonene were obtained by sulfuric acid hydrolysis of olive oil; and cetene (1-hexadecene, $C_{16}H_{32}$) was prepared from spermaceti wax found in sperm whales. All of these compounds reacted in the same way with chlorine, and were therefore placed in the same family with ethylene.

Some chemists reasoned that carbon, like many other elements, should exhibit more than one valence, and postulated an ethylene structure with both divalent and tetravalent carbon atoms. The first "ball and stick" model of ethylene was constructed by von Hofmann, who took six croquet balls, painted two of them black and four white, and connected them with metal tubes. This model seemed to explain the reactivity of ethylene with chlorine, since it was obviously an "unfinished" molecule which could be completed by adding two chlorine atoms to the unused bonds on the divalent carbon. Unfortunately, von Hofmann's model predicted the wrong structure for ethylene chloride, CH_3CHCl_2 rather than CH_2ClCH_2Cl. Eventually Crum Brown and other chemists arrived at the concept of a chemical double bond uniting two carbon atoms, and proposed the symmetrical ethylene structure that is now universally accepted. With the development of the multiple bond concept came a fuller understanding of the properties of ethylene and the other alkenes.

METHODOLOGY

The reaction first used to prepare ethylene—dehydration of an alcohol—is still an important general method for preparing alkenes. In this experiment you will be carrying

$$\underset{\text{4-methyl-2-pentanol}}{\underset{\overset{|}{CH_3}}{CH_3CHCH_2\overset{\overset{\displaystyle OH}{|}}{CH}-CH_3}} \xrightarrow[-H_2O]{H^+} \underset{\text{4-methyl-2-pentyl cation}}{\underset{\overset{|}{CH_3}}{CH_3CHCH_2\overset{\oplus}{CH}-CH_3}}$$

out the dehydration of 4-methyl-2-pentanol, which is expected to yield the 4-methyl-2-pentyl cation as the initial intermediate. This carbonium ion can either lose a proton directly to form one of two different alkenes (what are they?) or it can rearrange by one or more hydride shifts to form other carbonium ions, which can in turn form other alkenes. The composition of the resulting product mixture depends on several factors, including the stability of the carbonium-ion intermediates, the relative stability of the alkenes, and the reaction conditions. If complete thermal equilibrium of the product mixture is attained, the most stable alkene should be the major product, with less stable alkenes present in proportion to their relative stabilities. If the reaction time or the temperature is not sufficient to result in complete equilibrium, then the more easily formed products (those that do not require rearrangement of the first carbonium-ion intermediate) may predominate.

Henne and Matuszak carried out their dehydration reactions using either concentrated sulfuric acid or phosphoric acid mixed with the pure alcohol as a catalyst. The mixture was heated in a fractional-distillation apparatus to a temperature slightly above the boiling range of the products, so that they distilled out of the reaction mixture soon after they were formed. By maintaining the reaction temperature below the boiling point of the alcohol and above that of the alkenes, Henne and Matuszak insured that the reactants would remain behind in the boiling flask and that the reaction would be forced to completion by the removal of alkenes and water from the reaction mixture. After washing the distillate with dilute sodium bicarbonate to remove any acidic impurities, they separated and dried the alkene mixture and fractionally distilled it to separate the alkenes.

You will follow their procedure closely except for the final fractionation. Rather than separating the alkenes at this stage, you will distill them together in one fraction using a simple distillation apparatus, and then analyze the mixture using gas chromatography. When a small sample is passed through a gas chromatograph, the mixture is separated into its components, with each individual alkene giving rise to a different peak on a graph. Compounds of similar structure sometimes (but not always) appear on a gas chromatogram in order of boiling point, with the lowest-boiling component appearing first. That should be true in this case, so that if one of the peaks can be definitely assigned to a given alkene, the identity of the rest can be deduced. Thus by obtaining two chromatograms, one of your original product mixture and the other of your product "spiked" with a little 2-methyl-2-pentene, you should be able to determine the identity and quantity of each alkene in the product mixture.

Le Chatelier's principle predicts that removing one or more of the products of an equilibrium reaction will shift the equilibrium to the right.

Cis- *and* trans-*4-methyl-2-pentene will give rise to a single peak with most ordinary gas chromatographic columns.*

PRELAB EXERCISES

1.　Read the procedural part of ⟨OP–27⟩, pp. 554–557, and all of ⟨OP–32⟩; read or review the other operations as needed.

2.　Using 0.67 g/ml as the approximate density of the product, estimate the volume of alkene expected to result from dehydrating 20 ml of 4-methyl-2-pentanol.

3.　Write a checklist for the experiment.

Reaction and Properties

$$CH_3CH—CH_2\overset{\displaystyle OH}{\underset{\displaystyle CH_3}{\overset{|}{\underset{|}{C}}}H}—CH_3 \xrightarrow{H_2SO_4} C_6H_{12} + H_2O$$
alkenes

Table 1.　Physical properties

	M.W.	b.p.	d.
4-methyl-2-pentanol	102.2	132	0.802
2-methyl-1-pentene	84.2	61	0.680
4-methyl-1-pentene	"	54	0.664
2-methyl-2-pentene	"	67	0.686
cis-4-methyl-2-pentene	"	56	0.669
trans-4-methyl-2-pentene	"	$58\frac{1}{2}$	0.669

sulfuric acid　　　4-methyl-2-pentanol

The column need not be insulated.

Hazards

Sulfuric acid is very corrosive and reacts violently with water; handle with great care. 4-methyl-2-pentanol is somewhat flammable and should not be inhaled or kept in contact with the skin. The alkenes are flammable and may irritate the skin; avoid skin or eye contact and inhalation.

PROCEDURE

Reaction.　Assemble a fractional distillation ⟨OP–27⟩ apparatus using an *unpacked* column and a 25-ml (or larger) graduated cylinder as the receiver. Weigh ⟨OP–5⟩ about 20 ml of 4-methyl-2-pentanol and mix it with 2 ml of concentrated

---

Do not use an open receiver in the distillation if a burner is used for heating.   **WARNING**

---

sulfuric acid. Place the mixture (with boiling chips) into the boiling flask and slowly distill it (according to the directions in ⟨OP–27⟩), at a rate of about 1–2 drops per second. The distillation temperature should be maintained between 60° and 75°C to insure that none of the reactant distills over. Estimate the volume of alkene in the distillate, and stop the distillation when it is close to the theoretical value calculated in the Prelab Exercises and no more distillate comes over below 90°C.

Separation.   Carefully discard the residue in the boiling flask and wash the distillate ⟨OP–19⟩ in succession with one portion each of 5% aqueous sodium bicarbonate and saturated aqueous sodium chloride. Dry the alkene mixture ⟨OP–20⟩.

Dilute the residue by pouring it into a large volume of water if it is to be washed down the sink.

Anhydrous calcium chloride is satisfactory for drying alkenes.

Purification.   Distill the product mixture using the apparatus for semimicro distillation ⟨OP–25a⟩, collecting all distillate boiling between 50°C and 70°C. Weigh the product ⟨OP–5⟩ in a tared vial and save it for analysis.

To prevent the loss of volatile alkenes, the receiver can be cooled in ice water.

Analysis.   Mix a few drops of the product mixture with an equal quantity of authentic 2-methyl-2-pentene. Analyze this mixture and the untreated product by gas chromatography ⟨OP–32⟩. Identify the alkenes responsible for each peak, measure the peak areas, and determine the percent composition of the mixture. Turn in the remaining product to your instructor.

At your instructor's request, compare your results with those obtained by Henne and Matuszak.

Experimental Variations

   1.   Concentrated phosphoric acid (about 4 ml) can be used as the catalyst. The resulting product mixture can then be compared with that obtained with a sulfuric acid catalyst.

   2.   The presence of unsaturation in the product mixture can be verified by means of the bromine test described on page 399 or the Baeyer (potassium permanganate) test on page 406.

MINILAB 1

Preparation and Properties of Ethylene

Mix 2 ml of 1,2-dibromoethane (ethylene bromide), 2 ml of 1-pentanol, and 2 g of granular zinc in a Pyrex test tube fitted with a bent delivery tube (see ⟨OP–4⟩) as shown in Figure 3. Heat the tube gently to start the reaction, fill several large test tubes with the gas by displacement of water, and stopper them tightly until you are ready to use them. The first test tube will probably contain mostly air and should not be used in the tests.

WARNING   *Ethylene bromide and carbon tetrachloride have been classed by OSHA as carcinogens.*

Figure 3. Apparatus for ethylene generation

carbon tetrachloride

bromine

ethylene bromide

Test the flammability of ethylene by lowering a burning wood splint into one of the tubes. Add a few drops of 1% aqueous potassium permanganate to the second tube, then stopper and shake it. Add a few drops of 0.2M bromine in carbon tetrachloride to a third tube, then stopper and shake it. Record and explain your observations.

Topics for Report

"Adjacent" means next to the OH carbon in this case.

1. A rule proposed by Alexander Saytzev in 1875 states, "In an elimination reaction, hydrogen is removed preferentially from the adjacent carbon atom that is poorer in hydrogen." ***(a)*** What, by Saytzev's rule, should be the product of dehydrating 4-methyl-2-pentanol? What was your major product? ***(b)*** What should be the major product if thermal equilibrium among the five alkenes listed in Table 1 is attained? ***(c)*** Write mechanisms leading to the formation of all the alkenes observed in your product mixture, and explain your results on the basis of the mechanisms and alkene stability.

2.   Write a flow diagram for your synthesis, following the format given in Figure 1, Experiment 5.

3.   Predict at least two possible by-products (other than alkenes) that could result from alternative reactions of the *unrearranged* 4-methyl-2-pentyl cation. At what stage (or stages) would these by-products have been separated from the alkenes?

4.   Calculate the ratio of the volume of water to the volume of alkenes in the initial distillate, and compare it with the ratio actually observed.

The volume of water and alkenes in the distillate should be a part of your observations and be recorded in your laboratory notebook.

5.   (*a*) Predict the major product alkene that would result from dehydrating each of the following alcohols, with no carbonium-ion rearrangements.

1.

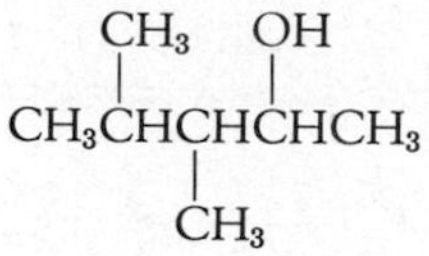

$$CH_3CHCHCHCH_3$$

with CH_3, OH groups above and CH_3 below

2.

structure with CH_3 and a cyclohexene ring bearing a $C-OH$ group with H_3C and CH_3

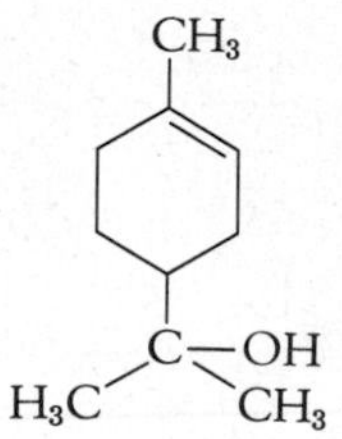

3.

$$CH_3C-CH-CH_3$$

with CH_3, OH above and CH_3 below

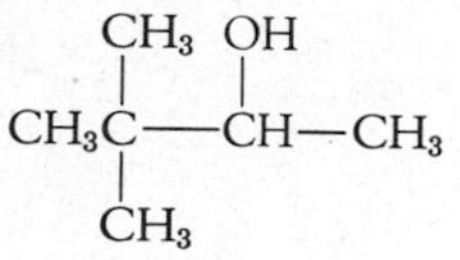

(*b*) In each case, predict the most stable dehydration product that could result after a single carbonium-ion rearrangement.

Library Topics

1.   (*a*) Read the paper by Henne and Matuszak cited in the Situation and report their conclusions regarding the mechanism of the dehydration reaction. (*b*) How does their preferred mechanism for carbonium-ion shifts differ from the currently accepted one?

2.   Report on some of the commercial uses of ethylene, giving equations for all reactions in which ethylene is converted to other useful products.

Separation of Petroleum Hydrocarbons by Fractional Distillation

Hydrocarbons. Separation Methods.

Operations

⟨OP–5⟩ Weighing

⟨OP–27⟩ Fractional Distillation

⟨OP–32⟩ Gas Chromatography

Experimental Objective

To separate a cyclohexane-toluene mixture by fractional distillation, determine the composition of the distillate, and measure the efficiency of the fractionating column.

Learning Objectives

To learn how to carry out a fractional distillation using a packed column.

To learn more about the laboratory practice of fractional distillation and some of its applications.

SITUATION

See Experiment 6 for the case of the carbonized code sheet.

CH₃

toluene

Having successfully reconstructed the missing code sheet so that your lab students could complete their last experiment, you have kept your job as a laboratory assistant at the Miskatonic Institute of Lucubration and are engaged in preparing reagents for the next experiment on aromatic substitution reactions. While transferring some toluene from a stock bottle to the reagent bottle, you run out of toluene and obtain more from the stockroom. After you have filled the rest of the reagent bottle with the new "toluene" you notice some light pencil marks forming an **X** over the original label on the stock bottle. Puzzled, you turn the bottle around and see, to your horror, a crudely lettered label reading CYCLOHEXANE. Hoping there is some mistake, you sniff the liquid remaining in the stock bottle. It doesn't smell like toluene.

  There is no more toluene in the laboratory, the stockroom has just closed, and your lab section starts in three hours. Cursing the unknown individual responsible for your predicament, you wait until your blood pressure has receded to a safe level, and then begin thinking of rational alternatives. Your only option is to separate the liquids and recover the toluene, but how? Fortunately you have access to an organic labkit that can be used to assemble a fractional distillation apparatus. . . .

cyclohexane

Possibly the same one that wiped up the sulfuric acid with your code sheet.

BACKGROUND

Distillation and Its Uses

Distillation has been used since antiquity to separate the components of mixtures—the ancient Egyptians were making an embalming fluid by distilling wood more than 3500 years ago. In the intervening years, the process of distillation has undergone many improvements and found numerous applications. In one form or another, distillation is used to manufacture perfumes, flavor ingredients, liquors, charcoal, coke, and a host of organic chemicals. Its most important modern application is its role in refining petroleum into fuels, lubricants, and petrochemicals. The first step in that process is the separation of petroleum into various hydocarbon fractions by distilling it through fractionating columns that are up to 200 feet high. Since components of different molecular weights and carbon structures usually have significantly different boiling points, this process separates the petroleum into portions containing hydrocarbons of similar carbon content and properties. Table 1 lists some typical petroleum fractions, along with their approximate boiling ranges and the number of carbon atoms in their constituent hydrocarbons.

Table 1.   Fractions from the distillation of petroleum

Name of Fraction	Carbon #	Boiling Range, °C
Natural gas	C_1–C_4	Below 20
Petroleum ether	C_5–C_6	20–60
Ligroin	C_6–C_7	60–100
Gasoline	C_6–C_{12}	50–200
Kerosene	C_{12}–C_{18}	175–275
Gas oil	Over C_{18}	Over 275

The names and boiling ranges of fractions may vary, depending on the producer and their anticipated uses. The "gasoline" fraction is straight-run gasoline and must be further refined before being used in gasoline engines.

Cyclohexane and other alicylic hydrocarbons are called naphthenes in the petroleum industry, while toluene is a high-octane aromatic that is a particularly important component of no-lead gasoline.

To obtain very pure toluene or cyclohexane would require more sophisticated equipment (and more time) than you will have available.

METHODOLOGY

In this experiment, you will separate two hydrocarbons, cyclohexane and toluene, that are commonly encountered in petroleum refining. In the process, you will also determine the efficiency of your fractionating column in terms of its HETP (height equivalent to a theoretical plate). For this purpose, it will be necessary to collect the initial few drops of distillate while distilling at a very slow rate, and to analyze the distillate by gas chromatography (or by refractometry—see Experimental Variation 1). During the separation, you will obtain four fractions of different boiling point ranges, each containing successively less of the lower boiling component (cyclohexane) and more of the toluene. The first fraction should be relatively pure cyclohexane; the residue remaining in the boiling flask after the third fraction has distilled should be fairly pure toluene. The intermediate fractions will be comparatively impure mixtures of the two components. The volumes of the first fraction and the residue you obtain should prove a good measure of your efficiency as compared with other students in your class, since better column packing and more careful distillation should result in larger volumes of the "pure" components (provided they are collected over the correct range). The purity of the fractions will be determined by gas chromatography or refractometry.

PRELAB EXERCISES

1. Read all of ⟨OP–27⟩ and review ⟨OP–32⟩ on gas chromatography.

2. Calculate the mass and volume of 0.20 mol of cyclohexane and 0.30 mol of toluene.

Properties

Table 2.   Physical properties

	M.W.	b.p.	d.
cyclohexane	84.2	81	0.774
toluene	92.2	111	0.867

Hazards

Cyclohexane is flammable and can cause narcosis or skin irritation after prolonged inhalation or contact. Toluene is flammable, a mild irritant, and can cause narcosis on inhalation. Avoid excessive contact with and inhalation of both liquids; keep them away from flames.

cyclohexane   toluene

PROCEDURE

Carefully measure out 0.20 mol of cyclohexane and 0.30 mol of toluene and combine the liquids in a 100-ml boiling flask. Pack a suitable distilling column with glass beads, steel wool, glass helices, or other packing materials provided, and assemble the apparatus for fractional distillation ⟨OP–27⟩. Measure the height of the packing in millimeters. Begin heating the pot, but just before the liquid begins to distill, slow the rate of heating and reflux the mixture for 5–10 minutes to establish equilibrium. The ring of condensing vapors should be kept at a level between the top of the packing and the sidearm of the still head. Then increase the rate of heating so that the liquid begins to distill very slowly. Discard the first few drops of distillate (which may contain traces of water) and collect five drops for gas chromatographic analysis; then quickly change the receiver and record the distillation temperature. At a normal distillation rate, collect the following numbered fractions in the ranges indicated (use labeled receivers):

A burner can be used if necessary, but great care must be taken to prevent fires—both hydrocarbons are very flammable.

Store all samples for gas chromatographic analysis in tightly capped vials or stoppered test tubes to prevent vaporization.

1. 81–85°C
2. 85°–97°C

If the distillation starts below 81°C, include all of the initial distillate in fraction 1.

3.   97°–107°C

4.   107°–111°C (residue in flask)

Let the apparatus cool before collecting the final fraction.

The last fraction need not be distilled, but can be collected from the boiling flask after the distillation temperature reaches 107°C. When the initial distillation is completed, pour fraction *4* into a labeled container and redistill all the fractions according to the following procedure:

Pour fraction *1* into the boiling flask and distill it, collecting the distillate in the labeled receiver you used previously for fraction *1*. When the temperature reaches 85° (or when the liquid has all distilled), remove the heat source and add fraction *2* to the boiling flask, then resume the distillation. Collect all the distillate coming over below 85° in receiver *1*, and collect the 85–97° fraction in receiver *2*. When the temperature reaches 97°, again remove the heat source and add fraction *3* to the boiling flask. Continue the distillation as before, collecting the 81–85° fraction (if any) in receiver *1*, the 85–97° fraction in receiver *2*, and the 97–107° fraction in receiver *3*. When the temperature reaches 107°, add fraction *4* and repeat the process, collecting fractions *1*, *2*, and *3* in the same temperature ranges as before. When the temperature reaches 107°, stop the distillation and, after cooling, retain the residue in the flask as fraction *4*. Weigh the four fractions ⟨OP–5⟩ and analyze them, along with the initial distillate, by gas chromatography ⟨OP–32⟩.

Calculations

Assume that the GLC peak areas are proportional to the masses of the two components.

1.   Calculate the percentage composition (by weight) of each fraction and the weight of cyclohexane and toluene in each fraction. Graph the weights of the components versus boiling temperature, using the midpoints of the temperature ranges specified above for the four fractions. Your graph should have two overlapping curves, one for cyclohexane and one for toluene.

The volatility factor, α, for 40 mol% cyclohexane in toluene is 2.43.

2.   Calculate the number of theoretical plates for your column, using the Fenske equation (see ⟨OP–27⟩). Remember that the pot contributes one theoretical plate, so the number of theoretical plates for the column is $n - 1$. Using the measured height of your column, calculate its HETP.

Experimental Variations

1. You may use refractometry rather than gas chromatography to analyze your distillation fractions. Construct a calibration curve by measuring the refractive index values of cyclohexane-toluene mixtures containing approximately 0, 20, 40, 60, 80, and 100 mol% cyclohexane, then graphing refractive index versus mol%. After the distillation is completed, measure the refractive index values for each of your fractions and determine their compositions from your calibration curve. If desired, 2–3 students can work together on the calibration curve, each contributing several determinations.

2. Somewhat more challenging distillations are *(a)* the separation of "isooctane" (2,2,4-trimethylpentane) and toluene, and *(b)* the separation of the two major components of turpentine, α-pinene and β-pinene. An efficient column (5–10 or more theoretical plates) is required.

Topics for Report

1. Explain why the procedure for redistilling the initial four fractions will result in higher yields of the "pure" components. For example, if the 85–97° fraction from the first distillation is distilled again, why will it yield some 81–85° fraction instead of distilling entirely at 85–97° as it did the first time?

2. Using the data in Table 3 for cyclohexane-toluene, and assuming a starting mixture containing 20 mol% cyclohexane, construct a temperature-composition diagram similar to that shown in Figure 3, ⟨OP–27⟩. From the graph, estimate the composition of the first drops of distillate *(a)* for a simple distillation; and *(b)* for fractional distillations with fractionating columns having 1, 2, 3, and 4 theoretical plates. It should not be necessary to use the Fenske equation.

3. Calculate the mol% of cyclohexane that your first few drops of distillate would have contained if the rating of your column had been six theoretical plates.

4. Suggest a chemical method that could effectively remove small amounts of toluene from cyclohexane.

Table 3.   Temperature-composition data for cyclohexane-toluene

Mol % Cyclohexane		T, °C
Vapor	Liquid	
0	0	110.7
10.2	4.1	108.3
21.2	9.1	105.5
26.4	11.8	103.9
34.8	16.4	101.8
42.2	21.7	99.5
49.2	27.3	97.4
54.7	32.3	95.5
59.9	37.9	93.8
66.2	45.2	91.9
72.4	53.3	89.8
77.4	59.9	88.0
81.1	67.2	86.6
86.4	76.3	84.8
89.5	81.4	83.8
92.6	87.4	82.7
97.3	96.4	81.1
100.0	100.0	80.7

Library Topic

Tell how the destructive distillation of wood is carried out, and indicate some of the products that can be obtained by this process.

Stereochemistry of the Addition of Bromine to Cinnamic Acid

Reactions of Alkenes. Electrophilic Addition to Carbon-carbon Double Bonds. Preparation of Stereoisomers.

Operations

⟨OP–5⟩ Weighing

⟨OP–7a⟩ Refluxing

⟨OP–10⟩ Addition

⟨OP–12⟩ Vacuum Filtration

⟨OP–28⟩ Melting Point

Experimental Objective

To carry out the addition of bromine to *trans*-cinnamic acid and determine the stereochemistry of the product.

Learning Objectives

To learn how to carry out a reaction requiring the addition of reactants under reflux.

To learn more about stereoselective and stereospecific reactions and their mechanisms.

To learn about some important compounds related to cinnamic acid.

SITUATION

An eminent Chaud-Froidian psychologist, Dr. Veronica Beccabunga, has recently published a book attempting to explain chemical reaction mechanisms using psychological principles of behavior. Dr. Beccabunga claims to be able to predict the outcome of chemical reactions on the basis of a complicated set of "molecular motivations" that can, in theory, be calculated from the appropriate molecular-orbital

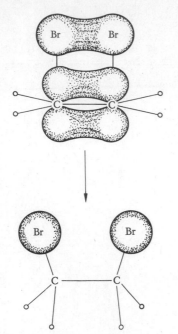

Figure 1. Dr. Beccabunga's proposed mechanism for bromine addition

—CH=CHCOOH

cinnamic acid

—CH=CHCHO

cinnamaldehyde

deamination: removal of an ammonia (NH_3) molecule.

wave functions. The theory predicts, for example, that a bromine molecule will approach parallel to a carbon-carbon double bond in order to exchange electrons and bond covalently to the carbon atoms. In the author's words (as translated by an impoverished foreign-languages major working his way through graduate school): "Therefore it is quite logical that the paired bromine atoms should lie down on the soft pi-electron cloud until, by the pleasure principle attracted to the waiting carbon nuclei, they bond with their new partners."

You are not persuaded by such arguments, but before rejecting them out of hand you decide to check some of the predictions in the laboratory. The addition of bromine to *trans*-cinnamic acid should provide an unambiguous test of this theory of bromine addition.

BACKGROUND

The Cinnamic Acid Connection

Cinnamic acid and its close relatives, cinnamaldehyde and cinnamyl alcohol, are naturally occurring compounds that are important as flavoring and perfume ingredients and as sources for pharmaceuticals. Cinnamaldehyde (the major component of cinnamon oil) is used to flavor many foods and beverages and to contribute a spicy, "oriental" note to perfumes. Esters of cinnamic acid are used in perfumery also, but of far greater significance is the role of cinnamic acid in secondary plant metabolism. As an intermediate in the shikimic acid pathway of plant biosynthesis, cinnamic acid is the source of an enormous number of natural substances that (to give only a few examples) contribute structural strength to wood, give flavor to cloves, nutmeg, and sassafras, and produce many of the brilliant colors of nature —the flower pigments that attract insects for pollination, the vivid and delicate shades of a butterfly's wings, and the radiant colors of leaves in autumn.

In nature, cinnamic acid is formed by enzymatic **deamination** of the important amino acid, phenylalanine. It can then be converted, by a wide variety of biosynthetic pathways, to coniferyl alcohol (a precursor of lignin) in sapwood, myristicin in nutmeg, safrole in sassafras bark, and to *flavonoids* in a wide variety of plant structures. The flavonoids are natural substances characterized by the 2-aryl-benzopyran structure found in flavanone, which itself is biosynthesized from cinnamic acid by a process involving

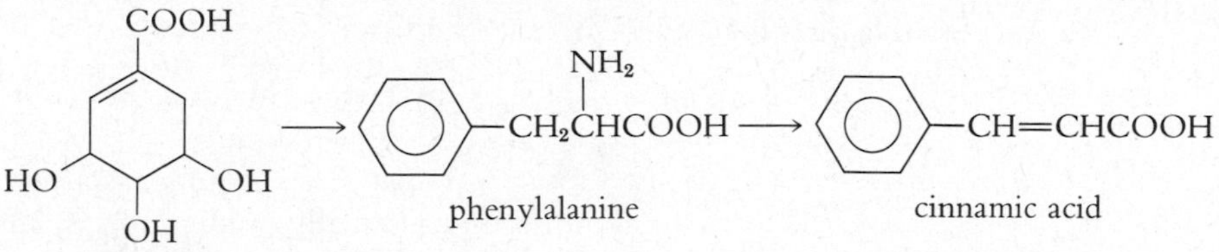

shikimic
acid

phenylalanine

cinnamic acid

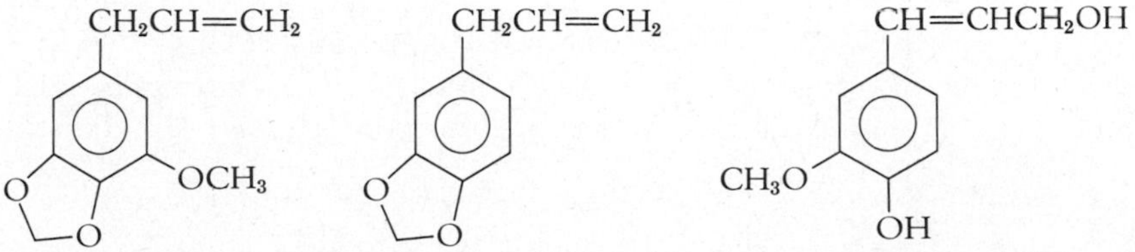

myristicin

safrole

coniferyl alcohol

Biosynthesis of flavanone from
cinnamic acid

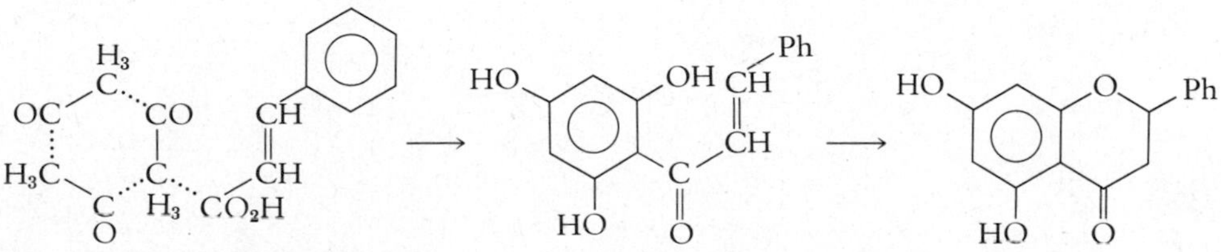

acetates   cinnamic acid

flavanone

the linkage of three acetate units to the carboxyl group of the acid. Flavonoids perform no single function in plants. Many are highly colored and attract insects for pollination or animals for seed dispersal, while others help to regulate seed germination and plant growth or protect a plant from fungal and bacterial diseases. Certain flavonoids contribute the bitter taste to lemons and the bracing astringency of cocoa, tea, and beer. Although flavonoids and the other derivatives of cinnamic acid have not furnished as many wonder drugs or "useful" chemicals to mankind as have the nitrogen-containing alkaloids, they provide much to delight the eye and stimulate the senses, and the world would be a drearier place without them.

METHODOLOGY

Reactions in which a given starting material preferentially forms one of several possible products are called *stereoselective*, while those in which stereoisomeric starting materials form different products (usually diastereomers) are called *stereospecific*. Many reactions involving additions to double bonds are stereoselective—a given alkene that is capable of forming several different products will nevertheless form one of them preferentially. A good example is the addition of methylene to *cis*-2-butene, which yields only *cis*-1,2-dimethylcyclopropane as the product. (This reaction is also stereospecific, since the stereoisomeric starting material, *trans*-2-butene, yields a different product, *trans*-1, 2-dimethylcyclopropane.)

Addition of methylene (CH_2) to *cis*- and *trans*-2-butene.

cis-2-butene

trans-2-butene

In this experiment, you will carry out the addition of bromine to *trans*-cinnamic acid and identify the product of the reaction from its melting point. The product (actually a mixture of enantiomers) could be either *threo*-2,3-dibromo-3-phenylpropanoic acid [whose enantiomers have the (2R, 3R) and (2S, 3S) configurations], the corresponding *erythro* compound [(2R, 3S) and (2S, 3R)], or a mixture of the two. The *erythro-threo* nomenclature is used to describe the configurations of compounds having two chiral centers but no plane of symmetry: it is derived from the structures of the two simple sugars, erythrose and threose. The *erythro* compound is analogous to a *meso* compound—when a plane is drawn separating the chiral centers, the maximum number of like substituents can be lined up directly across the plane from each other, with the remaining unlike substituents also being opposite. In the *threo* form, when one pair of like substituents is lined up, the other pair is not.

Stereoisomers of 2,3-dibromo-3-phenylpropanoic acid

threo
(2S,3S)

erythro
(2S,3R)

erythrose

threose

From the structure of your product, you should be able to deduce whether the addition of bromine to *trans*-cinnamic acid involves **syn** or **anti addition** (or both) and whether or not the reaction is stereoselective. You should then propose a mechanism to account for your results.

The reaction will be carried out by adding bromine (in carbon tetrachloride as the reaction solvent) to a solution of *trans*-cinnamic acid under reflux. It should be possible to carry out the addition in a quarter hour or less, and the product should begin to precipitate out of solution after about half of the bromine solution has been added. The crystalline product will be isolated by filtration, dried thoroughly, and its melting point obtained. Recrystallization is generally not necessary, since the product is sufficiently pure for most purposes; but the product may be recrystallized from a large volume of chloroform if desired (1 g dissolves in approximately 25 ml of hot chloroform).

Since bromine is a very corrosive liquid with a choking, poisonous vapor, it will be dispensed in solution unless your instructor specifies otherwise. Carbon tetrachloride is also a very hazardous liquid. Great care must be taken to minimize your exposure to both substances.

syn addition: entering groups add to the same side of the double bond.

anti addition: entering groups add to opposite sides of the double bond.

PRELAB EXERCISES

1. Read ⟨OP–10⟩ and review the other operations as needed.

2. Calculate (*a*) the mass of 20 mmol of *trans*-cinnamic acid, and (*b*) the volume of 2.0M bromine in carbon tetrachloride required to provide 20 mmol of bromine. $\frac{mole}{Liter}$

3. Write a checklist for the experiment.

Reaction and Properties

Table 1.   Physical properties

	M.W.	m.p.	b.p.	d.
trans-cinnamic acid	148.2	135–136		
bromine	159.8	−7	59	3.12
carbon tetrachloride	153.8	−23	76.5	1.59
erythro-2,3-dibromo-3-phenylpropanoic acid	308.0	202–4		
threo-2,3-dibromo-3-phenylpropanoic acid	308.0	93.5–95		

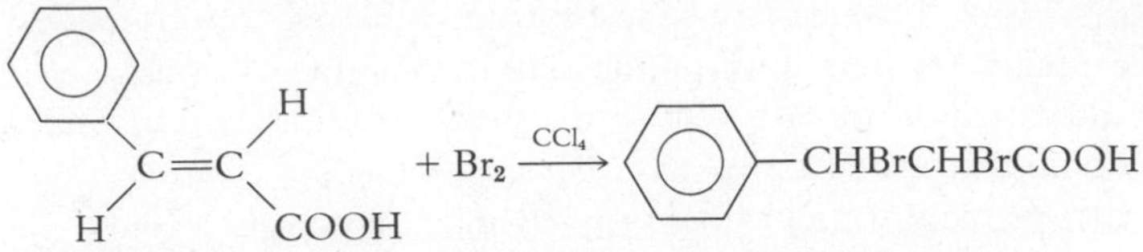

trans-cinnamic acid

2,3-dibromo-
3-phenylpropanoic acid

Hazards

bromine 　　carbon
tetrachloride

Bromine is a very corrosive liquid, causing serious burns to skin and eyes. Its vapors irritate the eyes and mucous membranes and can damage the respiratory tract. Even in solution, it should be handled with great care. Carbon tetrachloride vapors are very poisonous and cause permanent damage to the liver, kidneys, and other internal organs if inhaled or ingested. It has been classified as a carcinogen in animal tests. Avoid contact with skin, eyes, and clothing; do not breathe vapors; use only with adequate ventilation. It is recommended that rubber gloves be worn and that operations involving bromine and carbon tetrachloride be performed under an efficient hood. Cinnamic acid is a mild irritant; avoid skin or eye contact.

PROCEDURE

The cinnamic acid should dissolve after heating is begun.

Set up an apparatus for addition and reflux ⟨OP–10⟩, and combine 20 mmoles of *trans*-cinnamic acid with 20 ml of carbon tetrachloride in the boiling flask. Place the calcu-

---

WARNING

Carbon tetrachloride has been classified as a carcinogen based on tests with mammalian species. Appropriate protective measures should be taken to minimize exposure to this chemical.

---

*Reflux **gently** so that no bromine vapors escape out of the top of the condenser.*

lated volume of 2.0M Br₂/CCl₄ in the addition funnel and reflux the reactants ⟨OP–7a⟩ while adding the bromine solution portionwise ⟨OP–10⟩ over a period of 15 minutes or so. After each addition, wait until the bromine color fades to a pale orange before the next addition, and keep the funnel tightly stoppered between additions to keep bromine and carbon tetrachloride vapors out of the laboratory. When the addition is complete, continue refluxing for another 10–15 minutes, then cool the mixure in an ice bath and allow time for the product to crystallize completely.

Be careful in handling the reaction mixture—it may still contain some bromine.

Collect the product by vacuum filtration ⟨OP–12⟩ and wash it on the filter with three portions of cold methylene chloride. Obtain a melting point of the dry crystals ⟨OP–28⟩, weigh the product ⟨OP–5⟩, and turn it in to your instructor.

The filtrate should be shaken with a little 5% sodium bisulfite (to destroy any residual bromine) and then placed in a waste container.

Experimental Variations

2,3-Dibromo-3-phenylpropanoic acid can be dehydrohalogenated easily with alcoholic KOH to yield phenylpropiolic acid. A procedure is given in *J. Am. Chem. Soc. 64,* 2510 (1942). A somewhat more ambitious synthesis is the preparation of 2,3,3-triphenylpropanoic acid by Friedel-Crafts alkylation, described in *Organic Syntheses* (Bib–B1) Coll. Vol. 4, p. 960.

$$\text{PhCHBrCHBrCOOH} \xrightarrow{\text{KOH}} \text{PhC}\equiv\text{CCOOH}$$
phenylpropiolic acid

Ph₂CHCHCOOH
|
Ph
2,3,3-triphenylpropanoic
acid

Determination of Unsaturation in Commercial Products

Test samples of each of the following, using test C–7 from Part III (bromine/carbon tetrachloride), along with any additional products your instructor provides. In each case, record the amount of bromine solution needed to decolorize the sample. (Use approximately equal quantities of each sample.)

Turpentine

Mineral oil

Vegetable oil

Gasoline

Rubber cement

Linseed oil

Hydrogenated vegetable oil (such as Crisco)

Butter

Oleomargarine

Try to find out what kinds of unsaturated compounds are present in the samples that show a significant amount of unsaturation.

Topics for Report

Syn and *anti* addition are sometimes called *cis* and *trans* addition, respectively.

1. **(a)** Write the stereochemical formula of your product and tell whether the addition of bromine to *trans*-cinnamic acid involves *syn* or *anti* addition. Use models if necessary. **(b)** Write a mechanism that explains your results.

2. From your experimental results, can you conclude with certainty whether or not the reaction of bromine with cinnamic acid is stereospecific? What experiment would you have to carry out to determine this?

3. Write a flow diagram for your synthesis.

Assume the same addition stereochemistry as you observed in this experiment.

4. Predict the major product of the reaction of bromine **(a)** with *cis*-cinnamic acid; **(b)** with maleic acid; and **(c)** with fumaric acid. Draw stereochemical projections for the products in each case.

Library Topics

1. Find out how cinnamic acid can be prepared in the laboratory and how it is synthesized industrially, giving equations for the reactions used. Describe at least one industrial process that uses a reaction different from that used in the common laboratory synthesis.

2. Look up the structures and sources of some natural flavonoids and find out what functions they perform for the plants in which they occur.

Hydration of a Difunctional Alkyne

Alkynes. Electrophilic Addition to Carbon-carbon Triple Bonds.
Preparation of Carbonyl Compounds. Spectrometric Analysis.

Operations

⟨OP–5⟩ Weighing

⟨OP–10⟩ Addition

⟨OP–13⟩ Extraction

⟨OP–13c⟩ Salting Out

⟨OP–14⟩ Solvent Evaporation

⟨OP–15⟩ Codistillation

⟨OP–20⟩ Drying Liquids

⟨OP–25⟩ Simple Distillation

⟨OP–33⟩ Infrared Spectrometry

Experimental Objective

To carry out the hydration of 2-methyl-3-butyn-2-ol and obtain spectral evidence confirming that the expected transformation has taken place.

Learning Objectives

To learn how to carry out a simple codistillation, how to use the "salting out" phenomenon in separation procedures, and how to record an infrared spectrum.

To learn about the general procedures used in hydrating alkynes.

To learn about the synthetic sex hormones and other important compounds having alkyne functional groups.

SITUATION

As a graduate student at a major university, you are currently studying reactions in which a carbonyl and a hydroxyl group trade places, as in the isomerization of glyc-

Isomerization of glyceraldehyde

$$CH_2{-}CH{-}CH \overset{H^+}{\longrightarrow} CH_2{-}C{-}CH_2$$

glyceraldehyde   dihydroxyacetone

Anticipated isomerization of labeled
3-hydroxy-3-methyl-2-butanone

$$CH_3C{-}C{-}CH_3 \overset{H^+}{\longrightarrow} CH_3C{-}C{-}CH_3$$

Synthetic scheme

$$CH_3C{-}C{\equiv}CH \overset{H_2O}{\underset{Hg^{2+}}{\longrightarrow}} CH_3C{-}CCH_3$$

2-methyl-3-butyn-2-ol

perhydrocyclopentanophenanthrene

eraldehyde to dihydroxyacetone. You wish to find out whether the hydroxyketone 3-hydroxy-3-methyl-2-butanone undergoes a similar isomerization and, if so, what the mechanism is. Unfortunately, the product of the anticipated isomerization reaction (which involves a methyl shift) will be identical to the reactant if ordinary 3-hydroxy-3-methy-2-butanone is used. To get around this difficulty, you have decided to label the carbonyl oxygen atom (starred in the formula) by preparing the reactant with oxygen-18 enriched water. If analysis of the product mixture shows that some of the ^{18}O has been incorporated into a hydroxyl group, this will provide clear evidence that the rearrangement has occurred.

Since water enriched in ^{18}O is very expensive, you will first prepare some unlabeled 3-hydroxy-3-methyl-2-butanone using ordinary water so that you can develop a satisfactory procedure for synthesizing the labeled compound. One easy way of making 3-hydroxy-3-methyl-2-butanone is to add water to the hydroxy alkyne (alkynol) 2-methyl-3-butyn-2-ol.

BACKGROUND

From "The Pill" to Oblivon

Natural *estrogens* such as estrone and 17β-estradiol are responsible for promoting the development of secondary sex characteristics in females during puberty, while the natural *progestins* such as progesterone are necessary to maintain pregnancy in mammals. Both types of hormones belong to an important class of natural products called the steroids, which are characterized by a basic four-ring skeleton, the perhydrocyclopentanophenanthrene nucleus. Cholesterol is the most widely known of the steroids and, despite the "Mr. Cholesterol" TV commercials portraying him as a bad guy, is essential to the synthesis of the sex hormones and other body regulators.

progesterone

17β-estradiol

estrone

In addition to their natural function of promoting sexual maturity in females, the estrogens are very useful therapeutically. They help alleviate the mental and physical discomfort associated with menopause and are believed to prevent coronary atherosclerosis in younger women. Likewise, the progestins have been used to treat menstrual disorders and uterine bleeding and to prevent miscarriages. The natural hormones cannot be taken orally because they are rapidly deactivated in the liver, so the search for synthetic hormones with similar activity began soon after the natural ones were isolated and characterized. One way to prevent the deactivation of a steroid such as 17β-estradiol is to stabilize the C-17 hydroxy group with an appropriate substituent—the ethynyl (HC≡C—) function seems to fill the bill nicely. By treating estrone with potassium acetylide in liquid ammonia one can synthesize ethinyl estradiol, which is a potent estrogen that can be taken orally. Similarly, some synthetic progestins such as norethynodrel are more effective taken orally than progesterone is on injection.

cholesterol

ethinyl estradiol

norethynodrel

One of the most intriguing properties of the natural estrogens and progestins is their ability to inhibit ovulation in females. During the 1940s it was a common practice to treat menstrual disorders with these hormones, since preventing ovulation stops menstruation. Somewhere along the line it dawned upon a few investigators that preventing ovulation also prevents pregnancy. At that time, however, there were no orally active substances suitable for this application, and the idea of having frequent birth-control injections did not appeal to many women. By the early 1950s the picture had changed, since several effective progestins (19-norprogesterone, norethindrone, norethynodrel, etc.) had been synthesized by researchers for the Syntex and G. D. Searle companies, and early in 1953 these compounds were evaluated by Gregory Pincus and his colleagues for anti-ovulatory activity. It was soon discovered that the combination of a synthetic progestin with a small amount of synthetic estrogen

The first birth-control pill marketed, Enovid, contained 9.85 mg of norethynodrel and 0.15 mg of mestranol.

See J. Chem. Ed. 55, *591* (1978) for a more thorough discussion of the birth-control pill and its development.

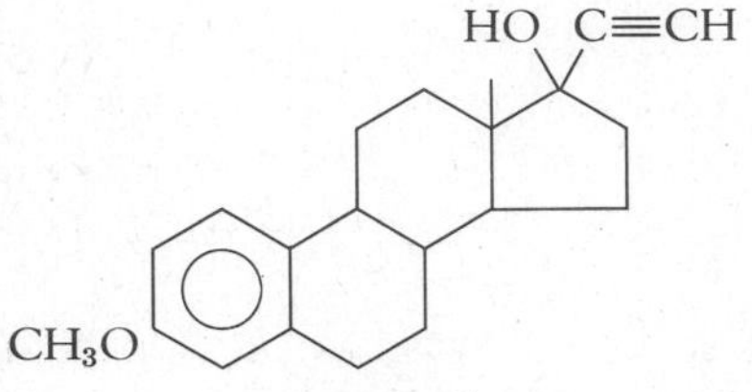

mestranol

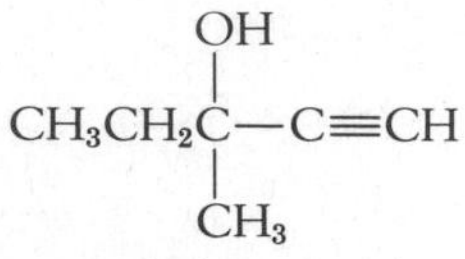

methylparafynol
(Oblivon)

provided the highest anti-ovulatory activity, and "the pill" was marketed soon afterward. In recent years more powerful synthetic hormones have been synthesized, and the mini-pill, the "morning-after" pill, and hormone implants have been developed. Although the long-term consequences of using oral contraceptives are not entirely known, they are currently the most reliable means of preventing pregnancy. Their impact on society has been enormous.

Alkynols having a hydroxy group adjacent to the triple bond (as in norethynodrel and mestranol) are comparatively easy to synthesize, and they appear in a number of pharmaceuticals. The 2-methyl-3-butyn-2-ol used in this experiment is prepared by combining acetone and acetylene with an alkali metal in liquid ammonia. A similar alkynol called methylparafynol has been used in sleeping pills under the trade name Oblivon. Apparently the tertiary alcohol portion of the molecule causes methylparafynol to act as sedative or depressant, while the acetylenic group gives it hypnotic (sleep-producing) properties.

Preparation of 2-methyl-3-butyn-2-ol

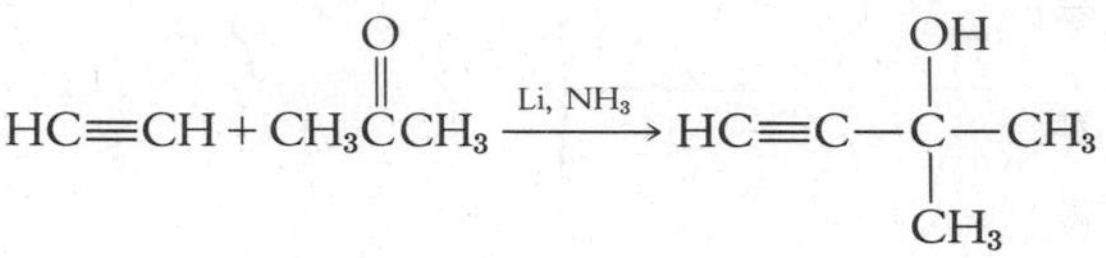

Hydration of an alkyne involves the initial formation of an enol, which rearranges to a keto form as illustrated below:

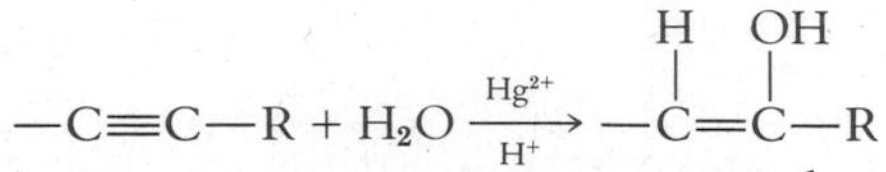

METHODOLOGY

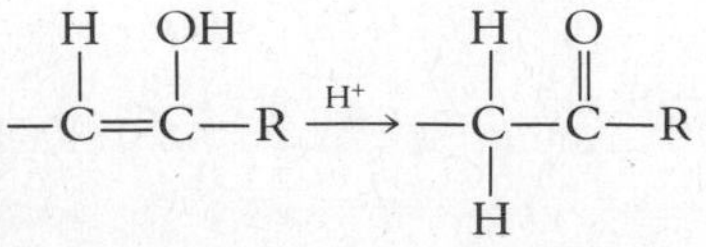

Hydration of alkynes is usually carried out in acidic solution in the presence of a mercuric salt such as mercuric sulfate or oxide. Since the reaction is strongly exothermic, it might become too violent if all the reactants were mixed at the start. For this reason, the alkynol is added over a period of time during the reaction by means of an apparatus for simultaneous addition and refluxing. The product will be isolated by codistillation with water, which is a convenient method for working up a product mixture containing a volatile organic compound and involatile impurities. The 3-hydroxy-3-methyl-2-butanone in the distillate can then be separated from the water (which codistills with it) by first adding some potassium carbonate and sodium chloride to "salt out" the organic layer and neutralize any acid that might be carried over in the distillate, then extracting the mixture with ether. A simple distillation is needed to purify the product after it has been recovered from the organic layer.

In chemical research, it is customary to analyze a synthetic product by some method that leaves little room for doubt about its identity. One of the simplest ways of doing this is to obtain an infrared spectrum, which is equivalent to a "fingerprint" of its molecules. Infrared spectroscopy is also useful for detecting functional groups in molecules, and with it one can trace a conversion from one functional group to another. In this experiment both reactant and product contain two functional groups. By obtaining infrared spectra of each compound, you should be able to identify the absorption bands that are characteristic of each functional group and so observe that a conversion has taken place.

PRELAB EXERCISES

1. Read ⟨OP–13c⟩, ⟨OP–15⟩, and ⟨OP–33⟩. Review the other operations as needed.

2. Calculate the mass and volume of 2-methyl-3-butyn-2-ol needed to provide 0.20 moles.

3. Write a checklist for the experiment.

Reaction and Properties

Table 1.   Physical properties

	M.W.	b.p.	d.
2-methyl-3-butyn-2-ol	84.1	104	0.868
mercuric sulfate	296.7		
3-hydroxy-3-methyl-2-butanone	102.1	140–41	0.953

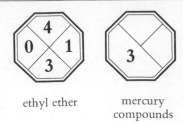

ethyl ether   mercury
compounds

Hazards

Mercuric sulfate is very poisonous if inhaled or ingested. Do not breathe dust or allow it in contact with skin or eyes. 2-Methyl-3-butyn-2-ol is toxic and flammable; avoid inhalation, ingestion, and skin and eye contact. Ethyl ether is extremely flammable and its vapor is harmful. Contact with the product should be avoided.

PROCEDURE

A precipitate may form when the alkynol is added, but it should dissolve immediately. The reflux should be just vigorous enough to keep the reactants well mixed.

Reaction.   Assemble an apparatus for addition and reflux ⟨OP–10⟩ using a 250-ml boiling flask. Place 75 ml of 3M sulfuric acid and 1.0 g of mercuric sulfate in the reaction flask, and put 0.20 mol of 2-methyl-3-butyn-2-ol in the addition funnel. Heat the reaction flask, with shaking, until the

WARNING   *Mercuric sulfate is poisonous, so handle with great care and wash hands afterwards.*

mercuric sulfate dissolves; then add the alkynol dropwise ⟨OP–10⟩ over about a ten-minute period while maintaining a gentle reflux. Continue the reflux for 30 minutes after all the alkynol has been added.

A graduated Erlenmeyer flask can be used as a receiver.

Separation.   Assemble the apparatus for codistillation ⟨OP–15⟩ by removing the reflux condenser and replacing it with the distillation assembly. Add 50 ml of water to the addition funnel and begin distilling the reaction mixture (*Caution:* foaming may occur). Add water as necessary to keep the water level in the flask approximately constant. Continue distilling until about 150 ml of distillate has been collected. Add to the distillate 25 grams of potassium carbonate dihydrate (or sesquihydrate) and enough sodium chloride to saturate the solution, and stir until the salts have dissolved ⟨OP–13c⟩. Transfer the solution to a separatory funnel (decant from any undissolved salt) and extract it ⟨OP–13⟩ twice with 25-ml portions of ethyl ether (*no flames!*). Combine the extracts, dry them ⟨OP–20⟩ with magnesium sulfate or another suitable drying agent, and remove the ether ⟨OP–14⟩ by simple distillation ⟨OP–25⟩ over a steam bath.

Purification.   Change receivers after the ether has distilled and purify the 3-hydroxy-3-methyl-2-butanone by simple distillation in the same apparatus ⟨OP–25⟩. Weigh the product ⟨OP–5⟩ in a tared, labeled vial.

The apparatus for small-scale distillation ⟨OP–25a⟩ can be used if desired.

Analysis.   Obtain infrared spectra ⟨OP–33⟩ of both the starting material and the purified product.

Experimental Variations

1.   2-methyl-3-butyn-2-ol can be used to demonstrate various tests for unsaturation, such as the bromine/carbon tetrachloride test and the potassium permanganate test. (See pages 399 and 406.)

2.   The NMR spectrum of the product can be obtained and each signal assigned to a given set of protons.

3.   The product can be characterized by means of a semicarbazone derivative. [See *J. Chem. Educ. 43,* 324 (1966) for experimental details.]

Topics for Report

1.   Interpret your infrared spectra as completely as possible, giving assignments for the major bands. Refer to the IR correlation chart (back endpaper) and to the discussion of infrared spectral interpretation in Part III for help. Explain clearly how the infrared spectral results show that the expected chemical reaction has occurred.

2.   Write a flow diagram for the synthesis you performed.

3.   The diol known as pinacol rearranges to pinacolone in the presence of acid by the mechanism shown. Propose a similar mechanism for the isomerization of 3-hydroxy-3-methyl-2-butanone described in the Situation. (*Hint:* look at some of the pinacol rearrangement steps in reverse.)

Pinacol rearrangement

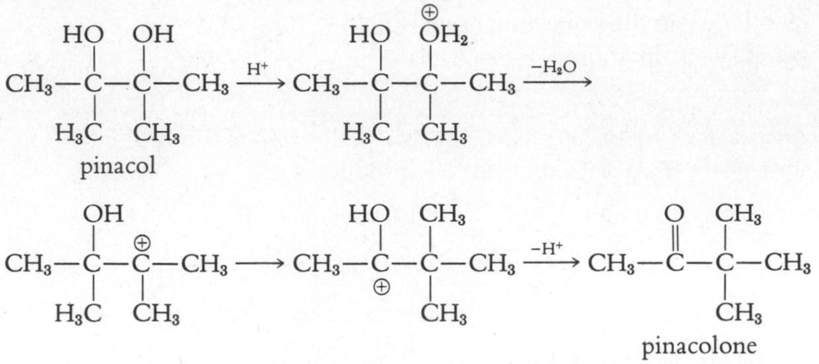

$$
\begin{array}{c}
\underset{|}{HO} \ \underset{|}{OH} \\
CH_3-C-C-CH_3 \xrightarrow{H^+} \\
\underset{|}{H_3C} \ \underset{|}{CH_3} \\
\text{pinacol}
\end{array}
\qquad
\begin{array}{c}
\underset{|}{HO} \ \overset{\oplus}{\underset{|}{OH_2}} \\
CH_3-C-C-CH_3 \xrightarrow{-H_2O} \\
\underset{|}{H_3C} \ \underset{|}{CH_3}
\end{array}
$$

$$
\begin{array}{c}
\underset{|}{OH} \\
CH_3-C-\overset{\oplus}{C}-CH_3 \longrightarrow \\
\underset{|}{H_3C} \ \underset{|}{CH_3}
\end{array}
\qquad
\begin{array}{c}
\underset{|}{HO} \ \underset{|}{CH_3} \\
CH_3-C-\underset{\oplus}{C}-CH_3 \xrightarrow{-H^+} \\
\underset{|}{CH_3}
\end{array}
\qquad
\begin{array}{c}
\overset{O}{\parallel} \ \underset{|}{CH_3} \\
CH_3-C-C-CH_3 \\
\underset{|}{CH_3} \\
\text{pinacolone}
\end{array}
$$

4. Draw structures for the major products obtained when the following alkynes undergo hydration: (a) acetylene; (b) 1-ethynylcyclohexanol; (c) 1-butyne; (d) 2-butyne.

Library Topic

"The pill" has been the subject of considerable controversy from both scientific and moral viewpoints since its introduction for contraception in the late 1950s. Find out about some of the beneficial and potentially harmful effects of oral contraceptives, describe some new developments in the field of contraception, and discuss the impact of modern birth-control methods on society.

The Synthesis of
7,7-Dichlorobicyclo[4.1.0]heptane

*Alicyclic Hydrocarbons. Carbene Addition Reactions of Alkenes.
Phase-transfer Catalysis. Cyclopropane-ring Compounds.*

Operations

⟨OP–5⟩ Weighing

⟨OP–7a⟩ Refluxing

⟨**OP–9**⟩ Mixing

⟨OP–13⟩ Extraction

⟨**OP–13b**⟩ Separation of Liquids

⟨OP–14⟩ Evaporation

⟨OP–20⟩ Drying Liquids

⟨OP–25a⟩ Semimicro Distillation

Experimental Objective

To prepare 7,7-dichlorobicyclo[4.1.0]heptane by adding dichlorocarbene to cyclohexene in the presence of a phase-transfer catalyst.

Learning Objectives

To learn how to conduct a reaction with magnetic stirring and how to purify a liquid by vacuum distillation.

To learn how phase-transfer catalysis works and how to apply it to an organic synthesis.

To learn about some natural and synthetic compounds having cyclopropane rings.

SITUATION

You are a research chemist employed by Phlogiston Chemicals, Inc. Your company has just been awarded a gov-

ernment contract by the Air Force to develop a highly volatile and powerful jet fuel for use in a new top-secret fighter
plane, the F–25 Starling.

A Monsanto research group that was working on a
similar contract found that highly strained hydrocarbons
show unusually high heats of combustion, and they synthesized a number of such compounds for possible use as jet
fuels. For example, starting with alkenes such as β-pinene
and bicyclohexylidene, they synthesized the tricyclic compounds shown by adding methylene to the double bonds.

Monsanto synthesis of strained
hydrocarbons

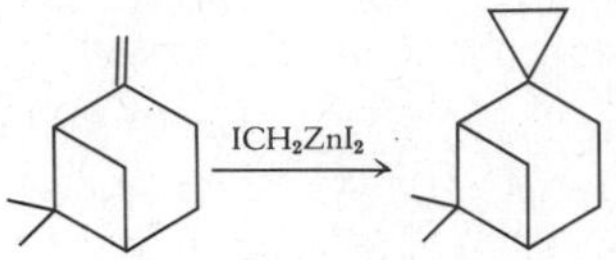

β-pinene

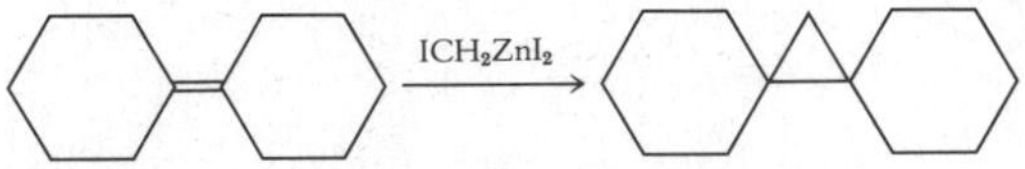

bicyclohexylidene

You recently came across a synthesis for an unusual hydrocarbon, tricyclo[4.1.0.0^{2,7}]heptane (Trich–41 for short),
that shows considerable promise as a high-energy fuel. It was
first prepared by W. R. Moore in 1961 using an intramolecu-

Moore's preparation of
tricyclo[4.1.0.0^{2,7}]heptane

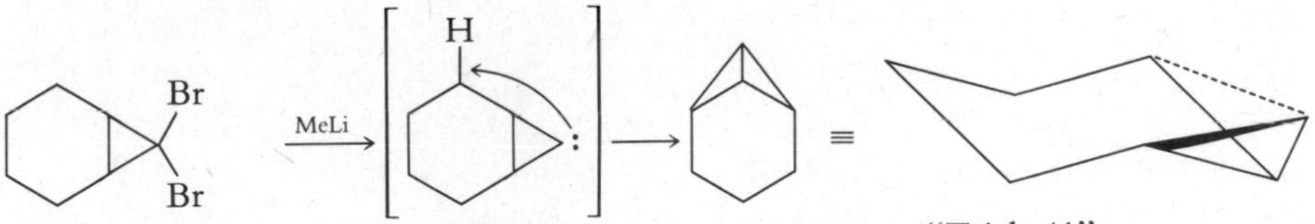

7,7-dibromonorcarane   carbene   "Trich–41"
                                intermediate

lar carbene-insertion reaction with 7,7-dibromonorcarane as
the starting material. Since 7,7-dichloronorcarane can be
synthesized more cheaply (from chloroform and cyclohexene) you plan to use it in place of the bromine compound. If
the conversion to Trich–41 is successful, you can then have
that hydrocarbon tested as a potential jet fuel. First you must
prepare 7,7-dichloronorcarane in good yield.

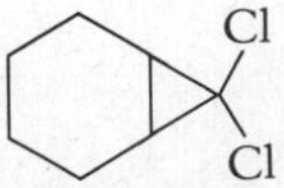

7,7-dichloronorcarane

BACKGROUND

The Ubiquitous Triangle

It often appears that Nature loves the hexagon, since so many natural compounds contain six-membered rings in their molecules. This might be expected, since both the aromatic benzene ring and the unstrained cyclohexane ring are unusually stable compared to other possible ring structures. So it is surprising to find numerous examples of the triangle —the comparatively unstable cyclopropane ring—in everything from arborvitae to water molds.

Henry David Thoreau reported that Northwoods lumbermen of the last century were accustomed to drink "a quart of arborvitae, to make (them) strong and mighty." An extract of the leaves of the arborvitae (white cedar) was thought to impart strength and prevent illness, particularly rheumatism. One of the main constituents of the oil from arborvitae and other *Thuja* species is the bicyclic terpenoid called thujone, which contains a three-membered ring fused onto a cyclopentane ring. Thujone is also found in the essential oils from tansy, sage, and wormwood. A 6:3 ring combination appears in 3-carene, which is obtained from various *Pinus* species and is a major constituent in turpentine oil obtained from Sweden and Finland. Another 6:3 bicyclic is found in the whimsically named sirenin which is a sperm attractant produced by the female gametes of a water mold, *Allomyces javanicus*. An unusual 7:3 combination occurs in ledol (ledum camphor), a major constituent of the essential oil from a northern shrub, Labrador tea. One of the more exotic sights in nature is a pumpkin-colored mushroom that glows in the dark around Halloween. This is the "Jack-O' Lantern" fungus, *Clitocybe illudens,* which has sometimes been mistaken for the edible chanterelle mushroom (with unpleasant consequences). It contains several antibiotic components, illudin-M and illudin-S, which have a cyclopropane ring fused by its apex to a six-membered ring.

Pyrethrin is a natural biodegradable insecticide, nontoxic to humans, obtained from the flowers of a daisy-like plant, *Chrysanthemum cineariaefolium*. It is made up of a mixture of esters like cinerin I that contain a cyclopropane ring in the carboxylic acid portion. In recent years, a number of pyrethroids, synthetic analogs of the natural compounds, have been developed in an effort to find a relatively safe but highly effective insecticide to replace the environmentally

hexagon

triangle

Euell Gibbons tasted arborvitae tea, and declared that he "would almost prefer rheumatism."

thujone

3-carene

sirenin

The sirens of Greek mythology were beautiful female creatures who lured mariners to destruction with their singing.

ledol

illudin-S

cinerin I

decamethrin

Renewed interest in "natural" pesticides was stimulated, in part, by Rachel Carson's book *Silent Spring*.

unsound "hard" pesticides. One of the most powerful of these is decamethrin, which is over sixty times more lethal to houseflies than parathion and over six hundred times more effective than DDT against certain mosquitoes. Because of their low toxicity and biodegradable properties, it seems quite possible that pyrethroids will become the insecticides of choice in the future.

METHODOLOGY

In this experiment, the unstable intermediate dichlorocarbene will be added to the double bond of cyclohexene to form dichloronorcarane, which has the systematic name 7,7-dichlorobicyclo[4.1.0]heptane. The carbene will be generated using the reaction of chloroform with aqueous sodium hydroxide, which is believed to occur by the alpha–elimination mechanism shown in the margin.

$:CCl_2$
dichlorocarbene

Formation of dichlorocarbene from chloroform

$$CHCl_3 + OH^- \longrightarrow CCl_3^- + H_2O$$
$$CCl_3^- \longrightarrow :CCl_2 + Cl^-$$

One of the drawbacks of this reaction is that the reactants are very different in polarity and will not dissolve in the same solvents under ordinary conditions. Sodium hydroxide is soluble in water and insoluble in the organic phase, while chloroform is only slightly water-soluble and cyclohexene almost completely insoluble. This means that the reactants will have a hard time "getting together;" as a consequence, the reaction may be very slow. An excellent way of getting around this difficulty is by use of a newly developed technique called phase-transfer catalysis, or PTC for short.

Reaction of alkyl halide with sodium cyanide

$$RCl + Na^+CN^- \longrightarrow RCN + Na^+Cl^-$$

Improved reaction with quaternary ammonium cyanide

$$RCl + Q^+CN^- \longrightarrow Q^+Cl^- + RCN$$
$(Q^+ = (CH_3CH_2CH_2CH_2)_4N^+)$

The principles of PTC can be readily explained with reference to the double-displacement reaction of an alkyl halide with sodium cyanide to form a nitrile. If a high-molecular-weight halide such as 1-chlorooctane is heated with aqueous sodium cyanide, there is no reaction—the cyanide stays in the aqueous layer, the alkyl halide stays in the organic layer, and "never the twain shall meet." If, instead of sodium cyanide, one uses a quaternary ammonium (Q) salt such as tetra-*n*-butylammonium cyanide, the reaction proceeds quite readily and gives a high yield of product. There are several reasons for this enhanced reactivity of the cyanide: First (and most important), the sixteen carbon atoms attached to the cation make it soluble in the organic phase and where the cation goes, the anion must follow. Second, the cyanide ion is more reactive in the organic phase than it would have been in the aqueous phase, since it is not solvated by water molecules that would shield it from the alkyl halide and decrease its reactivity. Finally, the large butyl groups around the positive nitrogen of the cation decrease the attractive forces between cation and anion and allow the cyanide ion more freedom to attack the alkyl halide. So by substituting a quaternary ammonium ion for the sodium ion in the cyanide salt, we have a good procedure for bringing about the desired transformation. The main disadvantage is that the quaternary ammonium cyanide is much more expensive than sodium cyanide.

The PTC technique gets around the high cost of quaternary ammonium salts by recycling them after each reaction step. If the quaternary ammonium cation in the product (tetrabutylammonium chloride) can be made to pick up some more cyanide ion to throw at another molecule of alkyl halide, it will function as a true catalyst, accelerating the reaction without being used up. All that is needed is a reservoir of cyanide ion and a small amount of the Q salt to keep the reaction going. The reservoir can be provided by an aqueous layer containing sodium cyanide.

Figure 1 diagrams the process, which occurs as follows: A catalytic amount of Q^+Cl^- combines with cyanide ion at the interface between the aqueous and organic layers, and the Q^+CN^- that forms dissolves in the organic layer. There it reacts with the alkyl halide to produce the nitrile (RCN) and form more Q^+Cl^-, which migrates to the interface, picks up some more cyanide, shuttles it back into the organic layer to react with the alkyl halide and form more product and Q^+Cl^-, and so on, ad infinitum.

The process involved in the PTC reaction of cyclohexene with dichlorocarbene is analogous, although not quite so simple. In principle, a quaternary ammonium salt could react with the sodium hydroxide, transferring OH^- to the organic layer to react with chloroform to form dichlorocarbene, which in turn would add to cyclohexene to form the product. However, Q^+OH^- is considerably less soluble in organic solvents than is Q^+CN^-. It seems likely that the reaction involves the generation of CCl_3^- at the interface between the two solvents, with the quaternary ammonium ion

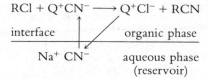

Figure 1.   Phase-transfer catalysis of halide displacement reaction

Reaction at interface

$$Q^+Cl^- + Na^+CN^- \longrightarrow Q^+CN^- + Na^+Cl^-$$

This process is somewhat different from the usual PTC route. Some investigators have suggested that it should be classified as a catalytic two-phase (CTP) reaction instead.

shuttling this ion between the aqueous and organic layers rather than the hydroxide ion.

The procedure used here for synthesizing dichloronorcarane is straightforward. Because the reaction mixture is heterogeneous, the two phases must be well mixed so that the reaction can proceed at a reasonable rate. A magnetic stirrer will be used for this purpose. It has been pointed out [*J. Chem. Educ. 51*, 216 (1974)] that some carbon monoxide may be generated, so the reaction should be carried out under a fume hood. (This is particularly important if the laboratory is not well ventilated.) After the product is isolated, it will be purified by semimicro distillation.

PRELAB EXERCISES

1. Read ⟨OP–9⟩ and ⟨OP–13*b*⟩. Review the other operations as needed.

2. Calculate the mass and volume of 0.10 mol of cyclohexene and the volume of 0.30 mol of chloroform.

3. Write a checklist for the experiment.

Reaction and Properties

cyclohexene   7,7-dichloronorcarane

Table 1.   Physical properties

	M.W.	m.p.	b.p.	d.
cyclohexene	82.15		83	0.810
chloroform	119.4		61.7	1.483
tetra-*n*-butylammonium bromide	332.4	102–4		
7,7-dichloronorcarane	165.1		$197-8^{760}$ 95^{35} 78^{15}	

Hazards

Chloroform is toxic by inhalation—prolonged breathing of the vapors can cause serious depression of body functions, or even death. It has been classified as a carcinogen. Avoid inhalation and skin and eye contact. Cyclohexene is flammable and its vapors should not be inhaled. Tetra-*n*-butylammonium bromide can react with oxidizing materials. Sodium hydroxide solutions can cause burns on skin and eyes;

  chloroform   sodium hydroxide   cyclohexene

spills should be flooded with water and (if necessary) neutralized with dilute acetic acid. Avoid inhalation and unnecessary contact with the dichloronorcarane, as with all other chlorinated hydrocarbons.

PROCEDURE

Reaction.   Assemble a 250-ml boiling flask for reflux ⟨OP–7a⟩ and place it through the rings in a steam bath so that it is just out of contact with the bottom of the bath. Add a magnetic stir bar, 300 mmol of chloroform, 100 mmol of cyclohexene, and 0.5 g of tetra-*n*-butylammonium bromide. Then add 75 ml of 10M sodium hydroxide (*care!*) through the reflux condenser, with stirring ⟨OP–9⟩. Heat the mixture gently, with vigorous stirring, for 45 minutes.

Heat so that the chloroform just barely refluxes.

Chloroform has recently been classified as a carcinogen based on tests in mammalian species that resulted in tumors upon ingestion. Although chloroform has been used safely in chemistry laboratories for many years without obvious harm, appropriate protective measures should be taken to reduce exposure to this chemical. The reaction must be carried out with adequate ventilation, since carbon monoxide may be evolved.

WARNING

WARNING *The aqueous layer contains sodium hydroxide and is very caustic — avoid contact, wash hands thoroughly after handling.*

Cooling with ice is suggested.

Separation. Cool the reactants to room temperature and separate the layers ⟨OP–13b⟩, saving the chloroform layer and returning the aqueous layer to the separatory funnel. Extract the aqueous layer ⟨OP–13⟩ thoroughly with two portions of methylene chloride, combine the extracts with the reserved chloroform, and dry the mixture ⟨OP–20⟩. Evaporate the solvents ⟨OP–14⟩ under vacuum on a steam bath, using a trap so that no chloroform is released into the atmosphere.

The layers should be shaken or stirred for several minutes to insure complete extraction.

Purification. Purify the product by semimicro distillation ⟨OP–25a⟩. Weigh the product ⟨OP–5⟩ and turn it in to your instructor.

Experimental Variations

See Vogel's Practical Organic Chemistry *4th ed., (Bib–B4), p. 869 for GLC conditions.*

1. A gas chromatogram of the product ⟨OP–32⟩ can be obtained to assess its purity, or an IR spectrum ⟨OP–33⟩ to establish its identity.

2. The distillation can be carried out under reduced pressure ⟨OP–26a⟩.

Relative rate = (decrease in alkene concentration) ÷ (decrease in cyclohexene concentration).

3. The relative rates of dichlorocarbene addition to alkenes can be determined by using 0.10 mol of an alkene along with 0.10 mol of cyclohexene and 0.10 mol of chloroform, then analyzing the reactant and product mixtures by GLC to measure the decrease in concentration of each alkene. The effect of substituents and ring size on rates of addition can be assessed by using such alkenes as 1-hexene, 2-methyl-2-butene, cyclopentene, and cyclooctene.

The alkyl groups in Adogen 464 and Aliquat 336 are a mixture of C_8–C_{10} alkyls.

4. Different phase-transfer catalysts can be used for this reaction, including benzyltrimethylammonium chloride and methyltrialkylammonium chlorides (Adogen 464 or Aliquat 336). The reaction time may need to be increased — particularly with the former — to obtain optimum yields.

Topics for Report

1. Predict the expected product of adding 1 mole of dichlorocarbene to each of the following alkenes: **(a)** 2-methyl-2-butene; **(b)** trans-2-hexene (show stereochemistry); **(c)** 1,5-cyclooctadiene; **(d)** α-pinene; and **(e)** camphene.

2. Write a flow diagram for the synthesis of 7,7-dichloronorcarane, showing how all reactants, catalysts, and by-products are separated from the major product during its isolation and purification.

3. Diagram the phase-transfer process that is apparently taking place in the synthesis of 7,7-dichlorobicyclo[4.1.0]heptane, showing all relevant reaction steps.

4. The carbon-carbon sigma bonds of cyclopropane rings exhibit some of the characteristics of pi bonds in other systems. For example, the cyclopropylmethyl cation undergoes reactions suggesting that it is best represented by the resonance structures shown. Explain the isomerization in acidic solution of chrysanthemyl alcohol to yomogi alcohol and artemisia alcohol by writing appropriate mechanisms.

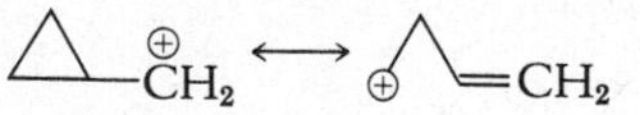

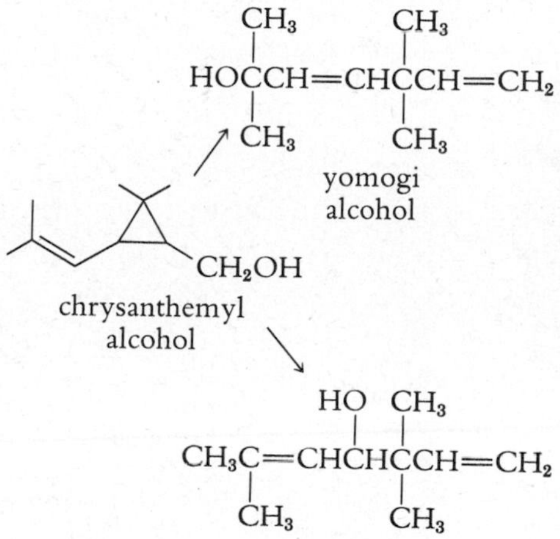

Library Topics

1. Look up the article in *Tetrahedron Letters* 3013 (1975) and find out which of the double bonds of limonene is attacked first by dichlorocarbene. Give the structure of the major product of this reaction and summarize the reaction conditions used, indicating how they differ from the conditions used in this experiment.

2. Research and describe some modern methods of insect control that have been proposed as alternatives to the use of chemical insecticides.

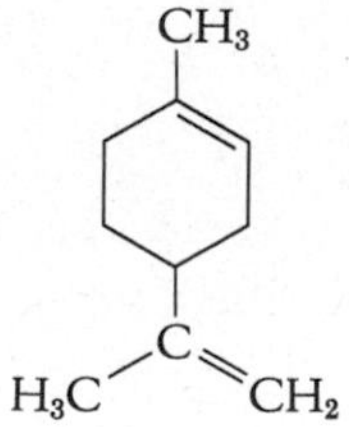

See The Merck Index (*Bib–A3*) *for structures if necessary.*

yomogi alcohol

chrysanthemyl alcohol

artemisia alcohol

limonene

The Preparation of Tropylium
Iodide from Cycloheptatriene

*Nonbenzenoid Aromatic Compounds.   Hydride-transfer Reactions.
Carbonium Ions.   Spectrometric Analysis.*

Operations

⟨OP–5⟩ Weighing

⟨**OP–8**⟩ Cooling

⟨OP–9⟩ Mixing

⟨OP–12⟩ Vacuum Filtration

⟨OP–21⟩ Drying Solids

⟨**OP–34**⟩ Nuclear Magnetic Resonance Spectrometry

Experimental Objective

To prepare tropylium fluoborate from cyclohepta-triene, confirm its aromaticity by NMR, and convert it to tropylium iodide.

Learning Objectives

To learn how to carry out a reaction with external cooling, and how to record an NMR spectrum.

To learn about some of the methods used for preparing long-lived carbonium ions.

To learn about some natural nonbenzenoid aromatic compounds.

SITUATION

Erich Hückel was a pioneer in applying quantum mechanical theories to organic molecules, especially to aromatic systems. His $4n + 2$ rule allowed him to predict which systems should be capable of aromaticity. He wrote in 1931: "One can expect that the seven-ring would have the tendency to give up its unpaired, non-bonding electrons, thus forming a positive ion. . . . Nothing is known about this." Hückel's prediction had already come true forty years earlier, but he had no way of knowing it because the unlucky discoverer of this "seven-ring ion" didn't know what to make of his discovery. The yellow salt that G. Merling isolated in 1891 (after brominating cycloheptatriene) had properties that made no sense at all according to the theories of his day; so he followed an all-too-human tendency to ignore what he couldn't explain and missed a golden opportunity to announce to the scientific world a new aromatic system and the first true carbonium ion. It was another sixty-three years before tropylium bromide was rediscovered and used to confirm Hückel's prediction.

Merling's preparation of tropylium bromide

cycloheptatriene   tropylium bromide

As a graduate student engaged in research on nonbenzenoid aromatic systems, you want to compare the properties of the tropylium halides and would like to develop a quick, convenient synthesis for tropylium iodide. Because iodine is not as reactive as bromine, Merling's method probably will not work for the iodide.

Molecular orbital calculations predict that tropylium ion should be very stable—even more so than the first stable carbonium ion known, the triphenylmethyl cation. If these calculations are correct, you should be able to prepare tropylium ion by a hydride-transfer reaction between cycloheptatriene and triphenylmethyl (trityl) fluoborate. The resulting tropylium fluoborate should then react with iodide ion to yield tropylium iodide. But before you can be sure that your "tropylium ion" is the real thing, you will have to prove that the product of the expected hydride transfer reaction is aromatic. The best way to do that is with nuclear magnetic resonance spectrometry.

See Experiment 17 for background on the triphenylmethyl cation:

BACKGROUND

Molds, Blue Oils, and King George III

Back in 1945, Michael J. S. Dewar puzzled over a strange substance called stipitatic acid, which had been isolated from a culture of the mold *Penicillum stipitatum*. Because stipitatic acid showed aromatic properties, such as undergoing substitution rather than addition reactions with bromine, other scientists had assigned it a benzene-type structure containing a six-membered aromatic ring. Dewar, however, disagreed with their reasoning. With little experimental evidence to go on, he proposed a seven-ring structure and declared that the compound was one of a previously unknown class of aromatic compounds which he named *tropolones*.

stipitatic acid

tropolone

resonance structure of tropolone

β-thujaplicin

Some essential oils yielding azulenes

araucaria
cajeput
camomile
elemi
galangal
ginger
hops
juniper
lemongrass
lovage
myrrh
niaouli
pimenta
rose
Siam wood
valerian
ylang-ylang
zdravetz

Dewar expected tropolone rings to show aromatic properties because one can draw resonance structures in which their six pi-electrons are delocalized over the seven-membered ring. His guess turned out to be correct, and before long other investigators were proposing tropolone ring structures for compounds with similar properties. One sample of an oil had been distilled from the heartwood of a western red cedar (*Thuja plicata*) and set aside, nearly forgotten, for sixteen years while it slowly crystallized. The crystals yielded β-thujaplicin, an effective fungicide and antibiotic, which was shown to be an isopropyl derivative of tropolone. This compound and two other fungus-destroying thujaplicins are believed to be responsible for the great durability of red cedar, which was used by American Indians to build canoes long before the white man discovered its value in fence posts, shingles, and cedar chests.

Soon after the discovery of distillation as a means of obtaining valuable "essences" from plants, it was found that many plants contained essential oils that turned a beautiful azure blue on distillation. At first it was believed that oxidation of the copper vessels used for distillation was responsible, but eventually the color was ascribed to certain decomposition products that were named *azulenes*. For example, a natural sesquiterpene alcohol from guaiacum wood, guaiol, can be decomposed to yield guaizulene; β-vetivone from vetiver oil gives rise to vetivazulene. These and other azulene precursors are found in everything from araucaria to zdravetz oil, as shown by the partial list in the margin. Sometimes the azulenes themselves occur naturally in small quantities. One group of scientists had to collect 160,000 liters of urine from pregnant mares in order to isolate a mere 20 milligrams of vetivazulene!

guaiaol   guaiazulene   β-vetivone   vetivazulene

Unquestionably the most important nonbenzenoid aromatic compounds are the *porphyrins*—without them there could be no plant or animal life as we know it, since they play a key role in both photosynthesis and respiration. Porphyrins are built around the system of four linked pyrrole rings that constitutes the parent compound, porphin. Tracing along the heavy line in the porphin molecule illustrated reveals an 18 pi-electron aromatic ring, to which each of two nitrogen atoms contributes one electron. Hemoglobin, which is a conjugated protein containing an iron-complexed porphyrin known as heme, is the pigment that shuttles oxygen through the bloodstream and keeps the respiratory process going. Hemoglobin itself is blue in color, but as it picks up oxygen in the lungs it is converted to bright red oxyhemoglobin, which gradually gives up its oxygen to the cells and is reduced to hemoglobin again for its trip back to the lungs.

Hemoglobin is an extremely complex protein whose structure differs slightly for different animal species; a typical empirical formula is $C_{738}H_{1166}O_{208}N_{203}S_2Fe$.

heme   porphin

Ordinarily the excess porphyrins in the body are metabolized by the liver into iron-free substances such as biliverdin and bilirubin, which collectively make up the bile pigments. If this metabolic process breaks down, a disease known as porphyria results, giving rise to agonizing attacks that may resemble psychotic episodes. It has been speculated that King George III suffered from porphyria, based on medical records that noted that his urine was red or discolored, a diagnostic symptom of the disease. If so, this must certainly have given rise to the episodes of "madness" that characterized the later years of his reign; and our image of George III as an evil king whose tyrannical acts precipitated the American revolution may undergo a revision.

See *Scientific American* (July 1969), p. 38 for more about George III's illness.

METHODOLOGY

One way to prepare a carbonium ion is by a hydride-transfer reaction. Just as a strong acid can transfer a proton to a strong base leaving behind a weaker (more stable) base, a hydrocarbon can transfer a hydride ion to a reactive carbonium ion leaving behind a less reactive (more stable) carbonium ion. In this experiment, the reactive ion will be trityl (triphenylmethyl) and the stable one tropylium. The hydride transfer takes place when trityl fluoborate reacts with the hydrocarbon cycloheptatriene, yielding tropylium fluoborate and triphenylmethane. Trityl fluoborate itself is prepared by treating the corresponding alcohol with 48% aqueous fluoboric acid in acetic anhydride. The anhydride is needed to take up the water present in the fluoboric acid, and also that resulting from the reaction—carbonium ions can react with water to yield alcohols or ethers. In the procedure used here, the trityl fluoborate will not be isolated but will be treated with cycloheptatriene directly. Tropylium fluoborate precipitates as the dark color of the trityl carbonium ion disappears; adding ether completes the precipitation. The iodide of tropylium is then easily formed by adding potassium iodide to an aqueous solution of tropylium fluoborate.

To determine whether or not the initial product is aromatic, an NMR spectrum will be obtained. If the product is not aromatic, most of its proton signals should occur in the same region as those of the starting material, which yields the NMR spectrum illustrated in Figure 1. If the product is aromatic, the deshielding effect of the so-called aromatic

Proton transfer

$$HB_1 + B_2^- \longrightarrow$$
stronger   stronger
acid   base

$$HB_2 + B_1^-$$
weaker   weaker
acid   base

Hydride transfer

$$R_1H + R_2^+ \longrightarrow$$
less stable
carbonium ion

$$R_2H + R_1^+$$
more stable
carbonium ion

trityl fluoborate

See Silverstein and Bassler (Bib–F8) for a qualitative discussion of the ring-current effect.

ring current (and of the ionic charge) should result in a spectrum that is shifted well downfield from that of the nonaromatic starting material.

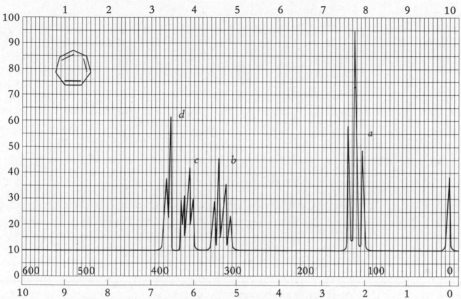

Figure 1. NMR spectrum of cycloheptatriene. (Reproduced from C. J. Pouchert and J. R. Campbell, *The Aldrich Library of NMR Spectra,* volume 1, by permission of Aldrich Chemical Company, Inc.)

PRELAB EXERCISES

1. Read ⟨OP–8⟩ and ⟨OP–34⟩. Review the other operations as needed.

2. Calculate the volume of 0.35 mol of acetic anhydride, the volume of 0.020 mol of 48% (by weight) fluoboric acid, the mass of 0.015 mol of triphenylmethanol, and the volume and mass of 0.017 mol of cycloheptatriene.

3. Write a brief checklist for the experiment.

Reactions and Properties

A. $Ph_3COH + HBF_4 \xrightarrow{Ac_2O} Ph_3C^+BF_4^- + H_2O$

$$Ph_3C^+BF_4^- + \text{⟨cycloheptatriene⟩} \longrightarrow Ph_3CH + \text{⟨tropylium⟩} BF_4^-$$

$$B. \quad \left(\!\!\! \bigcirc\!\!\!\right) BF_4^- + NaI \longrightarrow \left(\!\!\! \bigcirc\!\!\!\right) I^- + NaBF_4$$

Table 1.   Physical properties

	M.W.	m.p.	b.p.	d.
acetic anhydride	102.1	−73	140	1.082
fluoboric acid (48%)	87.8			1.41
triphenylmethanol	260.3	164	380	
cycloheptatriene	92.15	−80	117	0.887
potassium iodide	166.0			

Hazards

acetic anhydride

Acetic anhydride is extremely irritating to the skin, eyes, and respiratory system and reacts violently with water. Fluoboric acid is a highly toxic, very corrosive irritant; use extreme care to avoid inhalation or contact.

PROCEDURE

A. *Preparation of Tropylium Fluoborate*

Carry out this operation under the hood. Into a 125-ml Erlenmeyer flask measure about 0.35 mol of acetic anhydride.

CAUTION

Be very careful in handling fluoboric acid and acetic anhydride. Use rubber gloves.

Manual stirring is adequate.

A light yellow color may persist.

DMSO–d_6 is preferred, but ordinary protic DMSO can also be used as the solvent.

Place the flask in an ice bath ⟨OP–8⟩ and add dropwise, with swirling, 0.020 mol of fluoboric acid (48% aqueous). Then add 0.015 mol of triphenylmethanol in small portions, with continued cooling ⟨OP–8⟩ and mixing ⟨OP–9⟩, followed by 0.017 mol of cycloheptatriene, added dropwise until the dark color of the triphenylmethyl ion has faded. Precipitate the remainder of the tropylium fluoborate by adding 50 ml of anhydrous ethyl ether, then filter the product by suction ⟨OP–12⟩ and wash it with dry ether. Obtain an NMR spectrum of tropylium fluoborate in dimethyl sulfoxide ⟨OP–34⟩ and compare it with the spectrum of cycloheptatriene in Figure 1.

B. *Preparation of Tropylium Iodide*

Dissolve the tropylium fluoborate in the minimum amount of hot water, then add about 5 ml of saturated aqueous potassium iodide. Cool the solution in ice water to complete the precipitation, filter by suction ⟨OP–12⟩, and wash the crystals with a little cold methanol. Air dry ⟨OP–21⟩ and weigh ⟨OP–5⟩ the product.

The water should not be boiling.

Experimental Variations

1.   The ultraviolet-visible spectrum of a solution of tropylium fluoborate can be obtained in 0.1M HCl and compared to that of benzene in ethanol.

2.   An infrared spectrum of tropylium fluoborate can be obtained using a mull or a KBr pellet. The spectrum may be compared with that of other aromatic compounds, such as benzene.

Topics for Report

1.   (*a*) Compare the spectrum of your tropylium fluoborate with that of cycloheptatriene and tell what it reveals about the structure of tropylium fluoborate, giving reasons for your answer. (*b*) Assign as many NMR signals in the spectrum of cycloheptatriene as you can, indicating which set of equivalent protons is responsible for each signal.

Ignore minor peaks that may be due to impurities.

There are only four important signals, though each may be split into a number of smaller peaks.

2.   Trityl fluoborate should not be exposed to moist air for any length of time. Explain why this is so and give structures for some possible impurities in your tropylium iodide.

3.   Predict which of the following compounds should show aromatic properties.

1.

2.

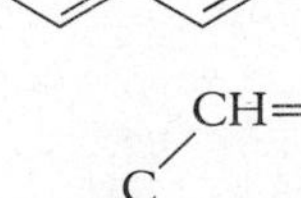

3.

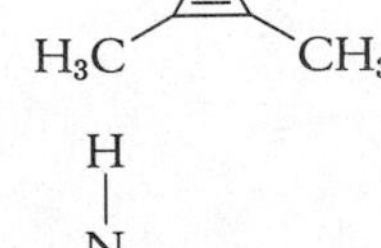

4.

5.

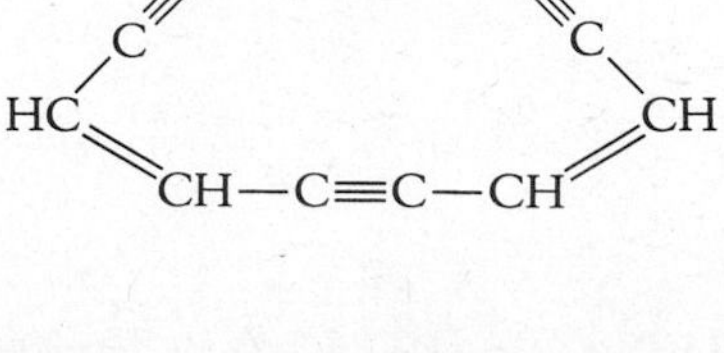

6.

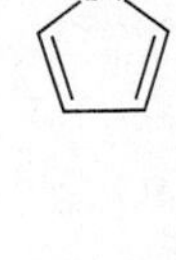

7.   **8.**   **9.**

[14]annulene

M. J. S. Dewar, *Nature 155,* 50 (1945).

cyclobutadiene

These metal complexes are "sandwich" compounds similar to ferrocene. One type is described as an "open-faced sandwich."

4.   When first prepared, [14]annulene was found to be far less stable than expected; it was considered nonaromatic until NMR and x-ray diffraction studies showed otherwise. Build a molecular model of [14]annulene in the conformation shown and attempt to explain its low stability.

Library Topics

1.   Read and summarize the article in which Dewar proposed a tropolone structure for stipitatic acid. Include the reasons he gave for believing the system to be aromatic.

2.   Cyclobutadiene, containing only four pi electrons on a ring, is not expected to be aromatic according to Hückel's $4n + 2$ rule. It is very unstable and has only recently been prepared in combination with certain metals. Look up and describe the preparation of cyclobutadiene complexes and discuss some recent speculations about the structure and properties of cyclobutadiene.

Preparation of Triphenylmethyl Bromide and the Trityl Free Radical

Arenes. Free-radical Substitution Reactions. Preparation of Halogenated Hydrocarbons. Properties of Free Radicals.

Operations

⟨OP–5⟩ Weighing

⟨OP–6⟩ Measuring Volume

⟨**OP–10a**⟩ Semimicro Addition

⟨OP–11⟩ Gravity Filtration

⟨**OP–12a**⟩ Semimicro Vacuum Filtration

⟨OP–14⟩ Solvent Evaporation

⟨OP–21⟩ Drying Solids

⟨**OP–22b**⟩ Trapping Gases

⟨**OP–23a**⟩ Semimicro Recrystallization

⟨OP–28⟩ Melting Point

Experimental Objectives

To prepare triphenylmethyl (trityl) bromide from triphenylmethane and convert it to the trityl free radical.

To observe the behavior of the trityl free radical in solution and interpret your observations.

Learning Objectives

To learn how to carry out an organic synthesis on a semimicro scale and how to isolate and purify small quantities of materials.

To learn about the nature and properties of free radicals and something of their history.

tris(4-*t*-butylphenyl)methyl

$$t\text{-Bu} = CH_3C-\begin{array}{c}CH_3\\|\\|\\CH_3\end{array}$$

SITUATION

You are one member of a team of organic chemists engaged in research on free-radical reactions. You are, of course, familiar with the ground-breaking work of Moses Gomberg, who prepared the first stable organic free radical, triphenylmethyl, at the turn of the century. At that time, Gomberg postulated that triphenylmethyl existed in equilibrium with its dimer, hexaphenylethane. However, a member of your research team recently prepared the very similar free radical, tris(4-*t*-butylphenyl)methyl, and discovered that it does not form a similar dimer. This is very puzzling, since there is no obvious reason why *t*-butyl groups at the *para* position on each benzene ring should prevent this kind of dimerization. Such an unexpected result has thrown some doubt on Gomberg's original work, and you have chosen to repeat his experiment in order to confirm or disprove the existence of a dimer.

Your laboratory does not have any triphenylmethyl bromide in stock, so you will prepare the compound from some triphenylmethane on hand. Then you can synthesize the radical, and presumably its mysterious dimer, by treating the triphenylmethyl bromide with metallic zinc.

Gomberg equilibrium

triphenylmethyl (trityl) free radical

hexaphenylethane (assumed)

BACKGROUND

The Strange Case of the Disappearing Dimer

In the mid-nineteenth century, many chemists were convinced that carbon could exist in a trivalent state in the form of "free" **radicals.** For example, it seemed reasonable to believe that if magnesium chloride could react with sodium metal to yield magnesium, then methyl halides should react with sodium to form "methyl." However, Charles Wurtz carried out the reaction of methyl iodide with sodium and

A **radical** is a group of atoms like methyl (CH_3-) that generally exists only in combination with other atoms or groups, such as in methyl bromide, CH_3Br.

obtained not methyl but ethane, a synthesis known today as the Wurtz reaction. After many similar attempts ended in failure, chemists became increasingly doubtful that truly "free" radicals could exist.

Moses Gomberg never meant to make a free radical. He was trying to synthesize hexaphenylethane to prove a point that, had he been successful, would be remembered today by only a handful of chemistry specialists. Gomberg's first attempts to prepare this compound using the Wurtz reaction were not productive, so he tried different metals such as silver and zinc, each time coming up with a snow-white solid melting at 185°C that gave the wrong analysis for hexaphenylethane. After repeated attempts to prepare this elusive compound it finally dawned on him that the product was reacting with oxygen in the air and forming triphenylmethyl peroxide. When Gomberg next ran the reaction, he was careful to exclude all air from the reaction mixture, and he obtained a white solid melting at 147°C that (at last) gave the correct analysis for hexaphenylethane. But this compound behaved very strangely for a hydrocarbon. It reacted with air in solution to form triphenylmethyl peroxide and it rapidly decolorized dilute halogen solutions—something no self-respecting hydrocarbon would think of doing.

Gomberg was eventually forced to the conclusion that he had synthesized the world's first stable free radical, triphenylmethyl. In solution it slowly equilibrated with its dimer (presumed to be hexaphenylethane). Recent research has shown that hexaphenylethane is *not* present in these solutions, and that it apparently does not even exist! After completing this experiment you may wish to read the article in the *Journal of Chemical Education* that shows the actual structure of the mysterious dimer and tells how it was established.

Reaction of methyl iodide and sodium

$$CH_3I + Na \longrightarrow \text{``}CH_3\text{''} + NaI$$
(expected)

$$2CH_3I + 2Na \longrightarrow CH_3CH_3 + 2NaI$$
(actual)

In 1896, Wilhelm Ostwald declared *"The very nature of the organic radicals is inherently such as to preclude the possibility of isolating them."* Ostwald was a first-rate chemist but a dismal prophet.

Attempted synthesis of hexaphenylethane

$$2Ph_3CBr + Zn \longrightarrow$$
$$Ph_3CCPh_3 + ZnBr_2$$

$$Ph_3COOCPh_3$$
triphenylmethyl peroxide

Triphenylmethyl is often called trityl for short.

J. Chem. Educ. **47,** 535 (1970).

METHODOLOGY

Triphenylmethane is more expensive than any of the starting materials you have used so far; consequently you will use less of it. When working with ten grams or more of starting material, it is possible to be a little sloppy about laboratory technique and still get respectable results. The loss of a few tens of milligrams in each transfer or a few tenths of a gram during washing or recrystallization may not be serious if you have plenty of material to work with. But in an experiment that starts with about a gram of starting material

semimicro: utilizing quantities of material on the order of a few grams or tenths of a gram.

Mechanism for the chlorination of methane

Initiation

1. $Cl_2 \xrightarrow{h\nu} 2Cl$

Propagation

2. $Cl\cdot + CH_4 \rightarrow CH_3\cdot + HCl$
3. $CH_3\cdot + Cl_2 \rightarrow CH_3Cl + Cl\cdot$
(steps 2 and 3 repeated indefinitely)

Termination

4. $Cl\cdot + Cl\cdot \rightarrow Cl_2$ (and other termination steps)

Gas trap reaction

$HBr + NaOH \longrightarrow H_2O + NaBr$

You will actually prepare the trityl free radical twice in this experiment—first as an intermediate in the bromination of triphenylmethane, and again by the reaction of zinc with trityl bromide.

or less, losses of that magnitude cut your yield drastically. In this experiment, it will be very important to avoid unnecessary transfers, to use minimal amounts of wash liquids and recrystallization solvents (well cooled), and in general to refine your technique so as to keep losses to a minimum. Another way of cutting down material losses is to reduce the surface area of glass or porcelain in contact with the chemicals (i.e., to use small-scale apparatus). So you will use **semimicro** apparatus and techniques in operations such as addition, filtration, and recrystallization.

The bromination of triphenylmethane proceeds by a chain mechanism very similar to that of the chlorination of methane. Both reactions involve the formation of a free radical intermediate, which then reacts with halogen to form the product. Before the reaction can begin, some bromine free radicals must be formed by irradiation from an appropriate light source; you can use an ordinary unfrosted light bulb for the purpose.

To cut down material losses, most of the operations will be carried out in the same vessel, a small sidearm test tube (see Experimental Variation 1 for an alternative setup). To avoid unwanted side reactions, the hydrocarbon will always be kept in large excess by adding bromine dropwise *to* the triphenylmethane solution. Since the reaction evolves hydrogen bromide, it will be necessary to trap this gas by passing it into a solution of dilute sodium hydroxide. When the reaction is complete, some excess bromine may remain, imparting color to the solution. This can be removed by adding a drop or two of cyclohexene.

Removal of excess bromine

cyclohexene   1,2-dibromocyclohexane
(a liquid)

To isolate the product, the solvent (CCl_4) can be removed by evaporation directly from the sidearm test tube. The product can be recrystallized in the same test tube and collected in a Hirsch funnel.

After you have prepared and purified trityl bromide, you will attempt to generate the trityl free radical by essentially the method Gomberg used—by adding zinc to a solution of the halide. You will be expected to observe the

resulting solution carefully so as to produce evidence supporting the existence of the radical and its dimer in solution.

*Free radicals are usually colored, hydro-
carbons colorless.*

PRELAB EXERCISES

1. Read ⟨OP–10a⟩, ⟨OP–12a⟩, ⟨OP–22b⟩, and
⟨OP–23a⟩. Review the other operations as needed.

2. Calculate the mass of 5.00 mmol of triphenylmethane, and calculate the volume (in ml) of 1.0M bromine in
carbon tetrachloride needed to provide 5.00 mmol of bromine.

3. Write a checklist for the experiment.

Reactions and Properties

A.

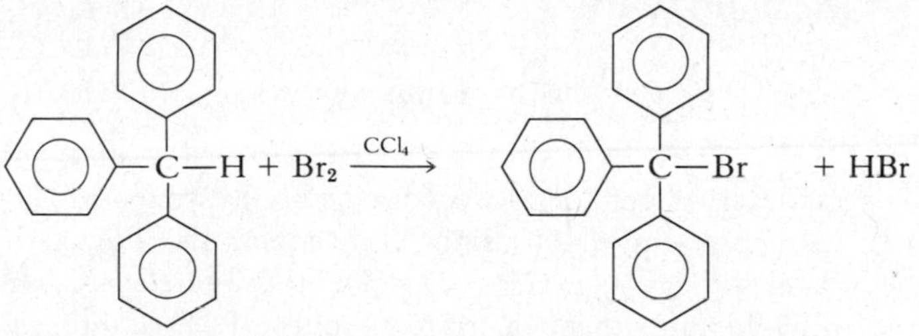

triphenylmethane   triphenylmethyl bromide

B.

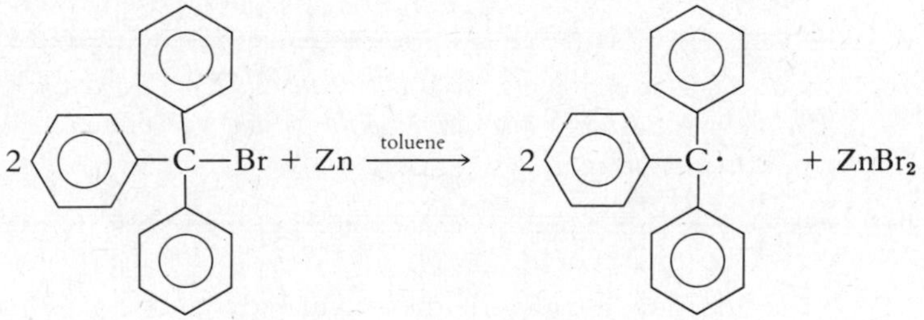

triphenylmethyl radical
(dimerizes in solution)

Table 1.   Physical properties

	M.W.	m.p.	b.p.	d.
triphenylmethane	244.3	94	359	
bromine	159.8	59		3.12
carbon tetrachloride	153.8	−23	76.5	1.594
triphenylmethyl bromide	323.2	152–54		

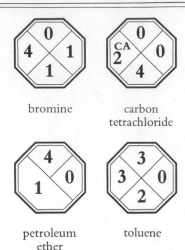

bromine   carbon
               tetrachloride

petroleum   toluene
ether

Hazards

Bromine is a very corrosive liquid, causing burns on skin and eyes. Its vapor irritates the eyes and mucous membranes and can damage the respiratory tract. Use great care in handling bromine or its solutions and do not breath its vapors.

Carbon tetrachloride has poisonous vapors whose inhalation can cause serious damage to internal organs; it is known to be carcinogenic to laboratory animals. Avoid contact and do not breathe vapors; use only with adequate ventilation. Petroleum ether is very flammable and should never be used when open flames are nearby. Triphenylmethyl bromide is corrosive and irritating; avoid contact with skin and eyes. Toluene is flammable and somewhat toxic; avoid contact and inhalation of vapors.

PROCEDURE

A. *Preparation of Triphenylmethyl Bromide*

Reaction.   Assemble the reaction setup illustrated in Figure 1 on a single ringstand, using dilute aqueous sodium hydroxide in the gas trap ⟨OP–22b⟩. Weigh the calculated amount of triphenylmethane on glazed paper ⟨OP–5⟩ and dissolve it in 10 ml of dry carbon tetrachloride in the reaction tube. Fill

If a sidearm test tube is not available, an ordinary 20-cm test tube with a two-hole stopper can be used.

WARNING   *Carbon tetrachloride has been designated as a carcinogen. Appropriate precautionary measures should be taken to minimize exposure to this chemical.*

The tip of the pipet should be several inches above the liquid surface.

the addition pipet with the calculated volume ⟨OP–6⟩ of 1.0M bromine/carbon tetrachloride. Arrange the assembled apparatus so that the reaction tube is about 10 cm from an unfrosted light bulb (several students can use the same bulb), then add the bromine solution slowly ⟨OP–10a⟩, with gentle shaking, so that the color is nearly discharged after each addition. When addition is complete, continue irradiation for about 10 minutes, or until the solution is nearly colorless. If an orange color persists, add a few drops of cyclohexene to remove the last traces of bromine.

Isolation.   Remove the dropping pipet and gas trap, attach the sidearm to a water trap and aspirator, insert a solid rubber stopper, and evaporate the solvent ⟨OP–14⟩ with gentle heating. Protect the product from moisture to prevent hydrolysis to triphenylmethanol.

Purification and Analysis.   Recrystallize the crude trityl bromide ⟨OP–23a⟩ from a little petroleum ether (*no flames!*), collecting the crystals by semimicro vacuum filtration ⟨OP–12a⟩. Let the product dry ⟨OP–21⟩ and obtain its weight ⟨OP–5⟩ and melting point ⟨OP–28⟩. Save enough for part **B** and turn in the rest of your product.

B. *Preparation and Reactions of the Trityl Free Radical*

Dissolve about 0.2 g of purified trityl bromide in 4 ml of toluene in a small test tube. Add 0.6 g of 30-mesh zinc, stopper immediately, and shake vigorously for ten minutes. Quickly filter the mixture by gravity ⟨OP–11⟩ through glass wool into a small test tube and stopper the tube immediately. Shake and allow the solution to stand for a few minutes, then open the tube to let air in, stopper, and shake again for a few minutes. Repeat this process until no more color changes occur. Record your observations in your laboratory notebook, and interpret them on the basis of the information given in the Background.

Experimental Variations

1.   At the instructor's option, the experiment can be carried out on a larger scale by increasing the quantities five-fold or more and scaling up the glassware accordingly. One arrangement uses a small round-bottomed flask fitted with a Claisen head as a reaction vessel. A gas trap is attached to the bent arm of the Claisen head; a separatory funnel is attached to the straight arm and used for adding bromine.

2.   The precipitate obtained in part **B** can be collected and washed with ether, and a melting point obtained to establish its identity.

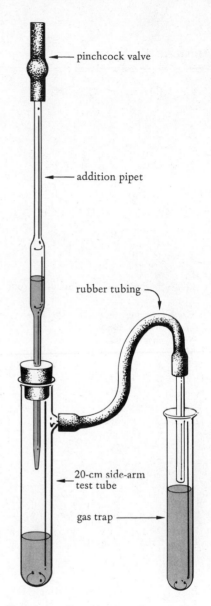

Figure 1.   Bromination apparatus

Free Radical Bromination of Hydrocarbons

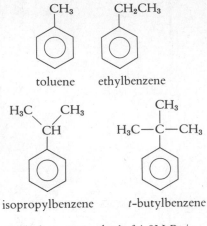

toluene   ethylbenzene

isopropylbenzene   *t*-butylbenzene

A blank containing ½ ml of 1.0M Br₂/ CCl₄ in 5 ml of carbon tetra- chloride can be prepared for color comparison.

In each of four 10-cm test tubes, place a solution of 1 ml of hydrocarbon in 4 ml of carbon tetrachloride. Use the following hydrocarbons: toluene, ethylbenzene, isopropylbenzene (cumene), and *t*-butylbenzene. Into each of four corresponding test tubes, measure 0.5 ml of 1.0M bromine in carbon tetrachloride. Add each bromine solution to the corresponding hydrocarbon solution as rapidly as possible (watch carefully for evidence of a quick reaction), stopper and shake the tubes, and place them near a window or strong light. Note and record the approximate time required for each solution to become decolorized. If any solutions are not completely decolorized at the end of the period, compare their relative color intensities. Arrange the four hydrocarbons in order of reactivity toward bromine, explain your results on the basis of free radical stability, and give the structure of the expected product in each case.

Topics for Report

1. Give equations for all reactions of the trityl free radical for which you saw evidence, and describe the evidence in each case.

2. Write a mechanism for the free radical bromination of triphenylmethane.

3. Construct a flow diagram for the synthesis of triphenylmethyl bromide.

4. Assuming a free-radical mechanism for the bromination of the following compounds, arrange them in order of their expected reactivity, most reactive first.

Library Projects

1.   Read the selections from Gomberg's work in the reference cited, describe the reasoning that led him to the conclusion that he had prepared the trityl free radical, and tell why he had tried to prepare hexaphenylethane in the first place.

2.   Read the article (cited in the Background) describing the discovery of a new structure for the dimer of triphenylmethyl, summarize the evidence for the proposed structure, and explain why tris(4-*t*-butylphenyl)methyl does not form a similar dimer.

H. M. Leicester and H. S. Klickstein, eds., *A Sourcebook in Chemistry, 1400–1900* (Boston: Harvard University Press, 1952), pp. 512–20.

The Isolation and Isomerization of Lycopene from Tomatoes

Carotenoids. Isolation of Natural Products. Isomerization about Carbon-carbon Double Bonds. Ultraviolet-visible Spectrometry

Operations

⟨OP–5⟩ Weighing

⟨OP–12⟩ Vacuum Filtration

⟨**OP–13d**⟩ Extraction of Solids

⟨OP–14⟩ Evaporation

⟨**OP–16**⟩ Column Chromatography

⟨OP–19⟩ Washing Liquids

⟨OP–20⟩ Drying Liquids

⟨**OP–35**⟩ Ultraviolet-visible Spectrometry

Experimental Objectives

To isolate the red pigment lycopene from tomato paste by column chromatography.

To convert lycopene to its geometric isomer neolycopene and verify the interconversion by ultraviolet-visible spectrometry.

Learning Objectives

To learn how to extract components from solid materials, how to separate the components of a mixture by column chromatography, and how to record an ultraviolet-visible spectrum.

To learn about the structures and occurrence of carotenoids and Vitamin A, and about the function of Vitamin A in vision.

SITUATION

As an analytical chemist working for Galactic Foods, Inc., you are responsible for testing new varieties of tomatoes for possible use in Galactic Tomato Catsup. One of your responsibilities is to assay all varieties for lycopene content, since the quality of a given lot can be correlated with the quantity of this red pigment present.

After working up an extract from a new variety called Galactic's Porkchop Hybrid VF, you ran an ultraviolet-visible spectrum of the red pigment in solution and came up with unexpected results. The spectrum bore a superficial resemblance to that of lycopene, but some peaks were not in the right location and their intensities were changed. Have your company's plant geneticists somehow come up with a mutant variety of tomato containing a new form of lycopene? Or did something happen during the treatment of the extract that isomerized its lycopene?

Thinking back over the events of the previous day, you recall that you assigned the job of extracting the pigment to an inexperienced laboratory technician who dropped a flask containing some extract and cut himself on the broken glass, then treated the cut with tincture of iodine. Considering his sloppy technique, it is not unlikely that traces of the iodine ended up in subsequent extracts. You decide to carry out the isolation yourself this time, starting with a tomato concentrate having the consistency of tomato paste. First you will find out if the red pigment is actually lycopene or an isomer. Then you will see if a little iodine can catalyze an isomerization reaction.

BACKGROUND

Carotenoids, Vitamin A, and Vision

Lycopene, with its 40 carbon atoms and 13 double bonds, is one of the most unsaturated compounds in nature. Because most of its double bonds are conjugated, lycopene absorbs radiation at long wavelengths in the 400–500-nm region of the visible spectrum. Its resulting deep yellow-red color is responsible for the redness of ripe tomatoes, rose hips, and many other fruits.

An even more important plant pigment is the yellow substance β-carotene, which is present not only in carrots but in all green leaves and many flowers as well. Both lyco-pene and β-carotene, along with most of the other natural carotenoids, exist in the stable all-*trans* form.

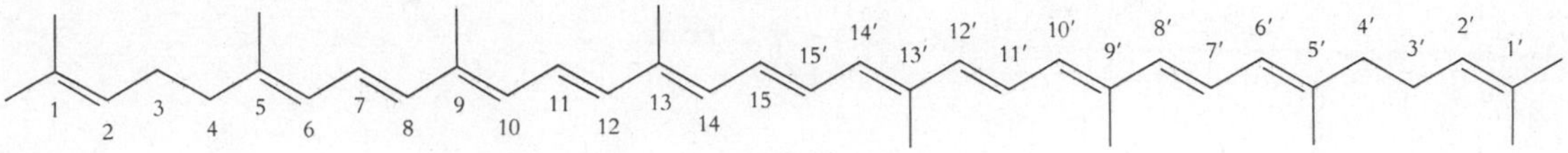

lycopene

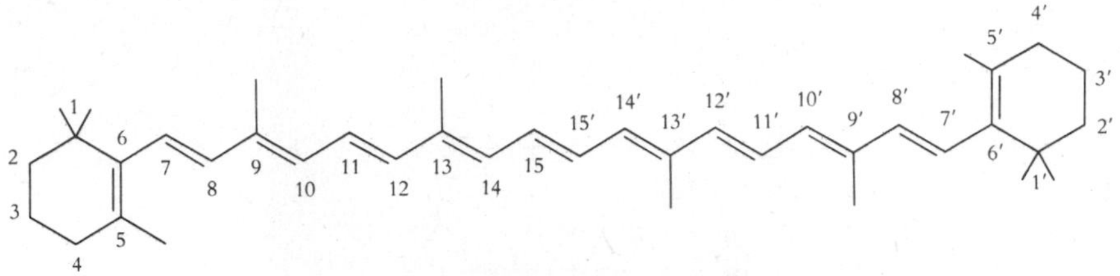

β-carotene

Figure 1. Structures of carotenoids. (*Note:* A single straight line branching off from a chain or ring stands for a methyl group in these and similar formulas. A carbon atom with the requisite number of hydrogens is at each bend of the chain.)

Although the main function of carotenoids in plants re-mains somewhat of a mystery, the importance of carotenes to animals is clear—β-carotene (and, to a lesser extent, its α- and γ-forms) act as provitamins, being converted in the in-testinal wall to Vitamin A, which is then esterified and stored in the liver. Generations of children have grown up

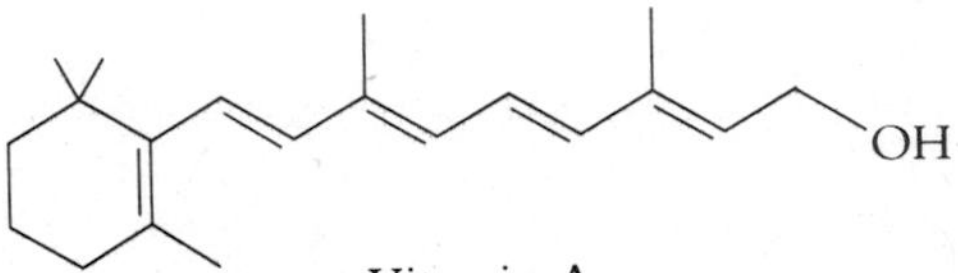

Vitamin A
(all-*trans*)

Cod liver oil is a more direct (if less palatable) source of Vitamin A than carrots and leafy green vegetables.

A similar process occurs in the cones of the retina, which are responsible for color vision.

to the mealtime refrain, "Eat your carrots—they're good for the eyes!" In fact, Vitamin A from carotenes and other sources does play an important role in vision, as well as acting to promote growth.

The process of vision, although extremely complex in its entirety, seems to be based on a simple isomerization of an oxidized form of Vitamin A called retinal. In the corneal rods that are responsible for night vision, retinal occurs in combination with the complex protein opsin to form rho-dopsin (visual purple). While bound to this protein, retinal must assume the shape shown in Figure 2A, with an 11-*cis* double bond and probably a cisoid conformation at the 12–13 single bond as well.

When a photon of light strikes a molecule of this 11-*cis*-retinal, it causes a configurational change that eventually results in the formation of all-*trans*-retinal. Now the *trans* form obviously has a different molecular shape than its *cis* counterpart, being straight and rigid where the *cis* is bent and twisted. As a result, it no longer fits well into its niche on the opsin molecule. Just as a mother kangaroo might react to a joey that is making a commotion in her pocket—shifting positions to minimize the discomfort and eventually ejecting her unruly passenger—the opsin undergoes a series of configurational and conformational changes that lead it to

A "joey" is a young kangaroo.

The mechanism by which a single photon can trigger the enormously amplified response of a receptor cell is not yet understood.

A. 11-*cis*-12-*s*-*cis*-retinal

B. all-*trans*-retinal

Figure 2.   Retinal isomers.

eject the transformed retinal. This dissociation of the rhodopsin molecule triggers the transmission of a visual message to the brain. Subsequently the *trans*-retinal is enzymatically reduced to all-*trans*-Vitamin A, which isomerizes to 11-*cis*-Vitamin A, which is oxidized back to 11-*cis*-retinal, which promptly combines with another molecule of opsin to regenerate more rhodopsin, which then waits for another light photon to come along and start the cycle all over again!

Although lycopene cannot be converted directly to Vitamin A and thence to retinal, it does undergo a configurational change analogous to those undergone by retinal. Adding a small amount of iodine to a solution of all-*trans*-lycopene can cause its partial conversion to 13-*cis*-lycopene (also called neolycopene A), with substantial changes in its properties. The conversion can easily be followed by spectroscopy, since isomerization causes significant changes in the ultraviolet-visible spectrum of this pigment.

The vision cycle

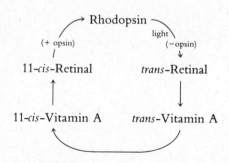

*Tangerine tomatoes contain a poly-*cis* isomer of lycopene called prolycopene; adding a drop of iodine to a pale yellow solution of prolycopene transforms it almost instantly to brilliant orange lycopene.*

METHODOLOGY

Although lycopene can be isolated directly from ripe tomatoes, the amount of water to be dealt with makes it more convenient to use a concentrated tomato preserve such as commercial tomato paste. The lycopene, along with other carotenoids (including the ubiquitous β-carotene), can be extracted with a mixture of acetone and hexane (or petroleum ether). The extract is washed to remove acetone and water-soluble constituents, dried, and reduced in volume. The lycopene is then separated from other pigments by chromatography on an alumina column. Since adsorption on such a column increases with the number of double bonds, lycopene with its 13 double bonds will remain on the column longer than γ-carotene with 12, or α- and β-carotene with 11 each. Carotenoids having polar functional groups, such as xanthophyll (with two OH groups per molecule), will remain on the column after the carotenes and lycopene have been eluted.

The isomerization of lycopene to neolycopene A can be observed directly in an ultraviolet-visible spectrophotometer by adding a drop of iodine solution to the cuvette after the spectrum of the lycopene has been recorded. If this is done while the wavelength control is set at the λ_{max} of lycopene, the reaction can be monitored by the change in absorbance. A spectrum of the equilibrium mixture can then be obtained when the absorbance reading stabilizes.

Fresh tomatoes contain about 20 mg of lycopene per kilogram, while tomato paste provides 150–300 mg/kg.

Lycopene is slowly oxidized or isomerized by heat, light, oxygen, contact with active surfaces (such as alumina), or simply by standing in solution; so it is important to avoid delays in working up the lycopene and obtaining its spectrum.

Light of a wavelength near the λ_{max} of lycopene accelerates the isomerization in the presence of iodine.

The equilibrium mixture is approximately 45% *cis*.

PRELAB EXERCISES

1. Read ⟨OP–13*d*⟩, ⟨OP–16⟩, and ⟨OP–35⟩. Review the other operations as needed.

2. Write a brief checklist for the experiment.

Properties

Table 1.   Physical properties

	M.W.	m.p.	b.p.	d.
lycopene	536.9	175		
hexane	86.2	−95	69	0.660
petroleum ether	variable		~60–75	~0.65
acetone	58.1	−94	56	0.791

Hazards

Acetone is flammable and irritates the eyes; inhalation can cause dizziness and narcosis. Petroleum ether is extremely flammable and should not be inhaled for prolonged periods. Inhalation of alumina dust can cause lung damage.

acetone   petroleum ether

PROCEDURE

Weigh 3–4 g of tomato paste ⟨OP–5⟩ into a small beaker and extract it ⟨OP–13*d*⟩ three times with 10-ml portions of 50:50 (by volume) acetone–hexane or acetone–

There should be no flames in the laboratory during this experiment.   ***WARNING***

---

petroleum ether. After each extraction, collect the extract by vacuum filtration ⟨OP–12⟩. Wash the residue on the filter with a little more of the solvent after the last extraction, and combine the wash liquid with the three extracts. Wash the combined extracts ⟨OP–19⟩ with 25 ml of saturated sodium chloride, 25 ml of 10% aqueous potassium carbonate, and 25 ml of water in succession, then dry the organic layer ⟨OP–20⟩ with magnesium sulfate or sodium sulfate.

Use low-boiling petroleum ether (approx. 60–75°).

The washing process removes the acetone and leaves the pigments dissolved in the petroleum solvent; potassium carbonate removes acidic impurities.

Remove most of the solvent under vacuum ⟨OP–14⟩ until only about 1–2 ml of liquid remain. Do *not* evaporate to dryness; use as little heat as possible to avoid isomerizing the lycopene.

Prepare an alumina column for chromatography ⟨OP–16⟩, using enough adsorbent to extend 15 cm or more in a 1-cm–diameter column. Elute the pigments using petroleum ether (or hexane) until the yellow (carotene) pigment begins to come off the column, then elute the orange-red (lycopene) band using 10% acetone in petroleum ether (or hexane). Collect a 5-ml sample of the lycopene solution from this band for spectrometric analysis. The remainder may be stored (in the dark) for later use, if desired.

Chromatography Grade II alumina (containing 3% water) is recommended. Use more alumina for a wider column or less for a narrower one, so that the length:diameter ratio is between 8:1 and 20:1.

The flow rate should be 1–2 drops/second. The column should be kept away from strong light or covered with aluminum foil to prevent isomerization.

Obtain an ultraviolet-visible spectrum ⟨OP–35⟩ of the lycopene eluate from 600 nm in the visible region to 250 nm in the ultraviolet. If necessary, adjust the instrument sensitivity or change the concentration of the lycopene solution (by adding or removing solvent) so that the strongest peak (at about 475 nm) covers 75–100% of the vertical scale.

Remember to change the light source when going from the visible to the ultraviolet region.

Tabulate all spectral data and include the spectra with your report.

After a good spectrum has been recorded, set the instrument on the 475-nm peak, add a drop of 0.025% iodine/hexane solution to the sample cuvette, shake to mix, and replace the cuvette in the instrument. When the absorbance reading appears to have stabilized (2–3 minutes), again record the spectrum in the 600–250 nm region.

Experimental Variations

1. The eluents containing lycopene may be concentrated until red lycopene crystals form on cooling, after which a melting point can be obtained. The melting point of crystals from the lycopene-neolycopene mixture may also be measured and compared with that of lycopene.

2. The carotene band from the chromatography of the tomato extract can be saved and its spectrum observed as was done with the lycopene.

3. Adding a drop or two of concentrated sulfuric acid to lycopene is said to yield a deep indigo blue color, due to the formation of carbonium salts.

These may be the 13-*cis* and 9-*cis* isomers, respectively.

4. Carotene can be isolated from strained carrots by a similar procedure and its ultraviolet-visual spectrum obtained. Treatment of β-carotene with iodine yields seven or more different *cis* forms, of which neo-β-carotene C and neo-β-carotene U predominate.

5. The carotenoid pigments can also be separated by TLC. See Experiment 26 in *Organic Chemistry: An Experimental Approach* by J. S. Swinehart for a procedure.

Published by Appleton-Century-Crofts, New York (1969).

Topics for Report

1. Calculate the concentration of your lycopene solution, given that its molar absorptivity (ϵ) at 473 nm is 1.86×10^4.

2. Describe any changes that occurred in the ultraviolet-visible spectrum when you added iodine, and discuss any evidence leading you to believe that a partial isomerization has occurred. Try to identify any absorption bands belonging to the *cis* isomer of lycopene.

3. Explain why hydrocarbons such as lycopene and β-carotene are colored, while most other hydrocarbons are not.

4. One step in the manufacture of synthetic Vitamin A involves the reaction of β-ionone with ethyl chloroacetate shown. Write a mechanism for this reaction. (*Hint:* Compare this reaction with the aldol condensation.)

5. Linus Pauling has predicted that the 7–*cis* and 11–*cis* forms of lycopene should be considerably less stable than any of its other mono–*cis* isomers. Draw structures for these isomers and explain why they should be less stable than 13–*cis*-lycopene.

Library Topics

1. An oxidized form of Vitamin A known as 13–*cis*-retinoic acid has recently been touted as a cure for acne. Give the structure of this compound and find out what you can about how it works, how effective it is, and what its potential side effects are.

2. Find a commercial synthesis of Vitamin A starting with β-ionone and outline all the synthetic steps.

3. Explain why it would be dangerous to eat a polar bear's liver (aside from the hazards of getting it from the polar bear).

β-ionone

The Effect of Catalyst Concentration in the Synthesis of Alkyl Bromides

EXPERIMENT 16

Alkyl Halides. Nucleophilic Aliphatic Substitution. Reactions of Alcohols. Acid Catalysis.

Operations

⟨OP–5⟩ Weighing

⟨OP–7a⟩ Refluxing

⟨OP–13b⟩ Separation of Liquids

⟨OP–15⟩ Codistillation

⟨OP–19⟩ Washing Liquids

⟨OP–20⟩ Drying Liquids

⟨OP–22b⟩ Trapping Gases

⟨OP–25⟩ Simple Distillation

Experimental Objective

To prepare either 1-bromobutane or 2-bromobutane and observe the effect of catalyst concentration on product yield.

Learning Objectives

To gain additional experience with procedures used in preparing organic liquids.

To learn how catalyst concentration can affect the rate and yield of chemical reactions.

To learn about some commercially important halogen compounds.

SITUATION

You are employed by the Westlady Organic Chemicals Co., which manufactures fine chemicals for research and for the preparation of commercially useful compounds. The company prepares large quantities of alkyl halides, which are in great demand as reaction intermediates, and is quite naturally interested in synthesizing them under conditions that will provide the maximum yield. Your research group's assignment is to determine the optimum reaction conditions for preparing 1-bromobutane and 2-bromobutane by varying the amount of sulfuric acid used as a catalyst. The optimum amount of catalyst should be the lowest quantity that provides the maximum yield in a given period of time. For example, if $\frac{1}{2}$ mole and 1 mole of sulfuric acid both result in a 90% yield of alkyl bromide, the smaller quantity would be preferred.

You and your coworkers will each prepare one of the two alkyl bromides from the corresponding alcohol, using amounts of sulfuric acid that vary between approximately 0.25 and 1 mol per mole of alcohol. A comparison of the yields obtained should indicate the best reaction conditions for the preparation of both halides.

BACKGROUND

The Ambivalent Halides

Alkyl halides are important as chemical intermediates because of the large number of reactions they undergo. The halide ion of an alkyl halide is a good leaving group—it can be easily displaced by more strongly nucleophilic ions or molecules. (Examples of such reactions are shown in the margin.) The reactivity of alkyl halides in such nucleophilic substitution reactions varies in the order RI > RBr > RCl; however, since bromides are reactive enough for most purposes and are cheaper than iodides, they are used most frequently in the laboratory.

Aliphatic halogen compounds find numerous commercial and medical uses as well. The first general anesthetic used successfully was chloroform (trichloromethane), which has been supplanted by less toxic compounds like cyclopropane and ethyl ether. Ethyl chloride is used as a local anesthetic; when sprayed on the skin it evaporates very rapidly, producing almost instantaneous freezing and temporary anesthesis. Chloral hydrate was the first synthetic sedative to supplant opium and alcohol; it is still used as a hypnotic. Its ability to quickly put a victim to sleep gave chloral hydrate the underworld sobriquet "Mickey Finn."

$$CH_3CH_2CH_2CH_2Br$$
1-bromobutane

$$CH_3CH_2\overset{\overset{\displaystyle Br}{\displaystyle |}}{C}HCH_3$$
2-bromobutane

Nucleophilic substitution reactions of alkyl bromides

$$RBr \xrightarrow[\text{or NaOH}]{H_2O} ROH \qquad \text{alcohol}$$

$$RBr \xrightarrow{NaOR'} ROR' \qquad \text{ether}$$

$$RBr \xrightarrow{NaSH} RSH \qquad \text{thiol}$$

$$RBr \xrightarrow{NaCN} RCN \qquad \text{nitrile}$$

$$RBr \xrightarrow[2.OH^-]{1.NH_3} RNH_2 \qquad \text{amine}$$

$$RBr \xrightarrow{R'C\equiv CNa} RC\equiv CR' \qquad \text{alkyne}$$

$$CHCl_3$$
chloroform

$$CH_3CH_2Cl$$
ethyl chloride

$$Cl_3CCH(OH)_2$$
chloral hydrate

trichloroethylene   tetrachloroethylene

$CH_2\!=\!CHCl \longrightarrow [CH_2\!-\!CHCl]_n$
vinyl chloride   PVC

$CH_2\!=\!CCl\!-\!CH\!=\!CH_2 \longrightarrow$
2-chloro-1,3-butadiene
$[CH_2\!-\!CCl\!=\!CH\!-\!CH_2]_n$
neoprene

hexachlorocyclopentadiene

Chlorinated hydrocarbons such as trichloroethylene, tetrachloroethylene, and methylene chloride are efficient industrial and dry-cleaning solvents with the ability to dissolve grease and oily grime and the advantage (over hydrocarbon solvents) of being virtually nonflammable. Unsaturated halides are widely used as monomers in forming plastics and rubbers. Vinyl chloride is the raw material for the important plastic polyvinyl chloride (PVC), and 2-chloro-1,3-butadiene polymerizes to form neoprene rubber.

Many of the so-called "hard" insecticides are alicyclic halogen compounds derived from hexachlorocyclopentadiene. The broad-spectrum pesticides chlordane, dieldrin, and aldrin are all synthesized by means of one or more Diels-Alder reactions, beginning with this compound. Mirex, a tetracyclic insecticide with an unusual cage structure, is prepared from the same starting material by a photochemical dimerization reaction. All of these polychlorinated insecticides show high mammalian toxicity and, since they are not easily broken down to simpler, nontoxic derivatives, tend to accumulate in the environment. For that reason, their use has been strictly limited in recent years.

chlordane

dieldrin

mirex

CCl_2F_2
Freon 12

Freon 114

Freon C318

The fluorine-containing compounds called freons find many applications as refrigerants, air conditioning fluids, and aerosol propellents. Their use in the latter application has been said to present a potential danger to the earth's ozone layer, which keeps most of the sun's ultraviolet light from reaching the ground. Photochemical reactions of compounds such as Freon 12 in the upper atmosphere may generate chlorine free radicals, which react in turn with ozone, O_3, and convert it to ordinary oxygen, O_2.

A number of other halogenated organic compounds, such as the polybrominated and polychlorinated biphenyls, have received a bad press recently because of their toxicity and suspected cancer potential. Their widespread use in fire retardants, coolants for electrical transformers, and other commercial products has resulted in their release into the environment as water pollutants and, in one case, as an accidental contaminant of cattle feed. Although some organic halogen compounds are probably quite safe, their reputation as a group has stimulated a search for safer alternatives to many of the commercially significant halides.

A polybrominated biphenyl occurring in PBB

METHODOLOGY

Catalysts are used to accelerate chemical reactions without themselves being consumed in the process. A catalyst can exert its effect by providing an alternative (low-energy) path to the products or by increasing the concentration of some crucial intermediate. In the reaction of an alcohol with HBr, for example, the sulfuric acid catalyst increases the concentration of the protonated alcohol, ROH_2^+, which is then converted to the product by an S_N1 or S_N2 mechanism. Since the reaction rate is proportional to the concentration of this intermediate (which in turn is proportional to the acid concentration), increasing the amount of the acid catalyst should increase the amount of product formed in a given period of time. However, using too high a catalyst concentration can lead to side reactions, such as alkene or ether formation, which lower the proportion of the desired product in the reaction mixture. Choosing an optimum catalyst concentration is therefore a matter of balancing these opposing factors—of using just the amount of catalyst that will give the highest yield of the desired product in a reasonable period of time.

Alcohol protonation equilibrium

$$ROH + H_2SO_4 \rightleftarrows ROH_2^+ + HSO_4^-$$

Many primary and secondary alkyl bromides can be prepared by treating the corresponding alcohol with about a 25% excess of 48% hydrobromic acid together with sulfuric acid, refluxing the mixture for an hour or more, and codistilling the alkyl bromide (with water) from the reaction mixture. The alkyl bromide layer, which may contain some alcohol, ether, or alkene impurities, is first washed with water and cold, concentrated sulfuric acid to remove the bulk of these impurities, and then washed again with sodium bicarbonate to remove the residual acid. After drying with calcium chloride, the halide is purified further by distillation.

A gas trap must be used to prevent the escape of corrosive hydrogen bromide gas and poisonous sulfur dioxide.

PRELAB EXERCISES

1. Review the operations as needed.

2. Calculate the quantity (mass and volume) of 0.20 mol each of 1-butanol and 2-butanol, and the volume of 48% HBr equivalent to 0.25 mol.

3. Write a checklist for the experiment.

Reactions and Properties

$$CH_3CH_2CH_2CH_2OH + HBr \xrightarrow{H_2SO_4} CH_3CH_2CH_2CH_2Br + H_2O$$

or

$$\underset{\substack{|\\ OH}}{CH_3CH_2CH}-CH_3 + HBr \xrightarrow{H_2SO_4} \underset{\substack{|\\ Br}}{CH_3CH_2CH}-CH_3 + H_2O$$

Table 1.   Physical properties

	M.W.	b.p.	d.
1-butanol	74.1	117	0.810
2-butanol	74.1	99.5	0.808
hydrobromic acid (48%)	80.9		1.49
1-bromobutane	137.0	102	1.276
2-bromobutane	137.0	91	1.259

48% HBr is approximately 8.9M.

Hazards

The butanols are flammable, narcotic, and may irritate the skin and mucous membranes; avoid unnecessary contact and breathing of vapors. Hydrobromic acid is toxic and can cause serious burns; its vapors irritate the eyes and mucous membranes. Handle with great care (preferably using rubber gloves) and avoid breathing the vapors. Sulfuric acid causes

1-butanol   2-butanol   2-bromobutane   sulfuric acid

burns and reacts violently with water; it must be handled with great care. The bromobutanes are somewhat flammable and may be narcotic in high concentrations; avoid breathing vapors and avoid contact with skin and eyes.

PROCEDURE

Each student should be assigned one of the two alcohols and either 3, 6, 9, or 12 ml of the sulfuric acid catalyst. Students in groups of eight can compare results, or the class results can be displayed somewhere in the laboratory.

Hydrobromic acid is corrosive and must be handled very carefully. Use gloves and a fume hood.

WARNING

Reaction.   Carefully measure 0.25 moles of 48% HBr by volume and add it to 0.20 mole of the assigned alcohol in a reaction flask. Slowly add the assigned volume of 18M sulfuric acid, with cooling. Reflux the mixture for one hour ⟨OP–7a⟩, using a gas trap ⟨OP–22b⟩ leading from the top of the reflux condenser to a container of dilute sodium hydroxide.

Separation.   Codistill the products ⟨OP–15⟩ with the water present until no more organic liquid distills over. Add 20 ml of water to the distillate, mix, and separate the two layers ⟨OP–13b⟩. Wash the alkyl bromide layer ⟨OP–19⟩ with about 10 ml of cold, concentrated sulfuric acid (*Take care! Possible violent reaction.*) followed by water and 5% sodium bicarbonate solution. Dry the butyl bromide ⟨OP–20⟩ with a small quantity of calcium chloride. Weigh the product ⟨OP–5⟩ at this point and report your crude yield to your coworkers or your instructor.

It will not be necessary to add more water to the pot, so the addition funnel in Figure 1, ⟨OP–15⟩ can be omitted.

Be sure to save the right layer!

Purification.   Distill the product ⟨OP–25⟩ over a 3–4°C boiling range, reweigh it in a tared, labeled vial, and turn it in.

Experimental Variations

1.   The experiment can be carried out using an equivalent amount of sodium bromide in place of the hydrobromic acid, and additional sulfuric acid (one mole per mole of NaBr) to form the HBr *in situ*. (The yields are generally slightly lower by this procedure.)

2. Other primary and secondary alcohols can be used in place of 1- and 2-butanol. Reflux times should be increased for most branched and high-molecular-weight alcohols.

MINILAB 4

Rates of Nucleophilic Substitution Reactions

Other halides can be used also, such as 1-chloro-2-butene and 1°, 2°, or 3° alkyl bromides.

It is not necessary to wait until one mixture has reacted before preparing the next one, as long as you record the starting times and observe all tubes frequently.

The silver nitrate solutions may be heated briefly to boiling if the water bath has no effect.

Measure 0.2 ml each of 1-chlorobutane, 2-chlorobutane, and 2-chloro-2-methylpropane into three clean, dry, labeled test tubes, and stopper the tubes. To each tube—one at a time—add 2 ml of a 15% sodium iodide–acetone solution. Stopper and shake the tube after each addition, and note the time of addition and the time when there is definite evidence of a precipitate. Repeat the process with new 0.2-ml samples of the three alkyl halides, using 2-ml portions of 1% ethanolic silver nitrate in place of the sodium iodide solution. If any tubes contain no precipitate at the end of the period, put them in a 50°C water bath for six minutes, allow them to cool, and look for evidence of a reaction.

On the basis of your results, decide whether each kind of reaction is S_N1 or S_N2, and write a general mechanism for each.

Use the yields before the final distillation for drawing your conclusions.

Topics for Report

1. From the class data, indicate the optimum quantity of sulfuric acid for each reaction and explain any differences between the optimum conditions for the primary and secondary compounds.

2. The reaction of 1-butanol with HBr is said to be predominantly S_N2, while that of 2-butanol is predominantly S_N1. (**a**) Write mechanisms for both reactions. (**b**) Which reaction should be accelerated by the addition of sodium bromide? Explain.

3. Give the structures of several different by-products that might be formed during the reaction of your alcohol with HBr. Write mechanistic pathways for their formation.

4. Construct a flow diagram for your synthesis, including the by-products from Topic 3 in the reaction mixture.

Library Topic

Research the effects of polybrominated biphenyls (PBB) and polychlorinated biphenyls (PCB) on the environment, giving specific examples of their release into the ecosystem and describing the consequences.

The Grignard Synthesis of Triphenylmethanol and Preparation of a Carbonium Ion

Preparation of Alcohols. Nucleophilic Addition. Reactions of Carbonyl Compounds. Long-lived Carbonium Ions.

Operations

⟨OP–1⟩ Drying Glassware

⟨OP–5⟩ Weighing

⟨OP–7a⟩ Refluxing

⟨OP–10⟩ Addition

⟨OP–12⟩ Vacuum Filtration

⟨OP–12a⟩ Semimicro Vacuum Filtration

⟨OP–13b⟩ Separation

⟨OP–14⟩ Evaporation

⟨OP–19⟩ Washing Liquids

⟨OP–20⟩ Drying Liquids

⟨OP–21⟩ Drying Solids

⟨**OP–22a**⟩ Excluding Moisture

⟨OP–23⟩ Recrystallization

⟨OP–28⟩ Melting Point

Experimental Objective

To synthesize triphenylmethanol by a Grignard reaction and use it to prepare the triphenylmethyl cation.

Learning Objectives

To learn how to carry out a reaction under anhydrous conditions.

To learn about some characteristics of Grignard reactions and the general experimental methods used in conducting them.

To learn about some long-lived carbonium ions related to the trityl cation, and about the related triphenylmethyl dyes.

SITUATION

Since the discovery of fuchsin in 1859, a controversy has raged over the structure of the triphenylmethane dyes — are they true carbonium ions, or do they exist primarily in the iminium ion forms with the positive charge on a nitrogen? The advent of modern instrumental methods such as carbon-13 magnetic resonance spectrometry now offers a means of testing these opposing theories. Because the position of a magnetic resonance signal depends on the electron density at the nucleus in question, a positively charged carbon-13 nucleus should be characterized by a low degree of diamagnetic shielding and, thus, by a downfield shift.

Refer to any modern organic chemistry textbook for an explanation of nuclear magnetic resonance terms.

Alternate structures for malachite green, a triphenylmethane dye

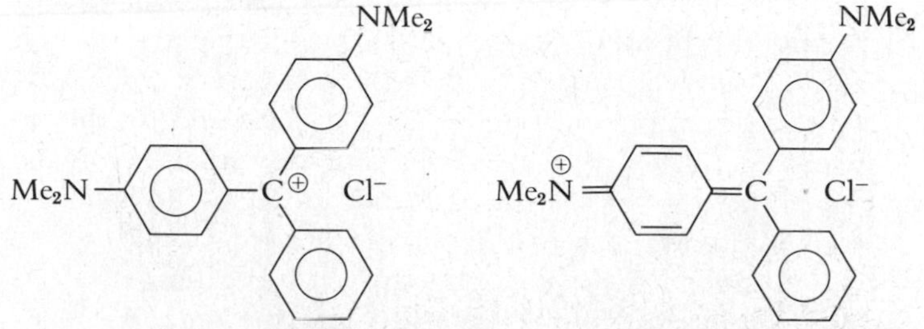

carbonium
ion form

iminium
ion form

Your research advisor has asked you to obtain carbon-13 magnetic resonance spectra of various triphenylmethane dyes (such as *p*-rosaniline, malachite green, and crystal violet) so as to shed some light on their structures. First you must prepare a model carbonium ion for comparison — one similar in structure to the dyes, but lacking the nitrogen-containing functions that may accommodate some of the positive charge. You have chosen the simplest member of the family, the triphenylmethyl cation itself, which you propose to synthesize as a fluoborate salt. This can be accomplished easily by treating triphenylmethanol with fluoboric acid. However no triphenylmethanol is available, so you must make it yourself.

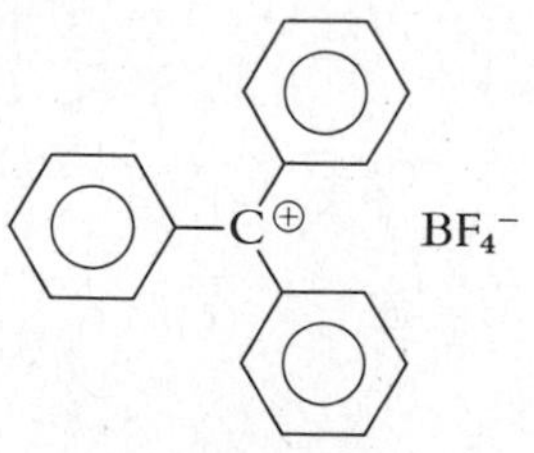

triphenylmethyl
fluoborate

BACKGROUND

The Colorful Career of the Triphenylmethanes

When W. H. Perkin prepared mauve from aniline in 1854, he started a mad race to develop other commercially profitable dyes from aniline and related compounds. Four years later somebody decided to heat the crude aniline of the day with stannic chloride, and produced a beautiful fuchsia-colored substance that was the first of the triphenylmethane dyes. Malachite green, crystal violet, and similar dyes followed fuchsin onto the scene not long afterward and stimulated research into the molecular basis of color. Why is malachite green, for example, brilliantly colored and its reduced form colorless?

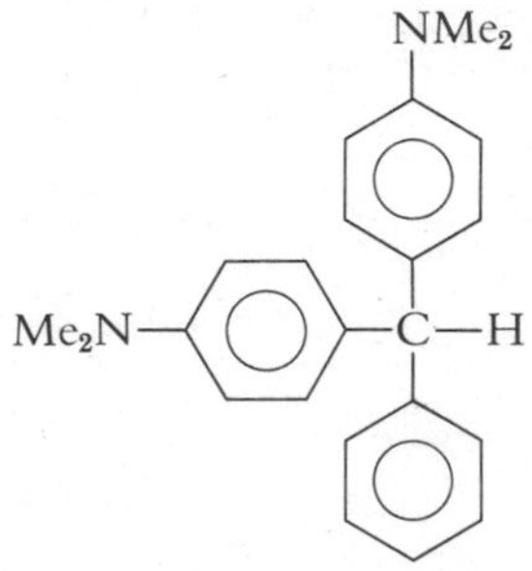

In the early days, the triphenylmethane dyes were represented by quinonoid (iminium ion) structures like the one shown above for malachite green. However when triphenylmethanol—which, having no nitrogen-containing substituents, cannot form iminium ions—was treated with strong acid, it also produced a colored solution. This led to the suspicion that the colored product, and by extension the triphenylmethane dyes as well, were actually carbonium ions having the positive charge located on the central carbon atom. Many heated arguments ensued between the supporters of carbonium and iminium ion structures for the triphenylmethane dyes. Such differences of opinion eventually paved the way for resonance theory, in which the opposing viewpoints were (at least in part) reconciled.

Although the triphenylmethyl (trityl) cation is not as stable as the tropylium ion prepared in Experiment 13, in association with anions such as fluoborate and hexafluoroantimonate it is stable enough to keep almost indefinitely under anhydrous conditions. It owes this unusual stability to delocalization of the positive charge about the three benzene rings—nine resonance structures of the trityl cation can be drawn with the positive charge on a ring atom. The cation is apparently shaped somewhat like a propeller with the "blades" (benzene rings) pitched at a 32° angle. (Steric interference between the *ortho* hydrogen atoms makes a planar configuration impossible.)

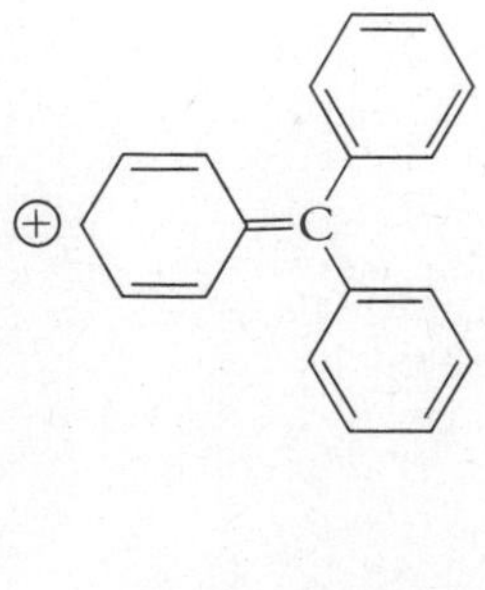

The year 1900 provided two significant milestones in organic chemistry. That was the year Moses Gomberg announced the discovery of the trityl free radical (see Experiment 14) and Victor Grignard reported on his discovery and use of Grignard reagents. Just a year later, trityl carbonium

---

Repeating this experiment with today's aniline would not produce fuchsin, because its formation was dependent on the presence of *p*-toluidine as an impurity.

See Library Topic 1.

Reduced form of malachite green

Johann Friedrich Wilhelm Adolf von Baeyer was the first to recognize the significance of these colored salts, which he christened "carbonium salts."

A resonance structure of the triphenylmethyl cation

Shape of triphenylmethyl cation

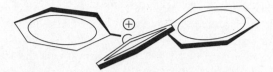

ions were being prepared from triphenylmethanol which, fortuitously, can be synthesized most readily from a Grignard reagent.

METHODOLOGY

Victor Grignard's original procedure for preparing a Grignard reagent was as follows: About 1 g-atom of magnesium metal was placed in a dry two-necked round-bottomed flask fitted with a reflux condenser and dropping funnel. One mole of the organic halide was dissolved in ethyl ether and 50 ml of this solution was added to the magnesium. When a white turbidity appeared at the metal surface and effervescence began, more ether was added in portions with cooling, followed by dropwise addition of the remainder of the halide-ether solution. The reaction was brought to completion by refluxing in a water bath, leaving a nearly colorless liquid. Essentially the same method is used today for making many Grignard reagents, although extensive studies of the reaction have led to some modifications of the reaction conditions.

It is extremely important that the reagents and apparatus be as dry as possible, since water not only reacts with Grignard reagents but appears to inhibit their formation. In a study with *n*-butyl bromide, it was found that the **induction period** for forming the Grignard reagent was $7\frac{1}{2}$ minutes using sodium-dried ether, 20 minutes using commercial absolute ether, and 2 hours using ether half-saturated with water. It is apparent that careful drying of the reaction apparatus and reagents saves time in the long run; this can increase the yield of Grignard reagent as well.

The type and quantity of reagents and solvents used is also important. For most purposes, an ordinary grade of well-dried magnesium turnings will suffice, if they are rubbed and crushed with a glass rod to remove some of the oxide coating and provide a fresh surface for reaction. A 5–10% excess of magnesium is commonly used to insure that the reaction goes to completion. Absolute ether is suit-

V. Grignard, Ann. chim. [7], 24, 433 (1901).

See Kharasch & Reinmuth (Bib–L46) for a very thorough discussion of Grignard reactions.

Reaction of water with Grignard reagents

$$RMgX + H_2O \longrightarrow RH + Mg(OH)X$$

induction period: the time elapsing between the combination of the reactants and the start of a noticeable reaction.

While water inhibits formation of Grignard reagents, iodine and other substances may accelerate it. For this reason it is a common practice to add a small crystal of iodine if the reaction does not start after 10–15 minutes.

Other solvents than ether can be used. Arylmagnesium chlorides form more readily in tetrahydrofuran.

tetrahydrofuran

Preparation of carbonium ions using fluoboric acid

$$ROH + HBF_4 \rightleftharpoons ROH_2^+ + BF_4^- \longrightarrow$$
$$R^+BF_4^- + H_2O$$

able for most routine preparations, but the optimum quantity of ether depends on the kind of Grignard reagent. One study showed that the highest yields of phenylmagnesium bromide were obtained with 5 moles of ethyl ether per mole of bromobenzene.

The addition rate of the halide also affects the yield of product, since rapid addition increases the reaction rate and may cause side reactions because of local superheating. Good yields are obtained when the halide is added over a period of 15–30 minutes with stirring or shaking rather than all at once. A 5–10-minute reflux period at the end of the addition should insure that the reaction has gone to completion.

When a Grignard reagent is to be used for preparing an alcohol, it is not isolated from the reaction mixture but is immediately treated with a solution of the appropriate carbonyl compound. The mixture is then refluxed to carry the reaction to completion. The intermediate magnesium salt is hydrolyzed by adding ice or cold water, and some mineral acid is used to dissolve any precipitated magnesium hydroxide. The product is then isolated from the ether layer and purified.

Carbonium ions can be prepared by treating alcohols with strong acids. Although many such ions are unstable and exist only as short-lived reaction intermediates, others like the trityl cation can be isolated if the associated anion is a weak nucleophile. Fluoboric acid (HBF_4) is a good choice since the fluoborate anion is an extremely weak nucleophile and will not bond covalently to the trityl cation or initiate side reactions.

PRELAB EXERCISES

1. Read ⟨OP–22*a*⟩ and review the other operations as needed.

2. Calculate the mass of 55 mmol of magnesium and 50 mmol of benzophenone, the mass and volume of 50 mmol of bromobenzene, and the volume of 250 mmol of ethyl ether.

3. Write a checklist for the experiment.

Reactions and Properties

A.

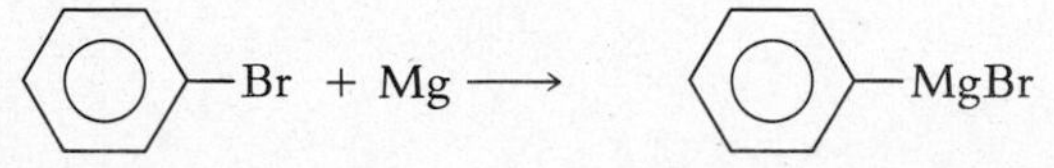

bromobenzene   phenylmagnesium bromide

B.

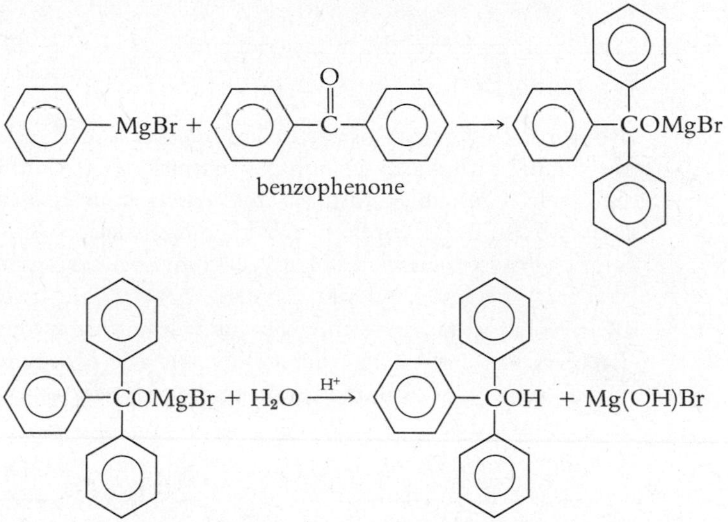

benzophenone

triphenylmethanol

C.

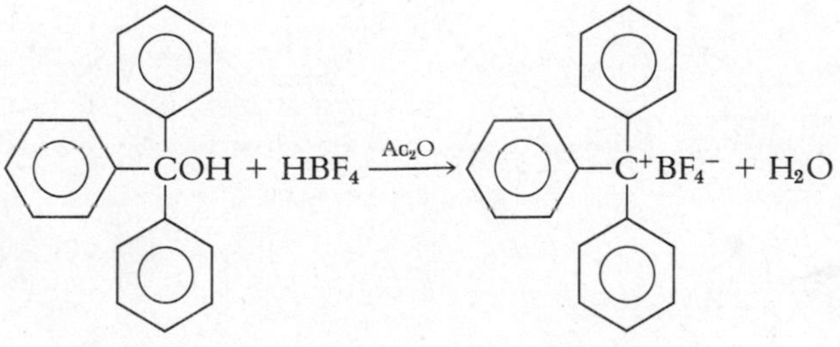

trityl fluoborate

Table 1.   Physical properties

	M.W.	m.p.	b.p.	d.
bromobenzene	157.0	−31	156	1.495
magnesium	24.3			
ethyl ether	74.1	−116	34.5	0.714
benzophenone	182.2	48	306	
triphenylmethanol	260.3	164	380	

Table 1.   Physical properties (*continued*)

	M.W.	m.p.	b.p.	d.
fluoboric acid (48%)	87.8			1.41
acetic anhydride	102.1	−73	140	1.082
trityl fluoborate	330.1			

Fluoboric acid is also called fluoro-
boric acid and tetrafluoroboric acid.

Hazards

Bromobenzene irritates the skin and may cause narcosis and
liver damage if inhaled in high concentration. Ethyl ether is
extremely flammable and may cause narcosis on inhalation;
keep away from heat or flames and avoid inhalation. Magnesium can cause dangerous fires (which are *not* extinguished
by water) if ignited. Acetic anhydride is very dangerous to
the eyes and irritating to the skin and respiratory system; it
also reacts violently with water. Fluoboric acid is very toxic
and can cause serious burns; use extreme care to avoid inhalation and contact.

    magnesium   acetic anhydride   fluoboric   bromobenzene   ethyl ether
                                          acid

PROCEDURE

A. *Preparation of the Grignard Reagent*

Thorough drying is essential.

Boiling chips are not necessary.

*The magnesium can be cleaned and dried
by rinsing it in ethyl ether and drying it
in an oven.*

Care: *Do not punch a hole in the flask!*

Clean and thoroughly dry ⟨OP–1⟩ all of the glassware
needed for addition and reflux ⟨OP–10⟩ and assemble the apparatus, using a drying tube to exclude moisture ⟨OP–22a⟩.
Weigh 55 mmol of clean, dry magnesium turnings ⟨OP–5⟩
and place them in the reaction flask. Mix 50 mmol of pure
bromobenzene with 250 mmol of anhydrous ethyl ether and
transfer the mixture to the addition funnel. Add 3–4 ml of
the mixture to the reaction flask, momentarily remove the
addition funnel, and carefully crush the magnesium turnings
with a flat-bottomed stirring rod. Replace the funnel and
observe the reactants for evidence of reaction. If no reaction
is evident after ten minutes or so, warm the flask with the

heat from your hand or add a small crystal of iodine. When the ether begins to boil gently with no external heating, add ⟨OP–10⟩ the rest of the bromobenzene solution slowly so as to keep the ether refluxing quietly. When all of the solution has been added and the reaction subsides, reflux ⟨OP–7a⟩ the mixture on a steam bath (turned very low) for ten minutes. Use the Grignard reagent in step **B** as soon as possible — it decomposes on standing.

Keep a beaker of cold water handy to moderate the reaction if necessary.

B. *Preparation of Triphenylmethanol*

Reaction.   Dissolve 50 mmol of benzophenone in 25 ml of anhydrous ethyl ether and transfer the mixture to the addition funnel. Add it ⟨OP–10⟩ to the Grignard reagent dropwise, with shaking, just fast enough to keep the mixture boiling gently. Reflux the reactants ⟨OP–7a⟩ on a steam bath for 20 minutes, cool them to room temperature, and pour them, with stirring, into a beaker containing about 50 ml of cracked ice and 20 ml of 3M sulfuric acid. Continue stirring until the solid material has dissolved.

The reflux may be omitted if the reactants are allowed to stand overnight or longer.

More ether can be added, if necessary, to dissolve all of the solid.

Separation.   Separate the layers ⟨OP–13b⟩ and wash the ether layer ⟨OP–19⟩ in succession with water and with 5% sodium bicarbonate. Dry the solution ⟨OP–20⟩ and evaporate the ethyl ether ⟨OP–14⟩, using a trap to recover it. Swirl and stir the residue with 25 ml of petroleum ether for a few minutes and collect the crystals by vacuum filtration ⟨OP–12⟩.

Washing the product with petroleum ether removes biphenyl, a by-product of the Grignard reaction.

Purification.   Recrystallize the crude triphenylmethanol ⟨OP–23⟩ from a 2:1 mixture of petroleum ether and ethanol. Collect the product by vacuum filtration ⟨OP–12⟩ and dry it ⟨OP–21⟩. Weigh the product ⟨OP–5⟩ and obtain its melting point ⟨OP–28⟩.

Use petroleum ether as the washing solvent.

C. *Preparation of Trityl Fluoborate*

Dissolve about 2 g of pure triphenylmethanol in 10 ml of acetic anhydride in a small Erlenmeyer flask, then add about 1.5 ml of 48% aqueous fluoboric acid (*Care!*). Stopper the flask and let it stand until the precipitate becomes crystalline

Acetic anhydride and fluoboric acid are very hazardous — use rubber gloves and a fume hood for this step.

WARNING

enough to filter. Collect the crystals by semimicro vacuum filtration ⟨OP–12a⟩, using a little ethyl ether as the wash solvent. Transfer them to a dry, tared vial. Turn in the weighed trityl fluoborate ⟨OP–5⟩ with the remainder of your triphenylmethanol.

Experimental Variations

1. Some properties of trityl fluoborate can be observed by **(a)** adding a few crystals of the salt to water or methanol, or **(b)** obtaining an ultraviolet-visible spectrum (300–600 nm) in acetone. The results should be discussed and interpreted.

2. Triphenylmethanol can also be prepared by treating phenylmagnesium bromide with ethyl benzoate. [See *Organic Syntheses, Coll. Vol. 3,* p. 839 (Bib–B1) for a procedure.]

3. Some experiments involving reactions of triphenylmethanol are described in L. F. Fieser and K. L. Williamson, *Organic Experiments,* 4th ed., (Lexington, Mass.: D.C. Heath, 1979), p. 123.

4. The infrared spectrum of triphenylmethanol can be obtained (using a KBr disc) and intepreted.

Topics for Report

1. Two possible by-products of the first step of this synthesis are benzene and biphenyl. Show how they might be formed, giving balanced equations for the reactions involved.

2. Construct a flow diagram for steps ***A*** and ***B*** of your synthesis, including the by-products (Topic 1) in the product mixture from step ***A***.

3. The reaction of phenylmagnesium bromide with benzophenone to form the salt of triphenylmethanol is an example of nucleophilic addition; its reaction with ethyl benzoate yielding the same product involves nucleophilic substitution followed by a nucleophilic addition step. Write reasonable mechanisms for both reactions.

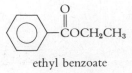

ethyl benzoate

4.   Outline a synthetic pathway for each of the following compounds, using the Grignard reaction and starting with benzene or toluene: **(a)** 1,1-diphenylethanol; **(b)** 1,2-diphenylethanol; **(c)** 2,2-diphenylethanol; and **(d)** 2,3-diphenyl-2-butanol.

5.   Draw all possible resonance structures for malachite green.

Library Topics

1.   Read about the chemistry of color and the structures of chromophoric groups, then explain why malachite green is colored while its reduced form (see Background) is colorless.

2.   Triphenylmethyl dyes such as malachite green can, in theory, exist in iminium-ion and carbonium-ion forms. According to current thinking, which best represents the true structure of such dyes? Give experimental evidence supporting your answer.

The Preparation of Synthetic Camphor from Camphene

Preparation and Reactions of Alcohols. Electrophilic Addition. Nucleophilic Acyl Substitution. Oxidation. Carbonium-ion Rearrangements. Reactions of Alkenes. Multistep Synthesis.

Operations

⟨OP–5⟩ Weighing

⟨OP–7⟩ Heating

⟨OP–7a⟩ Refluxing

⟨OP–10⟩ Addition

⟨OP–12⟩ Vacuum Filtration

⟨OP–13b⟩ Separation

⟨OP–19⟩ Washing Liquids

⟨OP–20⟩ Drying Liquids

⟨OP–21⟩ Drying Solids

⟨**OP–24**⟩ Sublimation

⟨OP–28⟩ Melting Point

Experimental Objective

To prepare synthetic camphor from camphene by a multistep synthesis.

Learning Objectives

To learn how to purify solids by sublimation.

To learn how to carry out a multistep synthesis with minimal losses of material.

To learn how to carry out a simple oxidation reaction and to gain more experience with nucleophilic addition and ester hydrolysis reactions.

To learn more about carbonium-ion rearrangements.

To learn about the history and significance of camphor.

SITUATION

(The following situation incorporates some genuine historical facts.)

You are a chemist working for the Acme Billiard Ball Company in the mid-1870s. Because of a worldwide shortage of elephants, your company had offered a $10,000 prize to anyone who could come up with a synthetic material to replace ivory in billiard balls. A New York printer named John Wesley Hyatt accepted the challenge and succeeded in mixing collodion (cellulose nitrate) with camphor to produce a hard, horn-like material he called "celluloid." After extensive testing of this substance, the board of directors of the Acme Company reluctantly concluded that celluloid was not suitable for use in billiard balls because of an unfortunate tendency to shatter on impact. However, the company envisages some commercial uses of celluloid in manufacturing combs, brush handles and the like, and has decided to branch out into these areas while the ivory shortage persists.

One of the factors limiting the manufacture of celluloid on a large scale is the scarcity of natural camphor, which is available in quantity only from the Chinese camphor tree, *Cinnamomum camphora*. A cheaper, more reliable source of camphor is necessary. While searching for such a source, you have come across a paper written by a Frenchman [Berthelot, *Compt. rend. 68*, 334 (1869)] who succeeded in converting camphene directly to camphor by chromic acid oxidation in acetic acid. Camphene itself can be made cheaply from the pinenes occurring in turpentine. Berthelot's reaction resulted in unacceptably low yields, however, and you feel that a multistep synthesis of camphor might be more productive.

Although the nature of the reactions occurring in Berthelot's synthesis is obscure, it is not unlikely that an alcohol is formed as an intermediate and is then oxidized to the ketone, camphor. The alcohol could conceivably result from hydrolysis of an acetate ester formed during the reaction of camphene with acetic acid. You have decided to test this theory by attempting to synthesize camphor in three steps, beginning with camphene.

Berthelot's reaction

camphene   camphor

Turpentine components

α-pinene   β-pinene

BACKGROUND

Camphor and Its Properties

The history of camphor is perhaps longer and more involved than that of any other natural product. Scientific

Perspective drawings of camphor structure

Compounds having camphoraceous odors

Construction of molecular models is necessary to demonstrate the "spherical" shape of these molecules.

This process should be differentiated from one in which intermediates are not isolated, either because they are too unstable or because the components of an intermediate product mixture are not detrimental to the subsequent reaction steps.

speculations about camphor have appeared in print since the time of Libavius (*Alchymia,* 1595), and its unusually complicated bicyclic structure stumped the scientists for sixty years after the correct molecular formula ($C_{10}H_{16}O$) was determined in 1833. More than thirty different structures were proposed, all wrong, before J. Bredt finally came up with the correct one, which was confirmed in 1903 with the total synthesis of camphor from simple open-chain esters.

Natural camphor is obtained from a tree that is native to Formosa, but which has been introduced into the southern United States and grows well in Florida and California. Most synthetic camphor is made from naturally occurring pinenes and consists of a racemic mixture of (+) and (−) forms, whereas the natural substance is mostly dextrorotatory. Besides its application as a plasticizer, camphor has been used in mothballs and for medical purposes.

The penetrating camphoraceous odor of camphor is shared by many compounds of similar molecular shape and size but vastly different composition. Compounds as diverse in structure as hexachloroethane, cyclooctane, dimethylpentamethylenesilicon, 2-nitroso-2-methylpropane, and thiophosphoric acid dichloride ethylamide all have the same distinctive odor. This correspondence has provided evidence in support of a stereochemical origin of odor. It has been suggested (Bib–L62, p. 150) that the olfactory receptor transmitting the sensation corresponding to a camphor-like odor has a hemispherical shape and a diameter of about 7 Ångstrom units, to accommodate roughly spherical molecules such as the ones named above. Although a completely satisfactory stereochemical theory of odors has not been developed, there seems little doubt that some relationship exists between molecular shape and odor quality.

METHODOLOGY

In a multistep synthesis, the starting material is first converted to one or more intermediate substances that are isolated, sometimes purified, and then transformed into the final product. Because the overall fractional yield of a multistep synthesis is equal to the product of the fractional yields of the individual reaction steps, it is essential to keep material losses to a minimum. For example, although 70% may be a respectable yield for a one-step synthesis, a five-step synthesis having 70% yields in each step will result in only a 17% yield of the final product ($.70^5 = .17$). Depending on the characteristics of the particular reaction sequence, it may be necessary to purify some or all of the intermediates, or only the final product. In this experiment only the camphor will be purified, although a somewhat better product could be obtained by distilling the intermediate isobornyl acetate under reduced pressure.

Three fundamental reaction types are represented in the synthesis of camphor from camphene. The first is an electrophilic addition reaction accompanied by a skeletal rearrangement. Adding a proton to the carbon-carbon double bond of camphene yields a carbonium ion, which can be represented either as the nonclassical carbonium ion shown, or as an equilibrium mixture of camphanyl and bornyl cations. Reaction of the "bornyl cation" with acetate ion could, in theory, yield either bornyl or isobornyl acetate, but only the latter is formed in quantity.

The second reaction is alkaline hydrolysis of the ester, isobornyl acetate, to yield isoborneol. This is a nucleophilic substitution reaction involving an attack by a hydroxide ion on the carbonyl carbon atom of the ester. Because no carbon atom on the bicyclic ring is attacked, the isobornyl configuration is retained in the product.

bornyl acetate

isobornyl acetate

camphene → nonclassical carbonium ion Equivalent to: camphanyl cation ⇌ bornyl cation

The final reaction is the chromic acid oxidation of the alcohol, isoborneol, to the ketone, camphor. The most likely mechanism for this reaction involves the formation of a chromate ester of the alcohol, followed by the elimination of $HCrO_3^-$ to form the carbon-oxygen double bond.

All three reactions are carried out using rather simple operations that have been practiced in previous experiments. The addition of acetic acid proceeds readily at steam-bath temperatures when catalyzed by sulfuric acid; it can be carried out in an Erlenmeyer flask. The alkaline hydrolysis requires heating at reflux temperatures. The oxidation reaction proceeds without external heating—Jones' reagent is simply

The chromium trioxide–sulfuric acid mixture used in this step is commonly known as Jones' reagent.

added to isoborneol dissolved in acetone, with stirring. The camphor is then precipitated in ice water and isolated by the usual methods.

Because of its compact, symmetrical structure, camphor can be converted directly from a solid to a vapor, which makes possible its purification by sublimation. This property makes it advisable to use a sealed tube in obtaining its melting point.

PRELAB EXERCISES

1. Read ⟨OP–24⟩ and review the other operations as needed.

2. Calculate the mass of 0.10 mol of camphene and the volume of 0.50 mol of glacial acetic acid.

3. Write a checklist for the experiment.

Reactions and Properties

A.

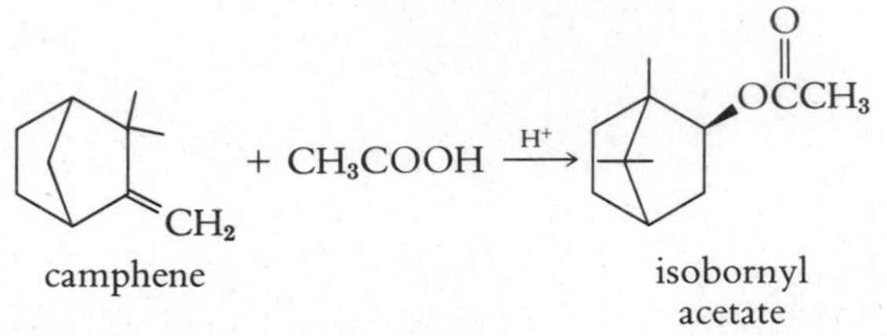

camphene + CH_3COOH $\xrightarrow{H^+}$ isobornyl acetate

B.

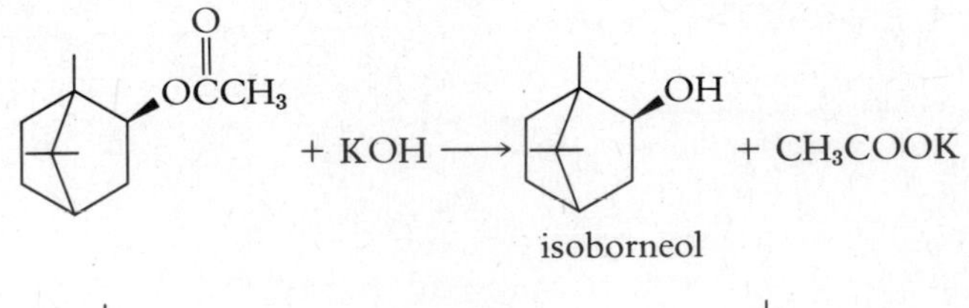

+ KOH ⟶ isoborneol + CH_3COOK

C.

3 (isoborneol) + $\underbrace{2CrO_3 + 3H_2SO_4}_{\text{"Jones' Reagent"}}$ ⟶ 3 camphor + $Cr_2(SO_4)_3 + 6H_2O$

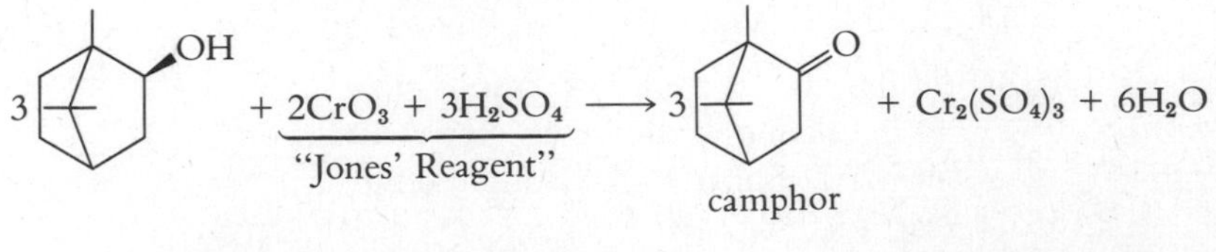

Table 1. Physical properties

	M.W.	m.p.	b.p.	d.
(±)-camphene	136.2	52	159	0.842[54]
acetic acid	60.05	17	118	1.049
(±)-isobornyl acetate	196.3		112[17]	

Table 1. Physical properties (*continued*)

	M.W.	m.p.	b.p.	d.
potassium hydroxide	56.1	360		
(±)-isoborneol	154.3	212	subl.	
(±)-camphor	152.2	179	subl.	

Hazards

Acetic acid can cause serious burns, particularly to the eyes; its vapors irritate the respiratory tract. Potassium hydroxide and its aqueous solutions cause severe burns. Jones' reagent is a strong acid and oxidant; it can cause serious burns and may react violently with oxidizable materials. The chromium trioxide in Jones's reagent has been found to be carcinogenic in animal tests. Camphor is an irritant, can damage the eyes, and is toxic if ingested. Unnecessary contact with camphor, camphene, isoborneol, and isobornyl acetate should be avoided.

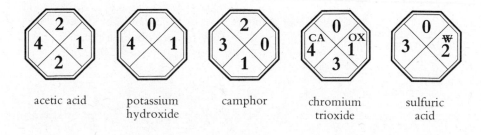

acetic acid   potassium hydroxide   camphor   chromium trioxide   sulfuric acid

PROCEDURE

A. *Preparation of Isobornyl Acetate*

Dissolve 0.10 mol of camphene in 0.50 mol of glacial acetic acid, then stir in 1.5 ml of 6M sulfuric acid. Heat the reactants ⟨OP–7⟩ for 15–20 minutes on a steam bath, then add 20 ml of cold water and cool the flask in an ice bath. Separate the ester layer ⟨OP–13b⟩, wash it ⟨OP–19⟩ with water and 10% aqueous sodium carbonate in sequence, then dry ⟨OP–20⟩ and weigh ⟨OP–5⟩ it.

A hood is advisable when using acetic acid.

Be sure to save the right liquid layer!

B. *Preparation of Isoborneol*

For every gram of isobornyl acetate, measure out 2.5 ml of a 2.5M solution of KOH in 75% ethanol. Combine the iso-

The KOH solution can be prepared from 8.25 g of 85% KOH pellets, made up to a volume of 50 ml with 75% ethanol.

Use additional water to transfer any precipitate remaining in the flask.

Jones' reagent can be prepared by dissolving 6.75 g of CrO_3 in 5.75 ml of conc. H_2SO_4 and diluting the mixture carefully to 25 ml with water.

bornyl acetate and the KOH solution in a boiling flask and reflux the mixture ⟨OP–7a⟩ for 45 minutes or more on a steam bath. Cool the reaction mixture to room temperature and pour it slowly, with stirring, onto 100 ml of cracked ice and water. Collect the precipitated isoborneol by vacuum filtration ⟨OP–12⟩, washing it well with cold water. Partly dry the crude solid ⟨OP–21⟩ and weigh it ⟨OP–5⟩. Do not use heat in drying the isoborneol, since it sublimes readily.

C. *Preparation of Camphor*

Reaction.   For every gram of crude, partly dried isoborneol, measure out 1.75 ml of Jones' reagent. Dissolve the isoborneol in 15 ml of acetone in an Erlenmeyer flask and add the Jones' reagent dropwise ⟨OP–10⟩ with shaking and swirling over about a ten minute period. Let the reactants stand, with frequent shaking or stirring, for 30 minutes.

Separation.   Pour the reaction mixture into 500 ml of ice and water, with stirring. Collect the camphor by vacuum filtration ⟨OP–12⟩ and dry it ⟨OP–21⟩ at room temperature.

Purification and Analysis.   Purify the camphor by sublimation ⟨OP–24⟩, weigh it ⟨OP–5⟩, and obtain a melting point ⟨OP–28⟩ in a sealed tube.

Experimental Variations

1.   A somewhat higher yield of camphor can be obtained if the separation is carried out as follows: Rinse the reaction mixture into a separatory funnel with 25 ml of ether followed by 25 ml of water, and extract the product into the ether layer. Extract the *water* layer twice with additional 25 ml portions of ether, combining all three ether extracts. Wash the ether solution with 5% sodium bicarbonate and with water in succession, then dry the solution and evaporate the ether.

2.   The experiment can be shortened by starting with borneol or isoborneol and performing only the oxidation step. A procedure for the oxidation of borneol is described in *Vogel's Textbook of Practical Organic Chemistry,* 4th ed., p. 427 (Bib–B4).

3.   A solution infrared spectrum of camphor in carbon tetrachloride can be obtained, compared with a standard spectrum (Aldrich or Sadtler), and interpreted.

See Bib–F22, F26.

4.   An experiment investigating the stereochemistry of borneol-camphor-isoborneol interconversions is described in *J. Chem. Educ.* **44**, 36 (1967). The isolation of *p*-dichlorobenzene from commercial air fresheners by sublimation is described in *J. Chem. Educ., 51,* 683 (1974).

Topics for Report

1.   The melting point depression constant (K_f) for camphor is about $40°C·kg·mol^{-1}$. Using the melting point of your product as the temperature at which the solid is completely liquefied, estimate the approximate mole percent of impurities present in your camphor.

Melting point (freezing point) depression equation

$$\Delta T = K_f m$$

ΔT = melting point depression
m = molal concentration of solute

2.   Both KOH and CrO_3 are used in excess for the hydrolysis and oxidation reactions in this experiment. Calculate the percentage by which each reagent is in excess of the stoichiometric quantity.

3.   Based on the discussion in the Methodology section, write reasonable mechanisms for (*a*) the hydrolysis of isobornyl acetate; and (*b*) the oxidation of isoborneol to camphor.

4.   (*a*) Show clearly how the camphanyl cation can be converted, by a single alkyl shift, to the bornyl cation. (*b*) Using the nonclassical carbonium-ion theory, explain why isobornyl acetate is obtained from the reaction of camphene with acetic acid, and not bornyl acetate.

Use models if necessary.

Library Topics

1.   Write all of the reaction steps in Gustav Komppa's total synthesis of camphor.

2.   Some monoterpenoids other than camphor that have been isolated from camphor oil are α-pinene, β-pinene, limonene, α-phellandrene, 1,8-cineole, borneol, α-terpineol, geraniol, and linalool. Such compounds are thought to be synthesized biochemically by way of geraniol pyrophosphate and the geranyl cation through a series of enzyme-

geranyl cation

catalyzed carbonium–ion reactions. (*a*) Give structures for all nine compounds. (*b*) Write mechanistic pathways showing how each compound might be formed by carbonium–ion reactions of the geranyl cation in a solution of aqueous mineral acid.

The Effect of Concentration on the Equilibrium Esterification of Acetic Acid

Carboxylic Acids. Nucleophilic Acyl Substitution. Preparation of Carboxylate Esters. Equilibrium Reactions.

Operations

⟨OP–6⟩ Measuring Volume

⟨OP–7a⟩ Refluxing

⟨**OP–25c**⟩ Water Separation

Experimental Objectives

To determine the equilibrium constant for the esterification of acetic acid with 1-butanol.

To measure the effect on reaction yield of increasing the concentration of the reactants and of removing a product from the mixture.

Learning Objectives

To learn how to use a water separator during a chemical reaction.

To learn about different methods for driving an equilibrium reaction to completion.

To learn about the significance of kinetic and thermodynamic factors in planning chemical conversions.

SITUATION

Your employer, United Ethyne Chemicals, specializes in the manufacture of chemicals from acetylene derived from limestone mined in Michigan's lower peninsula. Much of this acetylene is converted to acetaldehyde, which can in turn be processed into acetic acid and 1-butanol. Now

Chemicals from acetylene

$$HC \equiv CH \longrightarrow CH_3CHO$$

acetaldehyde

$$CH_3COOH \quad CH_3CH_2CH_2CH_2OH$$

acetic acid   1-butanol

$$\overset{O}{\underset{||}{CH_3COCH_2CH_2CH_2CH_3}}$$

n-butyl acetate

A **Fischer esterification** involves heating the carboxylic acid and ester together in the presence of an acidic catalyst.

Both batch and continuous processes are used in industrial processing. In the latter, reactants are fed in continuously at one level of the apparatus while products are withdrawn from another level.

United Ethyne wants to go one step further and prepare n-butyl acetate from these products. The ester is used in manufacturing automobile safety glass, and several companies have offered to purchase all of the butyl acetate you can produce if the price is right.

You and your coworkers have been assigned the task of designing an efficient industrial-scale process for manufacturing butyl acetate, using the standard **Fischer esterification** process.

The limitations imposed by your company's capital equipment inventory and anticipated production volume dictate a batch process operation. In this process, one batch of butyl acetate is prepared and purified before the reaction vessel is charged with reactants for the next one. Esterification is an equilibrium reaction, so you will have to find a suitable method for shifting the equilibrium to obtain nearly 100% yields of the ester—otherwise your process will not be cost-efficient and the competition can undersell you. Your plan is to take the reaction to 75% completion by refluxing the reactants until equilibrium is attained, then to shift the equilibrium by removing one of the products, water, in a low-boiling azeotrope. In order to calculate the initial mole ratios of reactants needed for 75% completion, you will have to know the equilibrium constant (K) for the reaction. Values for this constant have been published, but there is a distressing lack of unanimity among investigators about the equilibrium constants of esterification reactions. For instance, Poznanski found that the value of K for the reaction of acetic acid with ethanol varied from 1.0 to 6.8, depending on the proportions of reactants used. It is possible that the same phenomenon occurs in the butyl acetate esterification.

Your research team's major objectives, then, are to compare the yields of butyl acetate resulting from different initial mole ratios of 1-butanol to acetic acid, to compute the equilibrium constant for each mole ratio, and to find out how far the equilibrium can be shifted by removing water from the reaction mixture.

BACKGROUND

How Fast and How Far?

When a certain reaction is being considered for a chemical conversion, investigators must provide answers for two vital questions: *will it go,* and *how long will it take?* The first is a question of thermodynamics (reaction equilibria) and the second of kinetics (reaction rates). If the first question can be answered affirmatively, it may then be necessary to find out how far the reaction will go to completion—that is, to measure its equilibrium constant. This is usually done by monitoring the concentrations of reactants or products while the reaction proceeds. When they no longer change with time, the reaction has reached equilibrium. The equilibrium concentrations of one or more species can then be used to determine the extent of the reaction, from which the equilibrium constant can be calculated.

The value of the equilibrium constant is important because it provides a measure of how far a reaction will proceed under standard conditions. Table 1 illustrates this effect for a reaction of the esterification type (A + B $\rightleftharpoons$ C + D), showing the extent of reaction for various K values. For example, the maximum attainable yield from a reaction with $K = 2.25$ is 60% if equimolar concentrations of reactants are used and no attempt is made to shift the equilibrium. Since 60% yields are hardly acceptable in most industrial preparations, it is necessary to shift the equilibrium in favor of the products by using an excess of one reactant, or by removing one of the products, or both. Many ingenious ways of accomplishing the latter have been devised. Water can be removed by distillation—with or without an azeotroping agent—or by use of a dehydrating agent such as excess sulfuric acid or a column of alumina. Some esters that boil at a lower temperature than their component alcohols are removed by distillation; others can be removed by extracting them into a water-immiscible solvent such as ethylene chloride during the reaction.

Esterification provides some excellent examples of the principles of reaction equilibria and kinetics. The K values of many esterification reactions are close to unity, so the effect of structure on equilibria can be observed without difficulty. Reaction rates also vary considerably in response to changes in the structures of both the carboxylic acid and the hydroxylic reactant. Highly branched acids and alcohols react slowly because of steric interference in the transition states leading to products. Electronic effects can be observed by comparing the effects of different electron-withdrawing and donating substituents on the reaction rate.

The true thermodynamic equilibrium constant expression involves activities rather than concentrations, but the latter can be used as a first approximation.

Table 1. Extent of reaction as a function of K

K	Max. % Yield
1.00	50
2.25	60
5.44	70
16.0	80
81.0	90
361	95
9800	99
1.0×10^6	99.9

Note: Assume a reaction of the type A + B $\rightleftharpoons$ C + D, starting with equimolar quantities of reactants.

METHODOLOGY

The combination of acetic acid and 1-butanol to form butyl acetate is a typical esterification reaction, which is accelerated by an acidic catalyst. The effect of the catalyst is

Esterification mechanism

$$RC\overset{\displaystyle O}{\overset{\|}{-}}OH \xrightleftharpoons{H^+} RC\overset{\displaystyle OH}{\overset{\|}{\equiv}}\overset{\oplus}{O}H$$

$$\Bigg\updownarrow R'OH$$

$$R-\overset{\displaystyle OH}{\underset{\displaystyle OR}{\overset{|}{\underset{|}{C}}}}-\overset{\oplus}{O}H_2 \rightleftharpoons R-\overset{\displaystyle OH}{\underset{\oplus OHR'}{\overset{|}{\underset{|}{C}}}}-OH$$

$$\Bigg\updownarrow -H_2O$$

$$RC\overset{\displaystyle OH}{\overset{\|}{\equiv}}\overset{\oplus}{O}R' \xrightleftharpoons{-H^+} RC\overset{\displaystyle O}{\overset{\|}{-}}OR'$$

Calculation of the degree of completion:

$$\alpha = \frac{V_0 - V_e}{V_0} \qquad (1)$$

V_0 = volume of NaOH for initial titration

V_e = (volume of NaOH for equilibrium titration) $- V'$

V' (correction factor) = 9 ml × (aliquot volume ÷ initial vol. of reactants)

$$K = \frac{[BuOAc][H_2O]}{[BuOH][HOAc]} \qquad (2)$$

BuOAc = butyl acetate
BuOH = 1-butanol
HOAc = acetic acid

$$K = \frac{\alpha^2}{(n - \alpha)(1 - \alpha)} \qquad (3)$$

n = moles BuOH per mole HOAc (initially)

made clear by the reaction mechanism shown: protonation of the carboxylic acid makes it more susceptible to nucleophilic attack by the alcohol and greatly increases the reaction rate. The catalyst cannot, of course, increase the extent of reaction since it does not alter the relative free energies of reactant and product.

In this experiment, you will observe the effects on the extent of reaction of changing the initial quantity of one reactant (1-butanol) and of removing a product (water) from the reaction mixture. The progress of the reaction will be followed by titrating the reaction mixture with sodium hydroxide at intervals to measure the amount of acetic acid remaining. You can assume that equilibrium has been attained when the volume of the base used in successive titrations is constant (within about ± 0.2 ml). The degree of completion, α, is defined for this reaction as the amount of ester at equilibrium divided by the theoretical (100%) yield of ester—it can be calculated from the titration volumes using Equation 1 in the margin. Since the reaction mixture will contain a small amount of sulfuric acid catalyst, it will be necessary to subtract a correction factor, V', from the equilibrium titration volume to correct for the volume of sodium hydroxide used up in reacting with the sulfuric acid. Because the initial titration is performed before sulfuric acid is added, the correction factor is *not* subtracted from V_0.

The equilibrium constant for the reaction of 1-butanol with acetic acid is expressed in terms of the equilibrium concentrations of reactants and products by Equation 2. The equilibrium constant can also be expressed in terms of the degree of completion by Equation 3, which will be used to calculate K values in this experiment. It is important to recognize that Equation 3 cannot be used to calculate K from the value of α *after* water removal, since at that time the amount of water in the product mixture is no longer proportional to α.

To measure the effect of reactant concentrations on the degree of completion, each group of five students will use butanol/acetic acid mole ratios varying between 1 and 5. From the K values calculated using each mole ratio, you should be able to tell whether the equilibrium constant varies with the reactant concentrations, as Poznanski found for the reaction of ethanol with acetic acid. Then by removing water from the reaction mixture you can find out whether, and to what extent, this procedure drives the reaction toward completion.

Although the water could be removed by ordinary simple distillation, the use of a water separator will be demonstrated in this experiment. Water forms azeotropes with the ester and alcohol that boil close to 90°, so it should be possible to remove virtually all of the water around that temperature. The reported compositions of three possible azeotropes are shown in the margin. Since they all distill within a range of a few degrees it is not practical to separate them. Varying proportions of butyl acetate and 1-butanol will therefore distill with the water. The distillate will be combined with the product mixture before the final titration.

Azeotropes in the butyl acetate–1-butanol–water system
1. (b.p. 89.4°) 37.3% water, 35.3% butyl acetate, 27.4% 1-butanol.
2. (b.p. 90.2°) 71.3% butyl acetate, 28.7% water.
3. (b.p. 92.7°) 57.5% 1-butanol, 42.5% water.

PRELAB EXERCISES

1. Read ⟨OP–25c⟩ and review the other operations as needed. If necessary, review the procedure for titration in a general or analytical chemistry laboratory manual.

2. Calculate the mass and volume of 0.10 mol of acetic acid and 0.10 mol of 1-butanol.

3. Write a checklist for the experiment.

Reactions and Properties

$$CH_3COOH + CH_3CH_2CH_2CH_2OH \underset{}{\overset{H_2SO_4}{\rightleftharpoons}}$$
$$CH_3COOCH_2CH_2CH_2CH_3 + H_2O$$

Table 2.   Physical properties

	M.W.	m.p.	b.p.	d.
acetic acid	60.1	16.6	118	1.049
1-butanol	74.1	−89.5	117	0.810
butyl acetate	116.2	−78	126.5	0.883
sulfuric acid, 98%	98.1	10.4	338	1.84

Hazards

Acetic acid can cause serious burns and is very dangerous to the eyes; its vapors irritate the eyes and respiratory system. Sulfuric acid causes burns and reacts violently with water. 1-Butanol and butyl acetate are flammable and mild irritants; prolonged inhalation of their vapors can cause narcosis.

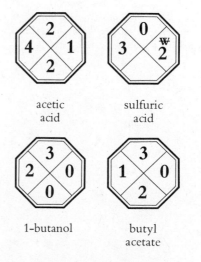
acetic acid   sulfuric acid   1-butanol   butyl acetate

PROCEDURE

Your instructor will assign each member of your group a quantity of 1-butanol, between 0.10 and 0.50 moles. Students assigned 0.30 mol or more should use 2.00-ml aliquots for titration; the others should use 1.00-ml aliquots.

Volumetric pipets should be used for withdrawing aliquots.

Attaining Equilibrium

Measure 0.10 mol of glacial acetic acid (*Care:* See Hazards.) into a 50- or 100-ml round-bottomed flask. Add your assigned quantity of 1-butanol and mix the reactants thoroughly. Pipet a 1- or 2-ml aliquot ⟨OP–6⟩ of this solution into a 125-ml Erlenmeyer flask containing 20 ml of water, add a drop or two of phenolphthalein solution, and titrate the solution with 0.40M sodium hydroxide to the pale-pink phenolphthalein end point.

The titration can be delayed until reflux is begun.

WARNING   *Do not use mouth suction for pipetting!*

Carefully add 0.10 ml of concentrated sulfuric acid to the round-bottomed flask and mix the reactants well. Insert a Claisen connecting tube into the flask, fitting the bent arm with a reflux condenser and stoppering the straight arm. Reflux the reaction mixture ⟨OP–7a⟩ gently for 30 minutes, then cool the reactants so that you can withdraw a 1- or 2-ml aliquot through the straight arm of the Claisen tube, and titrate the aliquot as before. Continue refluxing, removing and titrating aliquots every 15–20 minutes, until two successive titrations are within 0.2 ml of each other.

If an ester layer forms on top during the reflux period, be careful to remove the aliquot only from the lower *layer.*

Shifting the Equilibrium

Attach a device for water separation ⟨OP–25c⟩ to your reaction flask and reflux the reactants ⟨OP–7a⟩ gently until approximately 7 ml of liquid has passed into the separator. Cool the reaction mixture, return the contents of the water separator to the reaction flask, mix well, and titrate a 1- or 2-ml aliquot as before.

Do not add any water to the separator flask.

At some time during the experiment, record the α and K values determined by the other members of your group (or the data needed to calculate them).

Calculations

Calculate the total volume of your initial reaction mixture by assuming that it equals the sum of the volumes of the components, and use this to calculate the correction factor, V'. Then calculate the degree of completion of the reaction (*a*) when equilibrium is first attained; and (*b*) after the equilibrium is shifted by water removal. Use the first value of α to calculate the equilibrium constant for your reaction from Equation 3.

The degree of completion multiplied by 100 is equal to the percent yield of butyl acetate which would be attained if all the ester were recovered.

Experimental Variations

1. The butyl acetate can be isolated by washing the combined product mixture with dilute sodium hydroxide and water, drying the organic layer, and distilling it through an efficient fractionating column. (Collect the ester at 120–128°.) If a large excess of 1-butanol was used, this product may contain considerable alcohol. Its percentage composition can be determined by GLC.

2. The equilibrium constants for many other esters can be determined by essentially the same procedure as used here for butyl acetate, although esters of phenols and highly branched alcohols or acids may require too much time to reach equilibrium. Acetate esters of methanol, ethanol, 2-propanol, 2-methyl-1-propanol, and 3-methyl-1-butanol might be compared to determine the effect of branching on the equilibrium constant and the rate of attaining equilibrium.

Substituent Effects on
Acidity Constants
of Carboxylic Acids

MINILAB 5

Weigh out about 0.15 g of one (or more) of the acids listed, dissolve it in a solution made up by mixing 5 ml of standardized 0.1M sodium hydroxide with 15 ml of absolute ethanol, and measure the pH of the solution(s) with a pH meter. Calculate the equilibrium concentrations of the acid and its conjugate base (assuming that all of the sodium hy-

Table 3.   Carboxylic acids for Minilab 5

Acid	M.W.
p-acetamidobenzoic	179.2
p-anisic	152.2

Table 3.   Carboxylic acids for Mini-lab 5 (*continued*)

Acid	M.W.
benzoic	122.1
p-chlorobenzoic	156.6
p-hydroxybenzoic	138.1
p-nitrobenzoic	167.1
p-toluic	136.2

$$pK_a = pH + \log \frac{[HA]}{[A^-]} \qquad (4)$$

Report the α and K values obtained by your group in your discussion.

Random variations may be due to experimental error.

droxide is used up in a reaction of the type $HA + OH^- \rightarrow A^- + H_2O$) and use these and the pH to estimate the pK_a for the reaction (Equation 4). Compare your pK_a value(s) with those obtained by other students using different carboxylic acids, and explain the effect of the substituents on acidity. If desired, a Hammett plot of the data can be constructed and the value of ρ for 75% ethanol determined. (See Bib–H13 or other category H sources.)

Topics for Report

1. (*a*) Discuss the effect of increasing the butanol/acetic acid ratio, and the effect of removing water, on the maximum attainable yield of butyl acetate from the esterification reaction. (*b*) Were there any significant variations in the K values obtained for different butanol concentrations? Explain why, under certain conditions, an equilibrium "constant" would not remain constant.

2. Using your value for K, calculate the approximate molar ratio of 1-butanol–acetic acid that would result in 75% completion.

3. Besides the direct esterification procedure used in this experiment, there are many other synthetic methods used in preparing esters. Propose a high-yield synthesis for each of the following esters. Do not use the Fischer procedure; begin with any appropriate carboxylic acids, alcohols, or phenols.

4. (*a*) Write a mechanism for the reverse of the reaction used in this preparation—the acid-catalyzed hydrolysis of *n*-butyl acetate. (*b*) What reaction conditions could be used to drive the hydrolysis reaction to completion? (*c*) What is the value of K for this reaction?

5. Why was the separated liquid returned to the reaction flask before the last titration?

6. Write a reaction pathway showing all steps in the synthesis of *n*-butyl acetate, starting with acetylene. You may use any necessary solvents and inorganic reagents or catalysts, but no other carbon-containing reactants.

Library Topics

1.   Describe an industrial process for preparing butyl acetate by batch esterification. Include a flow diagram summarizing the process.

2.   The Fischer esterification proceeds by a mechanism that has been classified as $A_{AC}2$. Certain other esterification and hydrolysis reactions proceed by different mechanisms, such as the hydrolysis of methyl mesitoate in concentrated sulfuric acid ($A_{AC}1$), the hydrolysis of methyl salicylate in aqueous sodium hydroxide ($B_{AC}2$), and the transesterification of *t*-butyl benzoate with methanol ($A_{AL}1$). After consulting references on organic reaction mechanisms, explain what is meant by each symbol used in this classification system and write mechanisms for these reactions.

methyl
mesitoate

Identification of an Industrial Solvent

Carbonyl Compounds. Qualitative Analysis. Chromatographic Separations.

Operations

⟨OP–12*a*⟩ Semimicro Vacuum Filtration
⟨**OP–17**⟩ Thin-layer Chromatography
⟨OP–23*a*⟩ Semimicro Recrystallization
⟨OP–23*b*⟩ Recrystallization from Mixed Solvents
⟨OP–28⟩ Melting Point

Experimental Objective

To identify an unknown ketone from its physical properties and by preparing a derivative for analysis by TLC.

Learning Objectives

To learn how to carry out a thin-layer chromatographic separation and a small-scale mixed-solvent recrystallization.

To learn how to prepare derivatives for qualitative analysis, and to learn how they are used to identify organic unknowns.

To learn about some of the objectives and methods of forensic chemistry.

SITUATION

You are a forensic chemist investigating an "arson for profit" scheme. Several buildings in the quiet village of Brennstoff recently burned to the ground under highly suspicious circumstances shortly after being insured for large sums of money. All of the fires spread very rapidly and appeared to have more than one point of origin, which is almost certain evidence for the use of an **accelerant.** Police

An **arson accelerant** is a flammable substance such as gasoline, kerosene, turpentine, etc.

investigators collected samples of flooring and partly burned rags from one of the sites. From these samples your assistants isolated a small amount of volatile liquid. Infrared analysis of the liquid indicates that it is a ketone of a type widely used in industrial solvents.

See Table 1 for a list of these solvents.

The police have interrogated one suspect who works as a supply clerk for a large manufacturing firm and thus has access to a wide variety of solvents, several cans of which were discovered in his apartment. He is expected to confess and name his accomplices if evidence is found linking him to the scene of the crime. Your assignment is to identify the liquid found at the arson site and see if it matches one of the solvents in the suspect's possession.

BACKGROUND

Crime and Chemistry

Forensic chemistry is chemistry applied to the solution of crimes. It deals with the analysis of materials that were used in committing a crime or that were inadvertently left at the scene. Forensic chemistry may also involve the analysis of illicit drugs such as heroin or LSD and their **metabolites,** and the measurement of alcohol or other intoxicants in body fluids.

Drug metabolites are chemical compounds formed in the body as a result of the chemical or enzymatic breakdown of drugs.

Materials used in committing a crime might include a toxin used in a deliberate poisoning or an explosive used in a terrorist bombing. Such materials can be identified and sometimes traced to a particular source. Materials found at the scene of a crime might include contact traces, pieces of fiber from clothing, or particles of dust or soil. Contact traces, such as the chips of paint or glass found at the scene of a hit-and-run accident, can be analyzed chemically and under the microscope to determine the make and model of the car involved. Clothing can be traced by the dyes contained in fibers, and dust and soil particles may link a criminal to a particular occupation or location.

The conviction of a suspect requires that evidence be presented to establish, first, that a crime has actually been committed, and second, that the suspect is connected with it. In arson cases this ordinarily requires that the incendiary origin of the fire be established before introducing proof showing who did it. One way of establishing that a fire was deliberately set is to prove that an accelerant was used to intensify the fire and allow it to spread rapidly. Because fires burn from the point of origin upward, some accelerant may escape burning by soaking downward into flooring, rags, paper, or other porous materials. If the investigator traces a fire to its point of origin, he or she can often collect samples of material containing the accelerant, which are placed in airtight containers such as large paint cans. A forensic chemist can then separate the accelerant from the debris by, for instance, steam distillation or extraction. Once the accelerant has been isolated, it is usually classified according to chemical type (gasoline, turpentine, etc.) by an instrumental method such as gas chromatography.

In some cases, it may be necessary to match the sample with a liquid in the suspect's possession or with a commercial product. Since the more volatile components of an accelerant are evaporated and burned more rapidly in a fire, a sample of accelerant taken from the scene is not likely to have exactly the same composition as the control material. Therefore, the control is evaporated slowly and tested at various stages to see whether its composition at any point duplicates that of the recovered material. By this procedure it is often possible to determine the brand and grade of gasoline or other accelerant used. It may not be necessary to identify any specific component of a mixture if it can be shown to match a known sample. If the substance is a pure compound, however, its identity may be required. The analysis can be carried out using instrumental or (as in this experiment) chemical methods.

METHODOLOGY

The identity of an unknown organic compound can be deduced from its chemical and physical properties, its spectra, and its derivatives. In this experiment, the identity of an unknown ketone will be determined by preparing a derivative, the 2,4-dinitrophenylhydrazone, and carrying out a TLC separation. The identity of the derivative (and thus of the corresponding ketone) can be inferred by comparing its R_f value with those of known standards, and confirmed by obtaining its melting point.

A derivative preparation is actually a small-scale organic synthesis that is intended to provide a small quantity of very pure solid having a sharp melting point. An ideal derivative, then, is one that forms after a short reaction time and that can be isolated and purified with a minimum of

Since the molecular weight of an unknown is seldom known with certainty, no attempt is made to use equimolar quantities of reactants. Any excess reactant is removed during the workup and purification.

bother. 2,4-dinitrophenylhydrazones approach this ideal, since they usually form within 15 minutes at room temperature, can be separated by vacuum filtration, and can be purified by recrystallization from a solvent mixture.

PRELAB EXERCISES

1.   Read ⟨OP–17⟩ and review ⟨OP–23*b*⟩ and the other operations as needed.

2.   Write a brief checklist for the experiment.

Reaction and Properties

$$RC{=}O + H_2NNH{-}\bigcirc{-}NO_2 \xrightarrow{H^+} RC{=}NNH{-}\bigcirc{-}NO_2 + H_2O$$

$$\underset{R'}{|} \qquad\qquad \underset{O_2N}{} \qquad\qquad \underset{R'}{|} \qquad \underset{O_2N}{}$$

ketone   2,4-dinitrophenylhydrazine   2,4-dinitrophenylhydrazone

Table 1.   List of ketonic "industrial solvents" found in the suspect's possession

Ketone	m.p. of derivative
acetone	126
methyl ethyl ketone (2-butanone)	117
methyl propyl ketone (2-pentanone)	143
methyl butyl ketone (2-hexanone)	106

The common and functional class names are often used in industry.

Hazards

The 2,4-dinitrophenylhydrazine solution contains sulfuric acid and should be kept out of contact with skin and eyes. 2,4-dinitrophenylhydrazine as the dry solid is a high explosive that is sensitive to shock and must be handled with caution; it is less hazardous when moist or in solution.

sulfuric acid

PROCEDURE

Obtain a small (about 1 ml) amount of an unknown ketone from your instructor. Prepare derivatives of your unknown and of the known ketones specified by your instruc-

The spots should each contain about 0.3 μl of solution.

tor, following the instructions for procedure D–3, page 411. The derivatives should be separated by semimicro vacuum filtration ⟨OP–12a⟩ and recrystallized ⟨OP–23a⟩ from an ethanol-water mixture ⟨OP–23b⟩. Dissolve 10 mg of each derivative in 0.5 ml of ethyl acetate and carry out a TLC separation ⟨OP–17⟩ on silica gel using 3:1 toluene–petroleum ether (40–60°) as the developing solvent. Attempt to identify the unknown by comparing its R_f value with those of the known derivatives. Confirm your identification by obtaining a melting point ⟨OP–28⟩ of the unused derivative.

Experimental Variations

1. Additional unknowns may be used if desired, but they should be different enough in structure to allow a satisfactory chromatographic separation.

2. With the instructor's permission, the boiling point, refractive index, or infrared spectrum of the unknown can be obtained to assist in its identification.

3. An experiment involving the identification of an arson accelerant by GLC is described in *J. Chem. Educ. 51,* 549 (1974).

4. Experiments utilizing TLC for the separation and identification of drug components are described in *J. Chem. Educ. 49,* 834 (1972); *J. Chem. Educ. 50,* 852 (1973); and *J. Chem. Educ. 51,* 487 (1974).

Identification of a Spontaneous Reaction Product from Benzaldehyde

MINILAB 6

It is interesting to make a "tree" out of several pipe cleaners by twisting short lengths onto a main trunk to form branches.

Half fill a 30-ml beaker with dry sand and add enough benzaldehyde to just cover the surface of the sand. Push one or more pipe cleaners into the sand so that they extend to the bottom of the beaker, then let the apparatus stand for a week or more. Remove the white solid that forms on the pipe cleaners, purify it by recrystallizing it from boiling water, and identify it by any means at your disposal. Write an equation to account for the reaction.

Topics for Report

1. Describe and explain any trends you observed in R_f values during the TLC separation.

2. Write a mechanistic pathway for the reaction of your unknown with 2,4-dinitrophenylhydrazine.

3. Explain why the derivative of acetone melts at a higher temperature than that of methyl ethyl ketone, even though acetone has the lower boiling point and molecular weight.

4. (*a*) What was the limiting reactant in the 2,4-dinitrophenylhydrazone preparations? (*b*) By what percentages were the other reactants in excess?

Library Topics

1. Research some of the methods used to estimate the amount of alcohol consumed by persons suspected of being intoxicated and give some of the advantages and disadvantages of each. Include a balanced equation for the Breathalyzer reaction used to measure breath alcohol concentration.

2. Write a brief paper on the use of chemistry in enforcing laws against illicit drugs.

Preparation of the Insect Repellent
N,N-Diethyl-*meta*-toluamide

Functional Derivatives of Carboxylic Acids. Nucleophilic Acyl Substitution. Reactions of Carboxylic Acids. Schotten-Baumann Reaction.

Operations

⟨OP–5⟩ Weighing

⟨OP–7a⟩ Refluxing

⟨OP–8⟩ Cooling

⟨OP 9⟩ Mixing

⟨OP–10⟩ Addition

⟨OP–13⟩ Extraction

⟨OP–14⟩ Evaporation

⟨OP–20⟩ Drying Liquids

⟨OP–22b⟩ Trapping Gases

⟨**OP–26**⟩ Vacuum Distillation

⟨**OP–26a**⟩ Semimicro Vacuum Distillation

⟨OP–32⟩ Gas Chromatography

Experimental Objective

To prepare N,N-diethyl-*m*-toluamide from *m*-toluic acid.

Learning Objectives

To learn how to perform a small-scale vacuum distillation.

To learn about the methods used in carrying out a Schotten-Baumann reaction of an acyl chloride.

To learn about some insect repellents and how they work.

SITUATION

The government of the mosquito-infested tropical nation of Upper Miasma has spent millions on a crash program to eliminate that pest and the diseases it carries. As part of this effort, the country's National Health Foundation is subsidizing new approaches intended to prevent the transfer of mosquito-borne diseases from insect to human, including the development of more effective mosquito repellents. You have just received an NHF grant to study the mechanism of insect repellence and to come up with an effective substitute for the agents currently in use.

At present, the most popular insect repellent sold in Upper Miasma is Begone, a combination of citronella oil and gnu fat which, though effective, requires frequent re-application and tends to become rancid. Other repellents that discourage many species of mosquitoes are quite ineffective against your most troublesome pest, *Anopheles horridum*. However, you recently observed that certain American tourists visiting your country managed to avoid being bitten by *A. horridum*. On further investigation, you found that these tourists had brought along some spray cans of the insect repellent Off, whose major active component is a chemical named *N,N*-diethyl-*meta*-toluamide (Deet for short). You would like to find out whether Deet is effective enough to merit further consideration. Since the commercial preparation contains some isomers that might complicate the interpretation of your tests, you decide to prepare a pure sample of Deet for testing.

Deet
(*N,N*–diethyl-*m*-toluamide)

The purity of the sample will be checked by gas chromatography.

BACKGROUND

Chemical Mosquito Evasion

A recent *Scientific American* article theorized that mosquito "repellents" don't really repel mosquitoes in the same way that a disagreeable odor might repel a human from the source of the odor—instead, they appear to jam the insect's sensors so that it can't find its victim. Warm objects generate convection currents in the air around them; a warm living object also evolves carbon dioxide, which alerts the mosquito to its presence and starts it on its flight. This flight is initially random, but when the insect encounters a warm, moist stream of air it moves towards the source, which is generally a living object. Unless the object takes rapid eva-

R. H. Wright, "Why Mosquito Repellents Repel," *Scientific American*, July 1975, p. 104.

sive action, the mosquito is usually able to follow the convection current until it makes contact, or gets squashed, or both.

When you (as the intended victim) are protected by an effective insect repellent, the mosquito still knows you're around but is unable to find you. This is because the repellent prevents the insect's moisture sensors from responding normally to the high humidity of your convection current. Ordinarily, when a mosquito passes from warm, moist air into drier air, its moisture sensors send fewer signals to its central nervous system, causing it to turn back into the air stream. By blocking these sensors, the repellent reduces the signal frequency and convinces the mosquito that it is heading into drier rather than moister air; so it turns away just before landing.

The molecular features that make a compound a good repellent are not well understood at this time—repellents occur in nearly every chemical family and exhibit a large variety of molecular shapes. A number of N,N-disubstituted amides similar to Deet are represented (see Figure 1), including the diethylamide of thujic acid, a constituent of the western red cedar that may be partly responsible for that tree's resistance to insect attack. Some of the more effective repellents are esters like dimethyl phthalate and diols such as 2-ethyl-1,3-hexanediol. "Bug pills" that can be taken orally are already on the market. These claim to provide protection by causing a repellent chemical to be released through the skin.

Undoubtedly the phenomenon of insect repellence is more complicated than the explanation here might suggest; there must be individual variations in skin chemistry or physical characteristics to explain why some unlucky individuals are eaten alive by mosquitoes while others escape unscathed.

The diol 2-butyl-2-ethyl-1,3-propanediol is apparently more effective than Deet. In laboratory tests, one application has protected against mosquito bites for up to 196 days!

Figure 1.   Some typical insect repellents

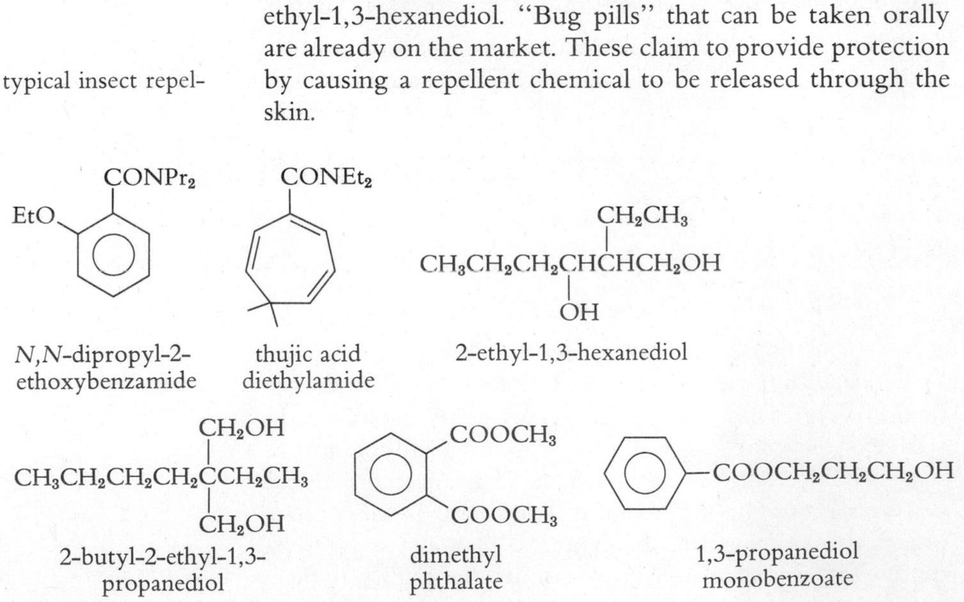

N,N-dipropyl-2-
ethoxybenzamide

thujic acid
diethylamide

2-ethyl-1,3-hexanediol

2-butyl-2-ethyl-1,3-
propanediol

dimethyl
phthalate

1,3-propanediol
monobenzoate

Et = CH_3CH_2—

Pr = $CH_3CH_2CH_2$—

When more is learned about the structural features contributing to repellency, it should be possible to develop even
more effective and convenient repellents for long-term protection against insect bites.

METHODOLOGY

Amides are usually prepared by treating a carboxylic
acid derivative with ammonia or with a primary or secondary amine. The acid derivative can be an acid anhydride or
even an ester, but acyl chlorides are most useful for preparing the widest variety of amides. Because of the high reactivity of acyl chlorides, the reactions are usually rapid and
exothermic—so much so that, in many cases, the rate must
be controlled by cooling or by using an appropriate solvent.
When the rection is carried out in an inert solvent such as
ethyl ether or benzene, it is necessary to use at least a 100%
excess of the nitrogen compound, because the reaction produces HCl that reacts with one equivalent of amine (or ammonia) to form the amine salt. This salt may also be difficult
to separate from the product. One way to avoid these disadvantages is to add a base such as pyridine, triethylamine, or
sodium hydroxide to react with the evolved HCl.

General reactions for preparing
amides

$$R-\overset{\overset{\text{O}}{\|}}{C}-Z + NH_3 \longrightarrow R\overset{\overset{\text{O}}{\|}}{C}NH_2 + HZ$$

$$R-\overset{\overset{\text{O}}{\|}}{C}-Z + R'NH_2 \longrightarrow R\overset{\overset{\text{O}}{\|}}{C}NHR' + HZ$$

$$R-\overset{\overset{\text{O}}{\|}}{C}-Z + R'-\underset{\underset{R''}{\|}}{N}H \longrightarrow R\overset{\overset{\text{O}}{\|}}{C}\underset{\underset{R''}{\|}}{N}R' + HZ$$

Z = Cl, OR, OCOR, etc.

Example of reaction in inert solvent

$$R\overset{\overset{\text{O}}{\|}}{C}Cl + 2R'NH_2 \longrightarrow R\overset{\overset{\text{O}}{\|}}{C}NHR'$$
$$+ R'NH_2^+Cl^-$$

Schotten-Baumann method for
amides

$$R\overset{\overset{\text{O}}{\|}}{C}Cl + R'NH_2 + NaOH \longrightarrow$$

$$R\overset{\overset{\text{O}}{\|}}{C}NHR' + NaCl + H_2O$$

The preparative method that uses an acyl chloride in the
presence of aqueous sodium or potassium hydroxide is
known as the Schotten-Baumann reaction. It has been used
to prepare esters and other acyl derivatives as well as amides.
Some acyl chlorides, particularly low-molecular-weight aliphatics, hydrolyze rapidly in water to form carboxylic acids

and thus are unsuitable reactants for this procedure. However, aromatic and long-chain aliphatic acyl chlorides, being relatively water-insoluble, hydrolyze slowly enough to make the Schotten-Baumann procedure practical for preparing their amides. The small amount of acyl chloride lost by hydrolysis can be compensated for by using an excess of this reactant. The presence of aqueous alkali in the reaction mixture also makes it possible to use an amine hydrochloride rather than the free amine, as the salt reacts with base to liberate the amine *in situ*. Since many amines are volatile, corrosive, and quite unpleasant to handle, the use of the relatively well-behaved amine salt can be a definite plus.

Acyl chlorides used in preparing amides are generally themselves prepared from the corresponding carboxylic acids. Several reagents are commonly used for this transformation, each with certain advantages and disadvantages. Thionyl chloride ($SOCl_2$) has the great advantage that all of the inorganic reaction products (HCl and SO_2) are gases and can be easily removed from the acyl chloride, which is often used for further reactions without isolation.

Reaction of an amine salt with a base

$$RNH_3^+Cl^- + NaOH \rightarrow$$
$$RNH_2 + NaCl + H_2O$$

Methods for preparing acyl chlorides

$$\underset{RCOH}{\overset{O}{\|}} + PCl_5 \longrightarrow \underset{RCCl}{\overset{O}{\|}} + POCl_3 + HCl$$

$$3\underset{RCOH}{\overset{O}{\|}} + 2PCl_3 \longrightarrow 3\underset{RCCl}{\overset{O}{\|}} + 3HCl + P_2O_3$$

$$\underset{RCOH}{\overset{O}{\|}} + SOCl_2 \longrightarrow \underset{RCCl}{\overset{O}{\|}} + SO_2 + HCl$$

The excess thionyl chloride from the first step is decomposed by its reaction with sodium hydroxide in the second.

In this experiment, you will prepare *m*-toluoyl chloride by treating *m*-toluic acid with excess thionyl chloride, then run a Schotten-Baumann reaction with diethylamine (from the hydrochloride) to prepare *N,N*-diethyl-*m*-toluamide. Because corrosive acidic gases are released in the reactions, a sodium hydroxide gas trap will be used to keep them out of the atmosphere. In the Schotten-Baumann reaction, stirring is necessary to provide adequate contact between the water-insoluble acyl chloride and the amine. Use of the surfactant sodium lauryl sulfate helps disperse the acyl chloride into smaller particles to increase the reaction rate. After the product is isolated by extracting it with ether, it can be purified by vacuum distillation and analyzed by gas chromatography.

PRELAB EXERCISES

1.   Read⟨OP–26⟩ and ⟨OP–26*a*⟩. Review the other operations as needed.

2.   Calculate the mass of 30 mmol *m*-toluic acid, the volume of 36 mmol thionyl chloride, the mass of 25 mmol diethylamine hydrochloride, and the volume of 100 mmol 3M sodium hydroxide.

3.   Write a checklist for the experiment.

Reactions and Properties

A.

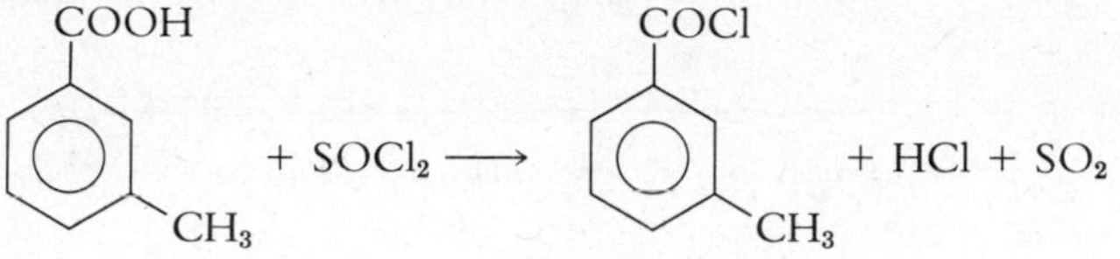

$+ SOCl_2 \longrightarrow$ $+ HCl + SO_2$

m-toluic acid   *m*-toluoyl chloride

B. $(CH_3CH_2)_2NH_2{}^+Cl^- \ + \ NaOH$
diethylamine hydrochloride

$$\longrightarrow (CH_3CH_2)_2NH + NaCl + H_2O$$
diethylamine

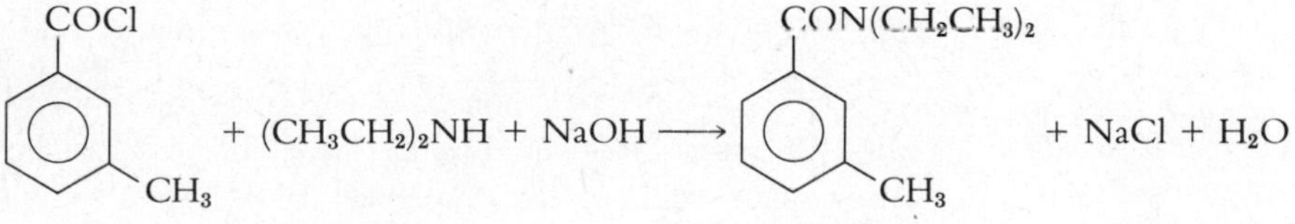

$+ (CH_3CH_2)_2NH + NaOH \longrightarrow$ $+ NaCl + H_2O$

Deet

Table 1.   Physical properties

	M.W.	m.p.	b.p.	d.
m-toluic acid	136.2	111–13		
thionyl chloride	119.0		79^{746}	1.65
diethylamine hydrochloride	109.6	227–30		
N,N-diethyl-*m*-toluamide	191.3		160^{19}	0.996

Hazards

Thionyl chloride causes severe burns and its vapors irritate the eyes and respiratory system; it decomposes on contact with moisture to produce HCl and SO_2. m-Toluoyl chloride strongly irritates skin, eyes, and mucous membranes. Diethylamine hydrochloride irritates skin and eyes; the free amine is toxic and very hazardous to the eyes and respiratory system. N,N-diethyl-m-toluamide can irritate the eyes and mucous membranes.

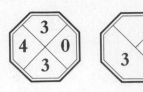

diethylamine thionyl chloride

PROCEDURE

A. *Preparation of* m-*Toluoyl Chloride*

Equip a 25-ml boiling flask with a reflux condenser and assemble a sodium hydroxide gas trap ⟨OP–22b⟩ to fit onto

WARNING *Thionyl chloride causes severe burns and respiratory problems — handle with great care.*

the top of the condenser. Combine 30 mmol m-toluic acid with 36 mmol of thionyl chloride (*Hood!*) in the boiling flask, and reflux ⟨OP–7a⟩ gently for 20 minutes or more, until evolution of gases stops. Cool to below room temperature (about 10°) in an ice bath.

B. *Preparation of* N,N-*Diethyl*-meta-*toluamide (Deet)*

Reaction. Set up an apparatus for addition and reflux ⟨OP–10⟩ using a 100-ml boiling flask and leaving the gas trap in place on the reflux condenser. Slowly combine, with stirring ⟨OP–9⟩ and cooling ⟨OP–8⟩, 25 mmol of diethylamine hydrochloride with 100 mmol of 3.0M aqueous sodium hydroxide in the boiling flask, then add 0.1 gram of sodium lauryl sulfate to the solution. Under the hood, carefully pour the cooled toluoyl chloride (which contains some

If magnetic stirrers are not available, the diethylamine hydrochloride–NaOH solution can be placed in a stoppered Erlenmeyer flask and the m-toluoyl chloride added in small portions with vigorous shaking (Hood!)

WARNING *Do not pour the toluoyl chloride into the reaction flask by mistake, as a violent reaction will result.*

excess thionyl chloride) into the dry addition funnel, stopper it immediately, and replace it on the reaction apparatus. Keeping an ice water bath handy to moderate the reaction if necessary, slowly add ⟨OP–10⟩ the acyl chloride to the aqueous amine solution with vigorous stirring ⟨OP–9⟩. When the addition is complete, warm the mixture on a steam bath for about 15 minutes (with continued stirring) to complete the reaction and hydrolyze the excess acid chloride. At this point, the odor of the acid chlorides should be gone and the solution should be basic to litmus.

Remove the stopper on the addition flask occasionally to break the vacuum.

Separation.   Transfer the contents of the boiling flask to a separatory funnel and extract the product ⟨OP–13⟩ with three 20-ml portions of ethyl ether. Combine the extracts, dry them ⟨OP–20⟩ with magnesium sulfate, and remove the solvent ⟨OP–14⟩ under aspirator vacuum on a steam bath to leave the crude liquid amide.

Purification and Analysis.   Purify the N,N-diethyl-m-toluamide by semimicro vacuum distillation ⟨OP–26a⟩, weigh it ⟨OP–5⟩, and analyze its purity by gas chromatography ⟨OP–32⟩.

Experimental Variations

1.   The product can also be purified by column chromatography ⟨OP–16⟩ on alumina, eluting with low-boiling petroleum ether. Deet should be the first major component to come off the column.

2.   A synthesis of Deet that does not use the Schotten-Baumann procedure is described in *J. Chem. Educ. 51*, 631 (1974).

3.   Interested students may wish to evaluate the effectiveness of their product as a mosquito repellent by making up a 15–20% solution (by weight) in isopropyl alcohol, applying measured quantities of this preparation and commercial mosquito repellents to each arm, and finding a convenient source of mosquitoes. As with any such preparation, care should be taken to avoid contact with eyes and mucous membranes.

One testing method is to place one's arm in a cage full of voracious mosquitos for three-minute periods at thirty-minute intervals until the arm is bitten twice during one three-minute period.

4.   An infrared spectrum of the product can be obtained and compared with the one found in the Aldrich Library of Infrared Spectra (Bib–F22) or in another collection of IR spectra.

Significant absorption bands should be identified.

Topics for Report

1. **(a)** Write equations for the reactions occurring in the gas trap. **(b)** Write equations for the reactions of excess thionyl chloride and *m*-toluenesulfonyl chloride with aqueous sodium hydroxide.

2. Construct a flow diagram for the preparation, including any by-products formed by the reactions in Topic **1(b)**.

3. Write reasonable mechanisms for: **(a)** the reaction of *m*-toluic acid with thionyl chloride; and **(b)** the reaction of *m*-toluoyl chloride with diethylamine.

4. The insect repellent *N*-butylacetanilide is used to treat clothing for fleas and ticks. Outline a synthesis of this compound, starting with aniline.

5. **(a)** Calculate the total mmoles of NaOH required for Part **B,** including the amount used up in combining with excess reactants. **(b)** Calculate the percentage excess of NaOH that was actually used.

Library Topics

1. Look up the structure of the antituberculosis drug isoniazide and relate the story of its discovery (Bib–L44).

2. Although Deet is an effective mosquito repellent, it actually helps attract certain types of insects such as the pink bollworm moth. Find out what function it performs for this species and what other chemical(s) occur in the attractant substance. Give their structures.

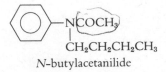

N-butylacetanilide

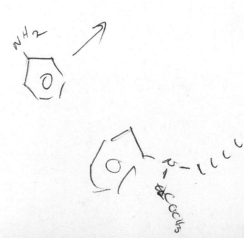

Isolation and Identification of an Essential Oil from Cloves

Phenols. Isolation of Natural Products. Spectrometric Analysis. Gas Chromatography.

Operations

⟨OP–5⟩ Weighing

⟨OP–13⟩ Extraction

⟨OP–14⟩ Evaporation

⟨**OP–15a**⟩ Steam Distillation

⟨OP–20⟩ Drying Liquids

⟨OP–32⟩ Gas Chromatography

⟨OP–33⟩ Infrared Spectrometry

Experimental Objective

To isolate and identify the major component of clove oil and to assess the purity of the oil.

Learning Objectives

To learn how to carry out a steam distillation.

To learn about some of the experimental methods used in working with natural products.

To learn about the use of plant sources in developing useful drugs.

SITUATION

For centuries, people have chewed on cloves to allay the pain of a toothache. Clove tea is said to provide relief from nausea, flatulence, dyspepsia, and languid indigestion; clove oil has been used as a local anesthetic and for driving out evil spirits.

You are one of a group of scientists exploring natural alternatives to some synthetic drugs, and the effectiveness of clove oil as a local anesthetic has made it a likely candidate for further research. You have already learned that the active principle of clove oil is a phenol with the molecular formula $C_{10}H_{12}O_2$; now you wish to isolate that component from cloves and determine its structure.

BACKGROUND

Plants and Healing

Herbal medicine, the use of plants for healing and for preventing disease, has long been associated with witch doctors, medicine men, and old wives' remedies. In this day of modern medical miracles, the arts of the herbalist are often considered mere superstition or quackery—and there may be an element of both in herbal medicine as it is sometimes practiced. Those who seek to cure a cancer by eating apricot pits rather than consulting a physician are taking a dangerous gamble, and the belief that "if it doesn't cure you, at least it won't hurt you" is not always valid—herbal preparations from *Datura, Aconitum,* and *Lobelia* species are considered dangerously poisonous. Nevertheless, some botanical preparations are undoubtedly effective, and many medicines used by the African witch doctor and the Indian medicine man have found their way onto the pharmacist's shelves.

 The widespread notion that all modern drugs and medicines come out of the test tube is wrong on several counts. Nearly half of all prescriptions written contain at least one drug of natural origin, and most synthetic drugs owe their existence to a natural substance whose molecules provided a blueprint for their development. Many drug companies employ explorers who search for promising botanicals, often basing their exploration on reports of plant usage by native peoples. Once a sufficient quantity of a plant has been collected, an extract is prepared and the active principles are isolated, analyzed, and tested. Some botanically derived compounds like reserpine are used in their natural form, but in many cases the molecules are altered to prepare even more effective drugs or drugs with fewer side effects.

 By making and testing numerous derivatives of a particular drug, scientists may be able to identify the parts of the molecule that are responsible for its action—specific groups of atoms whose location and configuration may be

Apricot pits, like cherry pits, apple seeds, and peach pits, can contain dangerous quantities of cyanide.

crucial. It is then often possible to prepare a purely synthetic compound displaying the same type of activity, or even a useful drug with a completely different therapeutic application, by reshuffling atoms and testing the resulting molecules.

The history of local anesthetics provides a good example of this approach. Certain South American Indian tribes have long been known to chew coca leaves to increase their endurance and allay hunger and thirst. The active principle of these leaves, cocaine, was isolated and found to be an effective local anesthetic, but it proved much too toxic and habit-forming to usc in routine applications. When it was discovered that the activity of cocaine depends on the presence of both a lipophilic portion and a hydrophilic portion separated by an intermediate structure containing an ester linkage, synthetic analogues were prepared in an effort to develop a drug as effective as cocaine without its disadvantages. This search led through β-eucaine to procaine (novocaine) and other modern local anesthetics such as tetracaine and lidocaine, whose structures bear little obvious resemblance to their naturally derived prototype.

Structural features of local anesthetics

Another native remedy that was used long before its "discovery" as a modern miracle drug is rauwolfia, a tranquilizer and hypertension preventive derived from the East Indian snakeroot, *Rauwolfia serpentina*. The botanical drug

The Indian snakeroot plant has been used for at least 3000 years as an antidote for snakebite, a cure for insanity and stomach ache, and to soothe grumpy babies.

The total synthesis of reserpine by R. B. Woodward in 1956 is considered one of the outstanding achievements of synthetic organic chemistry.

first attracted the attention of medical science when it was found to cure certain kinds of insomnia and to reduce blood pressure. Its use as a tranquilizer for mentally disturbed patients arose from the observation that heart patients treated with rauwolfia acted as if they hadn't a worry in the world. Analysis of the botanical drug led to the discovery that its most active component is the alkaloid now known as reserpine. No commercially feasible process for synthesizing reserpine has yet been developed, so the drug is still derived from rauwolfia extracts.

reserpine

Many plants that have been used for years by herbalists have attracted only superficial notice from medical science.

WARNING   *Like other drugs, botanical medicines may cause undesirable side effects or even death if used improperly; they should not be administered by those unskilled in their use without sound professional advice.*

The common North American weed called boneset (*Eupatorium perfoliatum*) provides a bitter-tasting tea that was a favorite early American remedy for reducing fevers; it is still used for that purpose by advocates of natural medicines. The closely related Joe Pye weed (*Eupatorium purpureum*) was named after an American Indian who gained fame using it to cure typhus. It is an effective diuretic for the treatment of kidney or bladder ailments. Euell Gibbons, in *Stalking the Healthful Herbs,* has called coltsfoot (*Tussilago farfara*) "one of the finest herbal cough medicines." It makes a pleasant-tasting tea with demulcent and expectorant properties, and has been used as an ingredient in cough drops. Legend has it that Achilles, during the seige of Troy, used yarrow to treat the wounded Greeks; its botanical name (*Achillea millefolium*) takes cognizance of that tradition. The bruised

In medieval times yarrow was said to help in preventing baldness, seducing maidens, and conjuring up the devil. Modern medical science has yet to experimentally verify these claims.

leaves of yarrow help to stop bleeding, heal cuts, and relieve the pain of a wound. The plant has been used in emergencies by backpackers and other outdoor adventurers.

Botanical medicine is far from a lost art, since there are more different kinds of molecules in the world's natural life forms than have been synthesized in all the world's chemistry laboratories. Most of them have never been identified, let alone used in medicine, so numerous botanicals should yet yield valuable drugs to researchers with the time and patience to carefully separate, analyze, and evaluate their constituents.

METHODOLOGY

Most plants contain a so-called essential oil that consists of relatively volatile and often highly odoriferous components. Although many plant components can be isolated from a plant by solvent extraction or other methods, the traditional way of separating essential oils is by steam distillation. This process is preferable to ordinary distillation because the volatile components distill at temperatures below their normal boiling points, and decomposition due to overheating is reduced or prevented. The volatile oil can then be separated from the water (which codistills with it) by solvent extraction. If, as is usually the case, the oil consists of a number of different components, these can be separated by various methods and then identified with the aid of gas chromatography and spectrometry.

In this experiment, the essential oil is unusual in that about 85–90% of it is a single compound—a phenol having the molecular formula $C_{10}H_{12}O_2$. The oil should thus yield an infrared spectrum that approximates the spectrum of the pure compound and that can be used in its identification. By consulting the formula index of a collection of standard infrared spectra, such as the *Aldrich Library of Infrared Spectra* (Bib−F23), you should be able to locate spectra for some common organic compounds having the correct formula. Then, by comparing the positions and relative intensities of the absorption bands, you should be able to identify the major component of clove oil as one of these compounds. Keep in mind that the presence of minor components may contribute additional absorption bands that are not present in the standard spectrum. The purity of the oil, along with the number and proportions of minor components, will be determined by gas chromatography.

Clove oil may also contain about 10% of a nonphenolic compound $C_{12}H_{14}O_3$, which can be separated and identified as described in Experimental Variation 1.

If standard spectra are not available in your library, use the CRC Atlas of Spectral Data and Physical Constants of Organic Compounds *(Bib−A4) or a comparable collection of spectral data.*

PRELAB EXERCISES

1. Read ⟨OP–15a⟩ and review the other operations as needed.

2. Write a brief checklist for the experiment.

PROCEDURE

A Moulinex coffee, spice, and nut grinder works very well.

The distillation usually takes ½–1 hour; about 200 ml of distillate should be collected.

methylene chloride

Weigh ⟨OP–5⟩ about 10 g of fresh whole cloves and grind them as finely as possible using a food mill or a mortar and pestle. Mix the cloves with 100 ml of water in a 500-ml boiling flask and steam distill the mixture ⟨OP–15a⟩ until a drop of the distillate shows no evidence of oily droplets when swirled on a watch glass. Extract the distillate ⟨OP–13⟩ with two 25-ml portions of methylene chloride, dry the extract ⟨OP–20⟩ with magnesium sulfate, evaporate the solvent ⟨OP–14⟩ and weigh the clove oil ⟨OP–5⟩. Analyze the oil by gas chromatography ⟨OP–32⟩ and compute the fraction of each component present. Obtain an infrared spectrum of the clove oil ⟨OP–33⟩ and identify its major component by matching the spectrum with a standard published spectrum.

Experimental Variations

1. The major and minor components of clove oil can be isolated by extracting the methylene chloride solution twice with 5% aqueous sodium hydroxide. Acidification of the aqueous layer with dilute HCl, followed by extraction with methylene chloride and evaporation of the solvent, yields the phenolic component; evaporation of the organic layer yields the nonphenolic component, which can also be identified by its infrared spectrum.

See the Sadtler Standard Spectra (Bib-F26) indexes.

2. The structure of the phenolic component can be confirmed by preparing a suitable derivative. See Part III for procedures.

3. A similar procedure can be used to isolate the essential oils from anise seed, caraway seed, and cumin. Anise oil contains 80–90% of a single compound, $C_{10}H_{12}O$, which can be identified from its infrared spectrum. Caraway seed contains carvone, which can be characterized by preparing the 2,4-dinitrophenylhydrazone; the cuminaldehyde in cumin can be characterized by preparing the semicarbazone. In each case, the percentage of the major component should be computed, assuming quantitative yields in the derivative preparations.

See Part III for procedures.

Preparation of Phenolphthalein

MINILAB 7

Mix 0.35 g of phenol with 0.20 g of phthalic anhydride in a test tube. Add 3 drops of concentrated sulfuric acid and heat, with stirring, at approximately 160°C for 2–3 minutes. Pour the hot melt into 50 ml of water and stir to dissolve. Make a portion of the solution alkaline with dilute sodium hydroxide and then re-acidify it with dilute hydrochloric acid. Describe and explain your observations, giving equations for the preparation and for the reaction of the product with base.

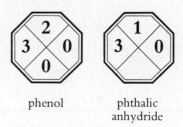

phenol   phthalic
              anhydride

Topics for Report

1. **(a)** Draw the structure of the major component of clove oil and give its common name. **(b)** Calculate the percentage of the oil in your cloves and compare your yield with the 15–18% reported in the literature.

2. Assign as many bands as you can in your infrared spectrum of clove oil and, by comparing it with the standard spectrum, indicate any bands that are probably due to minor components.

3. Outline a synthesis of vanillin from the major component of clove oil.

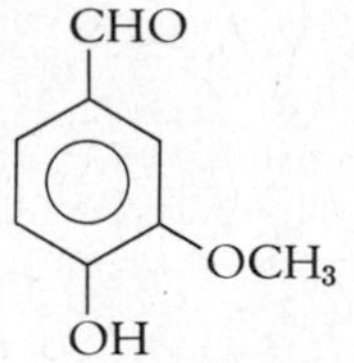

4. The leaves of the bearberry shrub (*Arctostaphylos uva-ursi*) contain a substance named arbutin, which is an effective disinfectant of the urinary tract. Arbutin has the molecular formula $C_{12}H_{16}O_7$ and can be hydrolyzed by emulsin or dilute HCl to glucose and compound X. Compound X, which shows germicidal activity, dissolves readily in dilute sodium hydroxide but not in sodium bicarbonate. Its NMR spectrum consists of two sharp singlets. Draw the structures of compound X and of arbutin.

CHO

OCH₃

OH

vanillin

hypericin

Library Topics

1. Some chemicals that have been derived from plants are podophyllotoxin from the May apple (*Podophyllum peltatum*), sanguinarine from bloodroot (*Sanguinaria canadensis*) and apocynin from American hemp (*Apocynum cannabinum*). (*a*) Find the structures for these compounds and describe their medical uses or physiological effects. (*b*) What president of the U.S. owed his life to *A. cannabinum* and why?

2. The common weed St. Johnswort (*Hypericum perforatum*) contains a photosensitizer, hypericin, that has caused skin lesions and death in livestock. Discuss the phenomenon of photosensitization, indicating the characteristics a molecule must have in order to act as a photosensitizer and describing the role of photosensitizers in chemical transformations such as photosynthesis.

Structure Determination of an Unknown Monosaccharide

Carbohydrates. Structure Determination. Polarimetry.

Operations

⟨OP–5⟩ Weighing

⟨OP–7⟩ Heating

⟨OP–11⟩ Gravity Filtration

⟨OP–12⟩ Vacuum Filtration

⟨OP–21⟩ Drying Solids

⟨OP–28⟩ Melting Point

⟨**OP–31**⟩ Optical Rotation

Experimental Objective

To determine the structure of an unknown monosaccharide by chemical methods.

Learning Objectives

To learn how to measure optical rotation with a polarimeter.

To learn about some of the experimental methods that are used to deduce the structures of carbohydrates.

To learn about simple sugars and the sweet taste sensation.

SITUATION

While analyzing an extract from the brain of the Eurasian chipmunk, you isolate and purify a monosaccharide having the same molecular formula as glucose. Tests by a panel of food tasting experts establishes that this monosaccharide, designated EC-25, is only half as sweet as glucose.

Intrigued by this result, you decide to investigate the structural features responsible for the sweet taste sensation and explain, if you can, the marked difference in sweetness. First you must establish the molecular structure of compound EC-25.

BACKGROUND

Sweet Molecules

One of the most obvious properties associated with the sugars is their sweetness. Fructose is the sweetest known sugar, being 1.8 times sweeter (in its crystalline form) than sucrose—but other substances are sweeter than any of the sugars. Cyclamates such as sodium cyclohexylsulfamate have been rated up to 30 times as sweet as sucrose, saccharin up to 350 times as sweet; and 1-n-propoxy-2-amino-4-nitrobenzene (P-4000) is estimated to be four thousand times as sweet as sucrose.

Many explanations have been advanced to explain the sweet taste sensation. One recent theory asserts that all sweet compounds contain a so-called AH,B couple that is necessary to bind it to the taste-bud receptor site. According to the theory, A and B are electronegative atoms such as oxygen or nitrogen (but sometimes halogen or even carbon) that must be 2.5–4.0 Å apart in order to interact (by hydrogen bonding) with the receptor site, which possesses a similar AH,B couple. In the sugars, AH and B are assumed to be an OH group and the oxygen atom of an adjacent OH

Fructose crystallizes as β-D-fructopyranose, which is the sweetest form; when dissolved in water it forms other tautomers, causing its sweetness to decrease soon after solution.

Proposed interaction of a sweet molecule with the receptor site

Gauche conformation in a sugar molecule

Proposed location of hydrophobic site on sweet molecules

group, respectively—ideally in a *gauche* conformation. In a *transoid* conformation they would be too far apart to interact and in a *cisoid* conformation they could hydrogen-bond intramolecularly rather than with the receptor site.

Another possible requirement for the sweet taste sensation is a hydrophobic or "greasy" site, γ, on the sweet molecule, which is approximately 3.5 Å from A and 5.5 Å from B in the triangular grouping illustrated. The dissymmetry of this arrangement could explain why enantiomers sometimes taste different. In β-D-fructopyranose the AH, the B, and the γ sites are considered to be the C-1 OH, the CH_2OH oxygen atom, and the ring methylene group, respectively. Proposed AH,B groupings for some other sweet compounds are illustrated in Figure 1.

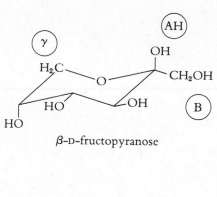

saccharin

chloroform

alanine

P-4000

β-D-fructopyranose

Figure 1.   AH,B Groupings in sweet molecules

METHODOLOGY

The structures of monosaccharides have been determined by a number of different methods, among them preparing osazones and oxidizing the sugars to aldaric acids. The formation of osazones from monosaccharides in effect "homogenizes" the molecules above carbon #3 by convert-

Effect of osazone formation

$$
\begin{array}{ccc}
CH{=}O & CH{=}NNHPh & CH_2OH \\
| & | & | \\
CHOH & \xrightarrow{\quad} \quad C{=}NNHPh \quad \xleftarrow{\quad} & C{=}O \\
| & | & |
\end{array}
$$

D-allose   D-altrose   D-allulose

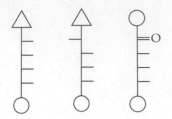

ing both —CO—CH$_2$OH and —CHOH—CHO to identical groupings. Thus aldoses and ketoses that are identical in configuration from C-3 on down the chain, but are different at C-1 or C-2, yield identical osazones. Considering only the D-hexoses (in their open-chain forms, for convenience) we can identify four C-3 to C-6 structural units that yield different osazones. For example, allose, altrose, and the ketohexose allulose, illustrated in the margin by means of Rosanoff symbols, all contain the structural unit **A** (Figure 2) and should yield the same osazone. The osazone of your unknown monosaccharide will be prepared by a simple test tube reaction, and its melting point will be measured. Because osazones decompose on melting, you will have to determine the melting point by a special technique that requires fast heating.

Figure 2.   C-3 to C-6 structural units of D-hexoses   Rosanoff symbols for functional groups

A   **B**   **C**   **D**

△ = CHO   ◺ = COOH

◯ = CH$_2$OH   ⊣ or ⊢ = OH

Example:

$$
\begin{array}{c}
\text{CHO} \\
\text{HO}-\!\!\!-\text{H} \\
\text{H}-\!\!\!-\text{OH} \\
\text{CH}_2\text{OH}
\end{array}
$$

Nitric acid oxidation of a monosaccharide converts both aldehyde and primary alcohol functional groups to carboxyl groups, forming an aldaric acid. Thus, glucose is oxidized to D-glucaric acid, which has COOH groups at each end of the molecule. (This acid actually exists as a dilactone, but for our purposes the dicarboxylic acid representation will be more instructive.) Since glucaric acid, like glucose, has no plane of symmetry, it is optically active and will rotate the plane of polarized light. The acid obtained by oxidizing D-ribose *does* have a plane of symmetry, and should therefore be an optically inactive *meso* compound. Thus the optical activity or inactivity of the aldaric acid formed from oxidation of a monosaccharide can yield valuable structural information.

Seliwanoff's reagent is prepared from 0.25 g of resorcinol dissolved in 500 ml of 6M HCl.

Seliwanoff's reagent, when heated with a dilute solution of a ketose, forms a deeply colored solution in 2½ minutes; aldoses react much more slowly with the reagent. From the outcome of the Seliwanoff test you should be able to decide whether your unknown monosaccharide is an aldose or a ketose. The melting point of its osazone should indicate which of the four C–3 to C–6 structural units it may contain. Then by determining whether or not the aldaric acid of EC–25 is optically active, you should be able to select the correct monosaccharide having that structural unit. As a check on the correctness of your conclusion, you can measure the optical rotation of the unknown monosaccharide and compare its specific rotation with published values.

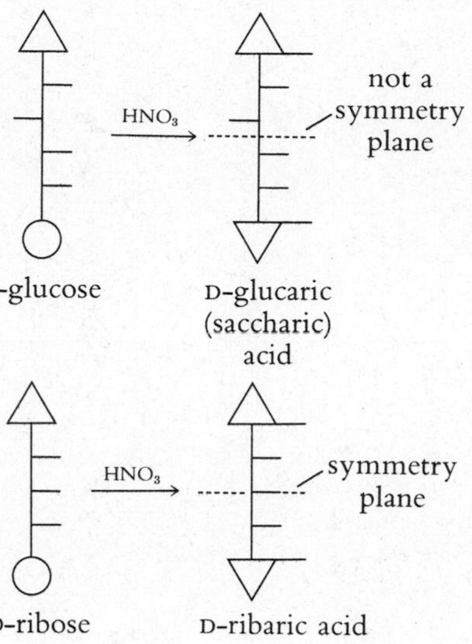

PRELAB EXERCISES

1.   Read ⟨OP–31⟩ and review the other operations as needed.

2.   Write a brief checklist for the experiment.

Nitric acid oxidation

D-glucose → (HNO₃) → D-glucaric (saccharic) acid — not a symmetry plane

D-ribose → (HNO₃) → D-ribaric acid — symmetry plane

==

Reactions and Properties

A.

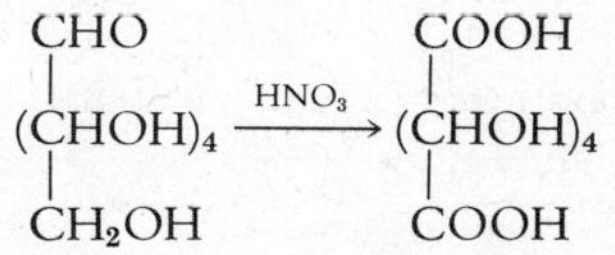

$$\text{CHO} \quad \xrightarrow{\text{HNO}_3} \quad \text{COOH}$$
$$(\text{CHOH})_4 \qquad (\text{CHOH})_4$$
$$\text{CH}_2\text{OH} \qquad \text{COOH}$$

B.

$$\begin{array}{c}\text{CHO}\\\text{CHOH}\\(\text{CHOH})_3\\\text{CH}_2\text{OH}\end{array} \quad \text{or} \quad \begin{array}{c}\text{CH}_2\text{OH}\\\text{C}=\text{O}\\(\text{CHOH})_3\\\text{CH}_2\text{OH}\end{array} \quad \xrightarrow{\text{PhNHNH}_2} \quad \begin{array}{c}\text{CH}=\text{NNHPh}\\\text{C}=\text{NNHPh}\\(\text{CHOH})_3\\\text{CH}_2\text{OH}\end{array}$$

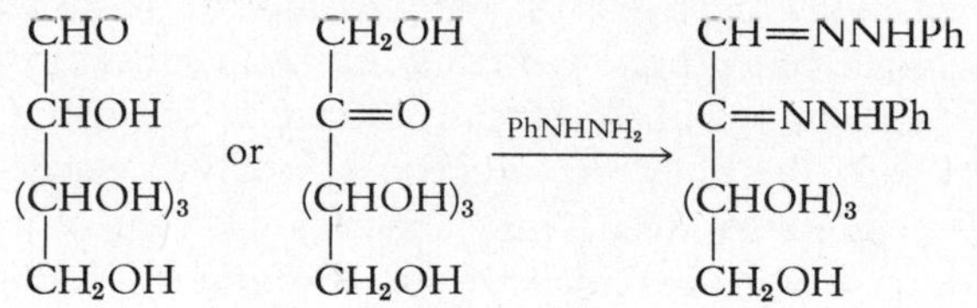

Table 1.   Osazone melting points

Structural unit	Osazone m.p.
A	178
B	205
C	173
D	201

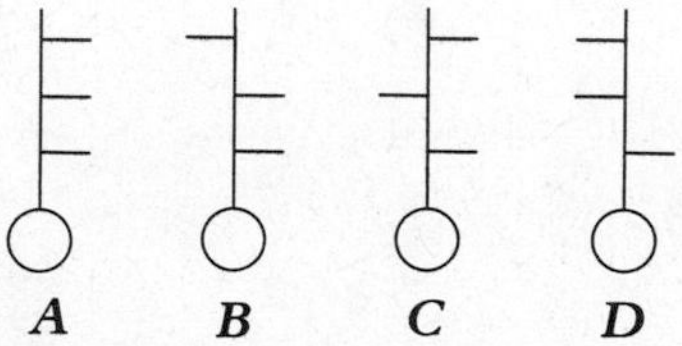

A   B   C   D

phenylhydrazine   nitric acid

resorcinol

Hazards

Nitric acid causes severe burns on skin and eyes; contact with combustible materials may cause fires. Phenylhydrazine and its hydrochloride are very toxic if ingested or allowed to contact the skin; the vapors are irritating to the eyes and should not be inhaled. Resorcinol (in Seliwanoff's reagent) is toxic and irritates the skin and eyes.

PROCEDURE

A. *Preparation of the Aldaric Acid of EC-25*

Dissolve 5.0 g of the unknown monosaccharide in 40 ml of water in a large evaporating dish (or a 250-ml beaker) and cautiously add 15 ml of concentrated nitric acid, with stirring, under an efficient fume hood. Heat the mixture gently

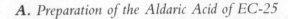

WARNING   *The NO_2 fumes evolved during this step are very toxic—do not inhale them.*

Some splattering or foaming may occur if too much heat is applied.

⟨OP–7⟩ over a boiling water bath (or on a steam bath) until the rapid evolution of red-brown fumes indicates the reaction has begun, then reduce the heat immediately. When the vigorous reaction has subsided, heat the mixture again, with occasional stirring, until most of the water has evaporated and a pasty mass remains. Cool the residue, add 20 ml of ice water, and collect the product by vacuum filtration ⟨OP–12⟩, washing it several times with cold water. Purify the product by dissolving it in 30 ml of ice cold 2M aqueous sodium hydroxide, filtering by gravity ⟨OP–11⟩ if necessary, and adding 12 ml of 6M hydrochloric acid slowly, with stirring and cooling. Let the solution stand until crystallization is complete (overnight if convenient) and collect the crystals by vacuum filtration ⟨OP–12⟩.

The solution should be strongly acidic after the HCl addition.

Keep the temperature below 25°C.

Accurate measurement of mass and volume is not necessary for the aldaric acid determination.

Dissolve about $\frac{1}{2}$ g of the air-dried aldaric acid in 25 ml of dilute aqueous sodium hydroxide (1 or 2M), filter the solution by gravity ⟨OP–11⟩ if necessary, and measure the optical rotation of the solution in a polarimeter ⟨OP–31⟩ (use the aqueous sodium hydroxide solution as a blank). Immediately rinse out the polarimeter tube with water, then measure the optical rotation of 1 gram (accurately weighed) of the original monosaccharide (EC-25) dissolved in enough water to make 25 ml of solution. The remainder of your product should be dried ⟨OP–21⟩, weighed ⟨OP–5⟩, and turned in to your instructor.

For best results, make the solution up to volume in a 25-ml volumetric flask and run a blank with pure water.

B. *Preparation of the Osazone of EC-25*

Dissolve 0.50 g of the unknown monosaccharide in 5 ml of water in a test tube and add 1.0 g of phenylhydrazine hydrochloride (*Care:* See Hazards), 1.5 g of crystalline sodium acetate, and 1.0 ml of saturated aqueous sodium bisulfite. Mix the reactants well, stopper the tube with a notched (or 1-hole) cork, and place it in a boiling water bath. Heat the tube for 30 minutes with occasional shaking, then set it in an ice bath to cool. Collect the product by vacuum filtration ⟨OP–12⟩, washing it on the filter with a milliliter or two of ice-cold methanol. Let the product air dry thoroughly on the filter, then pulverize it on a glass plate with a flat-bladed spatula. Prepare two samples for a melting point determination ⟨OP–28⟩ and record the temperature, T_1, at which the first sample melts with rapid heating (10–20° a minute). Let the temperature of the melting-point apparatus drop below T_1, then begin heating at a rate of 3–6° a minute and insert the second sample just as the temperature reaches T_1. Record the temperature when the second sample is completely melted as the melting point of the osazone.

The "melting point" is actually a decomposition temperature that varies with the rate of heating.

C. *Seliwanoff's Test for Ketoses*

Make up a few milliliters of a 1% aqueous solution of the unknown monosaccharide (EC-25) and mix together 0.5 ml of this solution, 1.5 ml of water, and 9 ml of Seliwanoff's reagent in a test tube. In another test tube, combine 0.5 ml of a 1% solution of D-fructose with 1.5 ml of water and 9 ml of Seliwanoff's reagent, and place both test tubes simultaneously in a beaker of boiling water. Record your observations after exactly $2\frac{1}{2}$ minutes of heating.

Fructose is a ketohexose.

Experimental Variations

1. Unknown carbohydrates can be classified using Benedict's test (Test C-6, Part III) to distinguish reducing from nonreducing sugars, Barfoed's test to differentiate

Procedures are found in R. M. Roberts et al., An Introduction to Modern Experimental Organic Chemistry, 2nd ed. (New York: Holt, Rinehart and Winston, 1974) pp. 397–98.

If a sugar does not form a precipitate after 30 minutes, its osazone is probably soluble.

monosaccharides and oligosaccharides, Bial's test to distinguish pentoses from hexoses, and Seliwanoff's test to distinguish aldoses from ketoses.

2. The phenylosazones of a number of different sugars can be prepared by the above procedure using 10 ml of water instead of 5 ml. The time required for the appearance of a precipitate should be noted in each case; the crystals can be studied and compared under a hand lens or microscope. Suggested sugars are glucose, fructose, galactose, mannose, maltose, and sucrose.

3. The preparation of 1-*n*-propoxy-2-amino-4-nitrobenzene (P-4000) is described in *J. Chem. Educ. 53,* 521 (1976).

4. An experiment involving the identification of monosaccharides by paper chromatography using an acetonitrile–ammonium acetate developer is described in *J. Chem. Educ. 50,* 562 (1973). Some other carbohydrate experiments in the same *Journal* deal with the analysis of glucose in blood serum [*53*, 126 (1976)], descending paper chromatography of oligosaccharides [*49*, 437 (1972)], and the preparation and NMR analysis of α- and β-D-glucose pentaacetates [*52*, 814 (1975)].

Preparation of Cellulose Acetate from Cotton

MINILAB 8

acetic acid   acetic anhydride   sulfuric acid

In a small flask, mix 20 ml of glacial acetic acid with 6 ml of acetic anhydride, then add 3 drops of concentrated sulfuric acid. Place 0.5 g of clean cotton in the flask and use a stirring rod to wet it as thoroughly as possible with the solution. Stopper the flask and let it stand until the next period. Pour the mixture in a fine stream into 500 ml of water and collect the precipitate by filtration. Dry the cellulose acetate, dissolve some of it in a little chloroform, and make a thin film as described in Experiment 43. Obtain an infrared spectrum of the film between two salt plates.

Topics for Report

1.   *(a)* Draw the structure of the monosaccharide EC-25 and write reaction paths for the formation of the osazone and the aldaric acid of EC-25. *(b)* Draw structures for all monosaccharides that should yield the same osazone as EC-25.

2.   Calculate the specific rotation of EC-25 and compare the value you obtained with that reported in the literature. See your instructor or an appropriate text or reference book to find out the name of the monosaccharide whose structure you have determined.

3.   *(a)* Make a cyclic (pyranose ring) model of the monosaccharide you identified and try to explain why it is not as sweet as glucose. *(b)* When a solution of this monosaccharide is heated, its sweetness approaches that of a glucose solution of the same concentration. Explain.

It has been suggested that the hydroxyl on C-4 is the "AH" group in most pyranose rings.

4.   Draw structures for the aldaric acids that would be obtained by the nitric acid oxidation of each of the eight D-aldohexoses. Which acids should be optically active?

Library Topics

1.   Write a short paper on the chemical basis of the four taste sensations, telling what chemical species are responsible for salty and sour tastes and illustrating (with examples) variations in sweetness and bitterness with minor changes in chemical structure.

2.   List some of the important natural sources of glucose, fructose, galactose, xylose, and ribose, and specify the uses or biological significance of each.

Most of these occur naturally as oligo- and polysaccharides or in other combinations.

Structure Determination of an Unknown Dipeptide

Amino Acids and Peptides.   N-Terminal Residue Analysis.
Chromatographic Separations.   Structure Determination.

Operations

⟨OP–4⟩ Glass Working

⟨OP–5⟩ Weighing

⟨OP–6⟩ Measuring Volume

⟨**OP–13a**⟩ Semimicro Extraction

⟨OP–14⟩ Evaporation

⟨**OP–18**⟩ Paper Chromatography

⟨**OP–22**⟩ Drying Gases

Experimental Objective

To determine the structure of an unknown dipeptide by hydrolysis and *N*-terminal residue analysis.

Learning Objectives

To learn how to identify substances using paper chromatography and to learn how to do a small-scale hydrolysis and extraction.

To learn how to conduct an *N*-terminal residue analysis using the Sanger method.

To learn about the nature and functions of some naturally occurring peptides.

SITUATION

Some weeks ago you received a shipment from Africa containing the venom of that extremely rare and poisonous snake, the mauve mamba. After isolating and purifying the major venom toxin, a polypeptide called Mamba-D toxin, you broke it down into 35 different dipeptides using the enzyme cathepsin C. You and your coworkers must now determine the structures of all 35 dipeptides so that this and other structural information can be used to piece together the structure of Mamba-D toxin.

BACKGROUND

Mushrooms, Black Mambas, and Memory Molecules

The naturally occurring polypeptides and proteins range in size from tripeptides such as glutathione to complex proteins having molecular weights on the order of one million; their variations in structure and function cover as broad a range. Some of them, such as the keratins of hair, horns, nails, and claws, act as biological building materials and have little or no biological activity; others exert a profound effect on biochemical reactions and function as hormones, enzymes, antibiotics, or toxins.

Note: Amino acids and most polypeptides actually exist as dipolar ions (zwitterions), such as $H_3\overset{+}{N}CH_2CO\overline{O}$ for glycine. For clarity, the neutral forms are used throughout this experiment.

$$H_2NCHCH_2CH_2CONHCHCONHCH_2COOH$$

COOH	CH$_2$SH

glutathione

Arg—Pro—Pro—Gly—Phe—Ser—Pro—Phe—Arg
bradykinin

Bradykinin, sometimes known as the "pain molecule," is released by enzymatic cleavage of plasma glycoprotein whenever tissues are damaged. It causes the sensation of pain by bonding to certain receptors on nerve endings. Peptides similar to bradykinin are present in wasp venom; other kinins act as hormones in stimulating a variety of physiological responses, such as the contraction or relaxation of smooth muscles and the dilation of blood-vessel walls. Analgesics of the aspirin type are believed to exert their pain-killing effects (at least in part) by suppressing the action of bradykinin.

Oxytocin and vasopressin are both secreted by the posterior pituitary gland. Their structures are identical except for two amino-acid units, but their functions are entirely different. Vasopressin increases retention of water in the kidneys and is used as an antidiuretic in a form of diabetes that is characterized by excessive urine flow. Oxytocin intensifies

(Consult your lecture text for the meaning of amino acid abbreviations.)

```
Cys—Tyr—Ile
 |
 S
 |
 S
 |
Cys—Asn—Gln
 |
Pro—Leu—GlyNH2
   oxytocin
```

```
Cys—Tyr—Phe
 |
 S
 |
 S
 |
Cys—Asn—Gln
 |
Pro—Arg—GlyNH2
   vasopressin
```

As little as one mushroom of the *Amanita phalloides* type can cause death in an adult. Some *Amanitas* (particularly *A. muscaria,* called "soma" or "fly agaric") have been eaten deliberately for their reported hallucinogenic properties, occasionally with fatal consequences.

uterine contractions during parturition and is used clinically to induce labor. Both are cyclic compounds (cyclopeptides) having a tripeptide side chain.

The stately mushrooms of the genus *Amanita* contain the cyclopeptides known as phallotoxins and amatoxins—the latter are believed to be the most virulent. These mushrooms are very deceptive, since the victim is usually not aware that he or she has been poisoned for ten hours or more following a mushroom feast. After a day or so of violent cramps, nausea, vomiting, and other symptoms, the patient appears to recover and may be sent home from the hospital. Death from kidney and liver failure often follows in several days. α-Amanitin and the other amatoxins are believed to attack the nuclei of liver and kidney cells, depleting their nuclear RNA and inhibiting the synthesis of more RNA, so that protein synthesis stops and the cells die.

α-amanitin

The venom toxin of the black mamba is probably the quickest—it can kill a mouse in less than five minutes.

Larger polypeptides, ranging from 60 to 74 amino-acid residues, are found in snake venoms like those of the African black mamba and the Indian cobra (Figure 1). Most of these venom toxins have long chains that are cross-linked by four or more cystine disulfide bridges.

One of the most exciting new areas of polypeptide biochemistry is that dealing with "memory molecules." Re-

Ser—Asp—Asn—Asn—Gln—Gln—Gly—Lys—Ser—Ala—Gln—Gln—Gly—Gly—TyrNH₂
scotophobin
(from the Greek *scotos,* dark; and *phobos,* fear)

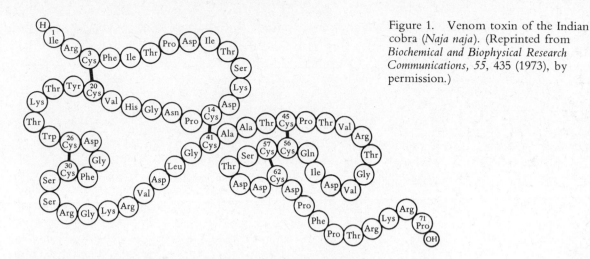

Figure 1.   Venom toxin of the Indian cobra (*Naja naja*). (Reprinted from *Biochemical and Biophysical Research Communications, 55*, 435 (1973), by permission.)

searchers have discovered that an animal's learned behavior can be forgotten when a substance interfering with peptide synthesis is injected into the animal's brain at a certain time, suggesting that peptide synthesis is involved in the consolidation of long-term memory. A peptide called scotophobin has actually been isolated from the brains of rats that had been trained to fear the dark. When scotophobin was injected into untrained rats, they also became afraid of the dark! This raises the intriguing possibility that breaking the chemical memory code could enable us to synthesize the peptides corresponding to any kind of learning experience. Perhaps someday it will be possible to get an injection of Biology 101 or Philosophy 416 rather than absorbing them in the classroom!

METHODOLOGY

You will hydrolyze your dipeptide with aqueous hydrochloric acid to obtain the component amino acids, and determine their identity by paper chromatography. (Their sequence will be determined by *N*-terminal residue analysis.) The chromatogram will be visualized with a cyclohexylamine-ninhydrin combination, which should allow you to identify most of the amino acids by their color reactions as well as by their R_f values. Since R_f values in these systems are not generally reproducible, it is advisable to chromatograph reference amino acids at the same time as the unknown mixture, so that the unknowns can be identified by comparison.

See Blackburn (Bib–L8), an organic chemistry textbook, or another appropriate source for a theoretical discussion of peptide sequence determination and N-terminal residue analysis.

2,4–dinitrofluorobenzene
(DNFB)

Table 1.   Approximate R_f values and colors for amino acids and DNP–amino acids

	1 Abbrev.	2 $R_f \times 100$	3 Color	4 DNP $R_f \times 100$
histidine	His	11	green	32 (bis)
lysine	Lys	12	purple	65 (bis)
serine	Ser	22	purple	29
aspartic acid	Asp	23	royal blue	8
glycine	Gly	23	red brown	*
threonine	Thr	26	gray-purple	41
alanine	Ala	30	purple	52
proline	Pro	34	yellow	*
tyrosine	Tyr	45	slate gray	75 (bis)
valine	Val	51	purple	69
phenylalanine	Phe	60	blue-gray	76
leucine	Leu	70	purple	78
(dinitrophenol. . . . . .62)				

Note: R_f values are multiplied by 100. Columns 2 and 3 give R_f values (from 12:3:5 1–butanol–acetic acid–water) and colors for the free amino acids; column 4 lists R_f values (from ammonia-saturated 1–butanol) for those DNP amino acids that can be identified by the method described. Amino acids forming *bis*-dinitrophenyl derivatives are so designated.
* DNP derivative destroyed under hydrolysis conditions.

2,4-dinitrophenyl group
(DNP)

Sanger won the Nobel Prize in 1958 for determining the structure of insulin.

Other amino acids, such as tryptophan, are decomposed or altered during the hydrolysis of the original peptide. They have been omitted for that reason.

Treatment of the dipeptide with 2,4–dinitrofluorobenzene under slightly alkaline conditions results in the attachment of a dinitrophenyl radical to the free amino group of the N-terminal amino acid. The dinitrophenylated dipeptide can then be hydrolyzed with HCl to yield the dinitrophenylated N-terminal amino acid. The original dinitrophenylation procedure developed by Frederick Sanger is still useful for most end groups. In this procedure, the peptide (or protein) is shaken with excess DNFB in 66% ethanol made alkaline with sodium bicarbonate. The excess reagent can be removed by extracting the mixture with ether at high pH; then the DNP-peptide is extracted at low pH and hydrolyzed in a sealed tube with constant-boiling HCl. There are several limitations to this method—the hydrolysis procedure destroys the dinitrophenylated derivatives of proline, glycine, and cystine under the conditions used, and dinitrophenylated arginine cannot be extracted from the reaction mixture with ether (it is water soluble). Therefore, these amino acids will not be included as N-terminal residues in the assigned dipeptides.

DNP–amino acids are light sensitive, so their chromatography must be carried out in the dark—for instance, inside a cabinet or drawer. Both DNP–amino acids and dinitrophenol (a by-product of the hydrolysis) form yellow spots on the chromatogram. The dinitrophenol spot can usually be distinguished by its R_f value or by holding the chromatogram over HCl vapor (this causes it to become colorless). Since the colors of the developed chromatograms may fade with time, spots should be circled as soon as possible after development or visualization.

This experiment requires long periods for the hydrolysis reactions and for developing the chromatograms, so it will be important to organize your time efficiently.

PRELAB EXERCISES

1. Read ⟨OP–13a⟩, ⟨OP–18⟩ and ⟨OP–22⟩, review ⟨OP–4⟩ on sealing glass tubing, and review the other operations as needed.

2. Write a brief checklist for the experiment.

Reactions and Properties

$$
\textbf{\textit{B.}} \quad \underset{\underset{R}{|}}{H_2NCHC}\overset{\overset{O}{\|}}{}-\underset{\underset{R'}{|}}{NHCHC}\overset{\overset{O}{\|}}{}-OH + H_2O \xrightarrow{HCl}
$$

$$
\underset{\underset{R}{|}}{H_2NCHC}\overset{\overset{O}{\|}}{}-OH + \underset{\underset{R'}{|}}{H_2NCHC}\overset{\overset{O}{\|}}{}-OH
$$

R, R′ = amino acid side chains

Table 2.   Physical properties

	M.W.	m.p.	b.p.
dinitrofluorobenzene	186.1	27.5–30	178[15]

Hazards

2,4-Dinitrofluorobenzene is an irritant and is probably toxic; avoid skin or eye contact and inhalation of dust or vapors. 1-Butanol is flammable and harmful to the eyes.

1-butanol

PROCEDURE

A. *Dinitrophenylation and Hydrolysis of the DNP-Dipeptide*

Use a pipet or a syringe for volume measurements.

The DNFB solution is prepared by dissolving 1 g of dinitrofluorobenzene in 19 ml of absolute ethanol.

*You should start on part **B** during the dinitrophenylation reaction.*

Approximately 0.1 ml of the HCl should be required.

Measure 2 mg of the dipeptide, 0.2 ml of water, 0.06 ml of 3.5% aqueous sodium bicarbonate, and 0.4 ml of stock dinitrofluorobenzene solution into a small, conical centrifuge tube (⟨OP–5⟩, ⟨OP–6⟩). Cover the tube with Parafilm and shake it intermittently over a one-hour period. Check the pH occasionally and add more of the sodium bicarbonate solution, if necessary, to keep it between pH 8 and 9 (use narrow-range pH paper).

When the reaction period is over, add 1 ml of water and 0.06 ml of 3.5% sodium bicarbonate, then extract the solution ⟨OP–13a⟩ twice with 3 ml portions of ethyl ether, removing the ether with a Pasteur pipet. Discard the ether extracts—they contain unreacted starting materials. Adjust the pH of the aqueous solution to about 1 using 6M HCl, and again extract the solution twice with 3 ml portions of ethyl ether. Remove the solvent ⟨OP–14⟩ from the combined ether extracts using a warm water bath and a dry air stream ⟨OP–22⟩ from a Pasteur pipet. Have ready a hydrolysis tube made of a 10-cm length of 5-mm O.D. soft glass tubing, sealed at one end ⟨OP–4⟩. Transfer the DNP-dipeptide into the tube using 0.2 ml of acetone, and evaporate the acetone in a dry air stream. Add 0.5 ml of 6M HCl, seal the tube ⟨OP–4⟩ (*Caution:* Be sure no flammable solvents are nearby), label it, and leave it in a 90°–100° oven for at least 12 hours. Open the (cooled) tube using a file, transfer the solution to a test tube with 2 ml of water, extract it ⟨OP–13*a*⟩ twice with 3-ml portions of ether, combine the extracts, and remove the ether as before. Add 0.5 ml of acetone and keep the solution stoppered in the dark until you are ready for part *C*.

All traces of ether must be removed before the hydrolysis step.

Save the aqueous layer in case it is needed for identifying the C-terminal amino acid (see Experimental Variation 1).

B. *Hydrolysis of the Dipeptide*

Dissolve 2 mg of the dipeptide in 0.5 ml of 6M HCl in a small test tube or centrifuge tube. Transfer the solution to another hydrolysis tube and seal it, then label the tube and heat it in a 100-110° oven for at least 24 hours. Open the (cooled) tube, transfer the solution to a small conical centrifuge tube, and evaporate all the solvent in a boiling water bath under a dry air stream. Add 0.03 ml of water and re-evaporate the solvent; then add 0.05 ml of water and keep the solution for chromatography.

Evaporation in a vacuum dessicator containing KOH and P_2O_5 is also suitable.

C. *Chromatographic Analysis*

Obtain a paper chromatogram ⟨OP–18⟩ of the DNP–amino acid solution (along with any reference solutions, if provided), using a piece of Whatman #1 chromatography paper measuring at least 12 cm along the direction of development. Use 1-butanol saturated with 0.1% aqueous ammonia as the developing solvent. Develop the chromatogram in the dark and use HCl vapors to locate the dinitrophenol spot. The yellow DNP–amino acid spot should be visible under ordinary light (ultraviolet light may help locate very faint spots).

    Obtain a paper chromatogram ⟨OP–18⟩ of the dipeptide hydrolysis solution and of any amino acid reference solutions provided. Use 12:3:5 1–butanol–acetic acid–water as the developing solvent. Visualize the spot by spraying the dry chromatogram with freshly prepared 5% cyclohexylamine in ethanol, drying it again, and dipping it in a freshly prepared solution of ninhydrin in acidic acetone.

Use only the organic phase.

The chromatogram should be partly dried before using HCl.

Allow the chromatography chamber to equilibrate for 10 minutes or longer before development.

The ninhydrin solution contains 0.25 g ninhydrin for every 7 ml acetic acid and 93 ml acetone.

After the chromatogram has air dried, heat it at 65° for five minutes and record the colors and R_f values of all of the known and unknown amino acids.

Identify the unknown amino acids and DNP-peptide by comparing them with the knowns and by referring to Table 1.

Experimental Variations

Evaporate the solution by the procedure described in part B.

1.   The C-terminal amino acid can be identified by (a) evaporating the aqueous layer that remains after the ether extraction of the DNP-peptide hydrolysate; (b) adding 0.05 ml of water; and (c) chromatographing this solution.

2.   For more certain identification, the amino acids can be chromatographed in two dimensions. One method is described in *J. Chem. Educ.* 48, 275 (1971).

3.   Thin-layer chromatography has been used to separate DNP-amino acids. A simple procedure is described in R. M. Roberts et al., *An Introduction to Modern Experimental Organic Chemistry,* 2nd ed. (Holt, Rinehart and Winston, 1974), pp. 417–18.

MINILAB 9   Isolation of a Protein from Milk

Avoid an excess of HCl—the pH should be about 4.7 when enough HCl has been added.

The precipitation of casein also occurs during cheesemaking processes such as the curdling of milk by rennet.

The HNO_3 test is called the xanthoproteic reaction and is a good test for proteins. The Biuret reaction and Millon's test for proteins may also be applied.

In a large beaker, dilute 100 ml of *skim* milk with an equal volume of water and add 3M HCl dropwise until no more precipitate forms. Allow the precipitate to settle, decant the supernatant liquid, and press the crude protein (casein) with a clean cork to remove excess water. Decant the water, transfer the protein to a small beaker, and cover it with 95% ethanol. Using a flat-bottomed stirring rod, triturate (stir and crush) the casein until it is finely divided, and collect it by vacuum filtration. Repeat the trituration with fresh ethanol, vacuum filter, and wash the casein with two portions of acetone. Let it air dry on a sheet of filter paper. Place about 0.1 g of the casein in a test tube, add 3 drops of concentrated nitric acid, warm the tube slightly, and record your observations.

Topics for Report

1.   Give the name and structure of your dipeptide and justify your conclusion.

2.   (*a*) Write equations illustrating the reactions undergone by your dipeptide during the experiment. (*b*) Explain why the DNP-peptide can be extracted into ethyl ether at low pH but not at high pH.

3.   Write a reasonable mechanism for the reaction of your dipeptide with 2,4-dinitrofluorobenzene. What class of reactions does this illustrate? Why is the reaction carried out in slightly alkaline solution?

4.   (*a*) Some amino acids, such as lysine and tyrosine, yield dinitrophenylated derivatives even when they are not at the end of a peptide chain. Explain, giving structures for the DNP derivatives. (*b*) Such derivatives do not usually interfere with the identification of the DNP *N*-terminal amino acids, since they are not extracted from the aqueous hydrolysis mixture at pH 1 by ethyl ether. Explain.

5.   Draw a flow diagram for the preparation and hydrolysis of the DNP-peptide.

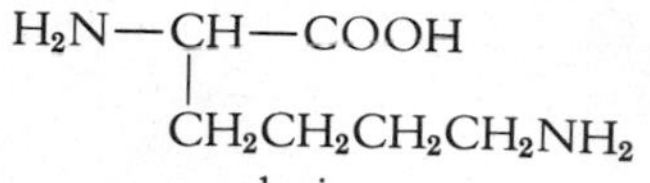

$$H_2N-CH-COOH$$
$$|$$
$$CH_2CH_2CH_2CH_2NH_2$$
lysine

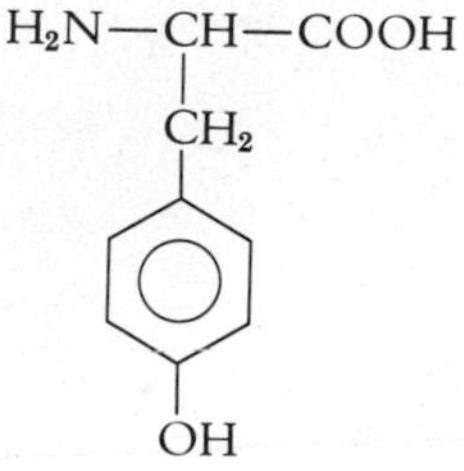

$$H_2N-CH-COOH$$
$$|$$
$$CH_2$$

OH
tyrosine

Library Topics

1.   Tell how each of the following enzymes or chemical reagents can be used to elucidate the structure of a pep-

Glu—Gln—Arg—Leu—Gly—Asn—Gln—Trp—Ala—Val—Gly—His—Lcu—Met—NH$_2$
bombesin

tide chain: trypsin, chymotrypsin, cathepsin C, and cyanogen bromide. Show the residues that would result if the peptide bombesin were treated in succession with trypsin and chymotrypsin. Show the residues that would result if bombesin were cleaved with cathepsin C, both before and after the *N*-terminal residue was removed with phenylisothiocyanate.

2.   Report on the structures, sources, and uses or biological functions of the following peptides: angiotensin II, gramicidin S, and bacitracin.

Bombesin is obtained from the skin of frogs of the genus *Bombina*.

PART II

Alternate and Supplementary Experiments

The Isolation of Carvone from Spearmint and Caraway Oils

Stereoisomerism. Separation of Natural-product Constituents.

Operations

⟨OP–5⟩ Weighing

⟨OP–14⟩ Evaporation

⟨OP–16⟩ Column Chromatography

⟨OP–31⟩ Optical Rotation

⟨OP–33⟩ Infrared Spectrometry

Experimental Objectives

To isolate carvone from either spearmint or caraway oil by column chromatography.

To observe and compare the physical and spectral properties of the carvones obtained from the two sources.

Learning Objectives

To learn about some of the methods used to separate the components of natural products.

To learn about the origin and biochemical significance of molecular dissymmetry.

SITUATION

Read an account of the Daedalus *project in* New Scientist 63, *522 (1974).*

You are a biochemist on the scientific team of the starship *Daedalus,* the first manned spacecraft from planet Earth to enter the planetary system of Barnard's Star. The third planet shows marked similarities to Earth; your mission is to explore its surface for signs of life and to describe and analyze any life forms you come across. As your shuttlecraft touches down on the planetary surface, the landing party is greeted by a small band of erect anthropoids whose resemblance to humans is astonishing—they even speak a barbaric-sounding language they call *Hsil'gne.* After your linguist has established a rudimentary means of communication with the inhabitants, you learn that the native name of the planet is *Arret,* and your party is invited to their village for a feast.

The meal is similar to an earthly dinner, including various boiled vegetables (overdone), a kind of bread covered with seeds, the roasted flesh of a domesticated animal called the *woc,* and an herb tea. There are some startled reactions from your team members as they begin eating, since the flavor of the food is quite unexpected; some dishes seem flat and nearly tasteless, while others exhibit a bitter aftertaste. The *yawarac* seeds on the bread taste like spearmint and the *t'nim* tea has an unappetizing caraway flavor. Shortly after dinner you experience a severe case of indigestion and are soon as hungry as if you had eaten nothing at all.

The Arretians hold these feasts every four years to promote their candidates for the post of Deah Oh'cnoh, and to celebrate the exploits of former tribal chieftains like B'nodnyl Nosn'hoj and D'lareg Drof.

After much thought, you have come up with a theory that you believe will account for your experience. It appears likely that molecular evolution on *Arret* has paralleled that on earth with one significant exception: all of the biologically active molecules on *Arret* are mirror image forms of the corresponding molecules on Earth. To test this theory, you have distilled an essential oil from *yawarac* seeds and another from the *t'nim* leaves used in brewing the herb tea. If your theory is correct, the Arretian *yawarac* oil should contain (*R*)-carvone instead of the (*S*)-carvone found in oil distilled from Earthly caraway seeds, and the *t'nim* oil should contain (*S*)-carvone rather than the (*R*)-carvone occurring in spearmint oil.

Due to the high cost of shipping the oils from Arret, you will use Earthly spearmint oil to simulate Arretian yawarac *oil, and caraway oil to simulate* t'nim *oil.*

BACKGROUND

Alice Through the Looking-glass: Dissymmetry and Life

One of the great unsolved mysteries of life concerns the origin of optically active compounds. Granted that molecular **dissymmetry** exists in living systems and is in fact necessary for life, how did such dissymmetry come about? Was there some "molecular Adam," a single dissymmetric molecule that gave rise to all molecular dissymmetry on Earth? Or is molecular dissymmetry a phenomenon that appeared independently at many different locations as a consequence of some kind of fundamental dissymmetry in our universe?

dissymmetry: The inability of an object to be superimposed on its mirror image.

chiral = dissymmetric

See Library Topic 2.

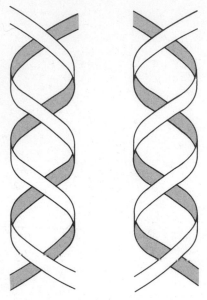

Figure 1.   Left-handed and right-handed double helices

The oil from both sources is approximately 55% carvone.

carvone

Many theories of the origin of **chiral** molecules have been proposed, some of them allowing at least the possibility that life forms on other planets may be made up of molecules that are mirror-images of our own. But convincing proof of such theories is hard to come by, and the answer may never be known.

Life on earth is intimately associated with molecular dissymmetry. Most of the molecules of life—proteins, carbohydrates, nucleic acids, and enzymes—are chiral and are built up of smaller units that are also chiral. A strand of DNA, for example, consists of two long chains each having a backbone of D-2-deoxyribose molecules twisted into a right-handed double helix. DNA and RNA regulate the synthesis of proteins from L-amino acids, which are combined in specific sequences inside the cellular structures called ribosomes. Some of these proteins make up the enzymes that assist in the digestion of carbohydrates, yielding D-glucose to be used by the body for fuel.

It is conceivable that life could be based on a mirror-image DNA containing L-2-deoxyribose and twisted into a left-handed double helix; but then protein synthesis could utilize only D-amino acids and the corresponding enzymes digest only L-carbohydrates. In other words, if the configuration of one link in the chain of life is reversed, all the rest must be reversed as well. If we could, by some magical contrivance, pass through the looking-glass as Alice did, all the people, plants, and other organic matter in the looking-glass world would presumably be constructed of these mirror-image molecules. We could not survive in such a world (though Alice did, in Lewis Carroll's imagination) since digestion, metabolism, reproduction, and other life processes involving chiral molecules would be inhibited or prevented entirely.

METHODOLOGY

Carvone is a natural ketone found in the essential oils of both caraway seeds and the spearmint plant, in association with other terpenoids such as limonene and the phellandrenes. Having a single chiral carbon atom, carvone exists in two enantiomeric forms that may exhibit different physical and physiological properties. The fact that caraway and spearmint exhibit distinctively different odors suggests that the carvones occurring in the two kinds of oils are stereoisomers. In this experiment, you will isolate the carvone from either spearmint or caraway oil so that you can compare the properties of both kinds of carvone and establish that they are in fact enantiomeric.

As described in Experiment 22, essential oils can be isolated from natural products by steam distillation, extraction, and other methods. The components of an essential oil can then be separated by any of a number of separation techniques, including vacuum fractional distillation, column chromatography, and preparative gas chromatography. In this experiment, the relatively polar carvone can be separated from the hydrocarbon components (mostly limonene) of spearmint or caraway oil by column chromatography. Carvone is adsorbed more strongly by silica gel than the hydrocarbons, which can be eluted from the column by a nonpolar solvent containing petroleum ether. The polarity of the eluent can then be increased, by adding methylene chloride in steps, to elute the carvone. After the carvone fractions are evaporated, the optical rotations and infrared spectra of the two kinds of carvone can be obtained and compared.

Enantiomers should, when pure, exhibit identical spectra and physical constants except for the direction in which they rotate polarized light.

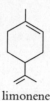

limonene

(S)-carvone should exhibit a positive optical rotation and *(R)*-carvone a negative rotation.

PRELAB EXERCISES

The Prelab assignment for all Part II experiments will include reading the experiment and understanding its objectives, reading or reviewing the operations to be used, and (at your instructor's request) preparing a checklist for the experiment. Only Prelab assignments in addition to these will be specified in future Part II experiments.

PROCEDURE

[Adapted from the *Journal of Chemical Education, 51,* 274 (1974), with permission.]

Prepare a chromatography column ⟨OP–16⟩ using silica gel in a slurry with petroleum ether (b.p. 65–90°). Place about 2 g of oil on the column and elute it with the following solvents in succession:

1. 25 ml petroleum ether;

2. 50 ml 10% methylene chloride–petroleum ether;

3. 25 ml 20% methylene chloride–petroleum ether;

4. 125 ml 50% methylene chloride–petroleum ether.

The volume of the last fraction should be more than 125 ml because of the petroleum ether initially present in the column.

Collect four fractions equal in volume to the eluent volumes, evaporate the solvent ⟨OP–14⟩ from fractions 2 and 4, and weigh the residues ⟨OP–5⟩ in these fractions. Obtain an infrared spectrum of the carvone ⟨OP–33⟩ in fraction 4 and measure its optical rotation ⟨OP–31⟩ in ethanol. Be certain to record the direction (+ or −) of the rotation as well as its magnitude. Compare your results with those obtained by students using the other essential oil.

Describe the odors of both kinds of carvone from fraction 4, and of both kinds of limonene from fraction 2.

Experimental Variations

A 30% GE SE-30–Chromosorb W column is suitable for GLC.

1.   Additional physical properties of the two carvones can be obtained and compared, including their boiling points ⟨OP–29a⟩ and refractive indices ⟨OP–30⟩. Their ultraviolet spectra ⟨OP–35⟩ can also be recorded and compared, and their purity analyzed by gas chromatography ⟨OP–32⟩.

2.   The hydrocarbons (predominantly limonene) in fraction 2 from both essential oils can be analyzed by GLC and their physical properties and spectra can be compared.

3.   Carvone can be isolated from caraway oil and spearmint oil by vacuum fractional distillation, as described in *J. Chem. Educ. 50*, 74 (1973).

4.   It has been reported that approximately 10% of the population is unable to distinguish between (*R*)- and (*S*)-carvone by odor. Students can test themselves or determine the percentage of students in each laboratory section having this characteristic.

MINILAB 10

Properties of Tartaric Acid Stereoisomers

Use water as the solvent for optical rotation measurements.

Obtain the melting point ⟨OP–28⟩ and optical rotation ⟨OP–31⟩ of one of the following: D-, L-, DL-, or *meso*-tartaric acid. Construct molecular models representing the molecule(s) present in each form and draw Fischer projec-

tions for each. Determine the Cahn-Ingold-Prelog configurations (R or S) of the chiral carbons in each case. Calculate
the specific rotation of your isomer and, using class data,
compare and discuss the properties of the different forms.

Topics for Report

1. **(a)** Calculate the specific rotation of your carvone
and interpret its infrared spectrum. **(b)** Using literature
values for the specific rotation of the carvones, calculate the
apparent optical purity of your product.

The *Merck Index* reports $-62.46°$ for
the specific rotation of (R)-carvone
and $+61.2°$ for (S)-carvone. In
theory, the two values should be
identical in magnitude.

2. Construct molecular models of both forms of carvone and draw stereochemical projections for each. Indicate
the chiral carbon atom, specify its configuration (R or S) in
each case, and tell which form is present in your essential oil.

cholic acid

3. The steroid cholic acid is said to have 2048 possible
configurations, of which only one occurs naturally. **(a)** Indicate each chiral carbon atom on the cholic acid molecule
with an asterisk, and perform a calculation to confirm this
isomer number. **(b)** Look up the stereochemical structure of
natural cholic acid in the *Merck Index* and give the R-S configuration at each chiral carbon where possible.

4. The Murchison meteorite, which fell near Murchison, Australia, in 1969, was found to contain traces of the
amino acid alanine. When a trifluoroacetyl derivative of this
alanine was treated with (S)-2-butanol and the resulting
esters were passed through the column of gas chromatograph, two separate peaks having identical peak areas were
observed. **(a)** Draw stereochemical structures of the two
compounds formed in the esterification reaction. **(b)** Discuss
the probability that the amino acid was biotic (derived from
once living matter).

trifluoroacetylalanine

2-butanol

Library Topics

1. Read the articles in *Science 172,* 1043 and 1044 (1971) describing how the odor difference between (*R*) and (*S*)-carvone was established. Summarize the experimental evidence that led to the conclusion that the difference was real and not due to residual impurities.

2. Read the articles in *J. Chem. Educ. 49,* 448 and 455 (1972) and write a research paper on the origin and consequences of optical activity. Use the bibliographies included with the articles to locate additional material.

Identification of an Unknown Diene by Preparing its Diels-Alder Adduct with Maleic Anhydride

Conjugated Dienes. 1,4-Addition. Diels-Alder Reaction. Dicarboxylic Acids.

Operations

⟨OP–5⟩ Weighing
⟨OP–7⟩ Heating
⟨OP–7a⟩ Refluxing
⟨OP–12⟩ Vacuum Filtration
⟨OP–21⟩ Drying Solids
⟨OP–23⟩ Recrystallization
⟨OP–28⟩ Melting Point

Experimental Objective

To characterize an unknown hydrocarbon by preparing its Diels-Alder adduct with maleic anhydride and hydrolyzing the adduct to a dicarboxylic acid.

Learning Objectives

To learn a procedure for conducting a Diels-Alder cycloaddition.

To learn about the characteristics and methodology of the Diels-Alder reaction.

To learn about some natural conjugated dienes and trienes.

SITUATION

You are a plant biochemist studying the relationship between the chemical constitution of plants and their botanical classification. On a recent botanical excursion you col-

maleic anhydride

β-myrcene and allo-*ocimene have an additional double bond, but only two of their three double bonds are involved in the Diels-Alder cycloaddition.*

$$CH_3C=CHCH=CHC=CHCH_3$$
$$\quad\ \ |CH_3 \qquad\qquad |CH_3$$

allo-ocimene

$$CH_3C=CHCH_2CH_2CCH=CH_2$$
$$\quad\ \ |CH_3 \qquad\qquad |CH_2$$

myrcene

$$CH_3C=CHCH_2CH=CCH=CH_2$$
$$\quad\ \ |CH_3 \qquad\qquad |CH_3$$

β-ocimene

α-terpinene

α-phellandrene   β-phellandrene

lected leaves from different kinds of eucalyptus trees for chemical analysis. When you steam distilled the leaves to obtain their essential oils, you discovered that the oils from most of the species yielded varying amounts of a hydrocarbon having the molecular formula $C_{10}H_{16}$. The ultraviolet spectrum of this hydrocarbon indicates that it is a conjugated diene, and you have narrowed down the possibilities to four naturally occurring conjugated hydrocarbons, α-terpinene, α-phellandrene, β-myrcene, and *allo*-ocimene. All four yield Diels-Alder adducts with maleic anhydride, and these adducts can be hydrolyzed to dicarboxylic acids. By preparing a Diels-Alder adduct and measuring its melting point, you should be able to identify your diene. Since information about the corresponding dicarboxylic acid is scarce, you can also hydrolyze your adduct to the acid and measure its melting point for future reference.

BACKGROUND

Dienes and Trienes in Nature

Hydrocarbons having two or three double bonds are found in many essential oils distilled from plants. If two or more of their double bonds are conjugated, they usually can form Diels-Alder adducts that can be used to identify them. Maleic anhydride is the dienophile most often used for this purpose, since it is reactive enough to combine with most dienes and forms an adduct that can be converted to a dicarboxylic acid for further characterization.

β-Myrcene is a pleasant-smelling liquid that was first obtained from bay leaves (*Myrcia acris*) and is also found in hops, verbena, and in certain lemongrass oils. β-Ocimene was first isolated from the Javanese oil of basil (*Ocimium basilicum*) and is usually found in combination with *allo*-ocimene. The latter is also obtained by pyrolysis of α-pinene, the most abundant component of turpentine, while myrcene is a major pyrolysis product from β-pinene. Among the common cyclic dienes are α-terpinene, found in the essential oils of cardamom, marjoram, and coriander; α-phellandrene, from oils of bitter fennel, ginger grass, star anise, cinnamon, and elemi; and β-phellandrene, from oils of water fennel, lemon, and Japanese peppermint.

METHODOLOGY

The Diels-Alder reaction is classified as a 4 + 2 cycloaddition reaction. One component, the *diene,* contributes four atoms to the six-membered product ring, or *adduct;* the other, the *dienophile,* contributes two. The four centers of the diene must be connected by two conjugated double bonds, while the dienophile must have a double or triple bond connecting the two carbon centers involved in the cycloaddition. The reaction can be viewed as a 1,4-addition to a conjugated diene in which the dienophile acts as the electrophile.

Diels-Alder reactions can be conducted by merely heating the diene and dienophile together and then isolating the product. A solvent is often used to moderate the reaction (which may be violently exothermic); catalysts may be useful for accelerating slow reactions. In this experiment, maleic anhydride will be refluxed with an excess of the diene in a small volume of ethyl ether. The adduct crystallizes from the reaction mixture on cooling, and can be recovered by vacuum filtration and recrystallized from methanol. Since anhydrides are hydrolyzed to dicarboxylic acids in the presence of water, it will be important to use dry glassware and to keep out moisture during the reaction and workup. Once the adduct has been purified, it can be converted to the dicarboxylic acid by boiling it with dilute aqueous sodium hydroxide.

A 4 + 2 cycloaddition reaction

diene   dienophile   adduct

transition state

Maleic anhydride must be protected from moisture, since it hydrolyzes readily to maleic acid.

PRELAB EXERCISE

Calculate the mass of 35 mmol of maleic anhydride and the mass of 50 mmol of the diene ($C_{10}H_{16}$).

Reactions and Properties

The actual structure of the adduct depends on the structure of the unknown diene.

Table 1.   Physical properties

	M.W.	m.p.	b.p.	d.
maleic anhydride	98.1	53	202	
ethyl ether	74.1	−116	34.5	0.714

Table 2.   Melting points of maleic anhydride adducts

β-myrcene	33–34°C
α-terpinene	60–61°C
allo-ocimene	83–84°C
α-phellandrene	126–27°C

Hazards

Maleic anhydride is a powerful irritant that can cause serious burns; inhalation can result in pulmonary edema. Avoid contact with skin, eyes or clothing; do not inhale dust or vapors. Ethyl ether and petroleum ether are both highly flammable; they must not be used anywhere near an open flame or allowed to contact hot surfaces.

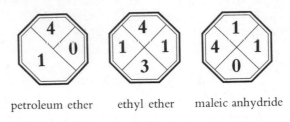

petroleum ether   ethyl ether   maleic anhydride

PROCEDURE

Preparation of a Diels-Alder Adduct.   Combine 35 mmol of maleic anhydride (keep dry!) with 50 mmol of the unknown

WARNING   *Maleic anhydride causes severe burns, so keep it off your skin and clothing.*

diene and 10 ml of ethyl ether in a 50 ml boiling flask, and reflux the mixture ⟨OP–7a⟩ on a steam bath ⟨OP–7⟩ for 45 minutes. Transfer the reaction mixture, while it is still hot, to an Erlenmeyer flask and let it cool to room temperature. Then cool the flask further in ice water to allow complete crystallization. Collect the adduct by vacuum filtration

Observe the nice display of crystal formation as the solution cools.

⟨OP–12⟩, washing the crystals on the filter with 10 ml of cold petroleum ether. Recrystallize ⟨OP–23⟩ the adduct from methanol (avoid prolonged boiling). Obtain the melting point ⟨OP–28⟩ of the dried product, and weigh it ⟨OP–5⟩.

Hydrolysis of the Adduct (Optional).   Reflux ⟨OP–7a⟩ about 1 gram of the adduct in 30 ml of water, containing 3 ml of 1 M sodium hydroxide, for 45 minutes. Cool the solution and acid-

Shake the reactants occasionally.
Filter the reaction mixture by gravity if some of the solid remains undissolved at the end of the reaction period.

Do not use a burner if ethyl ether or petroleum ether is in use nearby.

WARNING

---

ify it to pH 2 with about 1 ml of 3M hydrochloric acid. Allow the solution to stand until crystallization is complete, then collect the dicarboxylic acid by vacuum filtration ⟨OP–12⟩ and purify it by recrystallization ⟨OP–23⟩ from boiling water. Dry ⟨OP–21⟩ the product, measure its melting point ⟨OP–28⟩, and weigh it ⟨OP–5⟩.

Experimental Variations

Diels–Alder syntheses involving the following reactions can be carried out in addition to, or in place of, this experiment: **(a)** the reaction of cyclopentadiene with maleic anhydride [described in *J. Chem. Educ. 40*, 40 (1963)], **(b)** the reaction of anthracene with tetracyanoethylene [*J. Chem. Educ. 40*, 543 (1963)], or **(c)** the reaction of butadiene with maleic anhydride [*J. Chem. Educ. 45*, 55 (1968)].

Tetracyanoethylene is very toxic (evolves HCN) and must be handled with great care.

The Diels-Alder Reaction between Maleic Anhydride and Furan

MINILAB 11

Dissolve 2 g of maleic anhydride in 5 ml of dioxane in a test tube and add 1.5 ml of furan. Mix and set the tube aside

---

WARNING   *Avoid contact with dioxane and furan; do not breathe vapors.*

---

furan

Dimerization of 1,3-butadiene

The use of molecular models may be helpful.

for 24 hours or more, then recover the adduct by vacuum filtration, washing it with a little ethyl ether for quick drying. Obtain the melting point of the adduct and write an equation for the reaction.

Topics for Report

1.   Give the name and structure of the unknown diene, and draw structures for the adduct and its hydrolysis product.

2.   The side reaction most often encountered in Diels-Alder syntheses is dimerization or polymerization in which the diene also acts as a dienophile. For example, butadiene can react with itself to yield 4-vinylcyclohexene as shown. Draw the structures of four possible Diels-Alder dimers of your diene.

3.   Draw structures for all possible Diels-Alder adducts with maleic anhydride of the dienes named in the Background. Which diene will *not* form an adduct? Why?

4.   Write a flow diagram for the experiment.

5.   Write a mechanism for the 1,4-addition of bromine to butadiene and compare it with the mechanism for the Diels-Alder reaction of butadiene with maleic anhydride. Point out any similarities or differences.

Library Topic

A number of polychlorinated insecticides, such as dieldrin, aldrin, and chlordane, are synthesized using one or more Diels-Alder reactions. Report on the manufacture and use of these insecticides, giving equations for their synthesis from cyclopentadiene.

The Synthesis of 4-Methoxyacetophenone by Friedel-Crafts Acylation of Anisole

EXPERIMENT 27

Electrophilic Aromatic Substitution. Preparation of Aryl Ketones. Friedel-Crafts Reactions.

Operations

⟨OP–5⟩ Weighing

⟨OP–7a⟩ Refluxing

⟨OP–10⟩ Addition

⟨OP–12⟩ Vacuum Filtration

⟨OP–13b⟩ Separation

⟨OP–14⟩ Evaporation

⟨OP–19⟩ Washing Liquids

⟨OP–20⟩ Drying Liquids

⟨OP–22b⟩ Trapping Gases

⟨OP–25a⟩ Semimicro Distillation

⟨OP–28⟩ Melting Point

Experimental Objective

To prepare 4-methoxyacetophenone by the Friedel-Crafts acylation of anisole.

Learning Objectives

To learn about the characteristics and applications of the Friedel-Crafts reaction and about the experimental conditions for Friedel-Crafts acylations.

To learn about the properties and uses of some aromatic ketones.

"crataegon"

acetophenone   α-chloroacetophenone

The acronym MACE is derived from *M*ethylchloroform chloro*ACE*tophenone.

benzophenone

Zincke's attempted synthesis

$$PhCH_2Cl + ClCH_2COOH \xrightarrow[\text{benzene}]{\text{Ag}}$$

$$PhCH_2CH_2COOH$$

SITUATION

You are a chemist working for a perfume manufacturer, searching for useful perfume ingredients from natural sources. You have just isolated an odorous substance, *crataegon,* from hawthorn flowers (*Crataegus spp.*). Spectral analysis of crataegon indicates that it is the *para*-methoxy derivative of acetophenone, which has itself been used in perfumes. Now you wish to prepare a synthetic sample of 4-methoxyacetophenone to make certain that the odor is not due to traces of natural impurities, and to see if a commercial synthesis is feasible.

BACKGROUND

Friedel, Crafts and Phenones

Aromatic ketones having the carbonyl group adjacent to the benzene ring are frequently called phenones, the simplest member of this group being acetophenone. Acetophenone is a pleasant-smelling liquid that has been used to impart an "orange-blossom" odor to perfumes and is also prescribed as a sleep-producing drug under the generic name hypnone. Acetophenone was first prepared by Friedel in 1857—long before the discovery of the Friedel-Crafts reaction—by distilling a mixture of calcium benzoate and calcium acetate. When acetophenone is chlorinated on the side chain it yields α-chloroacetophenone, a liquid with distinctly unpleasant properties that make it a useful ingredient in MACE and other "tear gas" preparations. α-Chloroacetophenone is classed as a lachrymator, a chemical with tear-producing properties. Benzophenone is a white solid with a geranium-like odor that has been used as a fixative for perfumes and as a starting material for the manufacture of drugs and insecticides. It is the first aromatic ketone known to have been synthesized by a "Friedel-Crafts" reaction—Theodor Zincke prepared it accidentally, when he was trying to make something else, by heating benzoyl chloride with a metal in benzene.

Most of the phenones can be prepared by a Friedel-Crafts reaction of benzene or its derivatives with an appropriate acylating agent. The Friedel-Crafts reaction might well have been named the "Zincke reaction" if Theodor Zincke had understood the significance of an experiment that failed. In 1869, Zincke tried to synthesize 3-phenylpropanoic acid by combining benzyl chloride and chloroacetic acid in the presence of metallic silver (a variation of the Wurtz reaction). While carrying out the reaction with benzene as the solvent, Zincke observed, to his surprise, that a great deal of hydrogen chloride was evolved and that the major product was diphenylmethane instead of the expected carboxylic acid.

About four years later, a Frenchman named Charles Friedel was watching a student in Wurtz's laboratory perform a "Zincke reaction" using (appropriately) powdered zinc as the catalyst. When the reaction suddenly became violent, Friedel helped the student separate the solution from the zinc powder, thinking that removing the catalyst would moderate the reaction. To the astonishment of both, the reaction was just as violent in the absence of zinc! Although there is no record of his thought processes after this event, Friedel must have recognized its significance; in 1877, he and his collaborator, an American named Charles Mason Crafts, published a paper that marked the inception of the Friedel-Crafts reaction as one of the most important synthetic procedures in the history of organic chemistry. Friedel and Crafts' basic discovery was a simple one—it was a chloride of the metal, and not the metal itself, that catalyzed the reaction of organic halides with aromatic compounds. They found that anhydrous aluminum chloride was the most effective catalyst then available: it is still the catalyst of choice for most Friedel-Crafts reactions.

The "Zincke reaction"

$$PhCH_2Cl + PhH \text{ (benzene)} \xrightarrow{Ag}$$

$$PhCH_2Ph + HCl$$

See Experiment 14 for the story of another Wurtz reaction that "failed," with momentous consequences.

Crafts later returned to the U.S., where he became president of the Massachusetts Institute of Technology.

In Zincke's experiment, traces of metal chloride were formed during the reaction by the oxidation of the metal.

METHODOLOGY

The so-called "Friedel-Crafts reaction" is not a single reaction type, although the term has been most often applied to alkylations and acylations of aromatic compounds using aluminum chloride (or another Lewis acid catalyst) and a suitable alkylating or acylating agent.

A typical Friedel-Crafts acylation reaction uses a carboxylic acid chloride or anhydride as the acylating agent and anhydrous aluminum chloride as the catalyst. The acid anhydride usually provides better yields and a simpler workup than the acid chloride, but it requires more catalyst because aluminum chloride is tied up by complexing with the carboxylic acid formed during the reaction, as well as by reacting with the acylating agent and aryl ketone. Since acetic anhydride is readily available and gives high yields of product, it is often used instead of acetyl chloride for synthesizing acetophenone and its derivatives.

Many chemists regard any organic reaction catalyzed by aluminum chloride or a related catalyst as a Friedel-Crafts reaction.

Slightly more than one mole of $AlCl_3$ is used per mole of acyl chloride, while 2–3 moles may be needed per mole of anhydride.

Aluminum chloride complex with carbonyl compound

$$\overset{+}{O} - \overset{-}{Al}Cl_3$$
$$\parallel$$
$$R\overset{}{C}X$$

The reaction evolves gaseous HCl, so a gas trap is required.

In carrying out the acylation reaction, an excess of the aromatic hydrocarbon itself can be used as the solvent. Alternatively, another solvent such as methylene chloride, nitrobenzene, or carbon disulfide can be used. Since both of the latter have undesirable properties (nitrobenzene is high boiling and very toxic, CS_2 is toxic and extremely flammable), methylene chloride will be used as the solvent in this preparation. The acylation is highly exothermic, so it will be carried out by slowly adding acetic anhydride to the other reactants, then refluxing for a short time to complete the reaction. Pouring the product into ice water will decompose an aluminum chloride complex of the product and transfer inorganic salts to the aqueous phase. The product can then be recovered by evaporating the organic solvent and distilling the residue.

Polysubstitution should not occur to the extent that it does with Friedel–Crafts alkylations, since the acyl substituent deactivates the product toward further substitutions. Substitution on an aromatic ring with an *ortho, para*-directing substituent can, in theory, yield a mixture of *ortho* and *para* products, but unless the acylation reagent is extremely reactive (and therefore less selective), the *para* product usually predominates by a 20:1 ratio or more. Therefore, the reaction should be a relatively clean one and result in a good yield of the expected product.

PRELAB EXERCISES

Calculate the mass of 110 mmol of aluminum chloride and the mass and volume of 50 mmol of anisole and 50 mmol of acetic anhydride.

Reaction and Properties

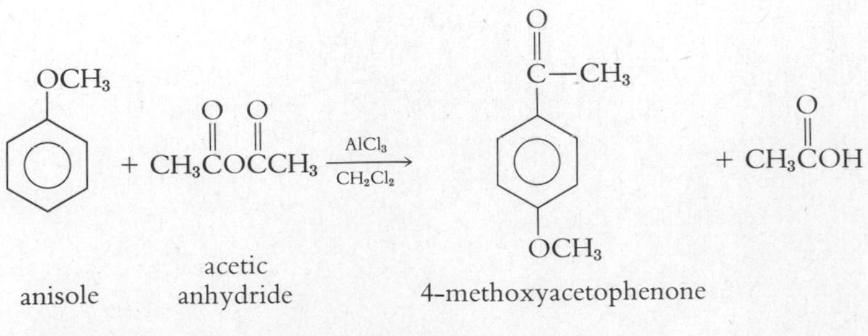

anisole

acetic anhydride

4-methoxyacetophenone

Table 1.   Physical properties

	M.W.	m.p.	b.p.	d.
anisole	108.2	−37.5	155	0.996
acetic anhydride	102.1	−73	140	1.082
aluminum chloride	133.3			
methylene chloride	84.9	−95	40	1.327
4-methoxyacetophenone	150.2	38−39	258	
			138−9^{15}	

Hazards

Acetic anhydride can cause severe burns to skin and eyes, its vapors are extremely irritating, and it reacts violently with water; avoid contact and do not breathe vapors. Aluminum chloride is a strong irritant and is especially dangerous to the eyes; it reacts with moisture to generate gaseous HCl. Do not breathe the dust, avoid contact with skin, eyes, and clothing. Methylene chloride is narcotic in high concentrations; avoid contact and inhalation.

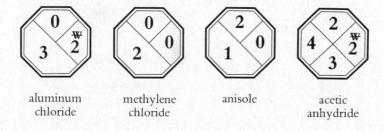

aluminum   methylene   anisole   acetic
chloride   chloride   anhydride

PROCEDURE

Reaction.   Assemble an apparatus for addition and reflux ⟨OP–10⟩, using a 100 ml round-bottomed flask, and fit it with a gas trap ⟨OP–22b⟩ attached at the top of the reflux condenser. Use dilute sodium hydroxide in the gas trap. Carefully weigh ⟨OP–5⟩ 110 mmol anhydrous aluminum chloride (*Care:* causes burns, may generate HCl gas) into a large, *dry* vial and immediately stopper the vial. Remove the reaction flask and add to it 35 ml of methylene chloride and 50 mmol of anisole; then add the aluminum chloride in *small* portions, through a dry powder funnel, shaking after each addition. Wash any adherent aluminum chloride into the

Clamp the Claisen connecting tube and the addition funnel securely so the entire apparatus can be shaken by moving the base of the ringstand back and forth. All glassware and other equipment must be dry.

Use a beaker of cold water to cool the mixture if necessary—it will boil if addition is too rapid.

The addition should take 10–15 minutes. A magnetic stirrer can be used if one is available.

Evolution of HCl should cease sometime during the reflux period.

You should be able to recover additional product by extracting the aqueous layer with methylene chloride ⟨OP–13⟩ and combining the extract with your organic layer.

The product is a solid at room temperature, but should remain liquid during the distillation. Semimicro vacuum distillation ⟨OP–26a⟩ can be used if desired.

flask with 5 ml of methylene chloride. Clamp the reaction flask in place at the Claisen connecting tube, measure 50 mmol of acetic anhydride into the addition funnel (be sure the stopcock is closed!), and turn on the water for the reflux condenser. Add ⟨OP–10⟩ the acetic anhydride slowly (20–30 drops a minute) to maintain gentle boiling while shaking the apparatus; use a beaker of cold water to moderate the reaction if necessary. When addition is complete, reflux ⟨OP–7a⟩ the reaction mixture gently for 30 minutes using a hot water bath or a steam bath.

Separation.   While the solution is still warm, pour it *slowly* (*Caution:* splattering) with vigorous stirring onto about 50 g of cracked ice in a beaker. Separate ⟨OP–13h⟩ the organic (bottom) layer, wash it ⟨OP–19⟩ in succession with 3M sodium hydroxide and saturated sodium chloride solutions, dry it ⟨OP–20⟩ over anhydrous magnesium sulfate, and evaporate ⟨OP–14⟩ the solvent.

Purification and Analysis.   Purify the product by semimicro distillation ⟨OP–25a⟩, using no cooling bath for the receiver. Transfer the distillate, while it is still liquid, to a watch glass or evaporating dish and let it crystallize. Wash the crystals on a vacuum filtration apparatus ⟨OP–12⟩ with some cold petroleum ether. Weigh the 4-methoxyacetophenone ⟨OP–5⟩ and measure its melting point ⟨OP–28⟩.

Experimental Variations

1.   p-Methoxyacetophenone can be used as a starting material for several multistep syntheses. D. A. Shirley (Bib–B3) gives procedures for nitrating and reducing it to yield 3-amino-4-methoxyacetophenone. One of these procedures is derived from a paper by Bogert and Curtin (*J. Am. Chem. Soc. 45,* 2161 (1923)), who converted the amino compound to the corresponding iodo and cyano derivatives via a diazotization procedure. (See figure on page 239.)

2.   The general method described in the Procedure can be used for acylating other aromatic compounds such as mesitylene and chlorobenzene.

Multistep syntheses with *p*-methoxyacetophenone

3.   A Friedel–Crafts alkylation of benzene with *t*-butyl alcohol is described in *J. Chem. Educ. 38,* 306 (1961).

4.   The infrared and NMR spectra of the product can be obtained and interpreted. A GLC of the product should show whether there is a significant amount of the *ortho* isomer present.

The melt technique described in
J. Chem. Educ. 50, *517 (1973) can be used for the IR spectrum.*

Topics for Report

1.   Write a mechanism for the reaction used in this experiment.

2.   Write a flow diagram for the synthesis of 4-methoxyacetophenone.

3.   When benzene reacts with *n*-propyl bromide in the presence of catalytic amounts of aluminum chloride, the major product is not *n*-propylbenzene but cumene (isopropylbenzene) instead. **(a)** Write a mechanism explaining this result. **(b)** Outline a Friedel–Crafts synthesis that can be used to prepare *n*-propylbenzene in high yield.

4.   Outline practical industrial syntheses for 4-methoxyacetophenone and for the phenones named in the Background, using benzene, ethylene, and methanol as organic raw materials.

Library Topics

1.   The Haworth synthesis of tetralones utilizes both intra- and intermolecular Friedel–Crafts acylation steps. De-

3,5,8-trimethyl-1-tetralone

scribe the Haworth synthesis and show how it can be used to prepare 3,5,8-trimethyl-1-tetralone from *p*-xylene.

2.   Write a short paper describing the uses of aldehydes and ketones in perfumery. Give examples illustrating the types of compound used and their synthesis from readily available starting materials.

Steric Effects on Orientation in the Nitration of Arenes

Arenes. Electrophilic Aromatic Substitution. Steric Effects of Alkyl Groups. Gas Chromatographic Analysis.

Operations

⟨OP–9⟩ Mixing

⟨OP–19⟩ Washing Liquids

⟨OP–20⟩ Drying Liquids

⟨OP–25⟩ Simple Distillation

⟨OP–32⟩ Gas Chromatography

Experimental Objectives

To carry out the nitration of an alkylbenzene, analyze the product mixture, and determine the ratio of *ortho* to *para* substitution.

To find out how the steric bulk of an alkyl group affects the *ortho/para* ratio.

Learning Objectives

To learn a simple procedure for nitrating aromatic hydrocarbons.

To learn about the occurrence and magnitude of steric effects in substitution reactions.

SITUATION

In 1902, Schultz and Flachslander carried out the nitration of ethylbenzene and painstakingly separated the mononitrated products by fractional distillation, isolating twice as much *ortho* as *para* product. Since there are twice as many *ortho* as *para* hydrogen atoms on each molecule, this 2/1 *ortho/para* ratio seems reasonable if one assumes that the

J. Prakt. Chem. [2] *66*, 160 (1902). The separation required 180 (!) successive fractionations.

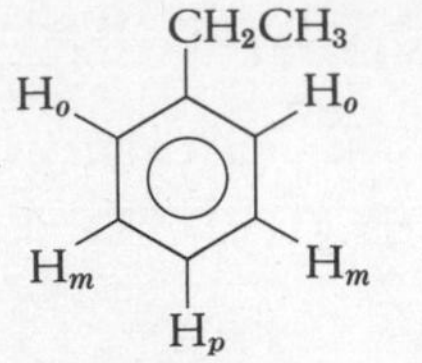

ethylbenzene

H_o = *ortho* hydrogens
H_m = *meta* hydrogens
H_p = *para* hydrogen

Mechanism of aromatic nitration

B: = some basic species

ethyl group exerts no steric effect on the outcome. However, some recent investigators (see Library Topic 1) have reported *ortho/para* ratios considerably less than 2—not only for ethylbenzene but for other alkylbenzenes as well.

Your "research group" will carry out the nitration of four different alkylbenzenes and measure the *ortho/para* ratio in each case to determine whether or not a steric effect is operating, and, if so, how the steric bulk of the alkyl group affects the results. In the process, you will be checking the validity of the results obtained by Schultz and Flachslander and other researchers who did not have the benefit of modern analytical instrumentation.

BACKGROUND

Steric Effects in Substitution Reactions

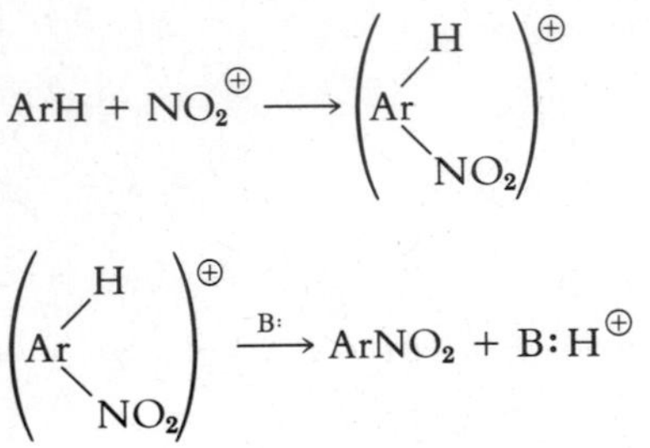

The nitration of benzene and benzene derivatives is an example of aromatic eletrophilic substitution, and proceeds by the initial formation of an electrophile, nitronium ion (NO_2^+), which attacks the benzene ring and displaces a hydrogen atom. Theories have been developed to account for the predominance of *ortho-para* substitution on rings containing electron–donating substituents and for *meta* substitution on rings with electron-withdrawing substituents. The reader may refer to any modern organic chemistry text for a discussion of such electronic effects of substituents.

In addition to their electronic effects, many substituents also exert *steric* effects in certain reactions—that is, effects that are caused by the size or "bulkiness" of a substituent. For example, 2-*t*-butylpyridine reacts with methyl iodide

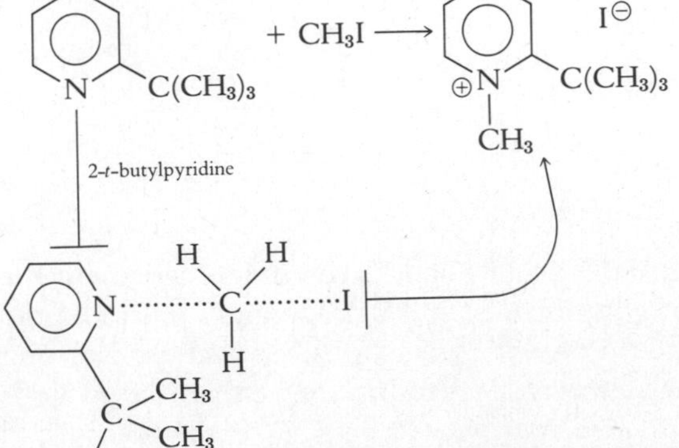

(Steric crowding destabilizes transition state.)

nearly 12,000 times more slowly than does 3-*t*-butylpyri-
dine. This is because the bulky *t*-butyl group in the 2 position interferes with the methyl group that is under attack by
the adjacent nitrogen atom. This crowding destabilizes the
transition state, and a less stable transition state means a
higher activation energy for the reaction, which results in a
lower rate. A *t*-butyl group at the 3 position, on the other
hand, is too far from the site of bond formation to exert a
significant steric effect on the reaction rate.

3-*t*-butylpyridine

A similar result might be expected for substitution on
aromatic rings; a bulky group already on the ring might retard the formation of an *ortho*-substituted product because
of crowding in the transition state leading to the intermediate aronium ion. Formation of the *meta* and *para* products
should not be affected by steric factors, since the substituent
is too far from the site of electrophilic attack to cause significant crowding. If such a steric effect does occur at the *ortho*
position, its magnitude should increase with the size of the
substituent—that is, the proportion of *ortho* product should
be smallest for a bulky group such as *t*-butyl and be greater
for successively smaller alkyl substituents.

aronium ion
for *ortho*
substitution

R = methyl, ethyl, isopropyl, *t*-butyl,
etc.

Merely predicting that such an effect *may* occur does
not insure that it *will* occur, however. Any such prediction
must be verified experimentally, and that is your objective
in this experiment. By determining the *ortho/para* ratio for
reactions involving methyl, ethyl, isopropyl, and *t*-butyl
substituents, you should be able to conclude whether or not
a steric effect is operating and to determine the relative magnitude of that effect for the different substituents.

METHODOLOGY

The nitration reaction will be carried out using a large
excess (about 5:1) of the aromatic hydrocarbon, which must
be removed by distillation after the reaction. The excess hydrocarbon will prevent the formation of di- and trinitrated
products by insuring that any nitronium ions formed will be
more likely to encounter (and react with) an arene molecule
than one of the nitroarenes. The reaction will be catalyzed
by acetic anhydride instead of by the more commonly used
sulfuric acid. Acetic anhydride reacts with nitric acid *in situ*
to form acetyl nitrate, which in turn decomposes to form the
nitronium ion and acetate ion. Because acetyl nitrate is dangerously explosive, it is important not to mix the acetic anhydride and nitric acid together, but to add them separately
to the alkylbenzene.

$$CH_3\overset{\overset{O}{\|}}{C}-ONO_2$$

acetyl
nitrate

Not all of the excess alkylbenzene will be removed during the distillation.

The gas chromatographic analysis can be carried out on a silicone column at 170°. The alkylbenzene peak should come off the column first, followed by the *ortho-, meta-,* and *para*-nitroalkylbenzenes. The *meta* peak should be very small and may not be observed.

PRELAB EXERCISES

If no alkylbenzene is assigned before the laboratory period, calculate the volumes for all four listed in Table 1.

Calculate the volume of 50 mmol of 15.9M nitric acid, 50 mmol of acetic anhydride, and 250 mmol of your assigned alkylbenzene.

Reaction and Properties

$R = CH_3—, CH_3CH_2—,$

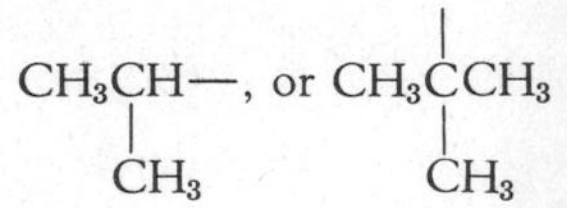

$CH_3\overset{\underset{\displaystyle CH_3}{|}}{CH}—,$ or $CH_3\overset{\underset{\displaystyle CH_3}{|}}{\underset{\displaystyle |}{C}}CH_3$

$Ac = CH_3CO—$

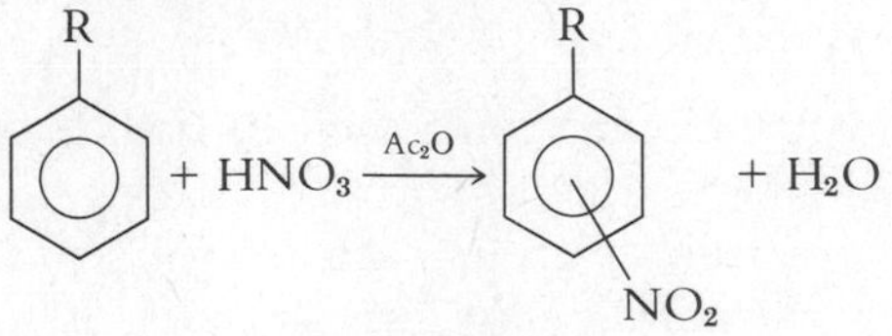

(mainly *ortho* and *para* products)

Table 1.   Physical properties

	M.W.	m.p.	b.p.	d.
acetic anhydride	102.1	−73	140	1.082
nitric acid (15.9M)	63.0			1.42
toluene	92.15	−95	111	0.867
ethylbenzene	106.1	−95	136	0.867
isopropylbenzene	120.2	−96	152	0.862
t-butylbenzene	134.2	−58	169	0.867

Concentrated (15.9M) nitric acid is about 70% HNO_3.

Hazards

Nitric acid causes severe burns on skin and eyes and is a strong oxidant; its vapors are very irritating and toxic. Avoid contact with skin, eyes, and clothing; do not breathe vapors; keep away from oxidizable substances. Acetic anhydride causes severe burns on skin and eyes and reacts violently with water. Avoid contact with skin, eyes, and clothing; do not breathe vapors; keep away from water. The

| nitric acid | acetic anhydride | toluene | ethylbenzene | isopropylbenzene |

alkylbenzenes are flammable and may irritate the skin; toluene is particularly hazardous to the eyes and its vapors are somewhat toxic. Aromatic nitro compounds are toxic if inhaled, ingested, or absorbed through the skin; they may explode if strongly heated. Avoid skin or eye contact with the product and do not inhale its vapors.

PROCEDURE

Each student in a group of four should be assigned one of the four alkylbenzenes listed in Table 1.

Reaction.   Measure 250 mmol of the assigned alkylbenzene into an Erlenmeyer flask and add *separately* (*Hood!*) 50 mmol

---

Do not mix acetic anhydride and nitric acid together; this may cause a violent reaction.

WARNING

---

of concentrated nitric acid and 50 mmol of acetic anhydride. Stopper the flask and let it stand at room temperature, with magnetic stirring or frequent shaking ⟨OP–9⟩, for 30 minutes.

Separation.   Wash the reaction mixture ⟨OP–19⟩ with water and then with two or more portions of 2M sodium carbonate (*Caution:* foaming) until no more carbon dioxide is evolved. Dry the organic layer ⟨OP–20⟩ and remove most of the excess alkylbenzene by simple distillation ⟨OP–25⟩.

Add the Na_2CO_3 solution slowly, with stirring or swirling.

It is not necessary to remove all of the alkylbenzene before the analysis. About 5–10 ml of liquid should remain in the reaction flask after distillation.

Avoid overheating near the end of the distillation —aromatic nitro compounds may explode when strongly heated.

WARNING

---

*Disregard the solvent (alkylbenzene)
peak in making your calculations.*

Analysis. Record a gas chromatogram ⟨OP–32⟩ of the product mixture, measure the peak areas, calculate the *ortho/para* ratio, and provide this information to the other students in your group.

Experimental Variations

1. A cooled mixture of nitrating acid containing 3 ml of concentrated nitric acid and 4 ml of concentrated sulfuric acid can be used in place of the nitric acid–acetic anhydride combination. The nitrating acid should be added to the alkylbenzene slowly, with cooling and swirling; the mixture should be stirred or shaken constantly during the reaction period.

2. A number of preparations for aromatic nitro compounds are described in *Vogel's Textbook of Practical Organic Chemistry,* 4th ed. (Bib–B4), pp. 624–629.

MINILAB 12

Nitration of Naphthalene

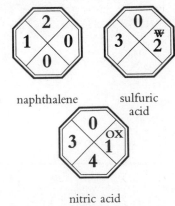

naphthalene   sulfuric
                   acid

nitric acid

Place 1 ml of concentrated nitric acid in a large test tube, cool it in ice, cautiously add 1 ml of concentrated sulfuric acid, and mix. Add (in small portions) 1 g of powdered naphthalene, shaking after each addition and keeping the temperature below 50°. Then shake the mixture in a 60° water bath for 20 minutes, stir it into 50 ml of ice water, and collect the precipitate by vacuum filtration. Purify the crude 1-nitronaphthalene by boiling it in 15 ml of water for 5 minutes (*Hood!*); then cool it in ice and collect the product by vacuum filtration. Write a mechanism for the reaction and explain why substitution takes place at the alpha position of naphthalene.

Topics for Report

1. Tabulate the percentage composition and *ortho/para* ratio obtained by nitrating each of the alkylbenzenes. Discuss and explain the results with reference to the Background material, and list the alkyl groups in order of their apparent steric size or "bulkiness."

2. Write mechanisms for the reactions leading to each product detected in your product mixture.

3. Write a flow diagram for the synthesis you carried out.

4. Predict the major product (or products) of the mononitration of (**a**) ethyl benzoate, (**b**) phenyl benzoate, (**c**) phenyl acetate, (**d**) *m*-nitrotoluene, and (**e**) anisaldehyde (*p*-methoxybenzaldehyde).

Library Topics

1. Read the article by Brown and Bonner on the nitration of alkylbenzenes, write a brief summary, and compare the class results and your conclusions with theirs.

J. Am. Chem. Soc. 76, 605 (1954).

2. Find a commercial method for preparing 2,4,6-trinitrotoluene (TNT) and describe the experimental conditions and apparatus used.

TNT

Structure Determination of a Natural Product by Side-chain Oxidation and Infrared Analysis

Oxidation of Aromatic Side Chains. Synthesis of Carboxylic Acids. Phase-transfer Catalysis. Spectrometric Analysis. Structure Determination.

Operations

⟨OP–5⟩ Weighing

⟨OP–7⟩ Heating

⟨OP–12⟩ Vacuum Filtration

⟨OP–13⟩ Extraction

⟨OP–13*b*⟩ Separation

⟨OP–21⟩ Drying Solids

⟨OP–23⟩ Recrystallization

⟨OP–28⟩ Melting Point

⟨OP–33⟩ Infrared Spectrometry

Experimental Objective

To determine the structure of an unknown aromatic compound by analyzing its infrared spectrum and by oxidizing it to a known carboxylic acid.

Learning Objectives

To learn how to carry out a side-chain oxidation with potassium permanganate using a phase-transfer catalyst.

To learn about some of the methods used in determining the structures of natural products.

To learn more about the interpretation of infrared spectra.

SITUATION

You are a natural-products chemist working in the artificial flavorings division of a large food manufacturer, looking for natural substances that may become useful flavoring ingredients. You have isolated several odorous substances from the essential oil of a pleasantly fragrant wild plant called sweet cicely and are attempting to determine the structure of one of them, an unsaturated aromatic compound you have tentatively named "cicelene". Preliminary analyses established that cicelene (molecular formula $C_{10}H_{12}O$) has a methoxy group and an unsaturated C_3H_5 side chain on a benzene ring. The position and structure of the side chain must be determined before the complete structure of cicelene is known.

Partial structure of "cicelene"

BACKGROUND

The Structure Puzzle — Taking
Molecules Apart and
Putting Them Back Together

Today, when a chemist can run an NMR or mass spectrum of an organic compound and often determine its structure in a matter of minutes, it is difficult to conceive of the time and effort that went into the structural analysis of even the simpler natural products. In a "classical" structure determination, the molecular formula is first found by elemental analysis and molecular-weight measurement. Then the compound is degraded (broken down) into smaller structural units that are isolated and, if possible, identified. Finding how the smaller units fit together to form the original molecule is an intellectual challenge that might be compared to putting together a jigsaw puzzle with some pieces missing, others that don't belong, and still others that have been chewed up by the family dog and are no longer recognizable. Finally, when enough information has been amassed to allow a reasonable structure to be proposed, that structure must usually be proven by an independent synthesis in which the compound is built up again, from known compounds, by reactions whose outcome can be reliably predicted.

In many cases, a structure determination may have involved the efforts of dozens or even hundreds of chemists over many decades, with the generation of much irrelevant or misleading information and many synthetic dead ends.

The advent of modern spectrometric methods has simplified the process enormously by providing detailed structural information that was not available to the chemists of earlier times. It is conceivable that someday one will only have to place a substance inside a "black box," press a button, and receive a computer printout listing the structures and quantities of all components present. Until that day comes, many of the traditional methods of structural analysis will remain useful.

METHODOLOGY

Strong oxidation of aliphatic side chains with $KMnO_4$ generally oxidizes them down to the ring, leaving a COOH group on the aromatic ring where the aliphatic group was originally located. In this experiment, the product of oxidizing cicelene will be one of three possible methoxybenzoic acids, all of which are solids whose melting points are recorded in the literature. Therefore, identifying the oxidation product as one of these three acids will establish the position of the cicelene side chain.

Recently it was discovered that many metal ions will dissolve in organic solvents if they are first complexed with an organic crown ether. This led to the use of "purple benzene" as an effective oxidizing agent. Potassium permanganate, when complexed with such an ether, will readily dissolve in organic solvents such as benzene; the resulting purple solution reacts rapidly with many oxidizable compounds, since the oxidant and the organic reactant are present in the same phase. Another way of obtaining purple benzene was reported in 1974 by Herriot and Picker, who added quaternary ammonium salts to dissolve the permanganate from an aqueous solution in a benzene phase, where it could oxidize water-insoluble organic compounds. The quaternary salt thus acts as a phase-transfer catalyst, transferring the permanganate ion from the aqueous to the organic phase where the reaction takes place. Its use allows the reaction to be carried out more rapidly (and under milder conditions) than would ordinarily be the case.

Because of the toxicity of benzene, the oxidation reaction will be carried out in toluene instead. As the reaction proceeds, the purple permanganate ion is reduced to a brown precipitate of manganese dioxide. The MnO_2, which would otherwise complicate the workup of the product, can be converted to soluble manganese(II) sulfate by adding so-

OCH₃

COOH

methoxybenzoic acid

(Substituents may be *ortho, meta* or *para*)

a crown ether with
solubilized $KMnO_4$

K⁺   MnO₄⁻

Tet. Lett. 1974, 1511.

See Experiment 12 for a discussion of phase-transfer catalysis.

Toluene is not readily attacked by quaternary ammonium permanganate solutions below 60°.

Removal of manganese dioxide

$$MnO_2 + NaHSO_3 + H^+ \rightarrow$$
$$MnSO_4 + H_2O + Na^+$$

dium bisulfite to the acidified solution. The product is extracted from the organic phase with aqueous sodium hydroxide (which converts it to the water-soluble sodium salt); then it is recovered by acidifying the aqueous layer and recrystallized from boiling water.

Infrared spectrometry is capable of distinguishing all possible C_3H_5 side chains, which must be either unsaturated or cyclic (a saturated aliphatic C-3 radical would have seven hydrogen atoms). The out-of-plane bending vibrations of vinylic protons are quite characteristic and can indicate the kind of substitution on a double bond, as shown in Table 1. Aromatic C—H out-of-plane bending vibrations also yield infrared bands in the same general region (Table 2), so it will be necessary to identify the aromatic bands before any conclusions can be drawn about the structure of the side chain. Once the position (*ortho, meta,* or *para*) of the side chain is established by oxidation, it should be an easy matter to identify any aromatic bands; then the remaining intense band(s) in the 1000–650 cm^{-1} region can be used, with the information from Table 1, to characterize the side chain.

Table 1.   Out-of-plane bending vibrations of vinylic C—H bonds

Structural type	Frequency range (cm^{-1})
$RCH{=}CH_2$	995–985 and 915–905
$RCH{=}CHR$ (*cis*)	730–665
$RCH{=}CHR$ (*trans*)	980–960
$R_2CH{=}CH_2$	895–885

R = alkyl or aryl

Table 2.   Out-of-plane bending vibrations of aromatic C—H bonds

Ring substitution	Frequency range (cm^{-1})
ortho	770–735
meta	810–750 and 710–690
para	840–810

PRELAB EXERCISE

Calculate the mass of 10 mmol of cicelene ($C_{10}H_{12}O$) and the mass of 40 mmol of potassium permanganate.

Reactions and Properties

$Q^+ \approx (CH_3(CH_2)_7)_3\overset{\oplus}{N}CH_3$

Table 3. Physical properties

	M.W.	m.p.	b.p.	d.
potassium permanganate	158.0			
o-methoxybenzoic acid	152.2	101		
m-methoxybenzoic acid	152.2	110		
p-methoxybenzoic acid	152.2	185		
toluene	92.2	−95	111	0.867

potassium permanganate   toluene

Hazards

Potassium permanganate is an irritant, and the solid can react violently with oxidizable substances; avoid contact of solution with skin or eyes. Toluene is flammable and is somewhat toxic on inhalation; the liquid is dangerous to the eyes. Sodium bisulfite reacts with acids to form sulfurous acid, which is toxic and corrosive; when strongly heated it yields poisonous sulfur dioxide gas.

PROCEDURE

Reaction. Mix 40 mmol of potassium permanganate with 50 ml of water in a 250 ml Erlenmeyer flask, then add 30 ml of toluene and 0.5 g of tricaprylmethylammonium chloride. Shake the mixture to dissolve the potassium permanganate. Add 10 mmol of cicelene in small portions (about 10 drops at a time), shaking vigorously and checking the temperature after each addition—do not let the temperature exceed 45°. When the addition is complete, shake the mixture vigorously in a 40–45° water bath ⟨OP–7⟩ for 30 minutes, then cool it to room temperature.

Use a magnetic stirrer for mixing if one is available.

Use a metal water bath or steam bath with the rings removed, and replenish the hot water periodically.

    Add solid sodium bisulfite, with shaking, until you can no longer detect the purple color of potassium permanganate. Then acidify the solution with 6M HCl and add solid sodium bisulfite in small portions with vigorous shaking until the brown precipitate of manganese dioxide has dissolved. Test the pH of the mixture after each bisulfite addition and, as necessary, add 6M HCl to keep it acidic. When the brown color has disappeared from both layers, adjust the pH of the aqueous layer to 2 with 6M HCl (use pH paper) and filter the mixture through a thin layer of glass wool into a separatory funnel.

Withdraw a little of the aqueous layer with a Pasteur pipet and test it with blue litmus paper or pH paper.

Approximately 6 g of sodium bisulfite and 10 ml or more of 6M HCl will be needed for the MnO_4^- and MnO_2 reductions. Avoid using a large excess of sodium bisulfite.

Separation and Purification. Separate the aqueous and organic layers ⟨OP–13*b*⟩ and extract the toluene upper layer ⟨OP–13⟩ with two 25-ml portions of 1M sodium hydroxide. Add 6M HCl (about 9 ml) to the combined sodium hydroxide extracts until the pH is about 2 and no more precipitate forms when another drop of HCl is added. Cool the mixture until precipitation is complete. Collect the product by vacuum filtration ⟨OP–12⟩ and recrystallize it ⟨OP–23⟩ from boiling water.

Analysis. Determine the mass ⟨OP–5⟩ and melting point ⟨OP–28⟩ of the dried ⟨OP–21⟩ methoxybenzoic acid. Obtain an infrared spectrum ⟨OP–33⟩ of a neat-liquid sample of cicelene sometime during the laboratory period.

Experimental Variations

1. To confirm the presence of unsaturation in the side chain of cicelene, a bromine/carbon tetrachloride test or a potassium permanganate test (C–7 or C–19, Part III) may be performed.

2. If a phase-transfer catalyst is not available, the oxidation can be carried out by the general procedure described in *Organicum*, p. 362 (Bib–B10) or in other sources of synthetic methods.

3. A procedure for oxidizing 1-decene to nonanoic acid using a phase-transfer catalyst is described in *J. Chem. Educ.* 55, 237 (1978) and in *J. Am. Chem. Soc.* 93, 199 (1971).

4. An interesting demonstration experiment illustrating the formation of purple benzene is described in *J. Chem. Educ.* 54, 229 (1977).

5. The NMR spectrum of cicelene in deuterochloroform can be obtained and interpreted.

Topics for Report

1. Draw the correct structure for cicelene and explain the reasoning that led to your conclusion. Include the possible structures for a C_3H_5 side chain (including geometric isomers) and tell how each incorrect structure was eliminated from consideration.

Excess $KMnO_4$ is needed, since some of it decomposes to MnO_2 while evolving oxygen.

2.   (*a*) Write a balanced equation for the reaction of cicelene with potassium permanganate. (*b*) Calculate the mass of potassium permanganate that is needed (in theory) to oxidize 10 mmol of cicelene, and the percent excess of permanganate that was actually used.

3.   Diagram the process of phase-transfer catalysis in permanganate oxidations as was done in Figure 1 of Experiment 12 for carbene reactions. Give equations for the reactions involved.

4.   Outline a synthesis of cicelene starting with benzene and using any necessary organic or inorganic reagents and solvents.

Such a synthesis, if successful, could definitively prove the structure you assigned to cicelene.

5.   The structure in the margin has been proposed for coniferyl alcohol, which can be obtained by the hydrolysis of coniferin, a natural product found in the sap of conifer trees. Assuming that coniferyl alcohol had not previously been reported in the literature, tell how you might go about proving its structure. Indicate what chemical tests and degradations might be carried out, describing the expected results and conclusions; and describe the information that could be derived from spectral analysis. Then show how it could be synthesized from readily available starting materials.

6.   Write a flow diagram for the procedure used in this experiment, showing how all of the reactants, catalysts, and by-products are separated from the product.

Library Topics

1.   (*a*) What is the traditional common name for cicelene? (*b*) Indicate the major natural sources of cicelene and give some of its uses.

2.   Capsaicin is a pungent substance found in tabasco, cayenne, and red pepper. Look up its structure and tell how the structure was determined, outlining the reaction steps involved.

Solvent Effects in the Hydrolysis of 1-Bromoadamantane

Alkyl Halides. Nucleophilic Aliphatic Substitution. Preparation of Alcohols. Reaction Kinetics.

Operations

⟨OP–5⟩ Weighing

⟨OP–6⟩ Measuring Volume

⟨OP–7⟩ Heating

Experimental Objective

To measure the hydrolysis rates of 1-bromoadamantane in different solvent mixtures and determine the relative ionizing power of the solvents.

Learning Objectives

To learn how to obtain kinetic data by titration, and how to use it to determine first-order rate constants.

To learn about the effect of molecular structure and solvent ionizing power on reaction rates.

To learn about some of the characteristics of adamantane and its derivatives.

SITUATION

Paul von R. Schleyer and his coworkers have suggested that 1-bromoadamantane (AdBr) is an ideal reference compound for determining solvent ionizing power (Y) in S_N1 reactions, since competing S_N2 and elimination reactions are ruled out by the unique cage structure of the adamantane nucleus. *t*-Butyl chloride is currently the most widely used reference compound.

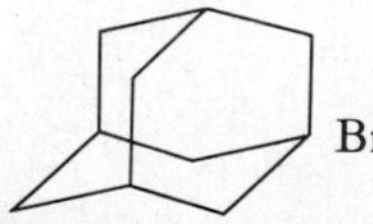

1-bromoadamantane
(1-adamantyl bromide, AdBr)

t-butyl chloride

adamantane

fragment of
diamond lattice

A **bridgehead** is the point at which
two or more fused rings are joined in
a bicyclic or polycyclic system.

Table 1.   Relative solvolysis rates of
tertiary halides

Substrate	Rel. Rate
	1
	10^{-3}
	10^{-6}

You have decided to test their theory by measuring hydrolysis rate of 1-bromoadamantane in different solvent mixtures. You can then assess the effectiveness of the solvents in promoting unimolecular substitution.

BACKGROUND

A Gem among Molecules

Adamantane, whose name is derived from a Greek word meaning "diamond," has molecules of elegant symmetry that can be regarded as fragments of a tetrahedral diamond lattice. Models of this unique molecule reveal that it consists of four interlocking chair-form cyclohexane rings, arranged somewhat like the four planes of a tetrahedron. A space-filling model is nearly spherical, and that molecular shape results in a particularly stable crystal lattice that is responsible for adamantane's melting point of 268°—unusually high for a hydrocarbon.

The extraordinary structure of adamantane has fascinated chemists for many years, because it has several features that make adamantyl systems almost ideal for the study of certain chemical phenomena. The rigid adamantane skeleton results in a system of known geometry with unstrained, tetrahedral bond angles; the cyclohexane rings making up the adamantane molecule come together at four points, forming a **bridgehead** at each junction; and the cage-like structure prevents certain kinds of interactions and reaction mechanisms, simplifying the analysis of reaction parameters.

In 1939, Paul D. Bartlett showed that most bridgehead positions are quite inert toward nucleophilic substitution reactions, and suggested that studies of bridgehead reactivity could yield valuable information about reaction mechanisms and the geometry of transition states. Solvolysis reactions are particularly interesting, since rear-side displacement by the solvent-nucleophile at a bridgehead is rendered impossible by the "cage" of carbon atoms behind it; the only feasible solvolysis mechanism is an S_N1 process involving a carbonium-ion intermediate. A comparison of the relative solvolysis rates in Table 1 shows that the formation of a carbonium ion (the rate-limiting step in S_N1 reactions) becomes more difficult as the atom holding the leaving group becomes more restricted in its movement. The t-butyl bromide molecule can easily go from a tetrahedral to a planar geometry during ionization to form the t-butyl carbonium ion. Formation of a planar 1-adamantyl carbonium ion, however, is prevented by the more-or-less rigid adamantane structure. As the tetrahedral bridgehead atom flattens out in attempting to form 120° trigonal planar bonds, the bond angles at the adjacent carbon atoms decrease from the tetrahedral angle toward 90°. As it turns out, the actual carbonium ion ends up somewhere in between, with bond angles of about 113° at the bridgehead carbon and 104° at the adjacent carbon atoms. Such a carbonium ion is less stable than a completely planar one, and is thus more difficult to form—by a rate factor of about 1000, in this case. As the rings fused at a bridgehead become more rigid or decrease in size, the ease of forming a carbonium ion also decreases, as shown by the other examples in Table 1.

In most solvolysis reactions, the role of the solvent is twofold. Some solvent molecules can act as nucleophiles and "push" the leaving group off from the rear, while other solvent molecules simultaneously "pull" it off from the front. In the adamantyl system, the solvent cannot act as a nucleophile because rear-side attack is forbidden. In this case, only the electrophilic strength (or ionizing power) of the solvent is important. Such solvolysis reactions often follow the Winstein-Grunwald equation

$$\log \frac{k}{k_0} = mY \qquad (1)$$

where Y is the ionizing power of the solvent and m is the sensitivity of the substrate to changes in Y. Upon measuring the reaction rate in solvents of known ionizing power, one can plot $\log k$ versus Y and derive the value of m for the substrate from the slope of the line. Alternatively, the ionizing powers of various solvents can be determined using a substrate having a known value of m.

METHODOLOGY

If the recommendation of von R. Schleyer and his coworkers were followed, the value of m for 1-bromoadamantane would arbitrarily be set at 1.00 and the Winstein-Grunwald equation would then take the form

$$\log \frac{k}{k_0} = Y_{Ad} \qquad (2)$$

where Y_{Ad} is the solvent's ionizing power measured with reference to AdBr. In this experiment, you will measure the

Table 1.   Relative solvolysis rates of tertiary halides (*continued*)

Substrate	Rel. Rate
Br	10^{-13}
113° ⊕ 104° 1-adamantyl carbonium ion	

Role of solvent in displacement reactions

solvent as nucleophile   solvent as electrophile

SOH = any hydroxylic solvent

k_0 is the rate constant for solvolysis in a reference solvent (such as 80% ethanol) at 25°C.

The value of m for AdBr calculated from Equation 1 is 1.20.

The rate constant for solvolysis of AdBr in 80% ethanol, k_0, is 5.10×10^{-7} sec^{-1}.

Solvolysis reactions of AdBr follow
the rate equation

$$-\frac{d[\text{AdBr}]}{dt} = k[\text{AdBr}]$$

solvolysis rates of 1–bromoadamantane in several aqueous–
organic mixed solvents to determine their Y_{Ad} values.

The reaction rate constants will be determined by measuring the time it takes for each reaction to reach the same
degree of completion and using an appropriate equation to
calculate k. For a first-order reaction, the integrated rate
equation is

$$-\ln \frac{c}{c_0} = kt \qquad (3)$$

where c_0 is the concentration of the reactant at the beginning
of the reaction and c is its concentration at time t. When the
reaction is halfway to completion, $c/c_0 = \frac{1}{2}$, and the equation
becomes

$$-\ln \tfrac{1}{2} = kt_{1/2}$$

or

$$k = \ln 2/t_{1/2}$$

where $t_{1/2}$ is called the half-life of the reaction. A more general equation can be derived by letting α represent the degree
of completion of the reaction and writing the ratio c/c_0 as
$1 - \alpha$. Rearranging Equation 3 then yields

$$k = \frac{-\ln(1 - \alpha)}{t_\alpha} \qquad (4)$$

where t_α is the time required to attain $(100 \times \alpha)\%$ completion. For example, suppose a reaction of 1-bromoadaman-
tane in a given solvent requires 125 seconds to reach 20%
completion. The first-order rate constant for the reaction is
then

$$k = \frac{-\ln 0.80}{125} = 1.79 \times 10^{-3} \text{ sec}^{-1}$$

From the value of k in a given solvent, the ionizing power of
the solvent can be calculated using Equation 2. In this example,

$$Y_{\text{Ad}} = \log \frac{1.79 \times 10^{-3}}{5.10 \times 10^{-7}} = 3.55.$$

The reaction times are measured by (in effect) titrating
the HBr evolved during the reaction with aqueous sodium
hydroxide. A specified volume of NaOH is added along
with bromothymol blue indicator to each AdBr–solvent
combination, and the time required for the indicator to
change color (t_α) is measured. To find the degree of comple-

tion, the volume of sodium hydroxide required at 100%
reaction (V_∞) is determined by warming AdBr in 40% ethanol until the reaction is complete, and titrating the mixture
with sodium hydroxide to the bromothymol blue end point.
In this experiment, 2 ml of the sodium hydroxide solution
will be used in each kinetic run, so the degree of completion
will be $\alpha = 2/V_\infty$.

PRELAB EXERCISE

Calculate the mass of 1-bromoadamantane needed to
prepare 25 ml of a 0.040M solution in ethanol.

Reactions and Properties

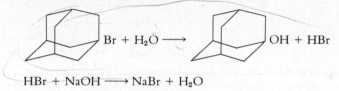

$$Br + H_2O \longrightarrow \qquad OH + HBr$$

$$HBr + NaOH \longrightarrow NaBr + H_2O$$

*Other solvolysis reactions are possible in
certain solvents, such as*

$$AdBr + EtOH \rightarrow AdOEt + HBr$$

*in aqueous ethanol; but these are gen-
erally unimportant compared to hydrolysis.*

Table 2. Physical properties

	M.W.	m.p.	b.p.	d.
1-bromoadamantane	215.1	116–18		
methanol	32.0	−94	65	0.791
ethanol	46.1	−117	78.5	0.789
2-propanol	60.1	−90	82	0.786

Hazards

The alcohols are flammable; their vapors are somewhat
toxic and should not be inhaled. Methanol can be absorbed
through the skin, so contact should be avoided.

methanol   ethanol   2-propanol

PROCEDURE

Preparation of Solutions.   Accurately weigh ⟨OP–5⟩ the calculated amount of 1-bromoadamantane and prepare 25 ml
of 0.04M AdBr solution in anhydrous ethanol, using a dry
volumetric flask. Prepare 100 ml of each of the following
solvent mixtures using distilled or deionized water: 40%
ethanol, 45% ethanol, 50% ethanol, 40% isopropyl alcohol,
and 50% methanol. Keep them in stoppered containers until
you are ready to use them.

*The solvents can be prepared in a 100 ml
graduated cylinder—measure their volumes as accurately as you can.*

Measuring Reaction Rates.   Into each of five clean, dry Erlenmeyer flasks measure 25 ml of each solvent mixture using a
graduated cylinder.   Then measure exactly 2 ml of
2×10^{-3}M sodium hydroxide into each of the flasks from a
25 ml buret, and add 2–3 drops of bromothymol blue indicator to each flask. Pipet ⟨OP-6⟩ exactly 1 ml of the 1-bro-
moadamantane solution into each flask, recording the time
when about half of the liquid has run into the flask, to the
nearest second, as the starting time for each addition. Shake
each flask to mix the reactants, and stopper it. Prepare an indicator blank in a sixth flask by adding 2–3 drops of the in
dicator to 25 ml of a phosphate buffer having a pH of 6.9.
For each reaction mixture, record the time when the
indicator color changes from blue to green (compare with
the blank). Save the reaction mixture containing 40%
ethanol for the determination of V_∞. Repeat the rate measurements once or twice using fresh 25 ml portions of each
solvent mixture.

*Be sure to label the flasks so you know
which solvent mixture each contains.*

*Use a stopwatch, digital timer, or a watch
that measures seconds for timing the reactions.*

*Set the flasks on a white sheet of paper so
the indicator change can be observed more
distinctly.*

*The indicator changes from blue to green
at the neutralization point, then to yellow
as HBr makes the reaction mixture more
acidic.*

Determination of V_∞.   After each 40% ethanol solution has
changed color, heat it in a 55–65° water bath for 30 minutes
then titrate it with 2×10^{-3} M sodium hydroxide until the indicator turns from yellow to green. If the indicator changes
back to yellow on standing, heat the solution in the water
bath a little longer, to make certain the reaction is complete,
and titrate it again.

Calculations

*Include the volume of NaOH used in the
kinetic runs when you calculate V_∞.*

1.   Calculate the degree of completion, α, from your
average value for V_∞ and use it to calculate the average value
of k for each solvent.

2.   Calculate the ionizing powers (Y_{Ad}) of the five solvents using Equation 2.

Experimental Variations

1.   The reaction rates for one or more solvents (50% ethanol is suitable) can be measured at several different temperatures, such as 25°, 38°, and 50°. Then the energy of activation for the reaction can be determined from an Arrhenius plot of ln k versus $1/T$.

Arrhenius equation

$$k = Ae^{-E_a/RT}$$

or

$$\ln k = -\frac{E_{act}}{RT} + \ln A$$

2.   An experiment on the solvolysis of different 1-haloadamantanes is described in *J. Chem. Educ. 54*, 773 (1977); another on bridgehead reactivity can be found in *J. Chem. Educ. 48*, 708 (1971). *J. Chem. Educ. 52*, 666 (1975) contains an interesting experiment on the hydrolysis of 3-chloro-3-methyl-1-butyne.

3.   This kinetic experiment can be combined with a synthetic one if 1-bromoadamantane is first prepared by brominating adamantane as described in *J. Chem. Educ. 48*, 708 (1971).

4.   The hydrolysis of *tert*-butyl chloride can be studied by essentially the same procedure as was used for 1-bromoadamantane.

Topics for Report

1.   Tabulate your results and arrange the solvents in order of ionizing power in the hydrolysis of 1-bromoadamantane.

2.   (*a*) Write a complete mechanism for the hydrolysis of 1-bromoadamantane, illustrating the activated complex for the rate-determining step. (*b*) Describe and explain (based on your mechanism) the effect of changing the water content of the solvent on the reaction rate. (*c*) Describe and explain the effect of using different alcohols on the reaction rate.

3.   Because some water and ethanol were added with the NaOH and AdBr solutions, the solvent mixtures were not of exactly the composition indicated. The presence of sodium hydroxide in the reaction mixture could also have a significant effect on the reaction rate. Tell how you would design an experiment to measure the hydrolysis rates as accurately as possible.

adamantane

Write a mechanism for E1 elimination from *t*-butyl chloride in aqueous solvents and explain why this is *not* a competing reaction in solvolysis reactions of 1-bromoadamantane.

5. After studying the nomenclature of bicyclic and polycyclic compounds and referring to the alternate structural representation in the margin, propose a systematic name for adamantane.

Library Topic

Read the article in *J. Chem. Educ. 50,* 780 (1973) and, using some of the references cited therein, write a short paper on the medical uses of adamantane derivatives.

The Oxidation of Menthol to Menthone

Reactions of Alcohols. Oxidation. Preparation of Carbonyl Compounds. Modifying Literature Procedures.

Operations

⟨OP–5⟩ Weighing

⟨OP–7c⟩ Temperature Monitoring

⟨OP–9⟩ Mixing

⟨OP–13⟩ Extraction

⟨OP–14⟩ Solvent Evaporation

⟨OP–19⟩ Washing Liquids

⟨OP–25⟩ Simple Distillation

Experimental Objective

To prepare menthone by the oxidation of menthol.

Learning Objectives

To learn how to carry out a dichromate oxidation.

To learn how to scale down a literature procedure.

To learn about some of the properties and uses of menthol and menthone.

SITUATION

You are a graduate student doing research on the rates of reactions involving *keto-enol* interconversions, and you are about to investigate the effect of various acid catalysts on the conversion of menthone to isomenthone. First you must prepare some menthone, but the *Organic Syntheses* procedure you have located is designed to yield about 75 grams and you have no use for that much menthone. So you must scale down the procedure to provide the 10 g or so of menthone that you estimate will be sufficient for the kinetic runs.

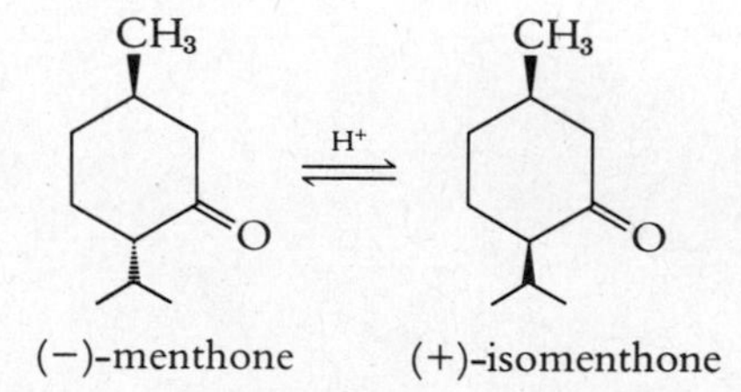

(−)-menthone   (+)-isomenthone

BACKGROUND

The Versatile Oil of Peppermint

According to Pliny, the Greeks and Romans crowned themselves with peppermint wreaths and used its sprays to adorn their tables.

(−)-menthol

thymol

The peppermint plant (*Mentha piperata*) yields a pleasantly pungent essential oil that has been used to treat a long list of human ailments, including bronchitis, cholera, colic, cramps, diarrhea, dyspepsia, flatulence, griping, headache, hysteria, influenza, laryngitis, lumbago, nausea, neuralgia, puerperal fever, rheumatism, seasickness, and toothache. Peppermint oil and its components have also been used to add a distinctive minty flavor or aroma to cigarettes, perfumes, liqueurs, and confections; to test for leaks in pipe joints; and to rid buildings of rats, which dislike its odor.

  The major component of peppermint oil is menthol, which makes up about 40–45% of the American oil. Peppermint oil also contains esters of menthol along with menthone, cineole, pinene, limonene, and other terpenoids. Natural menthol from peppermint oil exists primarily as the (−) enantiomer and is obtained by cultivating peppermint plants in Michigan and other Northern states. Synthetic menthol can be made by hydrogenating thymol, a process that yields the (±) racemate. The corresponding ketone, menthone, occurs in peppermint oil as the (−) enantiomer, but as the (+)-enantiomer in the oil of pennyroyal. (−)-Menthone isomerizes to (+)-isomenthone in the presence of acids. The extent of isomerization can easily be determined by optical rotation measurements, since the specific rotations of the two compounds are −30° and +85° respectively.

METHODOLOGY

  An organic chemist who wishes to prepare a certain compound will usually search the literature first for applicable procedures. Developing a practical laboratory synthesis from a literature procedure may require modifications in one or more of the following: (*1*) quantities of materials; (*2*) kinds of chemicals; (*3*) experimental apparatus and supplies; (*4*) experimental techniques; or (*5*) the time required for operations.

  Modifying the quantities of materials is usually just a matter of deciding how much of the major organic reactant you want to use, dividing this by the amount given in the procedure to get a scaling factor, and then multiplying all of the other quantities given by the same scaling factor. Obviously some common sense should be applied in the use of such scaling factors; it should not be necessary, for example, to painstakingly measure out 38.22 ml of ethyl ether for an extraction, even though 38.22 is the number you get when applying the scaling factor to the volume specified. Rounding the number off (to 40) is quite acceptable in this case and for other materials (solvents, drying agents, etc.) that are not involved in the reaction stoichiometry. The quantities of reactants and catalysts should, however, be determined with more precision so that their molar ratios remain essentially the same as in the literature procedure. This is particularly true for limiting reactants; more latitude can be allowed for catalysts and for reactants present in excess.

Substituting one chemical or solvent for another in a procedure may be risky, since it is not always possible to predict the result of such a substitution. Using a different reagent or reaction solvent may affect the outcome of a reaction or require substantial changes in the reaction conditions and workup procedure (unless the change is a minor one, such as substituting potassium dichromate for sodium dichromate in the procedure for this experiment.) Substituting one extraction solvent or drying agent for another during the workup is often permissible if the properties of the substitute are compatible with the system.

Scaling down an experiment will often require some changes in the equipment and methods used. The size of glassware will normally be reduced and sometimes the apparatus itself will be changed. Smaller quantities of materials require less powerful agitation, so magnetic stirring or even manual mixing can often be substituted for mechanical stirring. While mechanical stirring may require a three-necked flask fitted with a sleeve or bushing, magnetic stirring or manual agitation can be performed with a reaction vessel as simple as an Erlenmeyer flask.

Because material losses are more significant in small-scale syntheses, the experimenter must be more careful to minimize losses in such operations as material transfers, filtration, extraction, and distillation. Small vessels have a larger surface area to volume ratio than large vessels, so losses by surface adsorption will be proportionately greater if transfers are not quantitative. Therefore it is important to wash out the reaction flask and other vessels thoroughly during transfers.

The time required for various operations will usually decrease when a reaction is scaled down—it obviously takes

Surface area increases in proportion to the square of the radius; volume increases as its cube.

The reaction rate *may* vary with the scale of the reaction if the reaction is catalyzed by the inner surface of the reaction vessel, or if the efficiency of mixing and heat transfer is significantly lower in a larger reaction vessel.

less time to filter 10 g of solid than 100 g, or to distill 25 ml of liquid than 250 ml. The reaction time can also be reduced in some cases (even though a reaction in a small vessel will ordinarily proceed at the same rate as the same reaction in a larger one) because the smaller investment in chemicals and overall effort makes a lower percent yield acceptable. For example, a first-order reaction with a half-life of 30 minutes will be nearly 94% complete in 2 hours and 99% complete in $3\frac{1}{2}$ hours. The difference between 94% and 99% completion is ten grams in a synthesis having a theoretical yield of 200 grams, but only half a gram in a small-scale preparation yielding 10 grams. Unless the reactants are very costly, any reduction in the percent yield resulting from a shorter reaction time can be compensated for by using enough additional starting material to make up the difference.

PRELAB EXERCISE

Assume a 75% yield.

Write a detailed procedure for preparing about 10 g of menthone based on the *Organic Syntheses* procedure given. Specify the quantities of all reactants, solvents, drying agents, and other materials; the kind and size of glassware needed; and any changes in experimental methods. Substitute potassium dichromate for the sodium dichromate dihydrate specified in the procedure.

Reaction and Properties

$$3 \text{ (menthol)} + 4H_2SO_4 + K_2Cr_2O_7 \longrightarrow 3 \text{ (menthone)} + Cr_2(SO_4)_3 + K_2SO_4 + 7H_2O$$

Table 1. Physical properties

	M.W.	m.p.	b.p.	d.
(−)-menthol	156.3	44	216	
(±)-menthol	156.3	28–30	216	
(−)-menthone	154.3	−7	210	0.895
(±)-menthone	154.3		211	0.911[0]
potassium dichromate	294.2	398		
sodium dichromate (dihydrate)	298.0	357		

Hazards

Potassium dichromate is toxic and its dust irritates the respiratory tract and eyes; it may react violently with oxidizable materials. Some dichromates and chromates are considered carcinogenic, so exposure to this chemical should be minimized. Ethyl ether is highly flammable and inhalation can cause depression of central nervous system functions. Sulfuric acid causes severe burns to skin and eyes and reacts violently with water. Menthol and menthone are mild irritants, so excessive contact with the skin (and any contact with eyes) should be avoided.

ethyl ether   potassium dichromate

sulfuric acid

PROCEDURE

[Reprinted from a procedure by L. T. Sandborn appearing in *Organic Syntheses,* Collective Volume 1 (New York: John Wiley & Sons, 1941), pp. 340–41, by permission of the publisher and the Board of Editors of *Organic Syntheses.*]

In a 1-l. round-bottomed flask provided with a mechanical stirrer is placed 120 g. (0.4 mole) of crystallized sodium dichromate (or an equivalent amount of potassium dichromate), and to this is added a solution of 100 g. (54.3 cc., 0.97 mole) of concentrated sulfuric acid (sp. gr. 1.84) in 600 cc. of water. To this mixture 90 g. (0.58 mole) of menthol (crystals, m.p. 41–42°) is added in three or four portions and the mixture stirred (Note 1). Heat is evolved, and the temperature of the mixture rises to about 55° (Note 2). As soon as the reaction is complete the temperature falls. The oil is mixed with an equal volume of ether, separated in a separatory funnel, and washed with three 200-cc. portions of 5 per cent sodium hydroxide solution (Note 3). The ether is then removed by distillation and the residue distilled under reduced pressure, the menthone being collected at 98–100°/18 mm. If distilled under atmospheric pressure it boils at 204–207°. The yield is 74–76 g. (83–85 per cent of the theoretical amount).

Atmospheric pressure distillation may be used.

Notes

1. On addition of the menthol a black spongy mass forms which softens as the temperature rises and finally forms a dark brown oil.

2. The temperature may not reach 55°, in which case the mixture may be warmed gently with a small flame. If the reaction is slow in starting, gentle heating with a small flame is advantageous.

Avoid flames if ether is in use.

3. The oil, which is dark brown before washing with sodium hydroxide, becomes light yellow. If three washings are not sufficient to remove the dark color another portion of sodium hydroxide solution is used.

Experimental Variations

1. If optically active (natural) menthol is used as the starting material, the optical rotation of the product can be measured in alcohol and its purity calculated, assuming that any decrease in its specific rotation is due to its isomerization to isomenthone.

2. An infrared spectrum of the product can be obtained and interpreted.

3. Other oxidations of alcohols to ketones are described in Vogel, 4th ed., pp. 425–29 (Bib–B4).

MINILAB 13

Oxidation of Alcohols by Potassium Permanganate

When the reaction is complete, the solution should be colorless with a brown suspension of MnO₂.

When the reaction is complete, the solution should be colorless with a brown suspension of MnO_2.

Place 2 ml of 0.05% neutral potassium permanganate solution in each of five small test tubes. Using the first tube as a control, add to each of the other tubes 2 drops respectively of methanol, 1–butanol, 2–butanol, and 2–methyl-2–propanol. Shake each tube for 10 seconds and observe the color immediately, then after 5 and 10 minutes. To any solutions that have not completely reacted after ten minutes add 3 drops of 3M HCl, shake, and observe at the same intervals as before. To any solution that has not reacted (or has only begun to react) after ten more minutes add 3 drops of concentrated sulfuric acid (*Care:* causes severe burns), shake, and observe at the same intervals as before. Arrange the alcohols in order of oxidation rate and write balanced equations for their reactions with potassium permanganate. (*Hint*—what reactions of tertiary alcohols are induced by strong acids?)

Topics for Report

1. Calculate the theoretical amount of potassium dichromate needed for your oxidation and compare it with the amount actually used (compute the percent excess).

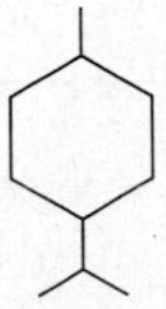

2. Give the R,S configuration for each of the chiral carbon atoms in (−)-menthol. Predict the number of possible stereoisomers having the basic menthol structure.

3. Cajeput oil contains an alcohol (**A**) with the p-menthane skeleton and the empirical formula $C_{10}H_{18}O$. Strong oxidation of **A** with potassium permanganate yields a compound **B** that readily dehydrates to form the γ-lactone **C**. Deduce the structures of **A** and **B** and explain your reasoning.

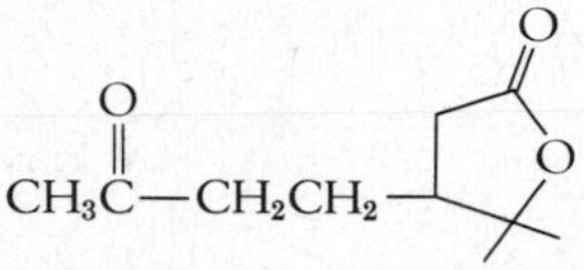

4. Write a flow diagram for the procedure, showing how all reactants and by-products are separated from the menthone.

5. Write a reasonable mechanism for the conversion of (−)-menthone to (+)-isomenthone.

6. Construct molecular models of (−)-menthol, (−)-menthone, and (+)-isomenthone. Draw the chair-form structure and give the equatorial-axial designation of each substituent in the most stable conformation of each compound.

Library Topic

1. Write a short paper describing the history, chemical composition, and uses of plants in the genus *Mentha*.

p-menthane

$$CH_3\overset{O}{\overset{\|}{C}}{-}CH_2CH_2{-}$$

C

Phase-transfer Catalysis in the
Williamson Synthesis of Phenetole

*Ethers. Nucleophilic Aliphatic Substitution. Reactions of Phenols.
Phase-transfer Catalysis.*

Operations

⟨OP–5⟩ Weighing

⟨OP–7a⟩ Refluxing

⟨OP–9⟩ Mixing

⟨OP–13⟩ Extraction

⟨OP–13b⟩ Separation

⟨OP–14⟩ Evaporation

⟨OP–19⟩ Washing Liquids

⟨OP–20⟩ Drying Liquids

⟨OP–25a⟩ Semimicro Simple Distillation

Experimental Objective

To prepare phenetole by a Williamson synthesis from
phenol using a phase-transfer catalyst.

Learning Objectives

To learn about the applications of the Williamson synthesis and the experimental conditions required for carrying
it out.

To learn more about the use of phase-transfer catalysts
in organic synthesis.

To learn about the sources and applications of some
useful aryl alkyl ethers.

SITUATION

The work of Makosza, Starks, and other investigators has led to the use of phase-transfer catalysts to accelerate reactions in two-phase systems. For example, the displacement of halide ions by nucleophiles such as cyanide can be catalyzed by quaternary ammonium ions, which transport cyanide ion from the aqueous to the organic phase where it reacts with the alkyl halide.

If the method works for cyanide ion, it should work for other nucleophiles as well. The old and familiar Williamson synthesis of ethers involves the displacement of halide ion (and other leaving groups) by nucleophiles such as alkoxide and aryloxide ions. You are convinced that phase-transfer catalysis can be used to provide a new twist to this old reaction. To test your idea, you decide to attempt the synthesis of phenetole (ethyl phenyl ether) by the reaction of sodium phenoxide with ethyl bromide using a phase-transfer catalyst.

See Experiment 12 for a review of the theory of PTC.

$$RBr + CN \xrightarrow{Q^+} RCN + Br^-$$

R = alkyl
Q^+ = quaternary ammonium ion

phenetole

BACKGROUND

Phenolic Ethers

Phenolic ethers frequently contribute a strong, pleasant odor to plant oils and extracts and are used in flavoring agents and perfumes. Although many common aromatic/aliphatic ethers like vanillin contain additional functional groups, monofunctional ethers are also of some importance.

Anisole has an agreeable odor and is used as an antioxidant for beer and as a stabilizer for vinyl polymers. Like many other compounds containing alkoxy groups, it helps to prevent autooxidation of compounds containing carbon-carbon double bonds. The closely related ether p-methoxytoluene has a powerful floral odor and is used in many perfumes. 2-Methoxynaphthalene (nerolin) and its 2-ethoxy homolog are perfume ingredients having delicate orange-blossom odors. Diphenyl ether has a strong, penetrating geranium odor and is used widely in cheap perfumes—especially for soaps—because of its low price. Mixed with biphenyl, it forms a eutectic mixture that solidifies at 12°C and is used as a heat-transfer fluid.

vanillin

anisole   p-methoxytoluene

nerolin

diphenyl ether

Williamson synthesis of ethers

$$NaOR + R'X \rightarrow ROR' + NaX$$

or

$$NaOAr + RX \rightarrow ArOR + NaX$$

METHODOLOGY

The traditional Williamson synthesis of ethers involves the reaction of an alkyl halide with the sodium salt of an alcohol or phenol. The alkoxide (or aryloxide) ion is strongly nucleophilic and displaces the halide by an S_N2 mechanism

Mechanism (S_N2)

$$RO^- + \overset{}{\underset{}{C}}{-}X \longrightarrow [RO\cdots\overset{\delta-}{\underset{}{C}}\cdots\overset{\delta-}{X}] \longrightarrow ROC + X^-$$

similar to that observed when preparing alcohols by the reaction of alkyl halides with hydroxide ion. Because alkoxide is also strongly basic, secondary or tertiary halides tend to undergo elimination as well as addition, and most aryl halides are so unreactive they cannot be used. Therefore the Williamson synthesis is most successful for combining a primary halide with a sodium alkoxide or aryloxide.

Preparation of alkoxide and aryloxide ions

$$ROH + Na \rightarrow NaOR + \tfrac{1}{2}H_2$$

$$ArOH + NaOH \rightarrow NaOAr + H_2O$$

Sodium alkoxides are usually prepared by adding bits of sodium metal to the corresponding alcohol. Then the alkyl halide is added and the mixture is refluxed for five hours or so to make the ether. Sodium aryloxides can be prepared by treating the corresponding phenol (which is much more acidic than an alcohol) with the less hazardous base, sodium hydroxide. Unfortunately, alkyl halides will not dissolve in an aqueous solution of NaOH, so the aryloxides are more often prepared by adding sodium to the phenol in an organic solvent such as ethanol or benzene, followed by the addition of the alkyl halide. When the phase-transfer-catalysis technique is used, however, sodium phenoxide is prepared in the aqueous phase by mixing the phenol with sodium hydroxide, while the alkyl halide is dissolved in an organic solvent such as methylene chloride. When the catalyst is added and the two-phase system is stirred, the phenoxide is carried into the organic phase by the quaternary ammonium cation (Q^+), where it can react with the halide to form a phenyl ether. Return of $Q+$ to the aqueous phase to pick up more $ArO-$ completes the cycle, as diagrammed in the margin.

Phase-transfer catalysis in the Williamson synthesis

$$ArOH + NaOH \longrightarrow NaOAr + H_2O$$

            OAr^- aqueous phase

―――――――――――――――――――

                organic phase (CH_2Cl_2)

$$RX + Q^+OAr^- \longrightarrow Q^+X^- + ArOR$$

$$X^- = halide$$
$$Q^+ = R'_4N^+$$

The catalyst used in this experiment, tricaprylmethylammonium chloride, is quite inexpensive and very effective for catalyzing certain S_N2 reactions. Its general formula is $R_3NCH_3^+$ Cl^-, where the R groups are C_8–C_{10} alkyls, with the C_8's predominating.

Commercial tricaprylmethylammonium chloride preparations include Aliquat 336 and Adogen 464.

After the reactants are refluxed for the required reaction period (with vigorous stirring to intermix the two phases), phenetole will be isolated by the usual procedures. The aqueous layer is extracted with methylene chloride to recover any additional phenetole dissolved in the water, and the organic layers are washed, dried, and evaporated to yield crude phenetole, which is then purified by semimicro distillation.

Because of the low boiling points of ethyl bromide and methylene chloride, two reflux condensers in series (one on top of the other) should be used on the reflux apparatus. Use a short length of tubing to connect the water outlet of the lower condenser to the inlet of the upper one.

PRELAB EXERCISE

Calculate the mass of crystalline phenol or 90% aqueous phenol needed to provide 25 mmol of phenol, and the volume of 50 mmol of ethyl bromide.

Your instructor will indicate the kind of phenol available.

Reactions and Properties

OH ————— O⁻Na⁺ ————— O⁻Na⁺ ————— OCH₂CH₃

phenol + NaOH ⟶ sodium phenoxide + H₂O

O⁻Na⁺ + CH₃CH₂Br $\xrightarrow{PTC}$ phenetole + NaBr

Table 1. Physical properties

	M.W.	m.p.	b.p.	d.
phenol	94.1	43	182	
ethyl bromide	109.0	−119	38	1.460
methylene chloride	84.9	−95	40	1.327
phenetole	122.2	−30	170	0.967

Hazards

Phenol is toxic and corrosive and can severely burn the skin and eyes; ingestion of as little as one gram has proven fatal. Avoid any contact with skin, eyes, and clothing. Ethyl bromide is toxic by ingestion, inhalation, and skin absorption; it is also flammable and the vapor is irritating to eyes and lungs. Avoid contact with the liquid, do not inhale vapors, and keep away from heat or flames. Methylene chloride vapors may be harmful and the liquid is dangerous to the eyes; avoid prolonged contact with the liquid and inhalation of vapors.

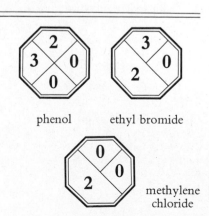

phenol   ethyl bromide

methylene chloride

PROCEDURE

Reaction. In a 100 ml round-bottomed flask, dissolve 25 mmol of phenol in 25 ml of 2M aqueous sodium hydroxide and add 25 ml of methylene chloride, 50 mmol of ethyl bromide, and 1 g of tricaprylmethylammonium chloride. Reflux the mixture ⟨OP-7a⟩ *gently* for two hours with vigorous magnetic stirring ⟨OP-9⟩, using two reflux condensers in series.

Set the reaction flask on the bottom of a non-ferric steam bath or hot water bath.

Separation and Purification. Separate ⟨OP-13b⟩ and save the organic (lower) layer, then extract the *aqueous* layer ⟨OP-13⟩ with two 20-ml portions of methylene chloride and combine the extracts with the reserved organic layer. Wash the organic solution ⟨OP-19⟩ with 25 ml of 2M aqueous sodium hydroxide, then with 25 ml of saturated aqueous sodium chloride. Dry the organic solution ⟨OP-20⟩ and evaporate the solvent ⟨OP-14⟩ using a solvent trap. Purify the crude phenetole by semimicro simple distillation ⟨OP-25a⟩ and weigh it ⟨OP-5⟩.

A dense white fog may form in the boiling flask near the end of the distillation —heating should be stopped at this point.

Experimental Variations

1. The purity of the product can be analyzed by gas chromatography, and infrared or NMR spectra can be obtained and interpreted.

2. The use of phase-transfer catalysis in the synthesis of methylenedioxybenzenes has been described in *Tetrahedron Letters,* 3849 (1975). Methylenedioxybenzene itself can be prepared from catechol and methylene bromide using Adogen 464 as the catalyst.

methylenedioxybenzene

3. The syntheses of phenetole, nerolin, and other alkyl phenyl ethers, using dialkyl sulfates as the alkylating agents, are described in *Vogel's Textbook of Practical Organic Chemistry,* 4th ed., pp. 755–56 (Bib–B4).

Topics for Report

1. Write a mechanism for the reaction used in this experiment, illustrating the role of the phase-transfer catalyst.

2. Write a flow diagram for the synthesis of phenetole.

3.   Phenetole can also be prepared by combining sodium phenoxide with any of the following compounds: ethyl chloride, ethyl *p*-toluenesulfonate, diethyl sulfate, and triethyl phosphate. **(a)** Write balanced equations for the reaction in each case. **(b)** What is the leaving group in the first alkylation step of each reaction?

4.   Outline a synthesis of each of the following ethers beginning with alcohols or phenols of the appropriate structure.

1. $OCH_2CH_2CH_3$ 2. $H_3C-\overset{\displaystyle CH_3}{\underset{\displaystyle CH_3}{\overset{|}{\underset{|}{C}}}}-O-CH_2CH_3$ 3.

4.   —OCH_2—   5. OCH_3 ... OCH_3

Library Topics

1.   Write a short paper describing the development of the phase-transfer catalysis method and giving some of its applications in organic synthesis.

2.   Report on the use of phenyl ethers as antioxidants, explaining how they function in this role and giving specific examples.

276

EXPERIMENT 33

The Borohydride Reduction of Vanillin to Vanillyl Alcohol

Carbonyl Compounds. Reduction. Preparation of Alcohols. Developing Synthetic Procedures.

Operations

The operations used will depend on the procedure you develop (see the Prelab Exercise).

Experimental Objective

To develop a suitable procedure for the sodium borohydride reduction of vanillin to vanillyl alcohol and to carry out the synthesis of vanillyl alcohol in the laboratory.

Learning Objectives

To learn about the uses of complex metal hydrides in synthesis and the procedures for carrying out sodium borohydride reductions.

To learn how to develop a specific synthetic procedure from a general description of reaction methods.

To learn about the characteristics and uses of vanillin and related aromatic compounds.

SITUATION

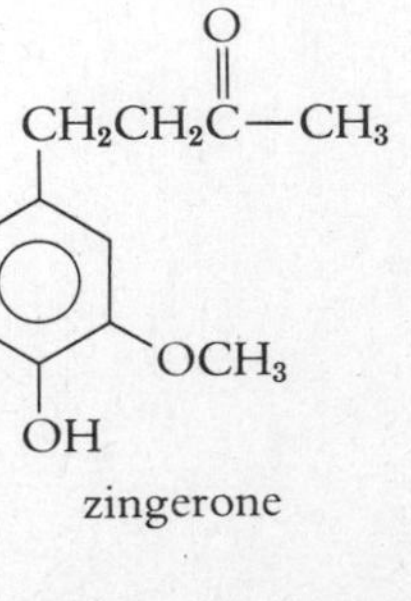

Your research group is trying to develop synthetic flavoring agents that may be used to replace or supplement some of the better known naturally derived flavor ingredients. One of these is zingerone, the pungent principle of ginger, which is prepared commercially by the condensation of vanillin with acetone followed by catalytic hydrogenation. While examining the structure of zingerone, it occurred to you that a zingerone analog having an oxygen atom in

zingerone

place of one methylene group could be created by treating vanillyl alcohol with an acetylating agent. Since this analog is structurally related to vanillin and (like many artificial flavors) is an ester, it should have some interesting flavor properties, although they may or may not resemble those of zingerone.

You have decided to prepare some vanillyl alcohol by reducing vanillin with the convenient reducing agent, sodium borohydride. A tentative literature search has turned up no directions for this specific reduction, so you will have to develop your own procedure from a general description of experimental conditions for borohydride reductions.

zingerone analog
(vanillyl alcohol acetate)

BACKGROUND

The Fragrant Aromatics

Chemists recognized at an early date that certain compounds obtained from natural sources showed a higher ratio of carbon to hydrogen than did typical aliphatic compounds. These compounds also had distinctly different chemical properties. Because many of them came from such pleasant-smelling sources as the essential oils of cloves, sassafras, cinnamon, anise, bitter almonds, and vanilla, they were called aromatic compounds. The name has stuck, although it is no longer associated with the odor of such compounds but rather with their structure and properties. Many aromatic compounds do justify the name, however, and among the most interesting and important of these are the chemical relatives of vanillin.

In 1520 the Spanish conquistador, Hernando Cortez, was served an exotic new drink by Montezuma II, emperor of the Aztecs. Cortez was pleased and soon the drink, a combination of chocolate and vanilla, found its way back to Europe. The vanilla plant (actually a climbing orchid, *Vanilla planifolia*) was also shipped back to the Old World in the hope that vanilla could be produced there successfully. The plant grew well but, mysteriously enough, would not fruit outside its native country. This mystery remained unsolved for more than 300 years, until someone discovered that the plant was pollinated by a native Mexican bee with an exceptionally long proboscis. A method of hand pollination was soon developed.

The vanilla flavoring comes from a long, narrow fruit that, after curing, looks somewhat like a dark brown string bean. The principal component, vanillin, does not exist as

Cortez later repaid Montezuma's hospitality by razing his capital to the ground.

such in the fresh "bean" but is formed by the enzymatic breakdown of a glucoside during the curing process.

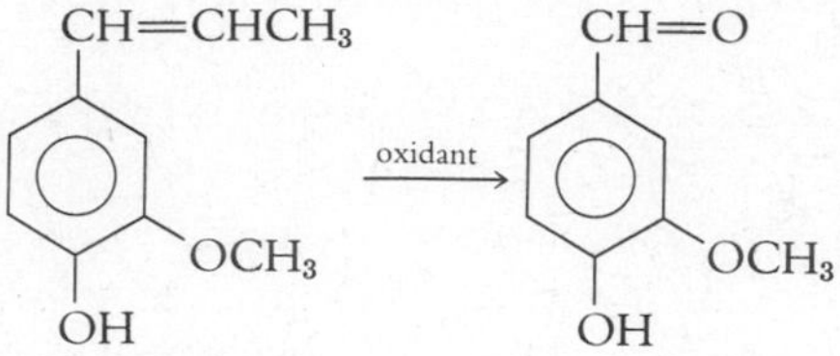

isoeugenol   vanillin

Although the finest vanilla flavoring is still obtained from natural vanilla, synthetic vanillin is far less costly. It is widely used as a component of flavorings, perfumes, and pharmaceutical products, and as a raw material for the synthesis of such drugs as L-dopa. At one time, most synthetic vanillin was made from isoeugenol, a naturally occurring compound with a fine carnation scent. Isoeugenol is still a widely used perfume ingredient.

Many other natural aromatics are related to vanillin. Bourbonal (3-ethoxy-4-hydroxybenzaldehyde) derives its trivial name from the island of Bourbon (now called Reunion), which produces a fine quality of vanilla. Although generally prepared from vanillin, bourbonal has been isolated in small quantities from vanilla beans and other natural sources. Safrole is a fragrant compound derived from sassafras and camphor oils. Once widely used as a flavoring in root beer, toothpaste, and chewing gum, it has since been banned for such uses because of its toxic, irritant qualities and because of its ability to produce liver tumors in rats and mice.

Piperonal can be derived from safrole by isomerization to isosafrole (which is analogous to isoeugenol) and oxidation. With its sweet "cherry pie" odor of heliotrope, piperonal is used widely in perfumes, cosmetics, and soaps.

The most important of all spices, black pepper, gets its "bite" from piperine, which contains the same methylenedioxy (—OCH_2O—) unit found in safrole and piperonal.

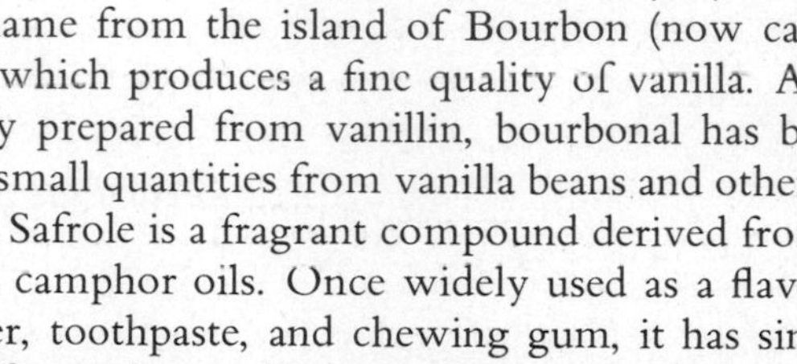

Today most vanillin is synthesized using lignin derived from wood pulp.

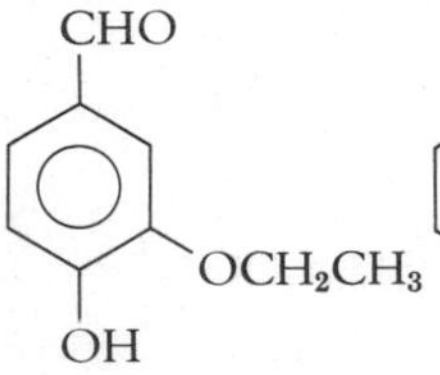

bourbonal   safrole

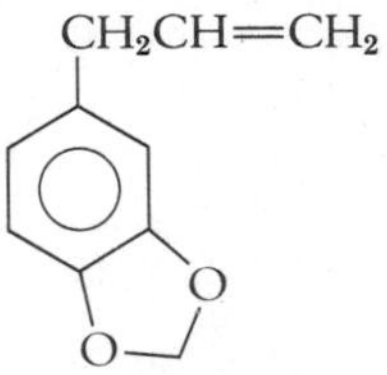

Bourbonal has a strong vanilla odor that is 3–4 times as powerful as that of vanillin itself.

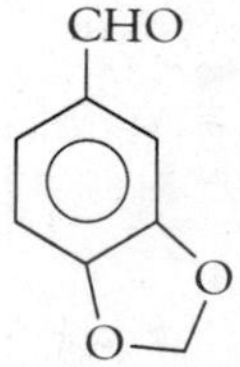

piperonal

piperine (*trans* double bonds)

Hydrolysis of piperine followed by oxidation yields both pi-peronal and the cyclic amine piperidine. Capsaicin, an even

$$\underset{\text{capsaicin}}{\text{HO}-\underset{\underset{\text{OCH}_3}{|}}{\bigcirc}-CH_2NH\overset{\overset{\displaystyle O}{\parallel}}{C}(CH_2)_4CH=CHCH\underset{\underset{CH_3}{|}}{C}H_3}$$

more pungent compound related to vanillin, is the fiery component of tabasco sauce, hot paprika, and cayenne pep-pers. Hydrolysis of capsaicin yields vanillylamine, which can easily be synthesized from vanillin. Another hot com-pound with the familiar 3-methoxy-4-hydroxy grouping is zingerone, the pungent principle of ginger. Apparently the same structural features that contribute to the pleasant odors of vanillin and piperonal produce quite a different effect in these fiery flavoring agents.

$$\underset{\text{zingerone}}{\text{HO}-\underset{\underset{\text{OCH}_3}{|}}{\bigcirc}-CH_2CH_2\overset{\overset{\displaystyle O}{\parallel}}{C}CH_3}$$

METHODOLOGY

When lithium aluminum hydride ($LiAlH_4$) was first in-troduced as a reducing agent in the late 1940s, it brought about a revolution in the preparation of alcohols by **reduc-tion.** Previously the two most popular methods for making alcohols from carbonyl compounds were reduction by so-dium metal in a hydroxylic solvent and catalytic reduction with gaseous hydrogen under pressure. The simplicity and convenience of the hydride reaction, however, soon made it the preferred method for a broad spectrum of chemical re-ductions. Lithium aluminum hydride is a powerful reducing agent with the ability to reduce aldehydes, ketones, acyl chlorides, lactones, epoxides, carboxylate esters, carboxylic acids, nitriles, and nitro compounds to alcohols or amines. Unlike catalytic hydrogenation, it does not reduce carbon-carbon multiple bonds (except in some α,β-unsaturated compounds). Its very reactivity, however, is a disadvantage in some applications. Since it reacts violently with water and other hydroxylic solvents to release hydrogen gas, it can only be used in aprotic solvents such as ethyl ether or di-glyme under strictly anhydrous conditions. It is also expen-sive and somewhat hazardous to use—even grinding it in a mortar can cause a fire.

By contrast, sodium borohydride ($NaBH_4$) is a much milder reducing agent that is safe to handle in the solid form;

Reduction in organic chemistry is defined roughly as the conversion of a compound in a higher "oxidation state" to a compound in a lower one. It is usually accompanied by a gain of hydrogen atoms or a loss of oxygen atoms, or both.

it can even be used in aqueous or alcoholic solutions. In such solutions, it is used primarily for reducing aldehydes and ke-

tones, since it is unreactive toward most of the other compounds affected by lithium aluminum hydride. Because of its much greater selectivity, sodium borohydride is widely used for reducing aldehydes and ketones that contain other functional groups. For example, a 3-keto bile acid ester (1) is reduced to the corresponding steroid alcohol without disturbing the ester function or the bromine atom.

Sodium borohydride reductions are usually carried out in water or in an alcohol such as methanol, ethanol, or 2-propanol. The reagent is not especially stable in pure water or dilute aqueous acid (it decomposes to the extent of about 4.5% per hour at 25° in neutral solution), so aqueous reactions are often run in dilute sodium hydroxide. When compounds having acidic functional groups are reduced, enough of the aqueous NaOH should be used to neutralize them and maintain a pH of 10 or higher.

Although sodium borohydride does react slowly with ethanol and methanol, these solvents are usually suitable when the reaction time is no more than 30 minutes at 25°; for longer reaction times or reactions at higher temperatures, isopropyl alcohol is a better solvent.

From the reaction stoichiometry, it can be seen that 1 mole of sodium borohydride will reduce 4 moles of aldehyde or ketone. In practice, it is wise to use a moderate excess of borohydride to compensate for any that reacts with the solvent or other hydroxylic materials. Since the reaction is first order in sodium borohydride (as well as the carbonyl compound), using an excess will also increase the reaction rate.

In most reactions with sodium borohydride, the aldehyde or ketone is dissolved in the reaction solvent and added to a solution of sodium borohydride, with external cooling if necessary, at a rate slow enough to keep the reaction tem-

1M sodium hydroxide is suitable for most reductions.

Unless neutralized, acidic functional groups may cause rapid decomposition of the sodium borohydride.

The rate of NaBH$_4$ decomposition in alcohols is in the order CH$_3$OH > C$_2$H$_5$OH > i-C$_3$H$_7$OH. Isopropyl alcohol is more difficult to remove during the workup, however.

Quantities of NaBH$_4$ that are 50–100% in excess of the stoichiometric amount are commonly used.

perature below 25°. The amount of solvent is not crucial, but enough should be used to readily dissolve each reactant and facilitate the workup of the reaction mixture. The solubility of sodium borohydride per 100 g solvent is reported to be 55 g in water at 25°, 16.4 g in methanol at 20°, and 4.0 g in ethanol at 20°.

Higher temperatures may decompose the hydride, especially in methanol or ethanol.

How much

Reaction conditions in borohydride reductions

Et = Ethyl
i-Pr = isopropyl
t-Bu = *t*-butyl
r.t. = room temperature

The time required to complete the reaction depends on the reaction temperature and the reactivity of the substrate. In kinetic studies of borohydride reduction in isopropyl alcohol, it was found that aldehydes are considerably more reactive than ketones and that aliphatic carbonyl compounds are more reactive than aromatic ones. For example, the comparatively reactive ketone 4-*t*-butylcyclohexanone is completely reduced at room temperature in 20 minutes, but benzophenone is reduced only by heating it at the boiling point of isopropyl alcohol for 30 minutes. Most reactions of aldehydes and aliphatic ketones are complete in 30 minutes at room temperature, whereas aromatic ketones or particularly hindered ketones may require more time or higher reaction temperatures.

After the reaction is complete, the tetraalkoxyborate intermediate (and excess sodium borohydride) must be decomposed to liberate the product. This is usually done by acidifying the reaction mixture to about pH 6 (slowly and with stirring) with 3–6M HCl. Hydrogen gas is evolved during this process as the excess sodium borohydride decomposes, so there must be no flames in the vicinity. Depending on the properties of the product and the solvent used, the product can be isolated by filtration or extraction, or by the evaporation of excess solvent. If the product is a solid that crystallizes from the reaction mixture, it can be collected by vacuum filtration, followed by extraction of the aqueous solution to recover dissolved product. Liquids or water-soluble products are generally isolated from an aque-

- HCl

distillation!
continuous addition of water
distil off ether

dissolve in ether

wash with H₂O until neutral

evap under vacuum

$(RCH_2O)_4B^{\ominus}$
a tetraalkoxyborate ion

Addition of acid may also generate some diborane (B_2H_6), which can cause side reactions if other reducible groups (COOH, COOR, C–C) are present.

Vanillyl alcohol tends to form supersaturated solutions in water; it may be necessary to scratch the flask to get it to crystallize.

ous solution by ether extraction and recovered by evaporating the ether. If the reaction solvent is an alcohol, the
reaction mixture is usually concentrated before the extraction step by evaporating most of the alcohol. Water can then
be added to facilitate the extraction. The product can be
purified by any appropriate method.

PRELAB EXERCISE

Each student should start with the same amount of vanillin, about 50 mmol.

Students should get together in small groups (3–6)
before the laboratory period to work out different procedures
for reducing vanillin. Experimental parameters such as the
amount of sodium borohydride, the quantity and kind of
solvent, the reaction temperature, and the reaction time
can be varied; each member of the group should be responsible for carrying out a given procedure. Alternatively,
students can develop their procedures independently and
compare results after the laboratory period.

Reactions and Properties

Table 1. Physical properties

	M.W.	m.p.	b.p.
vanillin	152.2	79	285
sodium borohydride	37.83		400d
vanillyl alcohol	154.2	115	d

Vanillyl alcohol is reported to be soluble *cold* in alcohol and
ether, and soluble *hot* in water, alcohol, ether and benzene. It
is relatively insoluble in cold water and benzene.

Hazards

Sodium borohydride is harmful if taken internally and is irritating to skin, eyes, and the respiratory system. Avoid contact with skin, eyes, and clothing; avoid breathing dust.

sodium borohydride

PROCEDURE

Develop your own procedure for this experiment. Obtain the yield and melting point of the purified product.

Keep the NaBH₄ container tightly closed when not in use.

Experimental Variations

1.   An infrared spectrum of the product (mull or KBr pellet) can be obtained and analyzed.

2.   A reduction using sodium bis (2-methoxyethoxy)-aluminum hydride is described in *J. Chem. Educ. 50,* 154 (1973); an interesting experiment involving the potassium borohydride reduction of menthone is in *J. Chem Educ. 50,* 292 (1973). Procedures for other reductions are given in *Vogel's Textbook of Practical Organic Chemistry* 4th ed., pp. 355–63 (Bib–B4).

Topics for Report

1.   Tabulate the results for your group, summarizing the experimental conditions for the *reaction* step and giving the yield and melting point for each set of conditions. On the basis of these results (and those of other groups, if available to you), discuss the apparent effect of changing various reaction parameters and state your conclusions with regard to the optimum reaction conditions.

Take into consideration the fact that the laboratory technique of students (and thus their results) can vary widely.

2.   Write a balanced equation for the acid-catalyzed decomposition of sodium borohydride in water.

3.   Write a flow diagram for the procedure you used in preparing vanillyl alcohol.

4.   Sodium borohydride is a strong base as well as a reducing agent. In the reduction of alkali-sensitive compounds, should the substrate be added to the borohydride solution or the borohydride to the substrate? Explain.

5. Write a reasonable mechanism for the reduction of vanillin with sodium borohydride.

Library Projects

1. Give structures of the recently developed hydride reducing agents lithium tri-*t*-butoxyaluminum hydride, sodium bis(2-methoxyethoxy)-aluminum hydride, lithium tri-*sec*-butylborohydride, and sodium cyanoborohydride. Describe some of their synthetic applications.

2. Write a short paper on the sources, nature, and uses of lignin, including a description of a commercial process for producing vanillin from the by-products of papermaking.

Lignin is second only to cellulose as the most abundant organic substance on earth.

The Multistep Synthesis of Benzilic Acid from Benzaldehyde

Carbonyl Compounds. Nucleophilic Addition. Oxidation. Rearrangement. Preparation of Carboxylic Acids. Multistep Synthesis.

Operations

⟨OP–5⟩ Weighing

⟨OP–7a⟩ Refluxing

⟨OP–9⟩ Mixing

⟨OP–11⟩ Gravity Filtration

⟨OP–12⟩ Vacuum Filtration

⟨OP–21⟩ Drying Solids

⟨OP–23⟩ Recrystallization

⟨OP–28⟩ Melting Point

Experimental Objective

To prepare benzilic acid by a three-step synthesis starting with benzaldehyde.

Learning Objectives

To learn about the history and experimental methodology of the benzoin condensation and the benzilic acid rearrangement.

To learn about the role of mandelonitrile and its glycosides in cyanogenetic organisms.

SITUATION

You are doing research on carbanion reactions involving the abstraction of protons from the α-carbon atoms of various esters. One of your objectives is to measure the effect of adjacent phenyl groups on the acidity of an α-hydro-

diphenylacetic acid

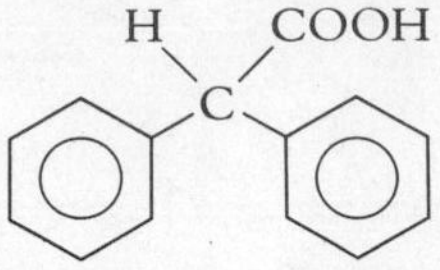

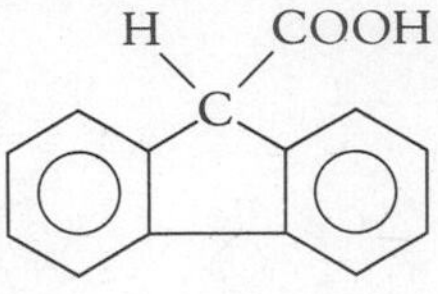

9-fluorenecarboxylic acid

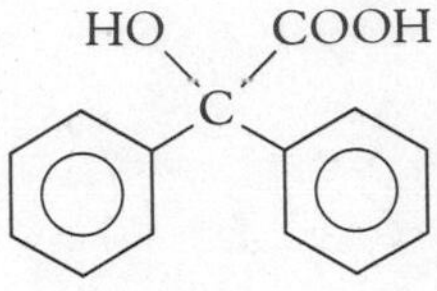

benzilic acid

One man who considered apple seeds a delicacy ate a cup of them at one sitting and died from cyanide poisoning.

gen atom and to find out if the orientation of a phenyl group has any bearing on carbanion stability. For this purpose, you would like to compare the acidity of the α-hydrogen in ethyl diphenylacetate, (where the two phenyl groups are free to rotate) with that of ethyl 9-fluorenecarboxylate (where they are linked to form the rigid fluorene ring and are thus locked into one plane). Both esters can be prepared from the corresponding carboxylic acids, diphenylacetic acid and 9-fluorenecarboxylic acid, which are in turn obtained from benzilic acid. You will prepare the benzilic acid using a familiar multistep synthesis from benzaldehyde.

BACKGROUND

Justus Liebig and the Bitter Almond Tree

Chemical warfare is generally thought of as a recent and uniquely human invention, but, as with most of our "discoveries," nature beat us to it. An otherwise unexceptional insect, the millipede *Apheloria corrugata,* discourages predators with a dose of "poison gas" powerful enough to kill a mouse. Many trees in the rose family (such as the cherry, apple, peach, plum, and apricot) insure their continued survival by protecting their seeds and foliage with cyanide-containing substances, and there are cases on record of human fatalities from eating the seeds of these species. The culprit in both of these examples is the cyanohydrin *mandelonitrile,* which can be decomposed by enzymes or stomach acids into benzaldehyde and lethal hydrogen cyanide.

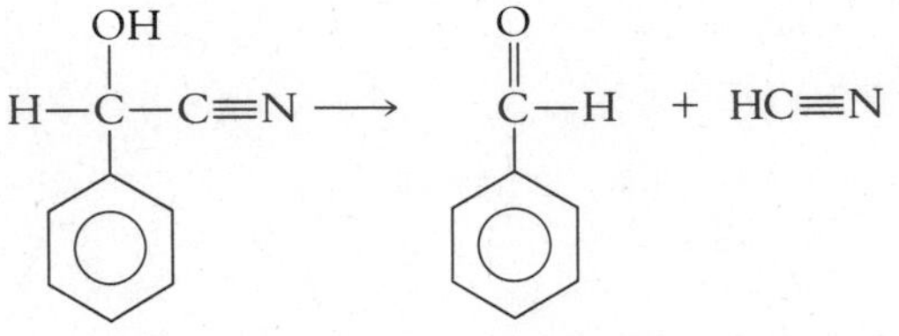

mandelonitrile   benzaldehyde   hydrogen cyanide

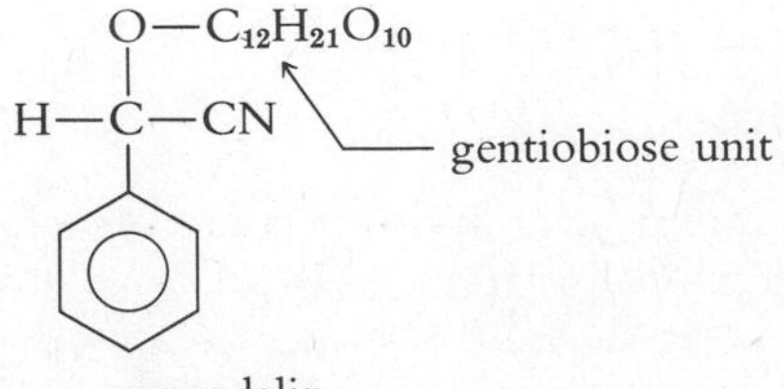

amygdalin

In most cyanogenetic (cyanide-forming) plants, mandelonitrile is present as a carbohydrate derivative called a glycoside. The most common of these is amygdalin, which is an acetal formed of mandelonitrile and the carbohydrate gentiobiose. Although amygdalin itself is not toxic, it can be broken down enzymatically under certain conditions to mandelonitrile (which *is* toxic) and its decomposition product, hydrogen cyanide. Closely related to amygdalin is the controversial cancer drug Laetrile, which allegedly kills malignant cells by releasing HCN (or mandelonitrile) at the cancer site.

One of the most prolific sources of amygdalin is the bitter almond, which (unlike the sweet varieties used for human consumption) is grown for the oil that can be pressed from its seed kernels. The characteristic odor of this oil (which is also used to preserve maraschino cherries) comes from benzaldehyde formed in the breakdown of amygdalin. In the early 1800s, workers often purified almond oil by washing it with aqueous alkali to extract acids, a process that resulted in the formation of a white solid later identified as benzoin. Friedrich Wöhler and Justus Liebig studied this reaction in 1832 and found that benzoin resulted from the catalytic action of sodium cyanide (formed when the HCN from amygdalin reacted with the alkali wash) on benzaldehyde. Surprisingly, cyanide ion is almost the only substance that catalyzes this reaction; so the early discovery of the benzoin condensation resulted from the lucky coincidence that both HCN and benzaldehyde were there when the bitter almond oil was washed with base.

Benzoin itself can be oxidized by a variety of oxidizing agents to the diketone benzil. It was not long after Liebig's work on the benzoin condensation that he discovered yet another unusual reaction, the benzilic acid rearrangement. When benzil is treated with hydroxide ion, it is converted to the benzilate ion which, upon acidification, forms benzilic acid. This reaction, the oldest known molecular rearrangement, is the prototype of a general class of rearrangements. Its mechanism has been the subject of much speculation over the years.

Laetrile

benzoin

Wöhler's interest in cyanides had previously resulted in his synthesis of urea from ammonium cyanate, an experiment that eventually helped demolish the "vital force" theory of organic chemistry.

benzil

benzilic acid

METHODOLOGY

Since this is a multistep synthesis, it will be important to minimize material losses at each stage to offset the multiplier effect of errors committed in sequence. Fortunately, all

The benzaldehyde must not contain any benzoic acid—it should be distilled if necessary. Fresh benzaldehyde from a previously unopened bottle should be usable without distillation.

The cupric ion is probably responsible for the oxidation of benzoin, being reduced to cuprous ion in the process. However, since it is continuously regenerated by a reaction with the ammonium nitrate, its overall effect is that of a catalyst.

three steps normally proceed in high yield, so it should be possible to obtain a reasonably good overall yield of benzilic acid from benzaldehyde.

The techniques involved in each reaction step are quite elementary and straightforward. The benzoin condensation is carried out by refluxing the reactants and recovering the benzoin (which crystallizes from the cooled solution) by vacuum filtration. This product need not be purified before its conversion to benzil, which is done by heating benzoin under reflux with ammonium nitrate and cupric acetate in acetic acid. The reaction mixture is diluted with water and the benzil is filtered and then recrystallized for use in the benzilic acid rearrangement. This reaction is carried out by refluxing benzil with alcoholic potassium hydroxide. The resulting suspension of potassium benzilate is dissolved by warming it in water, then filtered to remove solid impurities and acidified to precipitate benzilic acid, which is isolated by vacuum filtration and recrystallized.

The use of sodium cyanide makes the experiment potentially hazardous, since this substance is deadly poisonous if ingested and can generate lethal hydrogen cyanide in contact with acids. Consequently, it is extremely important that you read and understand the Hazards section of this experiment and carefully follow the directions given for disposing of the filtrate in the benzoin condensation.

PRELAB EXERCISE

Calculate the mass and volume of 150 mmol of benzaldehyde.

Reactions and Properties

A. Benzoin condensation

benzaldehyde　　　　　　　　　　　　benzoin

B.

benzil

$+ N_2 + 3H_2O$

C. Benzilic acid rearrangement

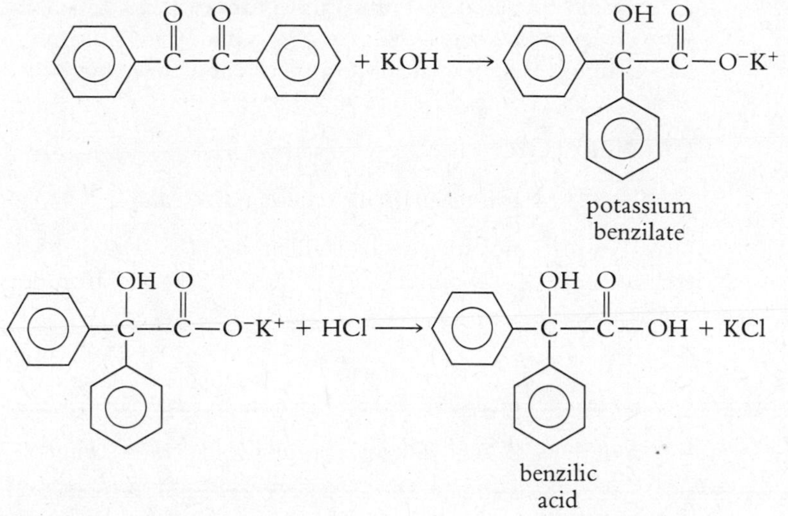

$$\text{(C}_6\text{H}_5)\text{C(=O)C(=O)(C}_6\text{H}_5) + KOH \longrightarrow \text{benzilate}$$

potassium
benzilate

$$\text{(C}_6\text{H}_5)\text{C(OH)(C}_6\text{H}_5)\text{C(=O)O}^-\text{K}^+ + HCl \longrightarrow \text{(C}_6\text{H}_5)\text{C(OH)(C}_6\text{H}_5)\text{C(=O)OH} + KCl$$

benzilic
acid

Table 1. Physical properties

	M.W.	m.p.	b.p.	d.
benzaldehyde	106.1	−26	178	1.042
benzoin	212.2	137		
benzil	210.2	95–6		
benzilic acid	228.2	151		
ammonium nitrate	80.0	170		
potassium hydroxide	56.1			

Potassium hydroxide pellets are about 85% KOH.

Hazards

Sodium cyanide is very poisonous, causing death through asphyxiation; it reacts with acids to produce poisonous HCN gas. Poisoning can occur through ingestion or skin contact, or by the inhalation of HCN gas. Avoid contact with skin or clothing, wash hands after use, and keep it away from acids. If necessary, cyanide solutions can be disposed of by washing them down a drain with large quantities of cold water.

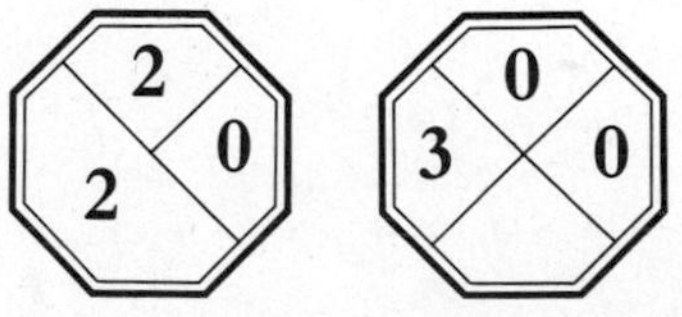

Benzaldehyde can irritate the skin and eyes, so unnecessary contact should be avoided. Ammonium nitrate is potentially explosive and should be kept cool, unconfined, and away from combustible materials. Copper (II) acetate is poi-

benzaldehyde

sodium cyanide

potassium
hydroxide

ammonium
nitrate

sonous if ingested; inhalation of dust and unnecessary contact should be avoided. Potassium hydroxide can cause severe burns to eyes and skin; prevent contact with skin, eyes, and clothing, and prepare solutions under a fume hood.

PROCEDURE

A.   *Preparation of Benzoin from Benzaldehyde*

Dissolve 0.15 mol of *pure* benzaldehyde in 30 ml of 95% ethanol in a 250-ml boiling flask and carefully add 10 ml of 2M aqueous sodium cyanide (*Poison!*). Reflux ⟨OP–7a⟩ the

---

WARNING   *Be certain to wash your hands after handling the cyanide solution.*

---

solution gently on a steam bath for 30 minutes, then let it cool to room temperature and cool it in an ice bath to complete crystallization. Collect the crystals by vacuum filtration ⟨OP–12⟩ and wash them first with 25 ml of cold water, then twice with 10-ml portions of ice-cold methanol. Press the product between two filter papers and weigh it ⟨OP–5⟩ when it is reasonably dry.

Dispose of the cyanide-containing filtrate properly (see Hazards).

B.   *Preparation of Benzil from Benzoin*

Combine 0.15 g of copper(II) acetate, 7.5 g of ammonium nitrate, and your crude benzoin in 35 ml of 80% (v/v) aqueous acetic acid in a flask set up for reflux ⟨OP–7a⟩. Heat the flask gently, with occasional shaking, to start the reaction, which is accompanied by the vigorous evolution of nitrogen gas. When the evolution slows down, heat the solution to the boiling point and reflux it gently for an hour or more.

If you recovered less than 10 g of benzoin, calculate the stoichiometric amount of ammonium nitrate and use a 25% excess.

Cool the reaction mixture to 50° and pour it, with stirring, onto 75–100 ml of crushed ice in a beaker. Collect the benzil by vacuum filtration ⟨OP–12⟩, wash it twice with water, and recrystallize it ⟨OP–23⟩ from 95% ethanol. Dry ⟨OP–21⟩ and weigh ⟨OP–5⟩ the product and, at your instructor's request, determine its melting point ⟨OP–28⟩.

Be sure to discard the filtrate; ammonium nitrate solutions can explode if concentrated.

C.   *Preparation of Benzilic Acid from Benzil*

Combine the benzil from step **B** with 95% ethanol (about 3 ml per gram of benzil) and add approximately 2.5 ml of aqueous 6M potassium hydroxide per gram of benzil. Reflux the mixture ⟨OP–7a⟩ for 15 minutes on a steam bath, then pour the hot liquid, with stirring, into 100 ml of water and let it stand a few minutes.

To prepare the 6M KOH, use 10 g of KOH pellets for every 25 ml of solution.

Warm the mixture to 50°, with stirring, to dissolve the potassium benzilate. There may be a colloidal suspension of unreacted benzil or by-products at this point, but most of the precipitate should dissolve—if it does not, add more water. Add a gram of decolorizing carbon and 0.5 gram of filtering aid (Celite, etc.), stir ⟨OP–9⟩ at 50° for 2 minutes, and filter the hot solution by gravity ⟨OP–11⟩.

Magnetic stirring is convenient, but not essential.

Mix 15 ml of concentrated hydrochloric acid with 100 ml of crushed ice in a beaker. Add 10–15 ml of the potassium benzilate solution, with stirring, and scratch the sides of the beaker until crystals begin to form. Add the rest of the solution, with continuous stirring, and check the pH. Collect the benzilic acid by vacuum filtration ⟨OP–12⟩ and wash it on the filter with cold water. Recrystallize it ⟨OP–23⟩ from boiling water and obtain its mass ⟨OP–5⟩ and melting point ⟨OP–28⟩ after it is thoroughly dry ⟨OP–21⟩.

The pH should be down to 2–3. Add more HCl if it is higher.

Experimental Variations

1. The syntheses of diphenylacetic acid and 9-fluorenecarboxylic acid from benzilic acid are described in *Organic Syntheses,* Coll. Vol. I, p. 224, and Coll. Vol. IV, p. 482 (Bib–B1).

2. The benzoin condensation has also been carried out using thiamine as a catalyst, although the yields are lower. The procedure is described in J. R. Mohrig and D. C. Neckers, *Laboratory Experiments in Organic Chemistry,* 2nd ed. (New York: Van Nostrand, 1973), p. 184. A benzoin condensation using tetra-*n*-butylammonium cyanide as the catalyst is described in *J. Chem. Educ. 55,* 237 (1978).

3. Tetraphenylcyclopentadienone is a useful reagent for visualizing Diels-Alder reactions (the purple color of the dienone fades as the adduct is formed). The synthesis of this reagent from benzil is described in *Organic Syntheses,* Coll. Vol. III, p. 806.

tetraphenylcyclopentadienone

Ph = phenyl

4. The preparation of a number of derivatives from benzoin, benzil, and benzilic acid is described in R. Adams,

J. R. Johnson, and C. F. Wilcox, Jr., *Laboratory Experiments in Organic Chemistry,* 6th ed. (New York: Macmillan, 1970).

MINILAB 14

Synthesis of 2,3-Diphenylquinoxaline from Benzil

Avoid contact with o-phenylenediamine; it can cause dermatitis and serious eye damage.

Dissolve 1 g of benzil in 4 ml of warm 95% ethanol and add a solution of 0.5 g of *o*-phenylenediamine in 4 ml of 95% ethanol. Warm the solution in a 50°C water bath for 20–30 minutes, add water to just saturate the warm solution

o-phenylenediamine   2,3-diphenylquinoxaline

The pure compound should melt at 125–26°C.

(watch for a slight cloudiness) and cool it in ice until the crystallization is complete. Recover the 2,3-diphenylquinoxaline by vacuum filtration. The product can be recrystallized ⟨OP–23*b*⟩ from ethanol–water and a melting point obtained, if desired. Write a balanced equation and propose a mechanism for the reaction.

Topics for Report

1.   Why might the presence of benzoic acid in the benzaldehyde used in step *A* retard or prevent the benzoin condensation?

The generally accepted mechanism has five steps.

2.   (*a*) The crucial step in the benzoin condensation is a proton transfer from carbon to oxygen, activated by an adjacent cyano group. Using what you know about nucleophilic addition reactions of carbonyl compounds, write a reasonable mechanism for the benzoin condensation of benzaldehyde. (*b*) The accepted mechanism for the benzilic acid rearrangement also involves nucleophilic additions to carbonyl. Write a reasonable mechanism for this reaction, starting with benzil.

3.   Write a synthetic pathway illustrating the preparation of compound **1** from a suitable five-carbon starting material, using the reactions described in this experiment.

4.   Write a flow diagram for the synthesis of benzilic acid from benzaldehyde.

Library Topics

1.   Benzaldehyde FFC is a term used to designate a grade of benzaldehyde *free from* chlorine. Describe several commercial methods for preparing benzaldehyde and explain why some commercial benzaldehyde might not be "FFC."

2.   Write a short paper on the Laetrile controversy describing the production of Laetrile, how it is used in treating cancer, and some of the arguments for and against its use.

Isolation and Identification of a By-product in the Condensation of Furfural with Cyclopentanone

EXPERIMENT 35

Carbonyl Compounds. Carbanions. Nucleophilic Addition. Spectrometric Analysis. Phase-transfer Catalysis.

Operations

⟨OP–5⟩ Weighing

⟨OP–8⟩ Cooling

⟨OP–12⟩ Vacuum Filtration

⟨OP–13⟩ Extraction

⟨OP–13*d*⟩ Solid Extraction

⟨OP–14⟩ Evaporation

⟨OP–19⟩ Washing Liquids

⟨OP–20⟩ Drying Liquids

⟨OP–21⟩ Drying Solids

⟨OP–23⟩ Recrystallization

⟨OP–23*b*⟩ Recrystallization from Mixed Solvents

⟨**OP–25*b***⟩ Distillation of Solids

⟨OP–26*a*⟩ Semimicro Vacuum Distillation

⟨OP–28⟩ Melting Point

⟨OP–34⟩ NMR Spectrometry

Experimental Objective

To prepare 2-furfurylidenecyclopentanone by the Claisen-Schmidt condensation of furfural and cyclopentanone, and to isolate and identify a by-product formed during the reaction.

Learning Objectives

To learn about some characteristics of the Claisen-Schmidt condensation and the experimental conditions for its use.

To learn more about the use of NMR spectrometry in structure determination.

To learn about the sources, characteristics, and uses of furfural.

SITUATION

A literature procedure for preparing 2-furfurylidenecy-clopentanone notes that a precipitate forms during the reaction and is filtered off and discarded. The authors did not bother to confirm the identity of this by-product, which is described as a yellow solid. You plan to prepare 2-furfurylidenecyclopentanone for your research project on the stereochemistry of Claisen-Schmidt condensations, and you believe that the by-product could also be used in your study. Before you can study it, however, you must isolate some of it from the reaction mixture and determine its structure.

2-furfurylidenecyclopentanone

BACKGROUND

Furfural—The Oat-hull Aldehyde

Furfural, also known as 2-furaldehyde, is the most important member of the furan series of aromatic compounds. The furan ring is aromatic because it has six pi-electrons (two from the oxygen atom) distributed about a five-membered ring. However, its aromatic sextet is less stable than that of benzene, so the furan ring can undergo reactions such as electrophilic addition, cycloaddition, and cleavage more readily than the corresponding benzene compounds.

Furfural has been known since its isolation by Dobereiner in 1832. It can be prepared in large quantities by treating such "waste" materials as bran, oat hulls, corncobs, and peanut shells with dilute acids. These materials contain polysaccharides known as pentosans that are hydrolyzed to pentoses (simple five-carbon sugars). These in turn are converted to furfural by acid-catalyzed dehydration. The first

furfural

Conversion of a pentose to furfural (Pentoses also exist in a cyclic hemiacetal form.)

$$-3H_2O$$

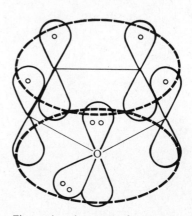

Figure 1.   Aromatic furan ring.

Furfural selectively removes alkenes and arenes from petroleum fractions, giving them better viscosity properties and greater resistance to oxidation.

commercial process for producing furfural was developed in 1922 by the Quaker Oats Company, which was trying to convert oat hulls into a better cattle feed at the time. Instead, it came up with a valuable commercial product that could be produced cheaply on a large scale. Among the many applications of furfural are the purification of lubricating oils, extractive distillation of butadiene used in the manufacture of rubber, synthesis of phenolic resins, and the manufacture of a large number of chemical intermediates.

Furfural behaves like a typical aromatic aldehyde in many of its reactions; it can be oxidized and reduced to the

Claisen-Schmidt condensation of furfural and acetone

$$\text{furyl}-\text{CHO} + \text{CH}_3\text{CCH}_3 \xrightarrow{\text{OH}^-} \text{furyl}-\text{CH}=\text{CHCCH}_3$$

furfurylideneacetone

furylacrylic acid

$$\text{furyl}-\text{CH}=\text{CHC}-\text{OH}$$

corresponding carboxylic acid and alcohol and, like benzaldehyde, undergoes the Cannizzaro reaction and the benzoin condensation. It also reacts with compounds having active methylene groups, such as aldehydes and ketones, to yield condensation products. For example, the reaction of furfural with acetone yields furfurylideneacetone by a Claisen-Schmidt condensation, and its condensation with acetic anhydride and sodium acetate forms furylacrylic acid by a Perkin reaction.

METHODOLOGY

The Claisen-Schmidt condensation is actually a kind of aldol condensation; it was discovered by Schmidt in 1880 and later improved by Claisen. It is usually regarded as a

Mechanism of a Claisen-Schmidt condensation

$$\text{CH}_3\text{CCH}_3 \xrightarrow{\text{OH}^-} \overset{\ominus}{\text{CH}_2}\text{CCH}_3 \xrightarrow{\text{PhCHO}}$$

$$\text{PhCH}-\text{CH}_2\text{CCH}_3 \xrightarrow{\text{H}_2\text{O}} \text{PhCHCH}_2\text{CCH}_3$$

$$\xrightarrow{\text{OH}^-} \text{PhCHCHCCH}_3 \xrightarrow{-\text{OH}^-} \text{PhCH}=\text{CHCCH}_3$$

reaction between an aromatic aldehyde and an aliphatic aldehyde or ketone that yields an α,β-unsaturated aldehyde or ketone under basic conditions. The mechanism is essentially the same as that for a typical aldol condensation (e.g., the self-condensation of acetaldehyde) up to the formation of a β-hydroxy carbonyl compound. In the Claisen-Schmidt condensation, this intermediate is then dehydrated by an E1cb mechanism, involving the formation of a resonance-stabilized carbanion, which loses hydroxide ion to form the unsaturated product. The overall mechanism is illustrated for the condensation of acetone and benzaldehyde.

E1cb refers to a unimolecular elimination reaction that proceeds via a carbanion.

Since cyclopentanone is a relatively reactive ketone in the Claisen-Schmidt condensation, the reaction can be carried out in very dilute base. The use of a phase-transfer catalyst such as tetrabutylammonium bromide, though not essential for a successful reaction, reduces the reaction time significantly. When this catalyst is used, the *by-product* will begin to precipitate shortly after the reaction begins and will be removed by filtration at the end of the reaction period. Extracting this solid with ether removes some 2-furfurylidenecyclopentanone, which is combined with that isolated from the reaction mixture. This product is a low-melting solid that can be separated from the unreacted starting materials by vacuum distillation and purified by recrystallization. It should be possible to deduce the structure of the by-product after studying its NMR spectrum and (if necessary) reviewing the aldol and related condensation reactions.

The reactivity of cyclic ketones decreases as ring size increases; cyclooctanone and higher homologs give very low yields.

Because furfural can be absorbed through the skin and the products will stain your hands yellow, it is advisable to wear rubber gloves during the reaction and much of the workup.

A comparison with the NMR spectrum of 2-furfurylidenecyclopentanone may also assist you in interpreting the by-product's spectrum.

PRELAB EXERCISE

Calculate the mass and volume of 100 mmol each of cyclopentanone and furfural.

Reaction and Properties

cyclopentanone   furfural   2-furfurylidenecyclopentanone

Table 1.   Physical properties

	M.W.	m.p.	b.p.	d.
furfural	96.1	−39	162	1.159
cyclopentanone	84.1	−51	131	0.949
2-furfurylidenecyclopentanone	162.2	60.5	154^{15}	

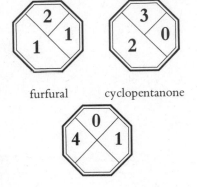

furfural   cyclopentanone

sodium hydroxide

Use a magnetic stirrer if one is available.

Cool the filter flask in ice water to retard evaporation of ether.

If solid matter remains in the ether after washing, it can be removed by gravity filtration.

The starting materials should distill below 95° with a reasonably good vacuum.

Hazards

Furfural irritates the skin, eyes, and respiratory system and its vapors are toxic; avoid inhalation and contact. Cyclopentanone is flammable and somewhat toxic; do not inhale its vapors, avoid unnecessary contact.

PROCEDURE

Reaction.   In a 250 ml Erlenmeyer flask combine 100 mmol cyclopentanone, 45 ml diethyl ether, 90 ml of 0.1M aqueous sodium hydroxide, and 1 g of tetrabutylammonium bromide. Cool the mixture to 5° in ice water ⟨OP–8⟩ and add 100 mmol of furfural (*Care:* see Hazards). Stopper the flask and shake it, vigorously and continuously, for 20 minutes. During this period, swirl the flask in the ice water bath frequently to prevent pressure buildup from vaporizing ether. Then let the reaction flask stand in an ice bath for 10 minutes or more.

Separation.   Isolate the solid by-product by vacuum filtration ⟨OP–12⟩ (save the filtrate!) and extract the solid ⟨OP–13d⟩ with 25 ml of ethyl ether. Combine this ether and the filtrate in a separatory funnel and use the ether to extract ⟨OP–13⟩ additional 2-furfurylidenecyclopentanone from the aqueous layer. Then extract the aqueous layer with a fresh 25 ml portion of ethyl ether and combine the extracts. Wash the ether solution ⟨OP–19⟩ twice with saturated aqueous sodium chloride, dry it over magnesium sulfate ⟨OP–20⟩, and evaporate the solvent ⟨OP–14⟩ until most of the ether has been removed. Transfer the residue to a 50 ml boiling flask and assemble an apparatus for semi-micro vacuum distillation ⟨OP–26a⟩. Distill off any residual ether and unreacted starting materials under reduced pressure, using a steam bath after the ether is gone. Then vacuum distill the 2-furfurylidenecyclopentanone with a heating mantle or oil bath. (See ⟨OP–25*b*⟩ for the procedure to be used when distilling solids.)

Flames should be avoided because of the ether used in this experiment.

Purification and Analysis. Purify the 2-furfurylidenecyclopentanone by recrystallizing it from aqueous ethanol ⟨OP–23*b*⟩, dry it ⟨OP–21⟩, and obtain its mass ⟨OP–5⟩ and melting point ⟨OP–28⟩.

Recrystallize ⟨OP–23⟩ the by-product from ethanol and obtain its mass ⟨OP–5⟩ and melting point ⟨OP–28⟩. Record its NMR spectrum ⟨OP–34⟩ in deuterochloroform or another suitable NMR solvent. An NMR spectrum of 2-furfurylidenecyclopentanone can be obtained for comparison, if desired.

A passable NMR spectrum can be obtained in ordinary methylene chloride.

Experimental Variations

1. See *J. Chem. Educ. 48,* 204 (1971) and *J. Chem. Educ. 55,* 339 (1978) for experiments using NMR and other methods to determine the stereochemistry of Claisen-Schmidt condensation products.

2. See *J. Chem. Educ. 49,* 836 (1972) for a procedure for isolating and characterizing furfural derived from natural products.

3. Other experiments involving aldol-type condensations are described in *J. Chem. Educ. 52,* 397 (1975) (synthesis of α- and β-ionone from citral); *J. Chem. Educ. 51,* 64 (1974) (synthesis of 9-benzalfluorene and its reduction to 9-benzylfluorene); and Vogel, 4th ed., pp. 793–96 (Bib–B4) (preparation of benzylideneacetone, dibenzylideneacetone, furfurylideneacetone, and benzylideneacetophenone).

Topics for Report

1. (*a*) Draw the structure of the by-product obtained in this reaction and state the reasoning that led you to your conclusion. (*b*) Assign the signals in the NMR spectrum of the by-product to specific protons.

There should be four downfield signals, two of them close together.

2. Write a mechanism for the reaction of furfural with cyclopentanone to form (*a*) 2-furfurylidenecyclopentanone, and (*b*) the by-product.

3. Write a flow diagram for your synthesis, showing how the product is separated from starting materials and by-products during isolation and purification.

Cis and *trans* are sometimes used rather than *Z* and *E*, although the terms are not synonymous.

4. Discuss the role of the phase-transfer catalyst in this reaction, using a diagram similar to that in Figure 1, Experiment 12 to illustrate a feasible process.

Library Topics

1. Write a short paper on the use of NMR to study geometric isomerism, telling how chemical shift values and coupling constants are used to determine whether a compound is the *Z* or *E* isomer.

2. Report on the ring-substitution reactions of furan derivatives, giving the preferred orientation of substitution and showing how they may differ mechanistically from the analogous reactions of benzene derivatives.

The Preparation of Azo Dyes

Reactions of Amines. Diazonium Salts. Electrophilic Aromatic Substitution. Dyes and Dyeing.

Operations

⟨OP–5⟩ Weighing

⟨OP–8⟩ Cooling

⟨OP–12⟩ Vacuum Filtration

⟨OP–21⟩ Drying Solids

Experimental Objective

To prepare Para Red and two or more additional azo dyes and use them to dye cloth.

Learning Objectives

To learn how to carry out diazotization and coupling reactions.

To learn about the history of the dye industry and some of the processes used in dyeing cloth.

SITUATION

You are a chemistry professor at that prestigious institution of higher learning, Kingston University. The student body has just voted unanimously to change the name of the university football team from the Kingston Artichokes to the Kingston Tigers, in the process changing the school colors from artichoke-green and white to orange and black. As a result, all of the old Kingston pennants, sweatshirts, and other paraphernalia have suddenly become obsolete. Since the change threatens to wipe out most of his inventory, the university bookstore manager has begun to suffer from acute hyperacidity. He has begged you to come up

"American Flag Red"
(Para Red)

Murphy's Law states: "If anything can possibly go wrong, it will." Serendipity refers to those rare occasions when things may go very wrong but everything turns out all right in the end.

See Bib–L45 for the fascinating story of the preparation of the "first" synthetic dye, pittacal.

with a solution to his problem—an orange dye that, when applied to green and white items, will turn them orange and black—before he runs out of Rolaids.

You recall from your undergraduate organic chemistry course that an azo dye known as Para Red was once used to dye the stripes in the American flag. By making minor modifications to the structure of the dye, you believe you can prepare an orange dye that will do the job. Your approach will be to first prepare "American Flag Red" and then, using the same general procedure but different components, produce more azo dyes until a suitable shade of orange is obtained.

BACKGROUND

Dyes and Serendipity

The ability of certain plant materials to dye cloth in different colors was probably known long before people began to keep written records. The rapid advancement of organic chemistry as a science is due, in large part, to the discovery that synthetic colors could be prepared that were in many ways superior to the natural ones. The first commercially important synthetic dye was made almost by accident when an 18-year-old research assistant, William Henry Perkin, naively tried to synthesize quinine by oxidizing allyltoluidine. Today we know that he had attempted the impossible—the structure of allyltoluidine bears hardly any resemblance to that of quinine, and the total synthesis of quinine was nearly ninety years in the future. But in 1856 the structural theory of organic chemistry was in a primitive stage of development and Perkin had only the molecular formulas to go on:

Perkin's idea

allyltoluidine   quinine

allyltoluidine's formula is $C_{10}H_{13}N$ while that of quinine is $C_{20}H_{24}N_2O_2$, so by adding a little oxygen and eliminating a molecule of water from two of allyltoluidine . . . at least it seemed like a good idea at the time! So Perkin painstakingly oxidized allyltoluidine with potassium dichromate and came up with a reddish-brown precipitate that was definitely not quinine. Most chemists would have thrown out the stuff and started over, but it had properties that interested him, so Perkin decided to try the same reaction with a simpler base, aniline. This time he obtained a black precipitate that, when extracted by ethanol, formed a beautiful purple solution that greatly impressed some of the local dyers. Perkin knew a good thing when he saw it, so he promptly gave up his study of chemistry and went into the business of manufacturing "aniline purple," or mauve as the dye soon came to be known.

While Perkin was getting the synthetic-dye industry under way, other chemists were experimenting with aniline and the vast array of other compounds that could be extracted from coal tar. One of these was a brewery chemist named Peter Griess, who took time off from the brewing of Alsopps' Pale Ale to discover the azo dyes. Undiscouraged by the fact that many of the diazo compounds he prepared had an unfortunate tendency to explode, Griess did some fundamental research into the diazotization of aromatic amines and went on to discover the coupling reaction by which virtually all azo colors are now synthesized. Aniline Yellow and Bismark Brown were developed in the 1860s; since then, the production of azo dyes has grown very rapidly. The number of azo dye formulations now outstrips that of all other dyes put together.

aniline

Perkin's dyestuffs plant was so successful that he was able to retire at the age of 36 and devote the rest of his life to pure research. Read an account of his remarkable career in Scientific American, Feb. 1957, p. 110.

The "aniline" of Perkin's time contained toluidine isomers that were incorporated into the structure of mauve.

mauve

An account of Griess's work can be found in J. Chem. Educ. 35, 187 (1958).

Aniline Yellow

Bismark Brown

Azo dyes are used to dye cloth by several different processes. In the *direct* process, the acidic or basic form of the dye is dissolved in water, the solution is heated, and the cloth to be dyed is immersed in the hot solution. The dye molecules attach themselves to the cloth fibers by direct chemical interactions. In the *disperse* dyeing process, a water-insoluble dye is suspended in water and a small amount of a carrier substance is added. The carrier dissolves the dye and carries it into the fibers, where it becomes trapped because of its hydrophobic nature. *Ingrain* (developed) dyes are synthesized right inside the fiber. Rather than mixing the diazonium salt and the coupling component directly, one immerses the cloth in solutions of the two separately. The relatively small molecules of the separate components can diffuse into the spaces between the fibers, but once they have combined within the fiber to form the dye, the larger dye molecules are trapped there.

METHODOLOGY

See the next section for the reactions used in preparing Para Red.

Diazotization

$$ArNH_2 \xrightarrow{\text{HONO}} Ar-N_2^+$$
$$\text{diazonium salt}$$

Coupling

$$Ar-N_2^+ \ + \ H-Ar' \xrightarrow{-H^+}$$
$$\text{coupling}$$
$$\text{component}$$

$$Ar-N=N-Ar'$$
$$\text{azo compound}$$

(Ar' must contain an activating group such as —OH or —NR$_2$.)

If the amine is insoluble, the diazotization is carried out in suspension, with stirring. To obtain as fine a suspension as possible, the amine is dissolved with heating and the solution is cooled rapidly with stirring.

The preparation of an azo compound involves two stages, known as *diazotization* and *coupling*. In the first, a primary aromatic amine reacts with nitrous acid (HONO), forming a diazonium salt. The nitrous acid is generated *in situ* from sodium nitrite and a mineral acid; the reaction is carried out at a low temperature so as not to decompose the diazonium salt. In the second stage, the diazonium salt is added to a weakly acidic or basic solution of the coupling component, which is usually a phenol or another amine. Phenols couple most readily in mildly alkaline solutions, whereas amines react best in acid. However, too low a pH will prevent an amine from reacting by causing protonation of the amino group, whereas too high a pH will cause the diazonium salt to change to a diazotate ion, which is incapable of coupling. The coupling reaction is basically an electrophilic aromatic substitution reaction with the diazonium salt acting as the electrophile. Because the diazo ($-\overset{+}{N}\equiv N$) group is only weakly electrophilic, the coupling component must contain strongly activating groups, such as OH or NR$_2$, in order for the reaction to occur.

A diazonium salt can be prepared by dissolving a primary amine in about 2.5 equivalents of dilute hydrochloric acid (or other suitable acid) in a flask, cooling the solution to 5°C or below in an ice-salt bath, and adding an equivalent amount of an aqueous solution of sodium nitrite while keeping the temperature at or below 10°C. Toward the end of the addition, the solution can be tested for excess nitrous acid using starch-iodide paper (a blue spot is a positive test). Nitrite is added until the test remains positive for a minute or so after the last addition. An excess of nitrous acid can be eliminated by adding a little urea to the solution.

Formation of unreactive species at low and high pH

Low pH: $ArNR_2 \underset{}{\overset{H^+}{\rightleftharpoons}} ArNHR_2^+$

High pH: $ArN_2^+ \underset{}{\overset{OH^-}{\rightleftharpoons}} ArN{=}N{-}O^-$
$$\text{diazotate}$$
$$\text{ion}$$

The coupling reaction is carried out by adding the diazo compound, with cooling and stirring, to a solution of a coupling component in dilute acid or base. If the coupling component is a phenol, it can be dissolved in about 2 equivalents of 1M sodium hydroxide and cooled before adding the diazonium salt solution. The pH should be adjusted, if necessary, after the addition; the azo dye crystallizes out on cooling. If the coupling component is an amine, it should be dissolved in 1 equivalent of 1M HCl; after the diazo component is added and coupling is complete (about 10–15 minutes) the solution is neutralized to litmus paper (with cooling) by adding 3M aqueous sodium carbonate.

A diamine requires 2 moles of HCl per mole of the amine.

The sodium carbonate solution should be added slowly to reduce foaming.

Since some dyes are more soluble than Para Red, it may be advisable to add about 2 g of sodium chloride to the mixture of diazo and coupling components before cooling.

The NaCl decreases the solubility of the dye by the salting-out effect (see ⟨OP–13c⟩).

PRELAB EXERCISE

Calculate the mass of 10 mmol each of *p*-nitroaniline and 2-naphthol. Calculate the volume of 25 mmol of 3M HCl, 20 mmol of 1M NaOH, and 10 mmol of 3M aqueous sodium nitrite.

Reactions and Properties

Table 1.   Diazo and coupling components used for preparing azo dyes

Coupling Component	M.W.	Diazo Component	M.W.
aniline	93.1	aniline	93.1
N-methylaniline	107	m-nitroaniline	138
N,N-dimethylaniline	121	p-nitroaniline	138
m-phenylenediamine	108	m-toluidine	107
		p-toluidine	107
phenol	94.1	m-anisidine	123
2-naphthol	144	p-anisidine	123

Equations are given for the preparation of Para Red only.

Diazotization

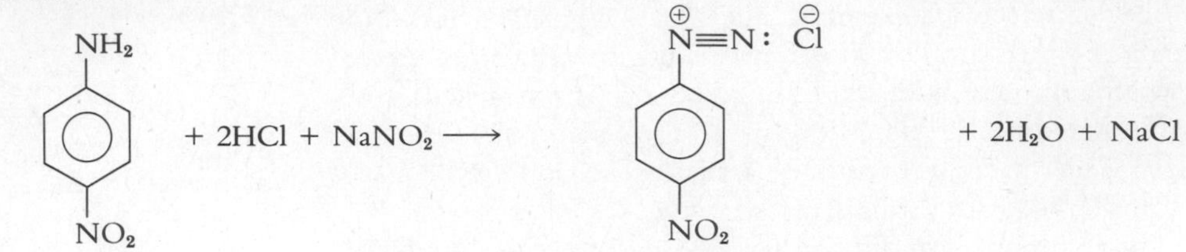

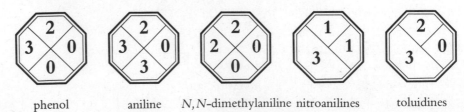

p-nitroaniline $+ 2HCl + NaNO_2 \longrightarrow$ p-nitrobenzenediazonium chloride $+ 2H_2O + NaCl$

Coupling

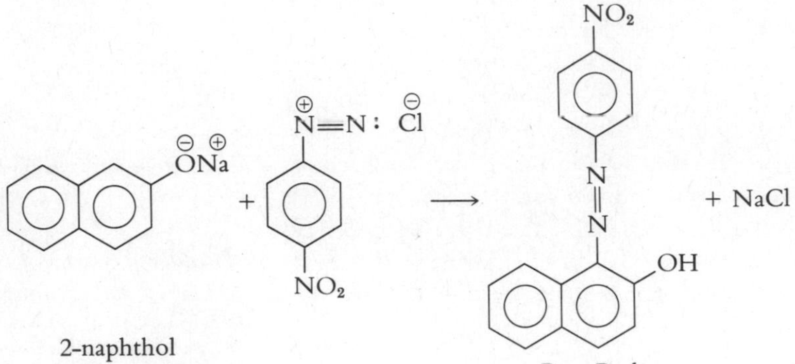

2-naphthol
(sodium salt) $+$ $\longrightarrow$ Para Red $+ NaCl$

Hazards

Aromatic amines such as aniline and the toluidines are toxic and can cause poisoning by inhalation, ingestion, or skin absorption. Prevent contact with skin, eyes, and clothing; do not inhale vapors. Phenols are toxic and corrosive; poisoning can result from ingestion or skin absorption. Prevent contact with skin, eyes, and clothing.

phenol 2 3 0 0   aniline 2 3 0 3   N,N-dimethylaniline 2 2 0 0   nitroanilines 1 3 1   toluidines 2 3 0

PROCEDURE

A. Preparation of Para Red

Stir 10 mmol of the *coupling component,* 2-naphthol, with about 20 mmol of 1M aqueous sodium hydroxide, warming the solution to dissolve the naphthol. Divide the solution into two equal parts, labeled 1c and 2c. Cool solution 1c in an ice bath ⟨OP–8⟩ and set it aside temporarily. Dilute solution 2c to 100 ml with water and prepare a piece of cloth (cotton or wool) for dyeing by soaking it in the solution for 2–3 minutes. Remove the excess solution by blotting the cloth between two towels and hang it up to dry.

The naphthol may not dissolve completely.

Students who are not careful to keep the solutions off their skin may be caught "red-handed." Rubber gloves are recommended.

Prepare a solution of the *diazo component,* p-nitrobenzenediazonium chloride, by dissolving 10 mmol of p-nitroaniline in 25 mmol of 3M aqueous hydrochloric acid (heating may be required). Cool ⟨OP–8⟩ the solution to 5°C in an ice bath and add 10 mmol of 3M sodium nitrite slowly enough to keep the temperature below 10°C during the addition. Divide the cold solution into two equal parts labeled 1d and 2d and use them in the following step as soon as possible.

Some precipitate may remain after the nitrite addition.

Add solution 1d, with stirring, to solution 1c and leave the mixture in an ice bath to allow complete crystallization. Dilute solution 2d to 100 ml with ice water and soak the treated cloth in it for a few minutes, then remove the cloth, rinse it with water, blot it, and hang it up to dry. Recover the Para Red from solution 1 by vacuum filtration ⟨OP–12⟩, wash it thoroughly with water, then dry ⟨OP–21⟩ and weigh ⟨OP–5⟩ it.

If a precipitate does not form, adjust the pH with dilute HCl or NaOH.

A follow-up washing with cold ethanol or methanol will speed up the drying of Para Red.

B. Preparation of Other Azo Dyes

From the list of diazo and coupling components in Table 1, choose enough combinations to prepare two or more different azo dyes. Follow the general directions given for Para Red, making any necessary changes described in the Methodology. Submit a dyed cloth and a dry crystalline product for each dye prepared.

Students may wish to work together in small groups (each student preparing a different set of dyes) and compare their results. Any of the coupling components in Table 1 can be used in combination with any diazo component.

Experimental Variations

1.   Para Red and other dyes can be used for direct dyeing by suspending $\frac{1}{2}$ g or more of the dye in 100 ml of hot water, acidifying the mixture with a few drops of concentrated sulfuric acid, and immersing pieces of wool or cotton in the mixture for about 5 minutes. The cloth is then removed and rinsed with water. With dyes other than Para Red, it may be necessary to vary the pH with acid or alkali to determine the optimum pH for dyeing.

2. Para Red and other water-insoluble dyes can be used for disperse dyeing by suspending $\frac{1}{2}$ g or more of the dye in 100 ml of hot water, stirring in 0.1 g of biphenyl (the carrier) and 2–3 drops of liquid detergent, then immersing a piece of Dacron or other polyester cloth in the mixture and heating on a steam bath for 15–20 minutes.

3. A special Multifiber Fabric can be used to test the effectiveness of a dye on different types of cloth. Multifiber Fabric 10A, for example, has the six fabrics wool, acrylic, polyester, nylon, cotton, and acetate rayon woven in sequence.

4. The preparation of other azo dyes is described in *Vogel's Textbook of Practical Organic Chemistry,* 4th ed., pp. 714–19 (Bib–B4).

MINILAB 15   # Preparation of *p*-Iodonitrobenzene

The reported melting point is 174°C.

Diazotize 10 mmol of *p*-nitroaniline as described above, filter the cold solution, and add the filtrate, with stirring, to a solution of 2.5 g of potassium iodide dissolved in 8 ml of water. Collect the product by vacuum filtration, recrystallize it from ethanol, and obtain the yield and melting point. Write equations for the reactions involved.

Topics for Report

1. Tabulate your results and those of the other members in your group (if applicable) and discuss the effect of azo dye structure on the colors obtained.

2. Write balanced equations and reaction mechanisms for each of the coupling reactions you carried out.

3. Why is it important to keep the temperature low during diazotization and coupling? Give the structure of the product that might form if the solution is heated during the diazotization of *p*-nitroaniline. Write an equation for its formation.

4. Why does the coupling of *p*-nitrobenzenediazon-
ium chloride occur mainly *para* to the dimethylamino group
of *N,N*-dimethylaniline, but *ortho* to the OH group of
2-naphthol?

5. Outline the syntheses of Aniline Yellow and Bismark Brown starting with benzene.

Library Topics

1. Tell how indigo is used in the vat dyeing of cloth,
describing a typical dyeing operation and giving equations
for the reactions involved.

2. Write a short paper on the molecular origin of
color, telling how chromophores and auxochromes are involved in the process of light absorption and showing how
changes in molecular structure affect the wavelength of light
absorbed.

Synthesis and Antibacterial Activity of Drugs Related to Sulfapyridine

Reactions of Amines. Electrophilic Aromatic Substitution. Nucleophilic Acyl Substitution. Sulfonic Acid Derivatives. Antibacterial Drugs. Multistep Synthesis.

Operations

⟨OP–5⟩ Weighing

⟨OP–7⟩ Heating

⟨OP–7a⟩ Refluxing

⟨OP–8⟩ Cooling

⟨OP–11⟩ Gravity Filtration

⟨OP–12⟩ Vacuum Filtration

⟨OP–20⟩ Drying Liquids

⟨OP–21⟩ Drying Solids

⟨OP–22b⟩ Trapping Gases

⟨OP–23⟩ Recrystallization

⟨OP–28⟩ Melting Point

Experimental Objective

To prepare a sulfanilamidopyridine from acetanilide and assess its antibacterial activity.

Learning Objectives

To learn how to carry out a chlorosulfonation reaction and the amidation of a sulfonyl chloride.

To learn about the discovery and applications of the sulfa drugs and about the general methods used in preparing them.

To learn how to evaluate the antibacterial activity of a drug.

SITUATION

Although antibiotics are used to alleviate most bacterial infections today, sulfa drugs are still applied to the treatment of urinary infections, bacillary dysentery, and meningitis. They are also important in veterinary medicine and for treating patients sensitive to antibiotics. One of the most widely used sulfa drugs is 2-sulfanilamidopyridine (sulfapyridine for short), which, though very effective, has some harmful side effects. Your employer, a pharmaceutical specialties firm, wants you to come up with a drug that might replace sulfapyridine in certain applications.

One of the techniques used in drug development is to take a successful drug and make minor changes in its molecular structure to see how the modifications affect its biological activity. While looking over the old chemical literature on sulfa drugs, you learn that a positional isomer of sulfapyridine, 3-sulfanilamidopyridine, was synthesized in the 1930s and found to show antibacterial activity comparable to that of sulfapyridine. It was not produced or extensively tested at that time because one of the starting materials was too expensive, but your company has assured you that if you come up with a suitable drug, they will find a way to manufacture it on a cost-effective basis.

sulfapyridine
(2-sulfanilamidopyridine)

3-Aminopyridine is about twice as costly as 2-aminopyridine today.

BACKGROUND

The First Wonder Drugs

The history of the sulfa drugs effectively began in 1932 when a German dye factory (I.G. Farbenindustrie) patented a rather ordinary azo dye called Prontosil, which incorporated a sulfamoyl ($-SO_2NH_2$) group to improve its re-

Prontosil

sistance to fading. Because a German chemist named Paul Ehrlich believed that some dyes might be useful as therapeutic agents, Prontosil was routinely tested for effectiveness against bacterial cultures but failed to show any antibacterial activity. Then Gerhard Domagk, a pharmacologist working for I. G. Farben, decided to test it on mice infected with he-

Domagk was later awarded a Nobel prize for his discovery, but Hitler did not permit him to accept it.

CO$_2$H   SO$_2$NH$_2$

NH$_2$ NH$_2$

PABA   sulfanilamide

molytic streptococci. When the mice got well, Domagk published his results, and the world had its first important antibacterial drug—or nearly so. It remained for a group of French chemists at the Pasteur Institute to prove that Prontosil breaks down in biological systems to form *sulfanilamide,* which is the true antibacterial agent. While Prontosil is effective only *in vivo* (in the body), sulfanilamide is biologically active both *in vivo* and *in vitro* (in cultures outside a living host).

Sulfanilamide does not actually kill bacteria; rather it inhibits their growth by depriving them of the essential nutrient folic acid and its derivatives. Mammals get their folic acid ready-made in the food they eat. Bacteria, however, must make their own in order to function, and sulfanilamide interrupts this process because of its close resemblance to *p*-aminobenzoic acid (PABA). A bacterium synthesizes folic acid by combining a molecule of pteridine with one each of PABA and glutamic acid. This process is directed by certain

H$_2$N

N   N

N

OH   N

CH$_2$ pteridine residue

NH

PABA residue

C=O

NH   glutamic acid residue

HOOCCH$_2$CH$_2$CHCOOH

folic acid

enzymes, including one called dihydropteroate synthetase that is responsible for attaching a PABA residue onto the pteridine. When a molecule of sulfanilamide is present, the enzyme apparently mistakes it for a molecule of PABA and uses it instead, taking the pteridine out of circulation and reducing folic acid synthesis to a critical level.

The discovery of the antibiotic effects of sulfanilamide led to an explosion of interest in these compounds, and over 5000 derivatives were prepared during the next ten years. Derivatives such as sulfapyridine and sulfathiazole were found to have many times the activity of sulfanilamide itself. These soon came into wide use for combating systemic dis-

Another theory of sulfa drug action holds that sulfanilamide merely binds to the active site on the enzyme and blocks out the PABA.

eases involving bacterial infections. However, these and other sulfa drugs had some undesirable characteristics, such as a tendency to crystallize out in the kidneys and block the renal passages, with sometimes fatal results. The "sulfonamide era," which extended roughly from 1935 to 1943, came to an abrupt end when penicillin was developed—production of sulfa drugs dropped from an all-time high of 10 million pounds in 1943 to less than half that amount the following year.

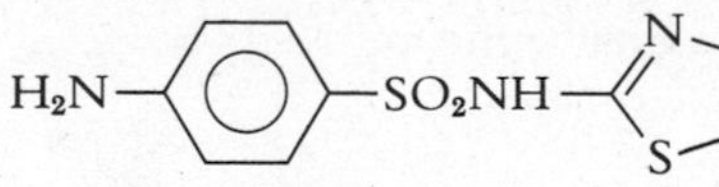

Although sulfa drugs are now seldom used—except in veterinary medicine—to combat most kinds of systemic bacterial infection, the sulfonamide era led to improvements in therapeutic techniques, stimulated research on antibiotic drugs, advanced the theory of drug design, and resulted in the development of many related drugs for specific illnesses.

METHODOLOGY

The usual synthesis of sulfanilamide and its derivatives is carried out by combining *p*-acetamidobenzenesulfonyl chloride (also called acetylsulfanilyl chloride or ASC) with ammonia or an amine, followed by hydrolysis to remove the acetyl group. Acetylsulfanilyl chloride itself is prepared by chlorosulfonating acetanilide, which is easily obtained from aniline. In this experiment, you will prepare a sulfanil-

sulfathiazole

Synthesis of sulfonamides

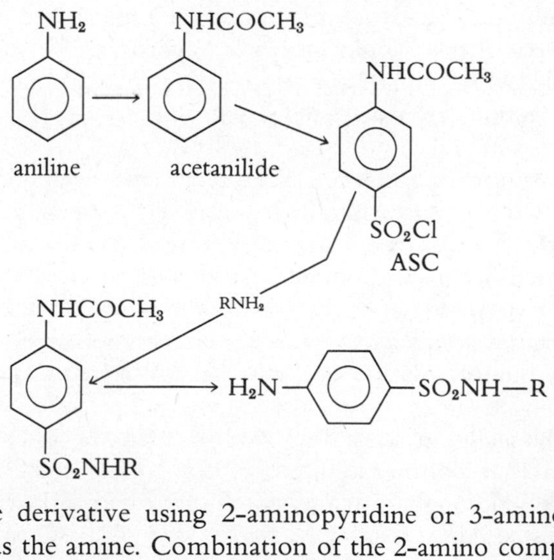

aniline   acetanilide

ASC

amide derivative using 2-aminopyridine or 3-aminopyridine as the amine. Combination of the 2-amino compound with ASC yields (after hydrolysis) 2-sulfanilamidopyridine,

The aminopyridines

2-aminopyridine

3-aminopyridine

pyridine

sulfonamide   carboxamide
linkage   linkage

which was the first sulfa drug to incorporate a heterocyclic ring into its structure; it is still one of the most effective. 3-Aminopyridine reacts with ASC to yield the isomeric 3-sulfanilamidopyridine. This compound was never used extensively as an antibacterial, perhaps because of the difficulty of synthesizing 3-aminopyridine.

Chlorosulfonation is an electrophilic aromatic substitution reaction that results in the transformation ArH → ArSO$_2$Cl. It is usually carried out by treating the aromatic compound with an excess of chlorosulfonic acid (ClSO$_3$H). Since gaseous HCl is evolved during the reaction, a gas trap must be used. After the reaction is complete, the reaction mixture is poured onto crushed ice to destroy the excess sulfonic acid and dissolve the inorganic products of the reaction. It is essential to remove all water from the crude acetylsulfanilyl chloride because this compound readily hydrolyzes to the corresponding sulfonic acid, which inhibits its reaction with the aminopyridines. This is accomplished by dissolving the partly dried ASC in acetone, filtering off the insoluble sulfonic acid, and drying the solution thoroughly with drying agents—magnesium sulfate to remove the bulk of the water and calcium sulfate to pick up the last traces.

Sulfonamides are synthesized by methods similar to those used in preparing carboxamides from carboxylic acid chlorides and amines. Unsubstituted sulfonamides are formed by adding the sulfonyl chloride solution to an excess of ammonia. The extra NH$_3$ ties up HCl (generated during the reaction) that would otherwise hydrolyze the sulfonyl chloride. Most amines are less reactive than ammonia, so their reactions are conducted using anhydrous solvents to keep the sulfonyl chloride from hydrolyzing under the more vigorous reaction conditions required. Since the aminopyridines used in this experiment are relatively expensive, keeping them in excess is not practical; instead, the reaction will be carried out in dry acetone (as the solvent) with a small amount of pyridine to react with the evolved HCl. The product, a 4-acetamido-N-pyridylbenzenesulfonamide, crystallizes out of solution and can be collected by vacuum filtration.

The acetyl group is removed by refluxing the substituted acetamidobenzenesulfonamide in dilute sodium hydroxide. Although both the sulfonamide and carboxamide linkages could, in theory, be hydrolyzed by base, sulfonamide groups are much more resistant to hydrolysis and are unaffected under the reaction conditions. After the reaction mixture is neutralized, the sulfonamide can be isolated by filtration and recrystallized from aqueous acetone.

Many sulfonamides are unaffected even by fusion at 250° with 80% sodium hydroxide.

The biological testing of the drugs is carried out on nutrient agar plates inoculated with bacteria by introducing discs soaked in different sulfa drugs, incubating the plates, and observing the effect of each drug on the growth of the bacterial colonies. Their relative effectiveness can be roughly evaluated by measuring the size of the "zone of inhibition" surrounding the antibacterial agent. Since the experiment requires sterile conditions and special equipment, it should be performed in a separate section of the laboratory or, if possible, in a biology lab equipped with the necessary materials and apparatus.

PRELAB EXERCISE

Calculate the mass of 25 mmol of acetanilide and the mass of 20 mmol of your aminopyridine.

Reactions and Properties

A.

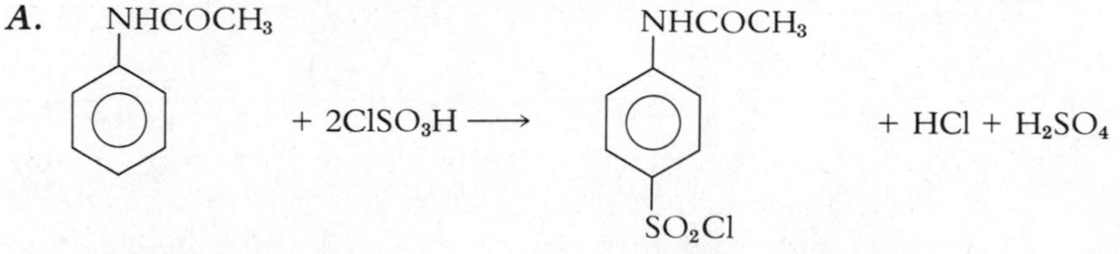
NHCOCH3
+ 2ClSO3H
NHCOCH3
SO2Cl
+ HCl + H2SO4

acetanilide — acetylsulfanilyl chloride

B.

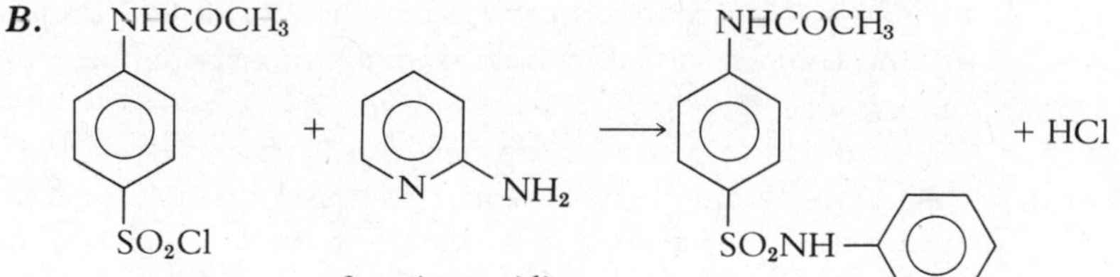
NHCOCH3
SO2Cl
+
N
NH2
NHCOCH3
SO2NH
N
+ HCl

2-aminopyridine

C.

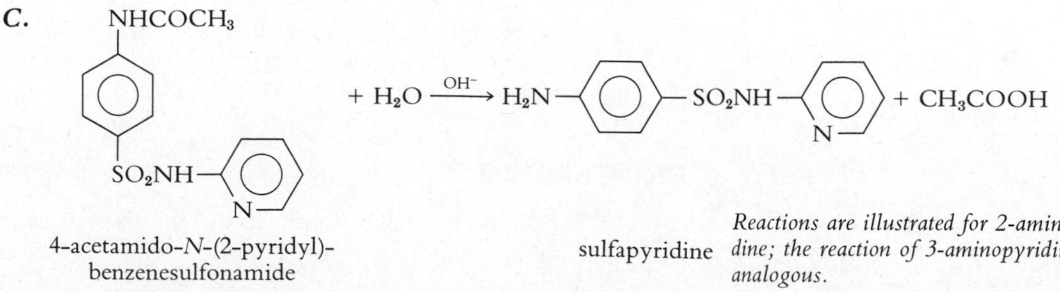
NHCOCH3
SO2NH
N
+ H2O
OH−
H2N
SO2NH
N
+ CH3COOH

4-acetamido-*N*-(2-pyridyl)-benzenesulfonamide — sulfapyridine

Reactions are illustrated for 2-aminopyridine; the reaction of 3-aminopyridine is analogous.

Table 1. Physical properties

	M.W.	m.p.	b.p.	d.
acetanilide	135.2	114	304	
chlorosulfonic acid	116.5	−80	158	1.766[18]
acetylsulfanilyl chloride	233.7	149		
2-aminopyridine	94.1	59–60	204	
3-aminopyridine	94.1	54–55	252	
pyridine	79.1	−42	115.5	0.982
2-sulfanilamido-pyridine	249.3	190–92		
3-sulfanilamido-pyridine	249.3	258–59		

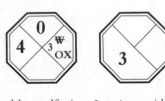

chlorosulfonic   2-aminopyridine
acid

pyridine

Hazards

Chlorosulfonic acid is a very corrosive liquid that causes severe burns and reacts violently with water to form hydrochloric and sulfuric acids. Its vapors are very irritating to the eyes and respiratory tract. Use gloves and a hood; avoid contact with skin, eyes, and clothing; do not breathe vapors; keep away from water. Acetylsulfanilyl chloride is a strong irritant; contact with skin, eyes, and clothing should be prevented. The aminopyridines are very poisonous—their action is similar to that of strychnine. Avoid contact with skin, eyes, and clothing; wash hands after use; do not breathe vapors. Acetone is very flammable and should not be used in the vicinity of open flames. Pyridine is toxic by inhalation, ingestion, and skin absorption. It irritates the skin and eyes and has a strong, unpleasant odor. It is also said to cause temporary impotence in males. Avoid inhalation and contact; handle under a fume hood.

PROCEDURE

Rubber gloves should be worn during this experiment, particularly while handling chlorosulfonic acid, the aminopyridines, and acetylsulfanilyl chloride. An efficient fume hood must be used during part **A**.

A. Preparation of Acetylsulfanilyl Chloride (ASC) from Acetanilide

Place 25 mmol of acetanilide in a dry 250 ml round-bottomed flask and melt it with a heat lamp ⟨OP–7⟩ or a small flame, then swirl to spread the acetanilide around the bottom half of the flask while it solidifies. Cool the acetanilide for ten minutes in an ice bath ⟨OP–8⟩ and attach a gas trap ⟨OP–22b⟩ directly to the reaction flask by means of a thermometer adapter or a rubber stopper. With the flask still in the ice bath, disconnect the gas trap and add 10 ml (about 150 mmol) of chlorosulfonic acid (*Hood!*) through a

*It is important that glassware used in parts **A** and **B** be thoroughly dried.*

Use dilute sodium hydroxide in the gas trap.

Chlorosulfonic acid is very corrosive and it reacts violently with water to release toxic gases. Handle with gloves under a hood and be prepared to take emergency action in case of spillage.

WARNING

---

dry funnel. Immediately connect the gas trap, remove the flask from the ice bath, and swirl it until the solid has completely dissolved; then heat the flask on a steam bath for ten minutes. Cool the reactants in an ice bath for ten minutes, then pour the liquid *cautiously*, in a thin stream with constant stirring, into a 150 ml beaker half full of cracked ice and water. Break up the resulting precipitate with a stirring rod and collect it by vacuum filtration ⟨OP–12⟩. Wash it with ice water and let it air dry on the filter for at least ten minutes. Press the crude ASC between large filter papers until it is friable (rubber gloves are recommended) then stir it with 20 ml of dry acetone until no more precipitate dissolves. Add 1 gram of anhydrous magnesium sulfate and swirl for 2–3 minutes, then filter the solution by gravity ⟨OP–11⟩ and wash the residue with a little dry acetone, adding the wash to the filtrate. Again dry the solution ⟨OP–20⟩ by the same procedure with another 1 g portion of magnesium sulfate, followed by a final drying with non-indicating calcium sulfate. Use enough wash acetone to keep the total volume of the ASC solution close to 20 ml.

The reaction is violent if the solution is added too fast.

Containers that held chlorosulfonic acid should be rinsed with acetone before adding water.

*The flocculent suspension of acetylsulfanilic acid forms by hydrolysis of ASC. Since 1-2% of this acid can prevent the reaction in Part **B**, it is important that the ASC solution be completely dry.*

The ASC solution should be used promptly in the next step.

B. *Reaction of Acetylsulfanilyl Chloride with 2- or 3-Aminopyridine*

The pyridine should be dried over KOH pellets and used under a hood.

Dissolve 20 mmol of 2- or 3-aminopyridine (*Caution:* see Hazards) in 5 ml of dry acetone in a small Erlenmeyer flask and add $2\frac{1}{2}$ ml of dry pyridine. Add the ASC solution from part *A*, swirl to mix, and watch the mixture for evidence of reaction. After any vigorous reaction has ceased, stopper the flask and set it aside for at least 24 hours. When the reaction is complete (any oil should have crystallized) cool the flask in an ice bath and collect the product by vacuum filtration ⟨OP–12⟩, washing it twice with a little ice cold 85% acetone.

If a precipitate forms immediately, break up any lumps by stirring and shaking.

To increase the yield of acetylated 2-sulfanilamidopyridine, it is advisable to concentrate the solution ⟨OP-14⟩ to about half its volume before filtering.

C. *Preparation of 2- or 3-Sulfanilamidopyridine*

If too much HCl is added the precipitate will dissolve; if this occurs, neutralize the solution with 3M NaOH.

Mix the product from part *B* with 15 ml of 3M sodium hydroxide and reflux the mixture ⟨OP–7a⟩ for 30 minutes. Transfer the cooled reaction mixture to a beaker and carefully make it *just* acidic (pH ~ 6.5) with 6M hydrochloric acid. Cool the mixture in an ice bath and collect the precipitate by vacuum filtration ⟨OP–12⟩, washing it with water. Recrystallize the air-dried sulfanilamidopyridine from 85% acetone ⟨OP–23⟩, using decolorizing carbon if necessary, filtering the hot solution when no more of the solid appears to dissolve. Dry the purified product ⟨OP–21⟩, weigh it ⟨OP–5⟩ and obtain its melting point ⟨OP–28⟩.

A rather large volume of the solvent may be needed, but be careful not to add too much.

D. *Evaluating the Antibacterial Activity of Sulfa Drugs (Optional)*

E. coli is suggested, but other safely handled bacteria may also be used.

Work in pairs, using a nutrient agar plate previously inoculated with a specific type of bacterium. Dissolve 0.10 g each of 2-sulfanilamidopyridine, 3-sulfanilamidopyridine, and sulfanilamide (as a reference compound) in separate flasks containing 30 ml of boiling 25% ethanol. Mark four quadrants on the bottom of the Petri dish with a grease pencil and number the quadrants 1–4; then write your names on the cover.

"Quadrant plates" having the quadrants already marked are available.

Throughout this process, care must be taken to protect the sample discs and the agar plate from contamination by undesirable bacteria. The filter paper and watch glass should be as sterile as possible.

Sterilize a metal-tipped forceps by dipping the ends in 70% ethanol and then placing them in a burner flame. Use it to dip a sterile paper disc into one of the solutions while it is still boiling. Transfer the disc to a piece of filter paper and cover it immediately with a watch glass. Repeat the process with the other solutions, sterilizing the forceps after each transfer. A fourth disc should be dipped into boiling distilled water for use as a control. When the discs are dry, again sterilize the forceps and transfer each disc to the appropriate quadrant inside the Petri dish: 1 for sulfanilamide, 2 for 2-sulfanilamidopyridine, etc.

The Petri dish should be opened by lifting the cover straight up, just high enough to allow room for the forceps, to keep out airborne bacteria.

Incubate the agar plate at 28–30°C in an incubator oven or temperature-controlled room and observe it after 24 and 48 hours of incubation. Holding the Petri dish up so that you can obseve it from the bottom, measure the diameter of the clear circle surrounding each sample tab and that surrounding the control. Calculate the area of each circle and subtract the area for the control from that for each sample.

The cloudiness on the plate is due to bacterial growth, so a clear area indicates inhibition.

Use a small metric ruler and measure the diameter in millimeters.

Compare **(a)** the relative effectiveness of the three sulfa drugs, as indicated by the areas of inhibition after 24 hours; and **(b)** their long-term persistence, as indicated by the decrease in size of the zone of inhibition between 24 and 48 hours. If the size of the zone does not decrease over that period, your product may contain bactericidal impurities that kill the bacteria rather than just inhibiting their growth.

It is a good idea to compare other students' results before drawing any conclusions about the relative effectiveness of the sulfapyridines, since the drug action is affected by impurities.

Experimental Variations

1. *p*-Acetamidobenzenesulfonyl chloride (ASC) can be used to prepare sulfanilamide according to the procedure in *Vogel's Textbook of Practical Organic Chemistry,* 4th ed., pp. 651–52 (Bib–B4). The activity of your sulfanilamide can then be tested according to the procedure given above or in Experimental Variation 2.

2. A procedure for biologically evaluating the purity of sulfanilamide preparations is described in *J. Chem. Educ. 52,* 676 (1975). Additional experiments with sulfa drugs on bacterial cultures are reported in *J. Chem. Educ. 70,* 76 (1973).

Topics for Report

1. Tabulate the test results of your group (and of other groups, if possible) and discuss the relative effectiveness of the three sulfa drugs in retarding the growth of bacteria.

2. Why was acetanilide, rather than aniline, used for the chlorosulfonation reaction in this synthesis, although its acetyl group had to be removed at the end of the procedure?

3. (*a*) Write equations for the decomposition of excess chlorosulfonic acid in water and for the formation of acetylsulfanilic acid as an impurity in step *A*. (*b*) Write equations to explain why your sulfanilamidopyridine is soluble in both acidic and basic solutions.

4. Write reasonable mechanisms for the reactions involved in the synthesis of your product.

5. Sulfapyridine has also been synthesized by the base-catalyzed reaction of 2-chloropyridine with sulfanilamide. Write a balanced equation and a mechanism for this reaction.

6. Write a flow diagram for the synthesis you carried out (see Topic 3*a* for by-products).

Library Topics

1. Some modern drugs like salicylazosulfapyridine, probenecid, tolbutamide, and dapsone bear structural relationships to the sulfa drugs. Give the structures of these drugs and report on their physiological action and applications.

2. Write a short paper on the early development and uses of the sulfa drugs, referring to *J. Chem. Educ. 19,* 167 (1942) for background and references.

Rate and Nucleophilicity in Substitution Reactions of 2,4-Dinitrochlorobenzene

EXPERIMENT 38

Aryl Halides. Nucleophilic Aromatic Substitution. Preparation of Amines. Reaction Kinetics. Colorimetric Analysis.

Operations

⟨OP–6⟩ Volume Measurement

⟨**OP–35***a*⟩ Colorimetric Analysis

Experimental Objectives

To measure the rate of the reaction between 2,4-dinitrochlorobenzene and a secondary amine and calculate the second-order rate constant for the reaction.

To determine the nucleophilic strengths of morpholine and piperidine in an S_NAr reaction.

Learning Objectives

To learn how to measure reaction rates by spectrophotometry and how to evaluate kinetic data for a second-order reaction.

To learn about nucleophilic substitution reactions of activated aryl halides.

SITUATION

Nair and Adams discovered a method of synthesizing substituted benzimidazoles starting with 2,4-dinitrochlorobenzene (or related aryl chlorides) and a cyclic secondary amine such as piperidine. While investigating the individual steps in this synthesis, you discovered that some cyclic amines react with 2,4-dinitrochlorobenzene more rapidly than others, and you concluded that this difference was

M. D. Nair and Roger Adams, *J. Am. Chem. Soc. 83,* 3518 (1961).

Synthesis of a piperidinobenzimidazole

2,4-dinitrochlorobenzene   piperidine   N-(2,4-dinitrophenyl)-
                                                    piperidine

caused by variations in the nucleophilic strengths of the amines. As part of your research project to optimize experimental conditions for the synthesis of benzimidazoles, you have decided to measure the reaction rates of some secondary amines with 2,4-dinitrochlorobenzene and determine their relative nucleophilicities with respect to this aryl halide.

BACKGROUND

The $S_N Ar$ Reaction

Nucleophilic aromatic substitution can occur (a) through an S_N1 mechanism, as in the substitution reactions of aryldiazonium salts; (b) by an elimination-addition mechanism involving an aryne intermediate; and (c) by a bimolecular addition-elimination route, often called the $S_N Ar$ reaction. In the bimolecular reaction (c), the nucleophile attacks an activated aromatic nucleus to form an intermediate complex, which then loses a leaving group to form the product.

The nature of the intermediate complex in the $S_N Ar$ reaction has long been the subject of speculation. The first evidence bearing on its structure was obtained in 1898 when Jackson and Boos mixed picryl chloride (2,4,6-trinitrochlorobenzene) with sodium methoxide in methanol and isolated a red salt, which was converted by ethanol to a second red salt. J. Meisenheimer prepared the second salt by two different methods, treating either 2,4,6-trinitroanisole with potassium ethoxide or 2,4,6-trinitrophenetole with potassium methoxide. He was then able to assign it the structure 1. Many other Meisenheimer complexes have since been prepared and characterized, and a large volume of experimental evidence indicates that the intermediates in $S_N Ar$ reactions are in fact σ-complexes of the Meisenheimer type. Their role in such reactions explains why only activated aromatic compounds undergo bimolecular substitution easily: nitro groups and similar electron-withdrawing substituents remove excess electron density from the aromatic nucleus, stabilizing the intermediate complex.

Preparation of Meisenheimer complexes

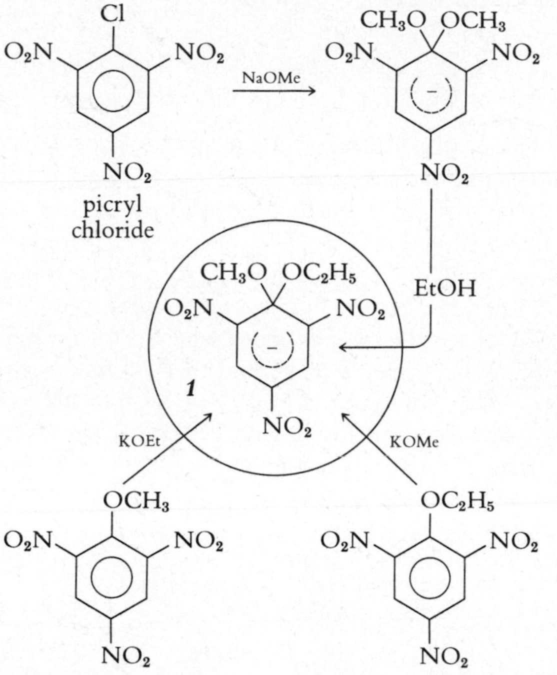

Cl

O_2N ⬡ NO_2

NO_2

picryl
chloride

NaOMe →

CH_3O OCH_3

O_2N ⬡ (−) NO_2

NO_2

EtOH

CH_3O OC_2H_5

O_2N ⬡ (−) NO_2

NO_2

1

KOEt ↗   ↖ KOMe

OCH_3

O_2N ⬡ NO_2

NO_2

2,4,6-trinitroanisole

OC_2H_5

O_2N ⬡ NO_2

NO_2

2,4,6-trinitrophenetole

The mechanism of most $S_N Ar$ reactions thus appears to be a simple two-step process involving the initial addition of the nucleophile to form a Meisenheimer complex, followed by the departure of the leaving group. If the first step of the reaction is rate limiting, the nucleophilic strength of the reactant—its ability to donate its electron pair to the substrate and form a sigma bond—should have a considerable effect on the reaction rate. The Swain-Scott equation

$$\log \frac{k}{k_0} = ns \qquad (1)$$

correlates reaction rates with respect to the strength of the nucleophile (n) and the sensitivity of the substrate to nucleophilic substitution (s). It has been applied widely in aliphatic systems, and some values of n for reactions of various nucleophiles with methyl iodide are given in Table 1. Attempts to use such correlations in $S_N Ar$ reactions have met

General $S_N Ar$ mechanism

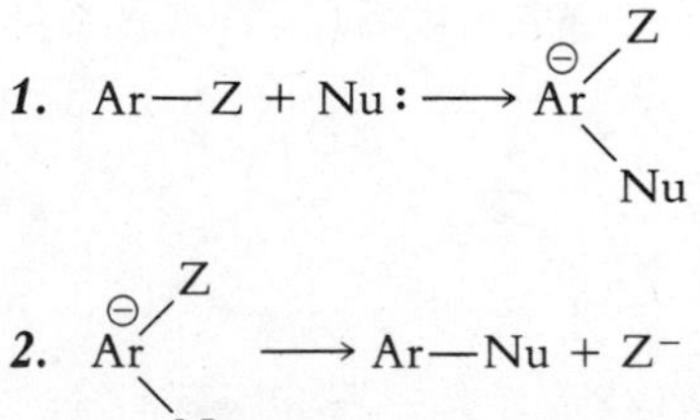

1. Ar—Z + Nu: ⟶ $Ar\overset{\ominus}{\underset{Nu}{\diagup}}^{Z}$

2. $Ar\overset{\ominus}{\underset{Nu}{\diagup}}^{Z}$ ⟶ Ar—Nu + Z⁻

(Ar must be activated by electron-withdrawing groups; Z is the leaving group and Nu: the nucleophile.)

Table 1. Nucleophilic strengths of various nucleophiles

Nucleophile	n
CH_3OH	0.00
F^-	2.7
Cl^-	4.37
pyridine	5.23
NH_3	5.50
aniline	5.70
Br^-	5.79
CH_3O^-	6.29
$(CH_3CH_2)_3N$	6.66
$(CH_3CH_2)_2NH$	7.0
pyrrolidine	7.23
piperidine	7.30
I^-	7.42

Note: n values are measured relative to methanol, with methyl iodide as the substrate.

In this experiment 2,4-dinitrochlorobenzene is the substrate, and the nucleophiles are piperidine and morpholine.

morpholine piperidine

The time should be measured in seconds.

with less success, since changing the substrate or solvent often changes the order of nucleophilic strength. However, an equation of this kind can be useful in comparing the nucleophilic strengths of various reactants with reference to the same class of substrates. For this experiment, we shall define a nucleophilicity parameter as follows:

$$n_{Ar} = \frac{1}{s} \log \frac{k}{k_0} \qquad (2)$$

The reaction of 2,4-dinitrochlorobenzene with ammonia in absolute ethanol, for which the rate constant (k_0) is 4.0×10^{-6} dm^{3}mol^{-1}s^{-1} at 25°C, is used as the reference reaction for Equation 2.

In this experiment, you will measure the reaction rates of the nucleophiles piperidine and morpholine with 2,4-dinitrochlorobenzene and use the rate constants to calculate their nucleophilic strengths. You should then be able to compare their n_{Ar} values with that of ammonia (for which n_{Ar} is zero) and to assess some of the factors that contribute to nucleophilic strength in S_NAr reactions.

METHODOLOGY

The reaction of 2,4-dinitrochlorobenzene with an amine is second order in both substrate and nucleophile and follows the general rate equation

$$\frac{dx}{dt} = k(S_0 - x)(N_0 - 2x) \qquad (3)$$

where x is the concentration of the product at time t and S_0 and N_0 are the initial concentrations of substrate and nucleophile, respectively. The computations can be simplified considerably if the experiment is carried out with the initial concentration of nucleophile being just twice that of the 2,4-dinitrochlorobenzene. Equation 3 then becomes $dx/dt = 2k(S_0 - x)^2$, and the integrated rate equation is

$$\frac{1}{(S_0 - x)} = 2kt + \frac{1}{S_0} \qquad (4)$$

By measuring the concentration of the product, x, at regular intervals during the reaction, one can calculate the term on the left side of Equation 4 which, when plotted versus time, should yield a straight line of slope $2k$.

The concentration of the product will be determined indirectly by measuring the absorbance of aliquots removed from the reaction mixture at various times. Each aliquot must first be quenched by adding dilute acid to stop the reaction; this is done so that the concentration will remain constant until you are ready to take the absorbance readings. The concentration term in Equation 4, $S_0 - x$, can be shown to be proportional to $A_\infty - A$, so that:

The N-(2,4-dinitrophenyl) amines are yellow-orange and absorb strongly in the visible region around 380 nm.

$$\frac{1}{(S_0 - x)} = \frac{q}{(A_\infty - A)} \qquad (5)$$

A_∞ is the absorbance when the reaction is 100% complete and A is the absorbance at time t.

The "infinity" value of the absorbance can be obtained by warming the reaction mixture to complete the reaction and measuring the absorbance of the resulting solution. The proportionality constant can then be calculated from the relationship $q = A_\infty / S_0$. Because of their different reaction rates, the amines will be used in different initial concentrations so that both reactions will be about half complete after 30 minutes. The absorbance values of the products at 380 nm are so high that, to obtain values in a convenient range, the solutions must be diluted before absorbance readings are taken.

Note: S_0 is the initial substrate concentration in the reaction mixture, not in the sample being analyzed spectrophotometrically.

PRELAB EXERCISE

Read ⟨OP–35*a*⟩ on the use of spectrophotometry for quantitative analysis.

Reaction and Properties

| 2,4-dinitrochlorobenzene | piperidine (Y = CH$_2$) or morpholine (Y = O) | dinitrophenylated amine | *The extra mole of amine combines with HCl liberated during the reaction.* |

Table 2. Physical properties

	M.W.	m.p.	b.p.	d.
2,4-dinitrochlorobenzene	202.6	53	315	
morpholine	87.1	−5	128	1.000
piperidine	85.2	−9	106	0.861

2,4-dinitrochlorobenzene

morpholine   piperidine

It is important that all glassware used in this experiment be clean and dry.

50 ml Erlenmeyer flasks are ideal, but large test tubes or other containers can also be used.

Rinse the pipet with the reaction mixture just before delivering each aliquot.

Swirl the reaction flask occasionally throughout the kinetic run.

Hazards

2,4-Dinitrochlorobenzene is poisonous by ingestion, inhalation of dust and skin contact. Prevent contact with skin, eyes, or clothing; wash hands after using. The secondary amines are toxic, flammable and irritating to the skin, eyes, and respiratory system. Do not breathe vapors; prevent contact with skin, eyes, and clothing.

PROCEDURE

Kinetic Runs

It may be necessary to work in groups of two or more in order to make the best use of equipment and facilitate the operations.

Using a small graduated cylinder, measure 18 ml of absolute ethanol into a 25-ml volumetric flask, then pipet in ⟨OP–6⟩ 4 ml of 0.10 M 2,4-dinitrochlorobenzene in ethanol.

Have ready seven numbered flasks with stoppers, each containing exactly 25 ml of the quenching solution (0.2 M sulfuric acid in 50% ethanol). Pipet 2 ml of 0.40M piperidine in ethanol into your volumetric flask, recording the time (or starting your timer) when about half of the solution has been delivered. Fill the flask to the mark with ethanol, stopper and shake it, and transfer the contents to a small Erlenmeyer flask. Pipet a 1 ml aliquot of this solution into the first quenching flask, recording the time when about half of the solution has drained out. Stopper both flasks and set the reaction flask aside until about five minutes have elapsed since the first measurement; then pipet another 1 ml aliquot into the second quenching flask, recording the time of addition as before. Repeat the process about every five minutes until six aliquots have been quenched. Warm the reaction flask in a 50° water bath for 2 hours or more to drive the reaction to completion. Then pipet a final 1 ml aliquot into quenching flask number seven to prepare the A_∞ solution.

Repeat the determination with morpholine, using 12 ml of ethanol, 10 ml of 0.10M 2,4-dinitrochlorobenzene/ethanol, and 2 ml of 1.0M morpholine/ethanol.

Absorbance Measurements

The ethanol should be measured with a volumetric pipet.

Dilute each solution by pipetting 1 ml into a numbered 15 cm test tube containing 10 ml of 95% ethanol, then stoppering and shaking the test tube. With the spectrophotometer set at 380 nm ⟨OP–35a⟩, adjust it for 100% transmittance with an ethanol blank. Then record the transmittance values for all of your solutions using the same instrument at the same wavelength. It should not be necessary to readjust the zero and 100% transmittance readings each time if the instrument has been warmed up and the readings are performed one after the other.

Use the same sample cuvette for all of your measurements, rinsing it after each reading with the next solution to be analyzed. Rinse it with ethanol after the last reading for each amine.

Calculations

1. Calculate the absorbance of each solution from $A = \log (1/T)$.

2. Calculate the value of the proportionality constant q for each amine from the absorbance of its "infinity" solution and the concentration of the substrate in its reaction mixture. Then calculate $1/(S_0 - x)$ for each aliquot from Equation 5.

3. Plot the data for each amine using Equation 4 and determine the value of k from the slope of the line.

4. Calculate the n_{Ar} values for both amines with Equation 2.

Since 2,4-dinitrochlorobenzene is the reference substrate, your reactions will have a sensitivity factor of 1.

Experimental Variations

1. The experimental infinity solution can also be obtained by leaving the reactants 48 hours or longer at about 25°.

2. Students can prepare their own stock solutions by weighing out the calculated amounts of amine and 2,4-dinitrochlorobenzene and diluting them with absolute ethanol. The amines should be weighed in closed containers, since they are hygroscopic.

3. Pure N-(2,4-dinitrophenyl)piperidine can be prepared by refluxing excess piperidine with 2,4-dinitrochlorobenzene for 15 minutes on a steam bath, stirring the reaction mixture into ice water, and recrystallizing the product from 95% ethanol. A solution of the "infinity" concentration can then be prepared and its absorbance compared with the A_∞ value from the kinetic measurements.

4. The reactions can also be followed by titrating the excess acid in the quenched solutions with sodium hydroxide. A procedure may be found in D. H. Rosenblatt and G. T. Davis, *Laboratory Course in Organic Chemistry* (Boston: Allyn & Bacon, 1971), p. 171.

MINILAB 16   # Synthesis of 2,4-Dinitrophenylhydrazine

Dissolve 0.5 g of 2,4-dinitrochlorobenzene in 7.5 ml of 95% ethanol in a test tube, heat it just to boiling, then add a

WARNING   *Hydrazine is very toxic and may be carcinogenic; 2,4-dinitrochlorobenzene is also toxic (see Hazards sections above), and contact with both chemicals must be prevented. 2,4-Dinitrophenylhydrazine is a high explosive and must be protected from heat or shock.*

solution containing 0.5 ml of 64% hydrazine dissolved in 2.5 ml of 95% ethanol. Let the solution cool 15–20 minutes, collect the product by semimicro vacuum filtration, and recrystallize it from ethyl acetate (about 50 ml per gram). Dry the product *at room temperature* and obtain the yield. Report your observations during the synthesis and write an equation and a mechanism for the reaction.

Topics for Report

1. Tabulate the k and n_{Ar} values obtained for the amines. Discuss the effect of amine structure on nucleophilicity, and attempt to correlate your results with amine basicity.

2. Write a mechanism for the reaction of 2,4-dinitrochlorobenzene with your amine.

3. (*a*) Derive Equation 5 using Beer's law and evaluate the constant q in terms of Beer's law parameters. (*b*) Derive Equation 4 from Equation 3.

pK_b values for ammonia, morpholine, and piperidine in aqueous solution are 4.75, 5.67, and 2.88 respectively.

4. Outline a synthesis of the spiro Meisenheimer complex shown, using picryl chloride and any appropriate inorganic and organic reactants.

5. (*a*) Calculate the concentrations of the dinitrophenylamines in the solutions used for the spectrophotometry infinity reading. (*b*) If you know the path length of the cuvette you used, calculate the molar absorptivities of these products at 380 nm.

6. Explain why "quenching" the reaction mixture with H_2SO_4 stops the reaction.

Library Topics

1. After reading the paper by Bunnett et al., tell what is meant by the "element effect" in reactions of 1-substituted-2,4-dinitrobenzenes, and explain how it was used to elucidate the mechanism of the S_NAr reaction. Explain why aryl fluorides showed anomalous behavior in the reactions studied.

2. Describe and compare some approaches toward correlating reactivities in nucleophilic substitution reactions, including the Winstein-Grunwald equation, the Swain-Scott equation, and the Edwards equation. Give some of the advantages and drawbacks of each approach.

a spiro Meisenheimer complex

Bunnett, Garbisch, and Pruitt, *J. Am. Chem. Soc. 79,* 385 (1957).

The Enamine Synthesis of 2-Propionylcyclohexanone

Carbonyl Compounds. Enamines. Carbanions. Nucleophilic Substitution. Preparation of Diketones. Multistep Synthesis.

Operations

⟨OP–5⟩ Weighing

⟨OP–7⟩ Heating

⟨OP–7a⟩ Refluxing

⟨OP–10⟩ Addition

⟨OP–13⟩ Extraction

⟨OP–13b⟩ Separation of Liquids

⟨OP–14⟩ Evaporation

⟨OP–20⟩ Drying Liquids

⟨OP–22a⟩ Exclusion of Moisture

⟨OP–25c⟩ Water Separation

⟨OP–26⟩ Vacuum Distillation

Experimental Objective

To prepare 2-propionylcyclohexanone by acylating an enamine.

Learning Objectives

To learn how to prepare, acylate, and hydrolyze an enamine.

To learn about the characteristics of enamines and some of their applications in organic synthesis.

SITUATION

Since the inception of the Stork reaction in 1954, many investigators have searched for new ways of using enamines in organic synthesis. For instance, Hunig, Lucke, and Benzing wrote an article entitled *"Kettenverlangerung von Carbonsauren um 6C-Atome"* in which they described the use of an enamine

Chem. Berichte 91, 129 (1958)

synthesis to increase the length of a carbon chain by six carbons. By this route acetic acid (as the anhydride or acid chloride) can be converted to octanoic acid, and other carboxylic acids can have their chains lengthened proportionately.

$$CH_3CH_2CH_2CH_2CH_2CH_2CH_2CH_2\overset{\displaystyle O}{\overset{\displaystyle \|}{C}}OH$$
pelargonic acid

Chain lengthening of acetic acid

$$CH_3\overset{\displaystyle O}{\overset{\displaystyle \|}{C}}OH \xrightarrow{\text{enamine synthesis}} CH_3CH_2CH_2CH_2CH_2CH_2CH_2\overset{\displaystyle O}{\overset{\displaystyle \|}{C}}OH$$
octanoic acid

Pelargonic acid (nonanoic acid) is a comparatively rare carboxylic acid that occurs naturally in Oil of Pelargonium and is used in manufacturing hydrotropic salts and other commercial products. You would like to develop a synthesis of pelargonic acid from propionic acid that might supplement the current industrial synthesis from oleic acid. First you must prepare 2-propionylcyclohexanone, an intermediate in the chain-lengthening process.

Hydrotropic salts increase the ability of water to dissolve sparingly soluble substances.

2-propionylcyclohexanone

BACKGROUND

Enamines and the Stork Reaction

In a brief "Communication to the Editor" in the 1954 *Journal of the American Chemical Society,* Gilbert Stork and his coworkers announced a new synthesis of 2-alkyl and 2-acyl ketones by way of intermediates known as enamines. Although enamines had been around for a long time (they are simply α-aminoalkenes, the nitrogen analogs of enols), Stork's publication was the first to suggest their vast potential in organic synthesis. The alkylation and acylation of carbonyl compounds through enamines (later to become known collectively as the Stork reaction) is only one application of enamines in synthesis, but it is still an important one.

Ketones can be alkylated by a base catalyzed reaction in which an α-proton is removed to form an enolate ion, which acts as a carbon nucleophile toward alkyl halides. However, multiple alkylation can also occur, with the second alkyl group usually ending up on the same α-carbon atom as the first. The requirement for a strongly basic catalyst (such as sodium methoxide) precludes the use of reactants having base-sensitive functional groups and can result in side reactions such as aldol condensations.

G. Stork, R. Terrell, and J. Szmuszkovicz, *J. Am. Chem. Soc.* **76**, 2029 (1954).

enol   enamine

Direct alkylation of a carbonyl compound

Formation of an enamine

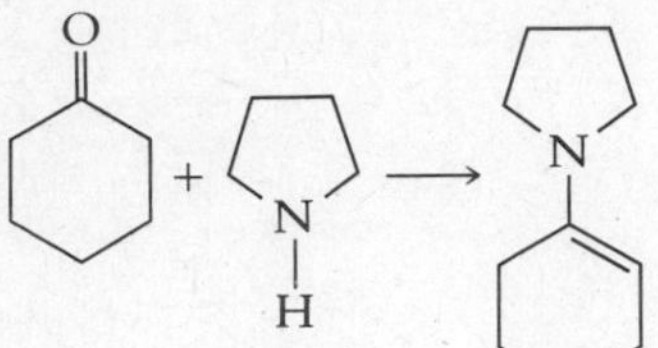

pyrrol-
idine   1

Alkylation and hydrolysis of an
enamine

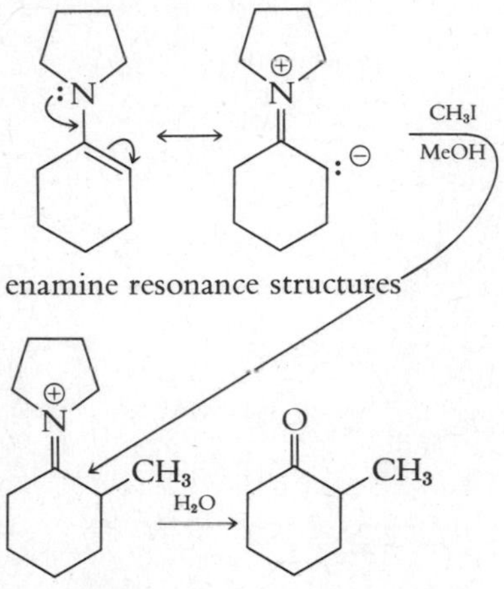

enamine resonance structures

iminium
salt

If a ketone such as cyclohexanone is condensed with a secondary amine such as pyrrolidine, it will, under the proper reaction conditions, yield the enamine 1-pyrrolidino-cyclohexene (1). This enamine can then be alkylated under neutral reaction conditions to yield a monoalkylated iminium salt, which can be hydrolyzed to the corresponding 2-alkyl ketone. Since one of the enamine's two resonance structures has a negative charge on the α-carbon atom, the alkylation step can be viewed as a nucleophilic substitution reaction with a carbanion acting as the nucleophile. Enamines thus provide an indirect method for preparing alkyl (and acyl) derivatives of carbonyl compounds, with the enamine functional group both activating the molecule and preventing unwanted side reactions.

Enamines can be acylated by acid chlorides and anhydrides; this provides a unique method for lengthening a carbon chain. Acetyl chloride reacts readily with 1-morpholinocyclohexene (2) to yield, after hydrolysis of the acylated enamine, 2-acetylcyclohexanone. Treatment of this diketone with strong base results in ring cleavage to form 7-oxooctanoic acid, which can be reduced by a Wolff-Kischner reaction to octanoic acid. Other carboxylic acids can also, via their acyl chlorides or anhydrides, undergo the same chain-lengthening process.

Enamine synthesis of octanoic acid

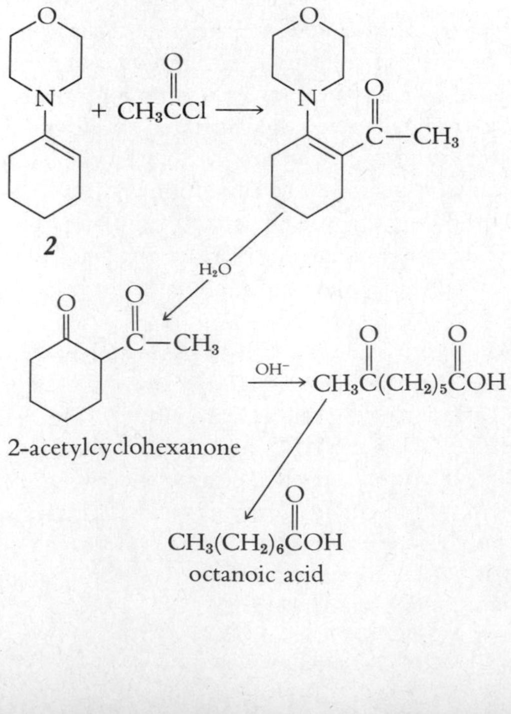

2

2-acetylcyclohexanone

$$CH_3C(CH_2)_5COH$$

$$CH_3(CH_2)_6COH$$
octanoic acid

METHODOLOGY

The formation of a ketone enamine is usually carried out in the presence of a catalyst such as *p*-toluenesulfonic acid; water is removed during the reaction to prevent the premature hydrolysis of the enamine. By using benzene or toluene as a reaction solvent and running the reaction under reflux, the water can be removed as an azeotrope through distillation into a water separator. Although pyrrolidine enamines form more rapidly than morpholine or piperidine enamines, morpholine compounds give higher yields in acylation reactions, so the morpholine enamine of cyclohexanone will be used in this preparation. The acylation of 1-morpholinocyclohexene will be carried out, using propionic anhydride as the acylating agent, by simply heating the reactants in methylene chloride and allowing the solution to stand. The acylated enamine is then hydrolyzed by refluxing it in water, and the product is purified by vacuum distillation.

Since this experiment involves 3 reflux periods and one reaction mixture that must stand overnight, it will be important to plan your time carefully so that you will be ready, if possible, to start the overnight reaction by the end of the first period.

$$SO_3H$$

p-toluenesulfonic acid

abbrev.: TsOH

A small excess of the amine is generally used, since some may be distilled off with the water.

Acylated enamines are too weakly basic to form iminium salts under the reaction conditions, so triethylamine is added to take up the propionic acid evolved during this step.

It is advisable to work on another experiment or minilab during the reflux periods.

PRELAB EXERCISE

Calculate the mass and volume of 0.10 mol of cyclohexanone, 0.13 mol of morpholine, 0.12 mol of triethylamine, and 0.11 mol of propionic anhydride.

Reactions and Properties

Table 1.   Physical properties

	M.W.	m.p.	b.p.	d.
cyclohexanone	98.2	−16	156	0.948
morpholine	87.1	−5	128	1.000
toluene	92.15	−95	111	0.867
p-toluenesulfonic acid	172.2	104–5	140^{20}	
1-morpholino-cyclohexene	167.3		119^{10}	
propionic anhydride	130.15	−45	168^{712}	1.011

Table 1.   Physical properties (*continued*)

	M.W.	m.p.	b.p.	d.
triethylamine	101.2	−115	89	0.728
2-propionylcyclo-hexanone	154.2		125^{20}	

A.

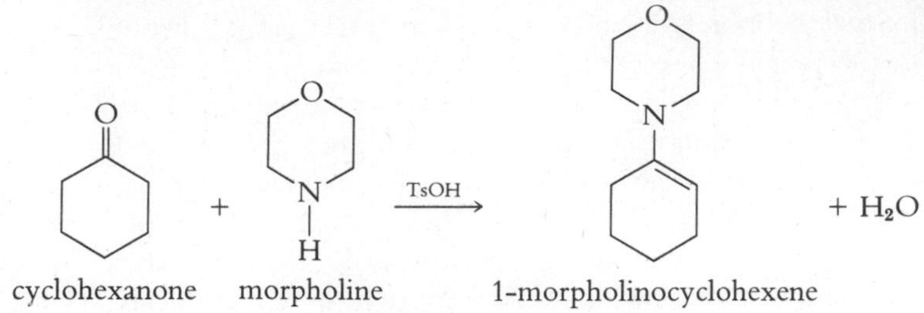

cyclohexanone   morpholine   1-morpholinocyclohexene

B.

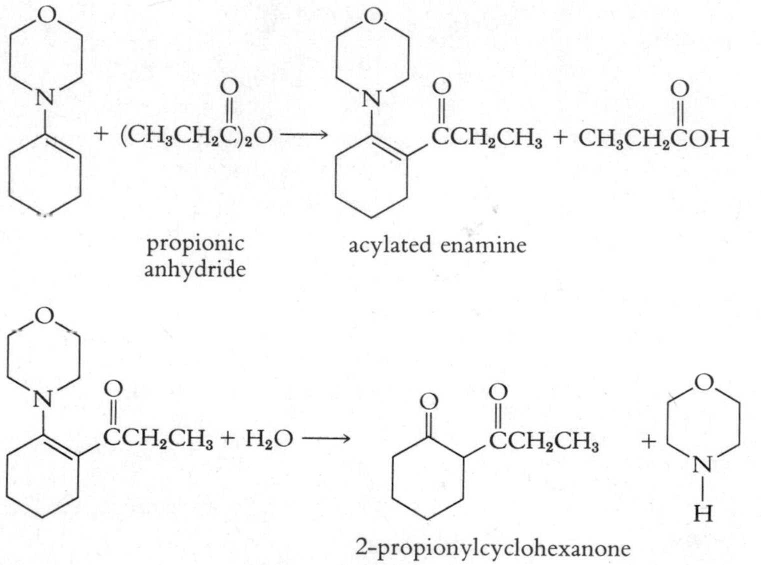

propionic
anhydride   acylated enamine

2-propionylcyclohexanone

Hazards

Morpholine is irritating to skin, eyes, and respiratory system. Avoid breathing vapors and contact with eyes or skin; keep away from flames. Cyclohexanone is flammable and irritant; avoid breathing vapors and contact with eyes or skin. *p*-Toluenesulfonic acid is caustic and can cause serious burns; prevent any contact with skin, eyes, and clothing. Propionic anhydride and triethylamine can cause burns on skin and eyes, their vapors irritate the eyes and respiratory system, and the amine is highly flammable. Measure under a hood, do not breathe vapors, prevent contact with skin, eyes, and clothing. Toluene and methylene chloride vapors are somewhat toxic, so do not breathe vapors and avoid unnecessary contact.

morpholine   *p*-toluenesulfonic   cyclohexanone   propionic   triethylamine
            acid   anhydride

PROCEDURE

A. *Preparation of 1-Morpholinocyclohexene*

Assemble an apparatus for reflux with water separation ⟨OP –25*c*⟩ and exclusion of moisture ⟨OP–22*a*⟩. Fill the water separator with toluene. Combine in the reaction flask 0.10 mol of cyclohexanone, 0.13 mol of morpholine, 0.10 g of *p*-toluenesulfonic acid, and 20 ml of dry toluene. Reflux the reactants ⟨OP–7*a*⟩ for an hour or more, then disassemble the water trap carefully, separate ⟨OP–13*b*⟩ the water, and measure its volume. Remove the toluene and excess morpholine by vacuum distillation ⟨OP–26⟩ on a water bath ⟨OP–7⟩. If the morpholinocyclohexene cannot be used immediately in the next step, it should be kept tightly stoppered in a refrigerator.

Apparatus used in this experiment must be thoroughly dried—preferably in an oven.

The water volume should be about 2 ml; if it is much less than this, the reflux should be resumed.

Use aspirator vacuum. Toluene boils at about 20°C and morpholine at 35°C at 20 torr.

B. *Preparation of 2-Propionylcyclohexanone*

Set up the reaction flask containing the 1-morpholinocyclohexene for addition and reflux ⟨OP–10⟩ with exclusion of moisture ⟨OP–22*a*⟩. Add a solution containing 0.12 mol of triethylamine (*Caution:* see Hazards) in 100 ml of dry methylene chloride to the reaction flask. Put 0.11 mol of propionic anhydride dissolved in 50 ml of dry methylene chloride in the addition funnel. Reflux the mixture ⟨OP–7*a*⟩ using a 50° water bath and add the propionic anhydride solution dropwise over a period of about 30 minutes. Disassemble the apparatus, place the drying tube directly on the flask, and let it stand (preferably in a warm place) for 24 hours or longer.

Prepare the triethylamine and propionic anhydride solutions under a hood.

Evaporate the methylene chloride ⟨OP–14⟩ on a water bath using a trap, add 25 ml of water to the residue, and reflux the mixture ⟨OP–7a⟩ for 30 minutes to hydrolyze the acylated enamine. Extract the reaction mixture ⟨OP–13⟩ with three portions of methylene chloride, dry the combined organic layers ⟨OP–20⟩, evaporate the solvent ⟨OP–14⟩, purify the product by vacuum distillation ⟨OP–26⟩ and obtain the yield ⟨OP–5⟩.

Experimental Variations

1. The infrared and NMR spectra of the product can be obtained and examined for evidence of enol forms. See Experiment 40 for information on the interpretation of enol spectra.

2. The product can be converted to 7-oxononanoic acid by the method described in *Organicum*, p. 505 (Bib–B10). The keto acid can be reduced to pelargonic acid by the method of Huang-Minlon [*J. Am. Chem. Soc. 68,* 2487 (1946)], which is a modification of the Wolff-Kischner reduction.

3. Many procedures for alkylating and acylating enamines may be found in *J. Am. Chem. Soc. 85,* 207 (1963).

Topics for Report

1. Why is it important to keep the reaction apparatus moisture-free in step *A*? Write an equation for the reaction that would occur if water were present.

2. (*a*) Write a reasonable mechanism for the acylation of 1-morpholinocyclohexene with propionic anhydride. (*b*) Write a reasonable mechanism for the hydrolysis of the iminium salt in step *B*.

3. There are four possible enol forms of 2-propionylcyclohexanone. (*a*) Draw their structures. (*b*) Which enol form would you expect to be most abundant? Explain.

4. Write a flow diagram for the synthesis, showing how the products in each step are separated from starting materials, catalysts, etc.

Library Topics

1.   *Queen substance* is a pheromone that inhibits the development of ovaries in worker bees and thus prevents the birth of a competing queen. Find the structure of queen substance and show how it can be synthesized from 7-oxooctanoic acid.

$$CH_3\overset{\displaystyle O}{\overset{\|}{C}}(CH_2)_5\overset{\displaystyle O}{\overset{\|}{C}}-OH$$
7-oxooctanoic acid

2.   Write a short paper on the applications of enamines in organic synthesis.

The Synthesis of Dimedone and the Determination of Its Tautomeric Equilibrium Constant

α, β-Unsaturated Carbonyl Compounds. Nucleophilic Addition by Carbanions. Conjugate Addition. Decarboxylation. Keto-Enol Equilibria. Spectrometric Analysis.

Operations

⟨OP–5⟩ Weighing

⟨OP–7a⟩ Refluxing

⟨OP–12⟩ Vacuum Filtration

⟨OP–14⟩ Evaporation

⟨OP–21⟩ Drying Solids

⟨OP–22a⟩ Exclusion of Moisture

⟨OP–23⟩ Recrystallization

⟨OP–28⟩ Melting Point

⟨OP–34⟩ NMR Spectrometry

Experimental Objective

To prepare dimedone starting with mesityl oxide, and determine its keto-enol composition in chloroform by NMR spectrometry.

Learning Objectives

To learn a method for carrying out the Michael addition and to learn how to decarboxylate a carboxylic acid.

To learn more about conjugate addition reactions, Dieckmann-type condensations, and decarboxylation reactions.

To learn how to determine a keto-enol equilibrium constant by NMR.

To learn more about compounds having stable enolic forms.

SITUATION

Dimedone is depicted as a diketone in most organic chemistry textbooks and in a recent edition of the *Merck Index*. Some other "diketones" like acetylacetone exist mainly in the enolic form, and you suspect that the textbook structure for dimedone may be wrong, or at least incomplete. You will synthesize dimedone from dimethyl malonate and mesityl oxide, and then use NMR spectrometry to find out whether your dimedone exists primarily in the keto or enol form in solution.

Possible dimedone structures

diketone   enolic
                    ketone

BACKGROUND

Stable and Unstable Enols

Enols have long been known as unstable intermediates in carbonyl-compound reactions such as the bromination of acetone. Although acetone exists overwhelmingly in the keto form, it is the minute amount of enol present that actually reacts with bromine under acidic conditions. Other enols are considerably more stable than that of acetone—ethyl acetoacetate contains about 7.5% enol at equilibrium and acetylacetone (2,4-pentanedione) contains about 80% enol. Certain natural compounds like Vitamin C (an enediol) exist almost entirely in the enolic form. In a sense, even phenols are enolic tautomers that could, in theory, exist in equilibrium with keto forms. For example, phlorglucinol forms carbonyl-type derivatives with hydroxylamine and similar derivatizing agents, which suggests that it exists in equilibrium with a triketone.

acetone
~100% keto   0.00025% enol

acetylacetone
20% keto   80% enol

Vitamin C
(ascorbic acid)

phlorglucinol   phlorglucinol
                         oxime

Ordinarily the keto form of a phenolic compound is the less stable tautomer, since its formation involves a loss of resonance energy. However, it was realized in the early fifties that bases such as guanine and thymine should exist mainly in the keto form at the pH of physiological systems —previously most textbooks had illustrated only the enol forms. This fact provided the key to the structure of DNA,

guanine
(keto)

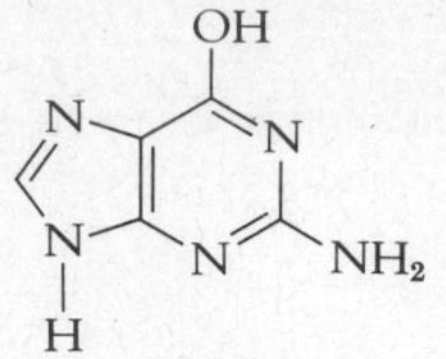

guanine
(enol)

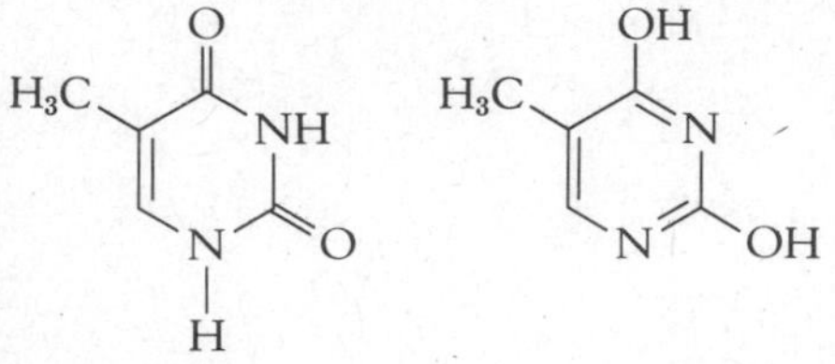

keto   enol

thymine

*The **Dieckmann condensation** is an intramolecular Claisen condensation of a diester that results in ring formation.*

Organic Syntheses Coll. Vol. II, p. 200.

Sodium methoxide can be purchased as the solid or as a 25% solution in methanol.

$$Na^+[CH(CO_2CH_3)_2]^-$$

sodiomalonic ester

The mesityl oxide should be freshly opened or recently distilled (Caution: see Hazards).

and thus to the genetic code. The bases, which are attached to the polyester backbone of a nucleic acid molecule, are responsible for the transmittance of genetic traits (by DNA) and the synthesis of protein (by RNA) in living beings. Only the keto form allows certain base pairs (such as adenine-thymine and guanine-cytosine) to form—a process that is essential to the operation of nucleic acid molecules and thus to life itself.

METHODOLOGY

The synthesis of dimedone from mesityl oxide involves four distinct reactions that take place in sequence. The first is a *Michael addition* of dimethyl malonate to mesityl oxide. The combination of sodium methoxide with dimethyl malonate generates a carbanion (sodiomalonic ester), which undergoes nucleophilic addition to the conjugated system of mesityl oxide. The next reaction is a *cyclization* which resembles the Dieckmann condensation, except that the attacking nucleophile is a carbanion adjacent to a ketonic carbonyl group rather than an ester carbonyl group. Both of these reaction steps occur spontaneously when dimethyl malonate and mesityl oxide are combined in a solution of the base in methanol. The resulting ester (see the Reactions section) is then *hydrolyzed* to yield a carboxylic acid, which is easily *decarboxylated* to form dimedone.

A standard synthesis of dimedone requires the *in situ* preparation of sodium ethoxide by adding metallic sodium to absolute ethanol. Since this operation is quite hazardous, commercial sodium methoxide in methanol will be used instead.

A sodiomalonic ester forms immediately when sodium methoxide is added to dimethyl malonate; it reacts readily with mesityl oxide, which is added in small portions. The Michael addition and the cyclization should go to completion during the subsequent reflux period. The cyclic ester is hydrolyzed with boiling sodium hydroxide and the solution acidified to form the carboxylic acid, which loses carbon dioxide on heating. The resulting dimedone is then isolated by vacuum filtration and recrystallized from aqueous acetone.

One of the best ways to detect and analyze enol content is by NMR spectrometry. The enolic OH protons of β-diketones absorb far downfield, with chemical shift values of about 10–16δ. Enols also show vinyl proton (H—C=C) absorption in the usual range for these signals, about 4.5–6.0δ. The keto form of a diketone can usually be recognized by the signal of the protons α to both carbonyl groups [H—C(C=O)$_2$]; these protons absorb near 3.5δ. Since there are two such protons for every enolic vinyl proton in the enol form of dimedone, the equilibrium constant for its keto-enol equilibrium can be determined from the integrated areas of these two signals using the formula

$$K = \frac{[\text{enol}]}{[\text{keto}]} = \frac{2 \times \text{area (H—C=C)}}{\text{area (H—C(C=O)}_2)}$$

PRELAB EXERCISE

Calculate the mass and volume of 50 mmol of dimethyl malonate, and of 52 mmol of mesityl oxide.

Reactions and Properties

dimedone

All of the cyclohexanedione derivatives can also be represented by enol forms.

Table 1.   Physical properties

	M.W.	m.p.	b.p.	d.
mesityl oxide	98.2	−52	129	0.858
dimethyl malonate	132.1	−62	181	1.154
sodium methoxide	54.0			
methanol	32.0	−94	65	0.791
dimedone	140.2	149–51		

25% sodium methoxide–methanol has a density of 0.945 g/ml.

mesityl oxide   methanol

Mesityl oxide should never be distilled to dryness; it can form explosive peroxides on standing.

Use a small amount of dry methanol for the transfers.

The solid that forms when sodium methoxide is added should dissolve on heating.

There may be considerable bumping during the evaporation; use gentle heating and add some fresh boiling chips.

The boiling should remove residual methanol as well as the CO₂ evolved during decarboxylation. Any oil that forms during this step should crystallize on cooling.

Hazards

Sodium methoxide is a strong alkali like sodium hydroxide and causes serious burns to skin and eyes. Prevent contact of the solution with skin, eyes, and clothing. Mesityl oxide is flammable, the vapors are irritant and toxic, and the liquid irritates the skin and eyes. Do not breathe vapors; avoid contact with skin, eyes, and clothing. Methanol is very flammable and is toxic by inhalation, ingestion, and skin absorption. Avoid breathing vapors, avoid contact with skin or eyes.

PROCEDURE

Reaction.   Equip a dry reflux apparatus ⟨OP–7a⟩ with a drying tube ⟨OP–22a⟩ and place 50 mmol of dimethyl malonate in the reaction flask. Add, with shaking and swirling, 12 ml of 25% sodium methoxide (∼53 mmol) in methanol. Attach the reflux condenser, heat the solution to boiling on a steam bath, then remove the heat source and add (through the condenser) 52 mmol of recently distilled mesityl oxide (*Caution:* see Hazards), in 1-2 ml portions. The addition should take about 2 minutes, and the apparatus should be shaken after each addition to mix the contents of the reaction flask. Reflux the reaction mixture over a steam bath for one hour.

Remove the reflux condenser, connect the flask to a trap and aspirator, and evaporate the methanol ⟨OP–14⟩ using a steam bath or hot water bath. Stop the evaporation when the solid residue is just moist but all standing liquid has been removed. Add 40 ml of 3M sodium hydroxide to the reaction flask and reflux ⟨OP–7a⟩ the mixture for another hour to hydrolyze the ester. Pour the warm reaction mixture, with stirring, into a 250 ml beaker containing 30 ml of 6M hydrochloric acid. Carefully (foaming!) boil the mixture for 10 minutes or more, until no more carbon dioxide is evolved. Cool the beaker in an ice bath until precipitation is complete.

Separation, Purification, and Analysis.   Collect the dimedone by vacuum filtration ⟨OP–12⟩ and recrystallize it ⟨OP–23⟩ from about 25 ml of 50% acetone. Dry it ⟨OP–21⟩, and obtain the melting point ⟨OP–28⟩ and yield ⟨OP–5⟩. Record an NMR spectrum ⟨OP–34⟩ of the dimedone in deuterochloroform.

Experimental Variations

1. The infrared spectrum of dimedone in Nujol can be obtained and interpreted. It should be possible to determine the dominant species present (keto or enol) by comparing the carbonyl and hydroxyl absorption bands.

An enolic O—H band absorbs at lower frequency than an alcohol O—H band.

2. The NMR spectrum of dimedone can also be obtained in other solvents (such as DMSO-d_6) and the spectra compared with that obtained in deuterochloroform.

3. The dimedone can be used to prepare a solid derivative of a low-boiling aldehyde such as butanal by the following procedure. Dissolve 0.3 g of dimedone in 4 ml of 50% ethanol, add 1 mmol of the aldehyde, and boil the solution gently for five minutes. Cool it and let it stand until the product has completely crystallized, then collect the crystals and recrystallize from an alcohol–water mixed solvent.

Up to four hours may be required for complete crystallization.

Conjugate Addition of Aniline to Chalcone

Dissolve 1.1 g of chalcone (benzalacetophenone; 1,3-diphenyl-2-propene-1-one) in 20 ml of absolute ethanol and add 0.5 ml of aniline. Shake until the mixture is homogeneous, stopper the flask, and let it stand until the next laboratory period. Collect the crystals by vacuum filtration, dry them, and obtain the yield. Write an equation and a mechanism for the reaction. The melting point (175°) can be determined if desired.

Caution: Avoid contact with aniline and chalcone; do not breathe aniline vapors.

It may be necessary to scratch the flask and cool it in ice to effect crystallization.

Topics for Report

1. Identify the enol OH, the enol H—C=C, and the keto H—C(C=O)$_2$ signals on your NMR spectrum and calculate the keto/enol equilibrium constant. What is the predominant form of dimedone in the NMR solvent you used?

The keto/enol ratio may depend on the solvent used.

2. **(a)** Write a mechanism for the addition of dimethyl malonate to mesityl oxide in the presence of sodium methoxide/methanol. **(b)** Write a mechanism for the cyclization step of your synthesis. **(c)** Write a mechanism for the decarboxylation step of your synthesis, illustrating the activated complex involved.

3. Give IUPAC names for all of the organic compounds (except methanol) illustrated in the Reactions and Properties section.

4. Construct a flow diagram for the synthesis of dimedone.

5. Draw the structure of the derivative formed when dimedone is condensed with butanal (see Experimental Variation 3).

Library Topic

Report on the history and applications of the Dieckmann condensation and tell what factors control the direction of ring closure when two or more products are possible.

The Oxidation of Anthracene to Anthraquinone

EXPERIMENT 41

Polynuclear Aromatic Hydrocarbons. Oxidation. Quinones.

Operations

⟨OP–5⟩ Weighing

⟨OP–7a⟩ Refluxing

⟨OP–10⟩ Addition

⟨OP–12⟩ Vacuum Filtration

⟨OP–21⟩ Drying Solids

⟨OP–24a⟩ Vacuum Sublimation

⟨OP–28⟩ Melting Point

Experimental Objective

To carry out the synthesis of anthraquinone from anthracene.

Learning Objectives

To learn a procedure for oxidizing aromatic hydrocarbons to quinones.

To learn about the history and uses of alizarin.

SITUATION

Your final assignment in a History of Chemistry course is to write a research paper on one of the milestones of synthetic organic chemistry, and then (as nearly as possible) to duplicate the original synthesis in the laboratory. Because of your interest in the history of the synthetic dye industry, you have decided to reproduce the Graebe-Liebermann synthesis of alizarin. First you must prepare anthraquinone by oxidizing anthracene.

alizarin

Alizarin itself is not a dye, but it forms different-colored lakes (metal complexes) with mordants such as aluminum (bluish-red), chromium (brown-violet), magnesium (violet), barium (blue), calcium (purple-red), and ferrous iron (black-violet). The aluminum-mordanted dye is known as Turkey Red.

Reduction of alizarin

anthracene

Such isomerizations can now be explained on the basis of a benzyne elimination-addition mechanism.

BACKGROUND

Madder and the Mad Race for Alizarin

The synthesis of alizarin by Graebe and Liebermann in 1868 was one of the milestones of organic chemistry, since it marked the first time that a natural dyestuff had been prepared synthetically. Along with it arose an important branch of the dye industry based on the synthesis and modification of natural dyes.

Alizarin and indigo were the two most important dyestuffs throughout the Middle Ages. Both were obtained from plant sources—woad or the Asian indigo plant for indigo, and madder (*Rubia tinctorum*) root in the case of alizarin. Considering the difficulties involved, it's a wonder that the art of dyeing with madder was ever developed. First the plant was fermented to convert the glycoside, ruberythric acid, into alizarin; then the cloth was impregnated with a mixture of lime and rancid olive oil, soaked in an aluminum sulfate solution, stirred with a suspension of the dye, steamed, and soaped. The whole process could take as long as four months, although later innovations cut the time down to about five days.

The story of the first alizarin synthesis is a record of diligent research crowned with incredible luck. Carl Graebe and Carl Liebermann were studying the dye just a year after Baeyer had discovered a method of reducing indoxyl (an indigo derivative) to indole. Graebe and Liebermann decided to try the technique (distillation with zinc dust) on alizarin and ended up with anthracene, a coal-tar derivative whose structure was not entirely worked out at that time. Having gotten that far, they reasoned that perhaps the process could be reversed. Knowing that anthracene could be oxidized to anthraquinone and that alizarin had the elemental composition of a dihydroxyanthraquinone, they brominated anthraquinone and fused the dibromide with alkali on the off-chance that the bromine atoms just *might* assume the proper orientation on the anthraquinone ring. To their amazement, it worked—the product of fusing the dibromide with alkali was identical to natural alizarin! But the amazing part of the whole synthesis was only revealed much later—the bromine atoms had not substituted at the 1,2 positions occupied in alizarin, but at the 2,3 positions instead. Then during alkali fusion, isomerization had occurred, leaving the hydroxyl groups in the correct 1,2 orientation of alizarin!

Unfortunately, the Graebe-Liebermann synthesis was not a practical one for commercial use because of the high

price of bromine. Undismayed, Graebe and Liebermann col-
laborated with Heinrich Caro of the Badische dye company
to develop a better synthesis involving the sulfonation of
anthraquinone followed by alkali fusion. They patented
their process on July 25, 1869—just one day before Perkin
patented the *same* process in England, having discovered it
independently! (Although the Germans won the race to the
patent office, the Englishman beat them to the draw in production and dominated the synthetic alizarin market for
several years.) Still another element of luck entered in here,
because the product of sulfonation was not the 1,2-disulfon-
ated compound as they thought, but the 2-monosulfonated
derivative. Under the vigorous conditions of the alkali fusion, oxygen in the air oxidized an intermediate and again
yielded alizarin.

Improved synthesis of alizarin

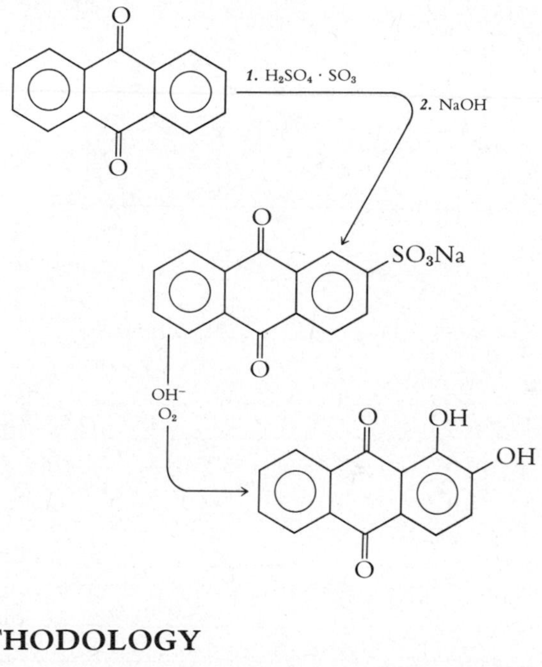

Graebe-Liebermann synthesis of
alizarin

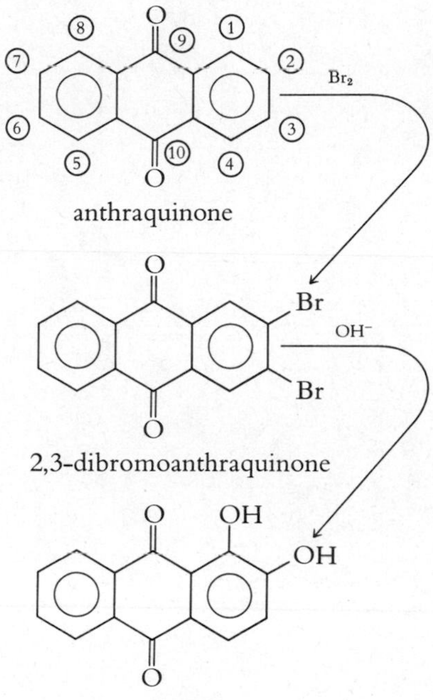

anthraquinone

2,3-dibromoanthraquinone

alizarin

METHODOLOGY

The starting material should be a good grade of anthracene—not only because it will give a better product, but because the impurities in commercial anthracene may be carcinogenic. The basic method used by Graebe and Liebermann
to prepare anthraquinone will be used here, with anthracene
being oxidized by chromium trioxide in glacial acetic acid.
The reaction is carried out by adding the chromic acid solution to a solution of anthracene in acetic acid at reflux, then

Mixtures of CrO_3 with acid and
water are often called "chromic acid"
solutions because they presumably
contain that acid as a result of the
reaction $CrO_3 + H_2O \rightarrow H_2CrO_4$.

refluxing the solution for a short time longer to complete the reaction. After workup, the anthraquinone can be purified by vacuum sublimation, forming beautiful yellow needles.

PRELAB EXERCISES

1.   Read ⟨OP–24a⟩ and review the other operations as needed.

2.   Calculate the mass of 20 mmol of anthracene and of 70 mmol of chromium trioxide.

Reaction and Properties

anthracene + 2CrO$_3$ + 6HOAc ⟶

anthraquinone + 2Cr(OAc)$_3$ + 4H$_2$O

Ac = acetyl (CH$_3$CO)

Table 1.   Physical properties

	M.W.	m.p.	b.p.	d.
anthracene	178.2	214–17	340	
chromium trioxide	100.0	196	d	
acetic acid	60.05	17	118	1.049
anthraquinone	208.2	286 (subl.)	380	

Very pure anthracene is colorless, but the usual reagent grade is pale yellow and technical grades are dark yellow or yellow-brown.

Hazards

Pure anthracene is not carcinogenic, but commercial anthracene often contains carcinogenic impurities. Avoid unnecessary contact with skin or clothing. Chromium trioxide is classed as a carcinogen and is a strong oxidant that can cause burns on skin and eyes. Avoid contact with eyes, skin, or

clothing; do not breathe dust; keep away from combusti-
bles. Acetic acid can cause serious burns and its vapors irritate the respiratory system. Avoid contact, do not breathe
vapors, handle under an efficient fume hood.

anthracene   chromium trioxide   acetic acid

PROCEDURE

Reaction.   Combine 20 mmol of pure anthracene and 35 ml
of glacial acetic acid (*Caution:* see Hazards) in an apparatus
for reflux and addition ⟨OP–10⟩. Dissolve 70 mmol of chromium trioxide (*Caution:* see Hazards) in 5 ml of water, stir
in 20 ml of glacial acetic acid, and transfer the mixture to the
addition funnel. Reflux the anthracene mixture ⟨OP–7a⟩ for
a minute or two to dissolve most of the anthracene, then remove the heat source. Without delay, begin to add the chromic acid solution, with shaking, just rapidly enough so that
the mixture continues to reflux without external heating.
This should take 5–10 minutes. Replace the heat source and
reflux the solution for an additional 10 minutes.

*Clamp the apparatus securely, since the
assembly will have to be shaken during
the addition.*

*Keep a cold-water bath handy to moderate
the reaction if necessary.*

Separation.   Cool the solution well in an ice bath, and carefully (Hood and gloves advised) collect the precipitated
anthraquinone by vacuum filtration, ⟨OP–12⟩ using water

*The solution contains acetic acid, which is irritant and corrosive,
and chromium salts, which may be carcinogenic. Avoid contact
with the liquid, do not breathe the vapors.*

WARNING

for the transfer. Wash the crude anthraquinone on the filter
with 35 ml of hot water, then with 35 ml of hot 1M sodium
hydroxide, and finally with cold water. Discard the filtrate
according to your instructor's directions.

*Important: Remove the filtrate
before washing.*

Purification and Analysis.   Dry the anthraquinone ⟨OP–21⟩ at
room temperature and purify it by vacuum sublimation

*It can be dried overnight in a dessicator.
A final wash with a little ice-cold 95%
ethanol will speed up the drying process.*

⟨OP–24a⟩. Weigh the anthraquinone ⟨OP–5⟩, and measure its melting point using a sealed capillary melting-point tube (see ⟨OP–28⟩ for directions).

Experimental Variations

1. The anthraquinone can also be purified by entrainer sublimation (see ⟨OP–24⟩) or by recrystallizing it from approximately 80 ml of boiling dioxane.

This compound is called "silver salt" because it separates as glistening silvery plates.

2. Sodium 2-anthraquinonesulfonate can be prepared from anthraquinone by a method described in *Vogel's Textbook of Practical Organic Chemistry,* 4th ed., p. 644 (Bib–B4).

Use approx. 1 × 10⁻⁴M anthracene and 2 × 10⁻⁵M anthraquinone solutions.

3. The ultraviolet spectra (200–400 nm) of anthracene and anthraquinone in ethanol can be recorded and compared.

4. The fluorescence of anthracene can be observed in a dark room using a "black light." A dilute alcoholic solution of very pure anthracene should fluoresce blue, while the ordinary reagent should fluoresce greenish-yellow.

Topics for Report

1. Calculate the percentage by which chromium trioxide and acetic acid were in excess in the oxidation.

phthalic anhydride

2. Anthraquinone can also be prepared by the reaction of phthalic anhydride with benzene in the presence of aluminum chloride, followed by cyclization with H_2SO_4. Write balanced equations and a mechanistic pathway for this synthesis. What is the name of the reaction involved?

3. Construct a flow diagram for the synthesis of anthraquinone.

4. Write a mechanism for the reaction of anthraquinone with fuming sulfuric acid to form 2-anthraquinonesulfonic acid.

Library Topics

1. Among the commercially significant derivatives of anthraquinone are Alizarin Orange, Alizarin Blue, carminic acid, and emodin. Report on the structures, sources, and applications of these compounds. Show how Alizarin Blue and Alizarin Orange can be prepared starting with anthraquinone.

2.   Describe a commercial process for preparing alizarin from sodium 2-anthraquinonesulfonate.

Preparation of 2-Phenylindole by the Fischer Indole Synthesis

EXPERIMENT 42

Heterocyclic Compounds. Fischer Indole Synthesis. Reactions of Carbonyl Compounds.

Operations

⟨OP–5⟩ Weighing

⟨OP–7⟩ Heating

⟨OP–12⟩ Vacuum Filtration

⟨OP–21⟩ Drying Solids

⟨OP–23⟩ Recrystallization

⟨OP–28⟩ Melting Point

Experimental Objective

To prepare 2-phenylindole from acetophenone by way of acetophenone phenylhydrazone.

Learning Objectives

To learn about the Fischer indole synthesis and the experimental conditions for its use.

To learn about the occurrence and significance of indole alkaloids.

SITUATION

The indole ring system is encountered frequently in molecules that exert profound effects on living systems. Indole derivatives occur in substances ranging from plant growth hormones to mind-altering drugs. Since most of the biologically active indoles have a substituent on the 3-position of the indole ring and none at the 2-position, it has been theorized that an open 2-position is necessary for binding at neurologically active sites. You would like to test this theory by preparing different 2- and 3-substituted indoles and observing their biological activity on laboratory animals. Among the compounds you intend to test are 2- and 3-phenylindole, both of which can be prepared by the Fischer indole synthesis. There are many different procedures reported for 2-phenylindole, so you have chosen a comparatively simple one utilizing polyphosphoric acid.

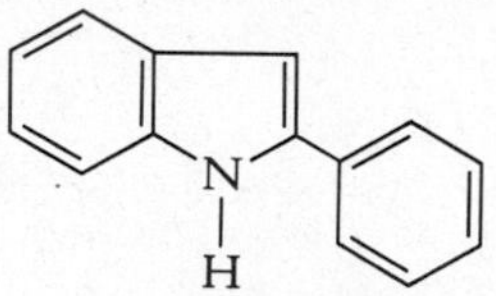

2-phenylindole

BACKGROUND

Of Toads and Toadstools

The indoles constitute a large group of natural and synthetic compounds with extraordinary properties and functions. Both benign and malevolent, indoles as a group have a Janus-faced aspect well illustrated by the parent compound, which has an odor of fine jasmine in dilute solutions but smells like feces undiluted. 3-Methylindole is a major product of the digestive putrefaction of proteins and is mainly responsible for the repulsive smell of feces, giving rise to its common name, skatole; but it is also found in cabbage sprouts, tea, and even lilies!

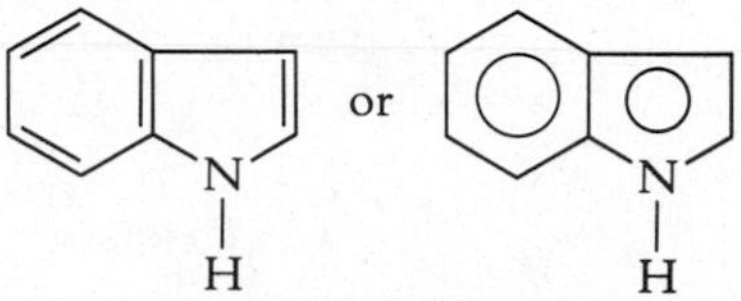

indole

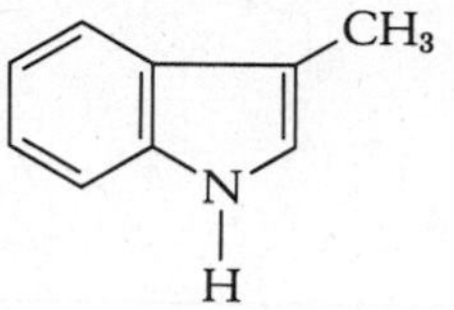

skatole

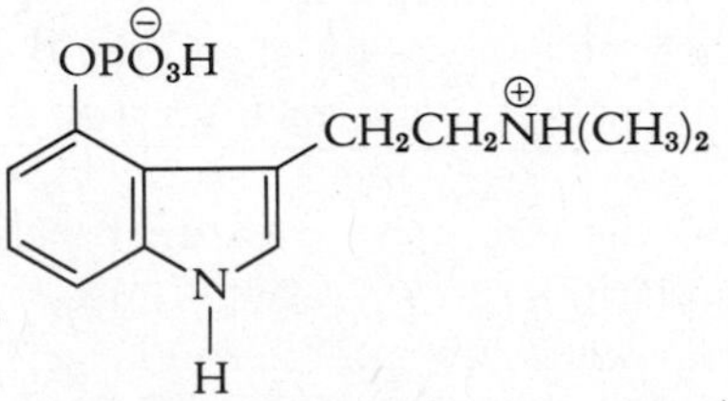

psilocybin

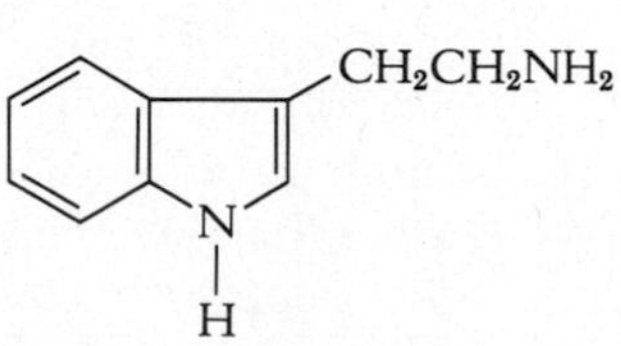

tryptamine

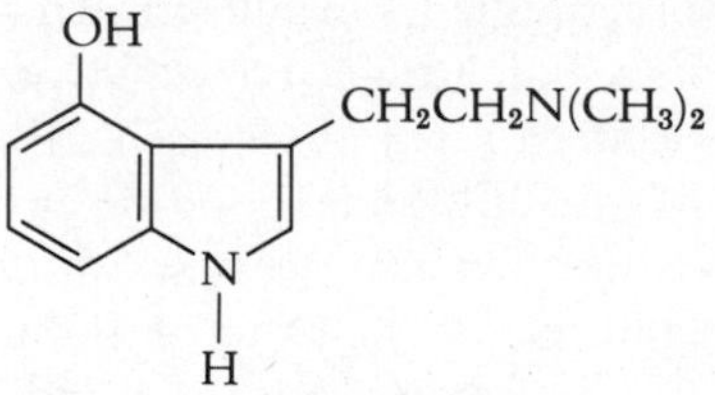

psilocin

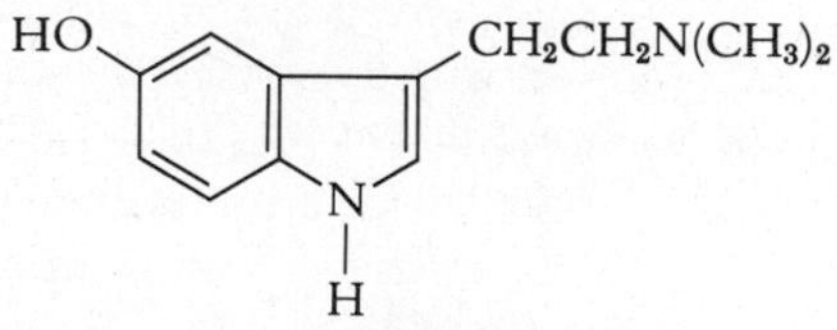

bufotenin

When the Spaniards conquered Meso-America in the sixteenth century, they observed the Aztecs using some small brown mushrooms called *teonanacatl* ("flesh of the gods") in their religious ceremonies. According to the Spanish friar, Bernardino de Sahagun, "They ate these little mushrooms with honey, and when they began to be excited by them, they began to dance, some singing, others weep-

Bufotenin has also been isolated from both toads (genus *Bufo*) and toadstools (*Amanita citrina*, etc.).

tryptophan

3-indoleacetic acid

indigo

serotonin

ing. . . . Some saw themselves dying in a vision and wept; others saw themselves being eaten by a wild beast; others imagined that they were capturing prisoners in battle, that they were rich, that they possessed many slaves, that they had committed adultery and were to have their heads crushed for the offense. . . ." Naturally the Spanish friars disapproved of these ceremonies, so they were driven underground and only rediscovered in the 1930s, later to be brought to public attention by R. Gordon Wasson. The little mushrooms actually came from a number of different species of the genus *Psilocybe;* their hallucinogenic constituents are two indoles of the tryptamine group, psilocin, and psilocybin. A similar tryptamine derivative, bufotenin, is the active principle of cohoba snuff, which is inhaled by Indians of South America and the Carribean to produce hallucinations and intoxication. Likewise, there is an indole ring in the complex molecule of lysergic acid, which can be isolated from the ergot fungus on rye and is the source of LSD.

Other indole derivatives are more consistently benign, such as the essential amino acid tryptophan, the beautiful dye indigo, the plant-growth hormone 3-indoleacetic acid (also called heteroauxin), and the bioregulator serotonin. *Tryptophan* is important as a source of serotonin and the B-vitamin nicotinamide, both of which are formed during its metabolism. Because the human body cannot generally biosynthesize aromatic compounds, tryptophan and the other aromatic amino acids (phenylalanine and tyrosine) must be obtained from the diet. *Indoleacetic acid* promotes the enlargement of plant cells and is the principal natural growth regulator for the higher plants. *Indigo* was one of the first natural dyes to be prepared synthetically. Its manufacture resulted from the brilliant researches of Adolph Baeyer, who confirmed its structure after eighteen years of research. Although the role of *serotonin* is not fully understood, it appears to be an important factor in mental processes. Some researchers see it as a mind stabilizer that helps to preserve sanity. The resemblance between serotonin and many mind-altering drugs is striking; it is possible that the psychological activity of these drugs is caused by their interference with the action of serotonin.

METHODOLOGY

The preparation of 2-phenylindole by the Fischer indole synthesis involves the acid-catalyzed cyclization of a phenyl-

Fischer indole synthesis

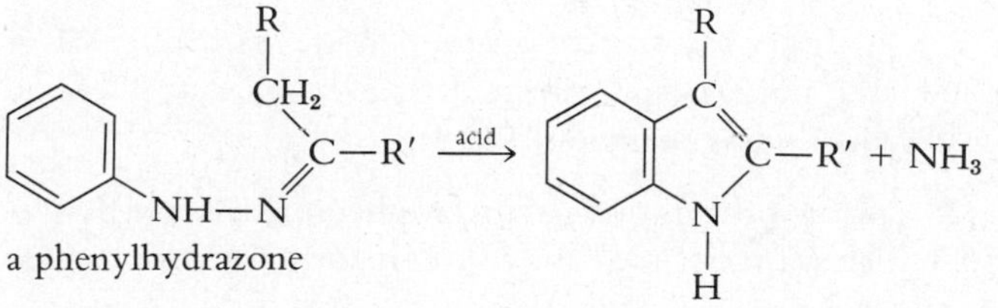

a phenylhydrazone

hydrazone with the loss of a molecule of ammonia. The phenylhydrazone can be prepared by combining phenylhydrazine (or a substituted phenylhydrazine) with an aldehyde or ketone of the structure RCH_2COR', where the R's can be alkyl, aryl, or hydrogen. The generally accepted mechanism for this *indolization* reaction is one proposed by Robinson and Robinson in 1918. It includes the following steps:

1.   A tautomeric shift of a proton from carbon to the β-nitrogen

2.   Protonation of that nitrogen

3.   A concerted electron shift that forms a C—C bond to the ring and breaks the N—N bond

4.   Another tautomeric shift from C to N

5.   Nucleophilic attack by nitrogen on a doubly-bonded carbon atom, followed by the loss of a proton

6.   Loss of ammonia to yield an aromatic pyrrole ring.

Robinson mechanism

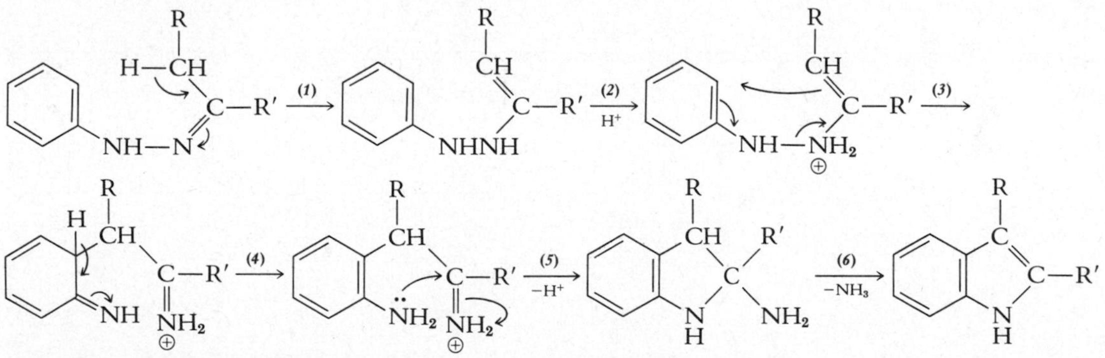

The later steps of the mechanism may vary somewhat depending on the reaction conditions, but the basic process is well established.

2-Phenylindole has been prepared from acetophenone by a variety of methods, including:

1. A direct reaction of acetophenone with phenylhydrazine and polyphosphoric acid in which acetophenone phenylhydrazone is formed *in situ;*

2. prior preparation of acetophenone phenylhydrazone, which is then stirred at 170° with anhydrous zinc chloride; and

3. heating acetophenone phenylhydrazone in 100% phosphoric acid.

In this experiment, you will use polyphosphoric acid as the catalyst, but will prepare the acetophenone phenylhydrazone beforehand to avoid the highly exothermic reaction that results from procedure 1.

The crude acetophenone phenylhydrazone, which results from heating the reactants in ethanol with a little acetic acid to catalyze the reaction, can be used directly in the second step of the synthesis. The reaction in polyphosphoric acid is exothermic and requires only a few minutes to go to completion, after which the reaction mixture is poured into water to dissolve the acid and precipitate crude 2-phenylindole. The recrystallization from ethanol is challenging, and good technique is required to get a high percent recovery.

PRELAB EXERCISE

Calculate the mass and volume of 25 mmol each of acetophenone and phenylhydrazine.

Organicum, p. 605 (Bib–B10).

Organic Syntheses, Coll. Vol. III, p. 125.

Grineva, Sadovskaya, and Ufimtsev, *J. Gen. Chem. USSR 33,* 545 (1963) (English translation).

This compound is light- and heat-sensitive. Do not use heat to dry it; store it in the dark.

Polyphosphoric acid (PPA) is made by heating H_3PO_4 with P_2O_5 and contains about 55% triphosphoric acid along with other polyphosphoric acids.

Reactions and Properties

Table 1.　Physical properties

	M.W.	m.p.	b.p.	d.
phenylhydrazine	108.15	20	243 (115[10])	1.099
acetophenone	120.2	20.5	202	1.028
2-phenylindole	193.25	188–89	250[10]	
acetophenone phenylhydrazone	210.35	105–06		

A.

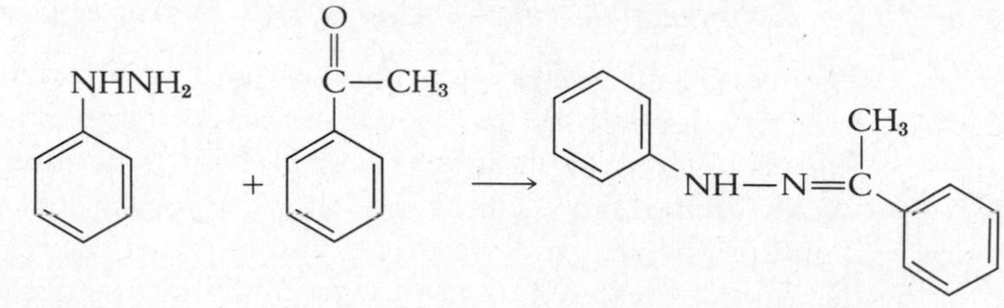

phenylhydrazine   acetophenone   acetophenone
phenylhydrazone

B.

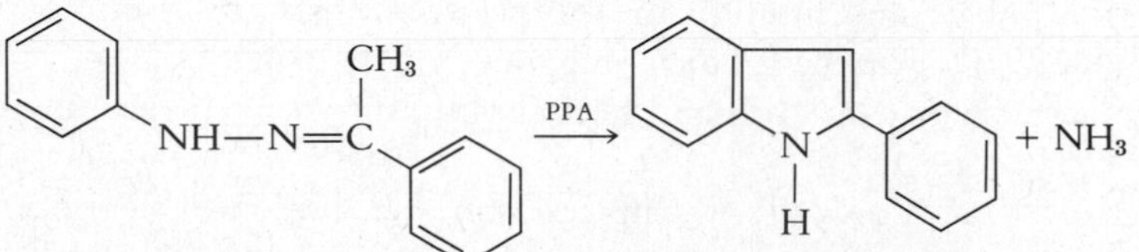

2-phenylindole

Hazards

Phenylhydrazine is highly poisonous and can cause severe
burns to skin and eyes. Some hydrazine derivatives (including phenylhydrazine hydrochloride) are classed as carcinogens. Avoid contact with skin, eyes, and clothing; do not
breathe vapors. In case of skin contact, wash off with 2%
acetic acid followed by soap and water. Acetophenone is
dangerous to the eyes; avoid contact with skin or eyes. Polyphosphoric acid is irritating to the skin and mucous membranes; avoid skin and eye contact. Some indoles are suspected carcinogens, so avoid unnecessary contact with
2-phenylindole.

phenylhydrazine   acetophenonone

PROCEDURE

A. *Preparation of Acetophenone Phenylhydrazone*

Combine 25 mmol of acetophenone with 25 mmol of recently distilled phenylhydrazine (*Caution:* See Hazards) in
10 ml of 95% ethanol and add five drops of glacial acetic
acid. Heat ⟨OP–7⟩ the reactants gently on a steam bath for 15
minutes, then cool the mixture and collect the precipitated
acetophenone phenylhydrazone by vacuum filtration ⟨OP–
12⟩. Wash it on the filter with 1M HCl followed by a little
cold 95% ethanol, let it air dry, and dry it further by blotting
it between filter papers (rubber gloves are advised).

*Impure phenylhydrazone is reddish-brown
and will give an inferior product.*

*Scratching the sides of the flask may help
promote crystallization.*

B. *Preparation of 2-Phenylindole*

Clamp a 50-ml beaker on a ringstand and position a steam bath on a support so that the bottom half of the beaker extends below the rings of the bath; then clamp a thermometer with its bulb inside the beaker. Half-fill the beaker with polyphosphoric acid and warm the viscous liquid enough so that you can readily mix in the acetophenone phenylhydrazone with a stirring rod. Heat the reactants, with stirring, until the temperature reaches 98°; then remove the steam bath and monitor the temperature. It should rise to 120° or so, then begin dropping; when it falls below 100°, replace the steam bath and heat the mixture for ten minutes with stirring.

If the temperature rises above 125°, apply a cold water bath to moderate the reaction.

Pour the warm reaction mixture into 70 ml of ice water, using more water for the transfer, and stir until all of the polyphosphoric acid has dissolved. Collect the product by vacuum filtration ⟨OP–12⟩ and wash it with water. Recrystallize the crude 2-phenylindole from 95% ethanol ⟨OP –23⟩, using about 0.2 g of decolorizing carbon. Dry the purified 2-phenylindole at room temperature ⟨OP–21⟩, weigh it ⟨OP–5⟩, and obtain its melting point ⟨OP–28⟩.

The 2-phenylindole dissolves slowly, so you may have to reflux it in 50–75 ml of ethanol for 15 minutes or more to get most of it in solution.

Experimental Variations

1. 2-Phenylindole can also be purified by vacuum sublimation; an IR spectrum in Nujol can be obtained.

2. A qualitative test for certain indoles is as follows: dissolve a little of the compound in ethanol and place a drop of the solution on a pine shaving moistened with HCl. Record your observations.

3. Part of the class can use one or more of the alternate procedures mentioned in the Methodology and compare the results with those of the rest of the class.

4. Some other experiments involving heterocyclic compounds may be found in the *Journal of Chemical Education 49,* 708 (1972) (isolation of caffeine from soft drinks); *51,* 22 (1974) (separation of drug alkaloids by TLC); and *53,* 256 (1976) (preparation of heterocyclic compounds from milk).

Synthesis of Indigo

Dissolve 0.5 g of *o*-nitrobenzaldehyde in 2 ml of acetone, add water dropwise until it turns cloudy, then add a drop or so of acetone to clear it up. Add 1M sodium hydroxide dropwise, with stirring, until the solution begins to heat up and turns dark brown. When the reaction is complete, cool the solution and collect the indigo by vacuum filtration. Wash the indigo with ethanol, then with ethyl ether, and let it air dry.

This is one of Baeyer's methods for synthesizing indigo. You can read his original paper (in German) in *Berichte 15,* 2856 (1882).

Topics for Report

1. Write a mechanistic pathway that illustrates the catalytic effect of acetic acid in preparing acetophenone phenylhydrazone.

2. Write a complete mechanism for your synthesis of 2-phenylindole, illustrating the transition states for steps 3 and 5.

3. Construct a flow diagram for your synthesis of 2-phenylindole.

4. R. B. Woodward's total synthesis of strychnine began with the preparation of 2-(3,4-dimethoxyphenyl) indole **(1)**. Show how this compound can be prepared from catechol and phenylhydrazine.

Library Topics

1. Write a short research paper on the sources, psychological effects, and social impact of hallucinogenic drugs.

2. Give several examples of natural and synthetic auxins (plant growth substances), describing their effects and applications.

1

Chain-Growth Polymerization of Styrene and Methyl Methacrylate

Addition Polymers. Free Radical Addition. Polymerization Methods.

Operations

⟨OP–5⟩ Weighing

⟨OP–7⟩ Heating

⟨OP–7*a*⟩ Refluxing

⟨OP–9⟩ Mixing

⟨OP–12⟩ Vacuum Filtration

⟨OP–21⟩ Drying Solids

⟨OP–25*a*⟩ Semimicro Distillation

⟨OP–33⟩ Infrared Spectrometry

Experimental Objectives

To synthesize polystyrene by solvent polymerization and emulsion polymerization methods and prepare a thin film for infrared analysis.

To synthesize methyl methacrylate by bulk polymerization.

Learning Objectives

To learn about and apply some of the techniques used in polymerizing unsaturated monomers.

To learn about the structure and properties of polymers and the process of chain-growth polymerization.

SITUATION

You are an amateur entomologist with a large collection of beetles that you wish to preserve for posterity. You know that biological specimens are sometimes preserved by embedding them in a layer of clear plastic, so you decide to experiment with different methods of forming plastics until you find one that meets your needs. One feasible method is to suspend each beetle in a liquid monomer that is then caused to polymerize, forming a rigid plastic matrix. Another possibility might be to prepare the polymer first by one of several possible techniques, then dissolve it in a volatile solvent, insert the beetle, and allow the solvent to evaporate. First you will prepare some polymers of polystyrene and poly(methyl methacrylate) and compare their properties.

BACKGROUND

Chain-growth Polymers

Polymers are of enormous importance to humankind—the meat, fruit, and vegetables we eat, the clothing we wear, and the wood we use for housing and furniture all consist partly or entirely of organic polymers. In fact, we are *made* of polymers—proteins in muscles and organs, blood cells and enzymes and protoplasm; lipids in nerve sheaths and cell walls and energy-storing fat tissues; nucleic acids in the chromosomes that control our heredity. Compared to these natural polymers, synthetic polymers are newcomers on the scene. Polystyrene was first synthesized in 1839—the same year Charles Goodyear learned how to vulcanize rubber—but its properties were not appreciated, and the first commercially useful synthetic polymer did not appear until 1907 when Leo Baekland synthesized "Bakelite" from phenol and formaldehyde. The discovery of Nylon 66 and polyethylene in 1939 and the development of synthetic rubbers during World War II (when natural rubber was unavailable) gave added impetus to the search for useful synthetic polymers. The more recent development by Ziegler and Natta of special catalysts for synthesizing stereoregular polymers has, with other advances, made it possible to design "tailor-made" polymers having almost any desired combination of properties.

Synthetic polymers have for many years been divided into the two broad categories of addition and condensation polymers. *Addition polymers* are built up by combining monomer units without eliminating any by-product molecules. The repeating unit of the polymer, therefore, has the same chemical constitution as the monomer. *Condensation polymers* are built up from monomer units containing two or more reactive functional groups that combine with the loss of a small molecule such as water or HCl. Thus the repeating unit of a condensation polymer does not have the same

In a sense, polystyrene is a "natural" polymer since it is found in styrax from the sweetgum tree. Some modified natural polymers, such as celluloid (produced by nitrating cellulose), were manufactured commercially before the advent of Bakelite, but these are not true synthetics.

Addition polymerization of ethylene

$$n\mathrm{CH_2}{=}\mathrm{CH_2} \longrightarrow -\!(\mathrm{CH_2CH_2})_{\overline{n}}-$$

ethylene polyethylene

(*n* is a large but indeterminate number.)

Formation of Nylon 6 by condensation polymerization

$$n\text{H}_2\text{N(CH}_2)_5\overset{\overset{\text{O}}{\|}}{\text{C}}\text{OH} \longrightarrow -(\text{NH(CH}_2)_5\overset{\overset{\text{O}}{\|}}{\text{C}})\overline{_n} + \text{H}_2\text{O}$$

6-aminohexanoic
acid

Nylon 6

Formation of Nylon 6 by addition polymerization

ω-caprolactam

$$n \longrightarrow -(\text{NH(CH}_2)_5\overset{\overset{\text{O}}{\|}}{\text{C}})\overline{_n}$$

Nylon 6

formula as the monomer(s), as illustrated for the Nylon 6 synthesized from 6-aminohexanoic acid. But when you consider that Nylon 6 can also be formed by an "addition polymerization" reaction from ω-caprolactam, it becomes apparent that the addition/condensation classification is somewhat arbitrary.

More recently, the terms *chain-growth* and *step-growth polymerization* have gained currency. The former refers to a process in which a few monomer molecules are activated by some initiator, after which a chain of repeating units builds up very rapidly. The latter refers to a process by which the chain length is built up more gradually. The polymerization

Mechanism of chain-growth polymerization of polystyrene

1. Initiator $\xrightarrow[\text{light}]{\text{heat or}}$ R · (free radical)

2. R· + CH$_2$=CH $\longrightarrow$ RCH$_2$CH·
　　　　　　　|　　　　　　　　|
　　　　　　　Ph　　　　　　　Ph

3. *etc.* R(CH$_2$CH)$_n$CH$_2$CH· + CH$_2$=CH
　　　　　　|　　　　　|　　　　　　|
　　　　　　Ph　　　　Ph　　　　　Ph

$\longrightarrow$ R(CH$_2$CH)$_n$CH$_2$CHCH$_2$CH·
　　　　　　　|　　　　　|　　　|
　　　　　　　Ph　　　　Ph　　Ph

Terminating steps:

2R(CH$_2$CH)$_n$CH$_2$CH· $\longrightarrow$
　　　　　|　　　　　|
　　　　　Ph　　　　Ph

R(CH$_2$CH)$_n$CH=CH + R(CH$_2$CH)$_n$CH$_2$CH$_2$
　　|　　　　　|　　　　　　|　　　　　|
　　Ph　　　　Ph　　　　　Ph　　　　Ph

(disproportionation)

or

R(CH$_2$CH)$_n$CH$_2$CH—CHCH$_2$(CHCH$_2$)$_n$R
　　|　　　　　|　　|　　　|
　　Ph　　　　Ph　Ph　　Ph

(radical coupling)

of styrene with a free-radical initiator is a typical example of chain-growth polymerization. An initiator such as benzoyl peroxide decomposes under the influence of heat or light to form free radicals, which add to styrene molecules according to Markovnikov's rule. Each "activated" styrene molecule then adds in similar fashion to another styrene molecule, leaving the unpaired electron at the end of the chain after each step. This process continues indefinitely until a reaction such as radical coupling or disproportionation occurs between chain ends (or with impurities) and deactivates the chain ends by forming stable products. Throughout a chain-growth polymerization, the bulk of the reaction mixture will consist of finished polymer molecules and unreacted monomers waiting to meet up with a reactive chain end. Because polymerization is so rapid once activation occurs, only one among many millions of molecules is actually involved in the growth process at any instant of time.

METHODOLOGY

There are a number of different methods used for chain-growth polymerization, each with its advantages and disadvantages. *Bulk polymerization* is the simplest; it is carried out by adding a suitable initiator and using heat or light to promote the reaction. The high heat of reaction makes bulk polymerization of vinyl monomers hard to control, and the method is seldom used commercially except for some polystyrene and poly(methyl methacrylate) products. The use of a solvent alleviates the heat removal problem, but it is often difficult (if not impossible) to entirely remove the solvent. *Solvent polymerization* thus works best for polymers that are commonly used in solution, such as acrylic finishes.

Suspension polymerization is carried out by mechanically dispersing the monomer in a liquid such as water, so that the polymer is obtained in the form of granular beads that can be easily isolated. *Emulsion polymerization* is similar to suspension polymerization in that the monomer is dispersed in water. In this method, however, the initiator is dissolved in the aqueous phase and the monomer is emulsified by detergent or some other **surfactant.** Polymerization starts in the surfactant micelles rather than in the monomer droplets. At some stage of the process, the polymer particles grow larger than the micelles and absorb all of the soap from solution, after which further polymerization occurs within the polymer particle itself. The monomer droplets provide a reservoir of monomer molecules that are continually fed into the growing polymer. The resulting dispersion resembles a rubber latex and can be used in that form, but it is more often coagulated and isolated as a finely divided powder.

In a simple demonstration of the bulk polymerization technique, a sample of methyl methacrylate will be mixed with a little benzoyl peroxide in a test tube and placed in the sunlight for a period of time. Since the inhibitor added to a commercial monomer would slow the reaction, the mono-

surfactant: surface-active agent.

Inhibitors are added to most vinyl monomers to stabilize them and prevent premature polymerization.

The polystyrene can be dried in an oven at 110°.

mer will be prepared by thermal depolymerization of poly(methyl methacrylate) in a distillation apparatus.

The emulsion polymerization of styrene will be carried out by dispersing the monomer in an aqueous detergent solution and heating the emulsion in the presence of a water-soluble initiator, potassium peroxydisulfate. The polymer forms as a rubbery latex that is broken up by adding an alum solution to precipitate polystyrene. As with most polymerization reactions, precautions must be taken to eliminate atmospheric oxygen.

Styrene can also be polymerized in solution, using an aromatic solvent such as toluene and a suitable initiator. The polystyrene is precipitated, in this experiment, by pouring the reaction mixture into methanol. The resulting polymer can then be cast into a thin film in order to obtain an infrared spectrum of polystyrene. It may require some experimentation with the quantities of polymer and solvent to obtain a film of appropriate thickness for IR analysis. Some thin films produce interference fringes in the IR spectrum due to interference between infrared rays reflected from the front and back of the film. This interference can be reduced or eliminated by smearing a little Nujol on one surface.

This method will, of course, result in additional bands where the Nujol absorbs.

Reactions and Properties

A,B.

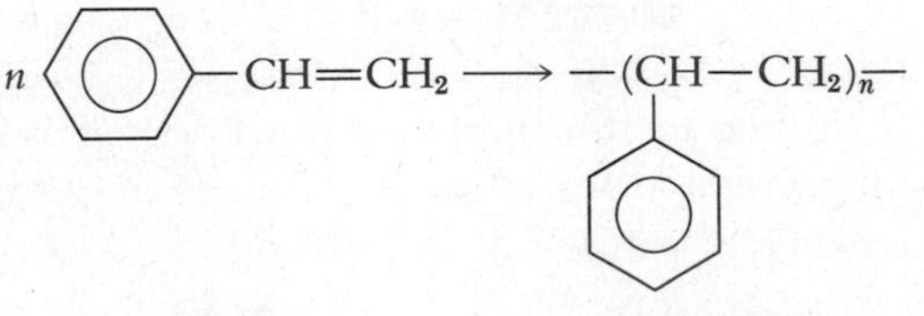

styrene   polystyrene

C.

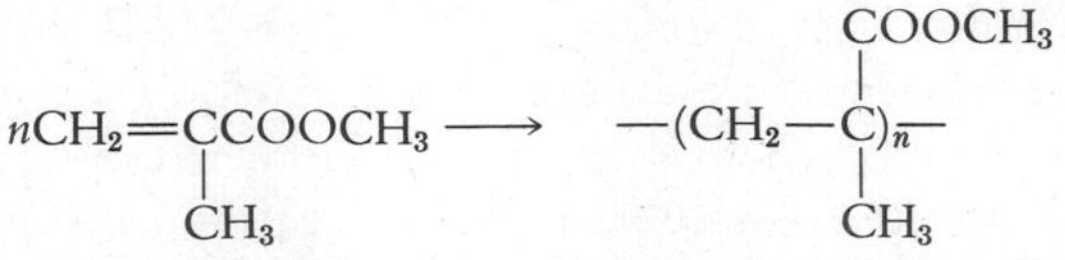

methyl methacrylate   poly(methyl methacrylate)

Table 1.   Physical properties

	M.W.	m.p.	b.p.	d.
styrene	104.15	−31	145–46	0.909
benzoyl peroxide	242.2	104	explodes	

Table 1.   Physical properties (*continued*)

	M.W.	m.p.	b.p.	d.
potassium persulfate	270.3	100d		
methyl methacrylate	100.1	−48	100	0.944

Hazards

Benzoyl peroxide, when dry, presents a serious risk of explosion from shock, friction, heat, or contact with certain metals; it can irritate skin and eyes. Do not grind; keep it away from oxidizable materials. The dry form should be weighed only on glassine paper using a nonmetallic (porcelain or glass) spatula. Methyl methacrylate and styrene are flammable and irritate the skin, eyes, and respiratory system; the unstabilized monomers are subject to rapid, uncontrolled polymerization that may be explosive due to the evolution of considerable heat. Do not breathe vapors; prevent contact with skin, eyes, and clothing; keep away from other chemicals. Potassium peroxydisulfate can react with oxidizable materials and alkalis.

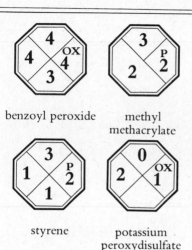

benzoyl peroxide   methyl
                        methacrylate

styrene   potassium
                    peroxydisulfate

PROCEDURE

A. *Emulsion Polymerization of Styrene*

Equip a 100-ml boiling flask with a reflux condenser ⟨OP–7a⟩ and magnetic stirbar. Place 20 ml of distilled water in the flask and dissolve in it 0.20 g of sodium lauryl sulfate and 0.05 g of potassium peroxydisulfate. Add 10.0 g of styrene and bubble nitrogen into the solution for a minute or two to displace air and disperse the styrene. Stir the mixture ⟨OP–9⟩ to maintain the emulsion while heating it in a 60° water bath ⟨OP–7⟩ for 2-3 hours.

Begin the other parts of the experiment during the heating period.

Precipitate the polymer by adding about 10 ml of 10% alum $(KAl(SO_4)_2 \cdot 12H_2O)$ solution and boiling it for a few minutes. Recover the polystyrene by vacuum filtration ⟨OP–12⟩ and wash it with methanol until it is no longer sticky. Dry the polystyrene ⟨OP–21⟩ and weigh it ⟨OP–5⟩.

If necessary, it can be transferred to a small beaker for washing.

B. *Solution Polymerization of Styrene*

Dissolve 1.0 g of styrene in 5 ml of toluene in a 15-cm test tube and add about 50 mg (0.050 g) of benzoyl peroxide

(*Caution:* see Hazards). Swirl to dissolve the peroxide, then heat the mixture on a steam bath for an hour or two, stirring occasionally. Pour the cooled reaction mixture, with stirring, into 25 ml of methanol and mix it well to coagulate the polystyrene. Decant or filter off the methanol, then add some fresh methanol and stir the mixture until the polymer is no longer sticky. Collect the polystyrene by vacuum filtration ⟨OP–12⟩, dry it ⟨OP–21⟩, and obtain the yield ⟨OP–5⟩.

Rinse out the test tube with a little methanol to recover all of the polystyrene.

Dissolve about 0.1 g of polystyrene in 1 ml of 2-butanone and, using a stirring rod, spread the mixture evenly over a clean microscope slide. Set the slide under the hood to evaporate the solvent, remove the film with a razor blade, and obtain its infrared spectrum ⟨OP–33⟩ between two salt plates.

An even film can be obtained by spreading the solution with a simple "doctor knife" constructed by wrapping two short strips of tape around a glass rod.

C. Bulk Polymerization of Methyl Methacrylate

Place 10 g of granulated poly(methyl methacrylate) in a 50-ml boiling flask set up for semimicro distillation and heat the flask with a small flame until the polymer softens and begins to depolymerize. Continue heating gently so that the methyl methacrylate distills over slowly ⟨OP–25a⟩. Stop before the residue in the flask turns black and tarry. Pour the distillate into a test tube, add 50 mg of benzoyl peroxide (*Caution:* see Hazards) and swirl to dissolve. Flush the tube with dry nitrogen, stopper it, and place it in direct sunlight until the next laboratory period. With your instructor's permission, you may break the tube to obtain the clear polymer.

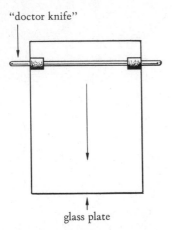

"doctor knife"

glass plate

Figure 1.   Polymer film apparatus

Flushing with nitrogen is not essential, but polymerization is slower without it.

Experimental Variations

1.   Small objects can be embedded in the bulk-polymerized methyl methacrylate by suspending them from a thread into the liquid. Metal objects should first be coated with clear enamel.

2. Try preparing a thin film of poly(methyl methacrylate) in a suitable solvent. Methods for preparing polymer films are described in *J. Chem. Educ. 50,* 228 (1973).

3. A multistep synthesis of polystyrene from benzene is described in *J. Chem. Educ. 38,* 305 (1961). A number of polymerization experiments may be found in *J. Chem. Educ. 42,* 10 (1965); a method for low-temperature polymerization of vinyl monomers is in *J. Chem. Educ. 33,* 231 (1956).

4. The preparation of isotactic polypropylene using a Ziegler-Natta catalyst is described in *Polymer Syntheses* Vol. I, p. 45 (Bib–B6).

Preparation of a Polyurethane Foam

MINILAB 19

Combine 5.5 ml of castor oil, 1.0 ml of glycerol, and 2 drops each of water, triethylamine, and Dow-Corning 200 silicone oil in a waxed paper cup and mix until smooth.

Do not breathe the vapors of the diisocyanate, or any of the vapors evolved during the reaction. Wash your hands thoroughly afterward.

WARNING

Take the cup under the hood and add 4.0 ml of tolylene-2,4-diisocyanate (*Caution:* very toxic), then stir vigorously until you have a creamy emulsion and bubbles start to form. Leave the mixture in the hood to complete the reaction and remove it from the cup after 24 hours.

triethylamine

Topics for Report

1. Write a mechanism for the free-radical-catalyzed polymerization of methyl methacrylate, using benzoyl peroxide as the initiator.

2. Indicate a monomer or pair of monomers that could be used to prepare polymers having the following repeating units:

Decomposition of benzoyl peroxide

$$\underset{\text{PhCOOCPh}}{\overset{\text{O}\quad\text{O}}{\overset{\parallel\quad\parallel}{}}} \xrightarrow[\text{light}]{\text{heat or}} 2\text{Ph}\cdot + 2\text{CO}_2$$

1. —(CHClCHCl)—
3. —(CH_2CH_2NH)—

2. —(CF_2CFCl)—
4. —(CH_2C=$CHCH_2$)—
             |
             CH_3

5. —(CH_2CH)—
      |
      CH_2CH_3

6. 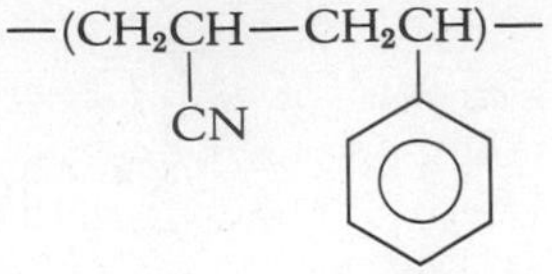 —(CH_2CH—CH_2CH)—
         |
         CN

3. Show how the two monomers illustrated could combine to form a Diels–Alder addition polymer, giving the structure of the polymeric repeating unit.

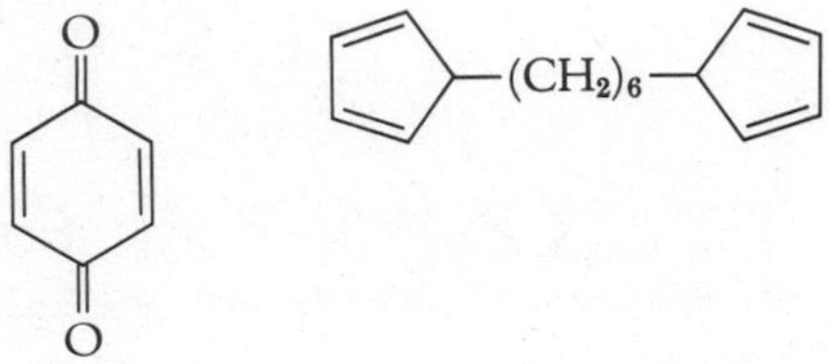

4. Poly(ethylene glycol) can be prepared from both ethylene glycol and ethylene oxide. Write a balanced equation for each reaction and classify each as an addition or condensation polymerization.

—(CH_2CH_2O)—

poly(ethylene glycol)

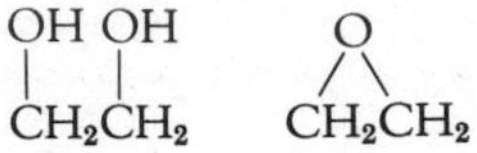

OH OH
 | |
CH_2CH_2
ethylene
glycol

    O
   / \
CH_2CH_2
ethylene
oxide

Library Topics

1. Write a short paper on the stereochemistry of chain-growth polymerization. Illustrate structures for atactic, isotactic, and syndiotactic polymers, tell how they are prepared, and compare their properties.

2. Look up the terms block copolymer, graft copolymer, cross-linked polymer, homopolymer, and alternating copolymer. Give definitions and examples for each.

Analysis of Fatty Acid Content in a Commercial Cooking Oil

Fats and Oils. Transesterification. Fatty Acid Esters. Gas Chromatrographic Analysis.

Operations

⟨OP–7⟩ Heating

⟨OP–11⟩ Gravity Filtration

⟨OP–13⟩ Extraction

⟨OP–14⟩ Evaporation

⟨OP–32⟩ Gas Chromatography

Experimental Objective

To analyze a commercial cooking oil for fatty acid content and determine the ratio of polyunsaturated to saturated fatty acids.

Learning Objectives

To learn how to carry out a small-scale transesterification of triglycerides, and how to identify fatty acid esters from their equivalent-chain-length values.

To learn about the history and significance of the triglycerides and their component fatty acids.

SITUATION

You are an analytical chemist serving as a consultant for the publishers of *Caveat Emptor,* a monthly magazine that rates consumer products and exposes inaccurate or misleading advertising. You have been asked to analyze the fat content of different brands of cooking oil to see how they compare in polyunsaturated and saturated fats and to find out if their advertised SFA and PUFA figures are accurate. Your

SFA: saturated fatty acid
PUFA: polyunsaturated fatty acid

approach will be to convert their component fatty acids to the corresponding methyl esters and analyze the resulting mixture of esters by gas chromatography.

BACKGROUND

Fatty Acids in Human Affairs

$$CH_3(CH_2)_{14}COOH$$

palmitic acid

$$CH_3(CH_2)_{16}COOH$$

stearic acid

Olive oil was so important to the medieval alchemists that they used this symbol, a circle within a cross, to represent it.

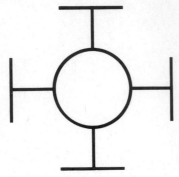

Some other fatty acids, as linolenic and arachidonic acids, have also been regarded as essential but probably are not. These acids have been lumped together under the name "Vitamin F."

$$CH_3CH_2CH=CH \ldots$$

The "ω" designation refers to the distance of the last double bond from the methyl end of the fatty acid chain, as illustrated.

Fats and oils have found many uses throughout human history. Excavation of an Egyptian tomb more than 5000 years old yielded several earthen vessels containing substances identified as palmitic acid and stearic acid, most likely formed by the breakdown of palm oil and beef or mutton tallow that were placed there as provisions for the deceased. The Egyptians used olive oil as a lubricant to move huge stones for their building projects, and a mixture of fat and lime as axle grease for their war chariots. The Romans used candles made of beeswax and tallow for illumination, and the Phoenicians of 600 B.C. were trading soap to the Gauls, who undoubtedly needed it. Pliny the Elder described a process for making soap by boiling goat fat with wood ashes, then treating the pasty mass with sea water to harden it; soap factories have been excavated at Pompeii, which was buried by a volcano in 79 A.D. The ancient art of painting with oils or waxes goes back at least to the ancient Egyptians, who used the encaustic technique (pigments mixed with natural waxes) to paint portraits on their mummy cases.

Modern research has revealed the importance of fats and oils in nutrition and is just beginning to clarify their role in human disease. One fatty acid, linoleic acid, is known to be an essential component of the human diet, since it is not synthesized in the body. Polyunsaturated fatty acids (PUFA) such as linoleic acid are believed to be involved in the bio-

$$CH_3(CH_2)_4\overset{\overset{\displaystyle H}{|}}{C}=\overset{\overset{\displaystyle H}{|}}{C}CH_2\overset{\overset{\displaystyle H}{|}}{C}=\overset{\overset{\displaystyle H}{|}}{C}(CH_2)_7COOH$$

linoleic acid

synthesis of the prostaglandins, which help control blood pressure and muscle contraction. They play a vital role in the functioning of biological membranes and have been implicated in the occurrence or prevention of such illnesses as atherosclerosis, cancer, and multiple sclerosis. Recent research seems to indicate that PUFA of the ω3 and ω6 families are especially effective in preventing the buildup of cholesterol in arteries, which is a major cause of heart disease. On the other hand, some PUFA may actually increase the growth rate of cancer tumors.

In recent years, polyunsaturation has become a selling point for many cooking oils and margarines because of the apparent relationship between certain kinds of fats and heart disease. The polyunsaturates are said to be generally beneficial because they reduce the buildup of cholesterol in the arteries by hastening its excretion; the saturated fats are in turn said to be harmful because they raise cholesterol levels in the blood. Monounsaturated fatty acids are assumed to be more or less neutral in this regard, having neither harmful nor beneficial effects.

Fatty acids can be named or symbolized by several different systems. Most of the common fatty acids have trivial names that have been used for years, partly because of their simplicity compared to the systematic names. Trivial names do not reveal the structures of the corresponding acids however, so several shorthand notations have been adopted. Oleic acid can be represented by c-9-18:1, where c means *cis,* 9 refers to the position of the double bond (numbering from the COOH group), 18 is the total carbon chain length, and 1 is the number of double bonds. Linoleic acid would be c,c-9, 12-18:2 under this system. Since all of the acids to be studied in this experiment are *cis,* the c prefixes will be omitted here.

Linoleic acid, for example, is named *cis*-9–*cis*-12-octadecadienoic acid under the IUPAC system.

$$CH_3(CH_2)_7-\overset{\overset{\displaystyle H}{|}}{C}=\overset{\overset{\displaystyle H}{|}}{C}-(CH)_7COOH$$
oleic acid (9-18:1)

METHODOLOGY

The fats and oils of the type to be analyzed in this experiment are triglyceride esters; that is, esters of the trihydroxy alcohol *glycerol* containing three fatty acid residues. The fatty acids can be obtained by hydrolyzing a triglyceride, usually in the presence of a base, to yield the corresponding acid salts. This is the process used in soap formation, with the acid salts comprising the "soap." Alternatively, a triglyceride can be transesterified with an alcohol such as methanol to yield a mixture of methyl esters. Since the esters are much more suitable for GLC analysis than the acids themselves, the second method will be used in this experiment; the fatty acid composition of each fat can then be derived from the percentages of the corresponding esters.

When saturated methyl esters of fatty acids are chromatographed on a suitable GLC column, their retention times (T_r) increase with the length of the carbon chain according to the relationship $\log T_r \propto$ carbon number. The

$$
\begin{array}{ll}
CH_3(CH_2)_{16}COOCH_2 & CH_2OH \\
\quad\quad\quad\quad\quad\quad\quad | & \quad\quad | \\
CH_3(CH_2)_{16}COOCH & CHOH \\
\quad\quad\quad\quad\quad\quad\quad | & \quad\quad | \\
CH_3(CH_2)_{16}COOCH_2 & CH_2OH
\end{array}
$$

a triglyceride,   glycerol
tristearin

$$CH_3(CH_2)_{16}COO^-Na^+$$

a soap, sodium stearate

$$CH_3(CH_2)_{16}COOCH_3$$

methyl ester of stearic acid
(methyl stearate)

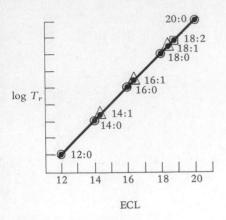

Figure 1. Plot of log T_r versus equivalent chain length

The graphing is easier if semilog paper is used.

Table 1.   ECL values for DEGS liquid phase

Acid (as Me ester)	Shorthand Notation	ECL Value
lauric	12:0	12.00
myristic	14:0	14.00
myristoleic	9-14:1	14.71
palmitic	16:0	16.00
palmitoleic	9-16:1	16.55
stearic	18:0	18.00
oleic	9-18:1	18.43
linoleic	9,12-18:2	19.22
arachidic	20:0	20.00
linolenic	9,12,15-18:3	20.12
behenic	22:0	22.00

Different students can analyze different kinds of oils (corn, cottonseed, safflower, peanut, sunflower, etc.) and compare their results.

Heptadecanoic acid (17:0) does not occur naturally, so the 17:0 peak will not interfere with other component peaks.

retention times of unsaturated fatty esters do not coincide with those of saturated esters; on a nonpolar liquid phase like Apiezon their retention times are shorter than those of the saturated esters of the same chain length, while on a polar liquid phase like diethylene glycol succinate (DEGS) they are longer. The actual value of T_r depends on the number and positions of the double bonds. By comparing the retention times of a large number of such esters, a series of equivalent-chain-length (ECL) values has been worked out for various esters on different liquid phases. The ECL of a *saturated* ester is the same as its actual chain length—for instance, 14.0 for myristic acid (as the methyl ester), 18.0 for stearic acid, etc. The ECL for an *unsaturated* ester on a polar liquid phase is greater than its actual chain length—for instance, 18.43 for methyl oleate (9-18:1) and 19.22 for methyl linoleate (9,12-18:2), both measured on DEGS. If the retention times for two or more saturated esters are known, a calibration curve can be constructed by plotting the logarithms of their retention times against their chain lengths. To identify an unknown fatty acid ester one calculates the logarithm of its retention time, reads its ECL value off the graph, and compares that value with the ECL values of known methyl esters (see Table 1).

In order for ECL values to provide an accurate means of identifying the esters, the same type of column and carrier gas should be used. Even then, the experimental values will not usually correspond exactly to literature values because of differences in such factors as the kind of column support, the amount of liquid phase, the age of the column, and the experimental conditions. In addition, some peaks may overlap in complex mixtures so that two or more columns must be used for their separation, and deviations from linearity in the log T_r plot may occur at low carbon numbers. The relatively simple mixtures of fatty acids that you will encounter in the commercial oils should not present any serious experimental difficulties, however.

The methanolysis (transesterification) of the oil will be carried out by saponifying the triglycerides in methanolic sodium hydroxide and esterifying the resulting fatty acids in methanol with a boron trifluoride catalyst. After the methyl esters have been isolated from the reaction mixture, they will be mixed with an internal standard, methyl heptadecanoate, and analyzed by gas chromatography. To minimize the effect of retention time drift due to changes in operating parameters, relative retention values should be calculated by dividing the retention time (or distance) for each peak by that of the 17:0 peak. Once the 17:0 peak has been located, those of the 16:0 and the 18:0 esters can be identified by the fact that the log T_r interval between their peaks and 17:0 is the same. Then a log T_r versus ECL plot can be prepared, from which the other esters can be identified. Since peak area should be nearly proportional to component mass for fatty acid esters having over ten carbon atoms, the percentage composition of the mixture can be determined from peak areas with good accuracy.

The 16:0 ester often gives the first strong peak on the chromatogram.

The ECL graph can be refined after several more esters have been identified by using their reported ECL values to plot a more accurate straight line.

Reaction and Properties

$$
\begin{array}{c}
\text{R}'\text{COOCH}_2 \\
| \\
\text{R}''\text{COOCH} \quad + \; 3\text{CH}_3\text{OH} \; \xrightarrow[\text{BF}_3]{\text{NaOH}} \\
| \\
\text{R}'''\text{COOCH}_2
\end{array}
\qquad
\begin{array}{c}
\text{R}'\text{COOCH}_3 \\
+ \\
\text{R}''\text{COOCH}_3 \\
+ \\
\text{R}'''\text{COOCH}_3
\end{array}
\; + \;
\begin{array}{c}
\overset{\text{H}}{\text{O}} \; \overset{\text{H}}{\text{O}} \; \overset{\text{H}}{\text{O}} \\
| \quad | \quad | \\
\text{CH}_2\text{CHCH}_2
\end{array}
$$

(R groups can be alike or different.)

Table 2.   Physical properties

	M.W.	m.p.	b.p.	d.
methanol	32.0	−94	65	0.791
boron trifluoride– methanol complex	131.9		59[4]	
boron trifluoride	67.8	−127	−100	

The complex has the composition $BF_3 \cdot 2CH_3OH$

Hazards

Methanol is toxic and can be absorbed through the skin. Avoid breathing vapors or contact with skin or eyes. Boron trifluoride is very toxic, corrosive on the skin, and irritating to eyes and mucous membranes. Avoid contact with skin, eyes, and clothing; do not breathe vapors. The methanolic sodium hydroxide should also be kept out of contact with skin and eyes.

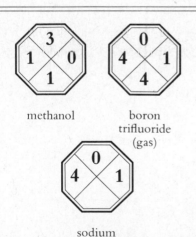

methanol   boron trifluoride (gas)

sodium hydroxide

PROCEDURE

[Adapted from a procedure in the *Journal of Chemical Education, 541,* 406 (1974), with permission.]

Reaction and Separation.   Combine 0.15 g of a commercial cooking oil with 5 ml of 0.5M methanolic sodium hydroxide in a 15 cm test tube and heat the tube ⟨OP−7⟩ on a steam

This usually requires 3–5 minutes.

The 12.5% boron trifluoride solution can be prepared by mixing 5 ml of methanol with 1 ml of BF₃–methanol complex.

Most of the methanol will be in the aqueous layer.

Use about 2 µl of sample. A 60/80 mesh Chromosorb W column coated with 15% DEGS is suitable for the analysis. The column temperature should be about 190°C.

bath until the fat globules have dissolved. Carefully (see Hazards) add 6 ml of 12.5% (weight/volume) boron trifluoride in methanol and boil the mixture for two minutes. Transfer the mixture to a small separatory funnel using 30 ml of 30–60° petroleum ether, add 20 ml of saturated aqueous sodium chloride, and extract the methyl esters ⟨OP–13⟩ from the aqueous layer. Filter the petroleum ether layer by gravity ⟨OP–11⟩ and evaporate the solvent ⟨OP–14⟩.

Analysis.   Add 0.1 ml of a 50% wt/v solution of methyl heptadecanoate in chloroform to the residue and analyze the mixture by gas chromatography ⟨OP–32⟩. Measure all retention values and peak areas, then identify the peaks and determine the composition of the mixture as described in the Methodology.

Experimental Variations

1.   Commercial oleomargarines can be analyzed by the method described if they are dried first. An ether solution of the oleomargarine can be dried with magnesium sulfate and then evaporated just before the hydrolysis step.

2.   The cooking oil can be tested for unsaturation using test C–7 or C–19 in Part III.

3.   The isolation of trimyristicin from nutmeg is described in R. Ikan, *Natural Products, a Laboratory Guide,* p. 25 (Bib–B5).

Topics for Report

The PUFA/SFA ratio has been used as one measure of the dietary quality of a fat or oil.

1.   (*a*) For your cooking oil, calculate the percentage of polyunsaturated fatty acids (linoleic and linolenic), the percentage of saturated fatty acids, and the PUFA/SFA ratio. (*b*) Compare the PUFA/SFA ratios of your product with those obtained by other students with other cooking oils and rank the oils in the order of their apparent dietary health benefit.

2.   Give structures, common names, IUPAC names, and shorthand notations for all of the fatty acids present in your cooking oil.

3.   Stearic acid makes up about 90% of the fatty acids present in hydrogenated vegetable shortenings made from corn and soybean oils. Explain.

4.   Explain why no glycerol peak was observed on the chromatogram, even though glycerol was a product of the hydrolysis.

5.   Outline a synthesis of the detergent sodium lauryl sulfate from the triglyceride of lauric acid, trilaurin.

$$CH_3(CH_2)_{10}CH_2OSO_2O^-Na^+$$
sodium lauryl sulfate

Library Topics

1.   Give the structures, major natural sources, and some uses of lauric, myristic, stearic, oleic, linoleic, linolenic, arachidic, and behenic acids.

The acids occur naturally in the form of triglycerides or other esters.

2.   Write a short paper on the health implications of saturated and polyunsaturated fatty acids and the apparent relationship between cholesterol and heart disease.

PART III

Systematic Organic Qualitative Analysis: Identification of Unknown Organic Compounds

Background: The Chemist as Detective

Professional organic chemists are often confronted with unknown compounds whose identity must be established. If a chemist carries out a reaction that is expected to produce a certain compound, he or she must prove that the product is, in fact, the expected one. This may require no more than a melting or boiling point and some spectral data. However, if the outcome of the reaction is *not* known, the problem of characterizing the products is a more difficult one, and the chemist will ordinarily have to go through some kind of systematic procedure for identifying them. The same is true of a chemist who has isolated unknown components from natural or man-made products.

The process of identifying an unknown can be compared to the approach used by a detective in identifying the perpetrator of a crime. The detective first looks for clues that help to characterize the criminal, narrowing down the range of suspects and perhaps suggesting the most productive areas of investigation. When one or more likely suspects has been tracked down, the detective has to find and analyze evidence to help eliminate some suspects from consideration and to establish the probable guilt of the prime suspect so that an indictment can be obtained. Finally, the detective has to amass enough additional evidence to convince a jury that the accused is, in fact, guilty of the crime.

In carrying out the identification of an organic compound you, like the detective, should be constantly on the lookout for clues to the identity of your unknown. Spectral and chemical data should allow you to confine your search to a particular chemical family. Additional physical and chemical evidence will help you narrow down the list of "suspects" and focus your attention on a few of the most probable compounds. Finally, the preparation of one or more derivatives should lead you to a definite conclusion and provide you with sufficient evidence to convince the "jury" (your instructor) that your compound is, in fact, what you believe it to be.

As in the solution of any other problem, you must first ask yourself the right questions before you can arrive at the correct answer. Some important questions to be answered with regard to an unknown compound are: *(1)* Is it pure? *(2)* What functional group(s) does it contain? *(3)* Are there any other significant structural features that might aid in its identification? Each bit of evidence that you obtain should, if interpreted correctly, help reveal the answer to one or more of these questions; and all of them combined should provide you with an answer to the ultimate question, "What is it?"

A detective trying to solve a case will almost invariably come upon clues that lead nowhere or, even worse, to false conclusions. The same is true in chemical problem-solving, so it is important to keep an open mind throughout your investigation and avoid jumping to conclusions before all the evidence is in. You may formulate tentative conclusions based on your initial observations (e.g., since it turns chromic anhydride reagent green, it may be an alcohol), but you should be ready to revise or discard such conclusions if they are not supported by subsequent observations (e.g., its IR spectrum shows a $C{=}O$ stretching band but no $O{-}H$ band, so it may be an aldehyde instead).

Chemical and physical evidence can be misleading for a variety of reasons:

1. Some compounds of a given family may be atypical in their reaction with a given reagent and yield either a false positive or a false negative result.

2. Some reagents give positive tests with more than one functional group.

3. Impurities may complicate or invalidate a test.

4. Spectral bands may occur outside the expected frequency ranges or may be incorrectly assigned.

Because of these and other possible sources of error, it is wise not to rely on a single piece of evidence in formulating a conclusion. For example, the classification of an unknown as a secondary alcohol is adequately established by a slow reaction with Lucas's reagent, a positive chromic anhydride test, *and* an infrared band in the 1100 cm^{-1} region; but not by any one of these alone.

There is no single "best" way to identify an organic compound and no one path to follow. Although a general outline and description of one useful approach is given in the Methodology, it is not necessary (or even desirable, in some cases) to follow it slavishly. You must use your own judgment and initiative in choosing which tests to perform, which physical properties to measure, which derivatives to prepare, and which spectra to record. By keeping your eyes and mind open at all times, you may find clues to the structure or identity of your compound that suggest a way to bypass some of the usual intermediate steps. At the same time you should avoid acting on "hunches" that may lead you on a wild-goose chase. For example, an unknown liquid with a distinct wintergreen odor might be methyl salicylate; it could be tested for ester and phenolic functional groups immediately after distillation, followed by the preparation of suitable derivatives if the tests are positive. On the other hand, a liquid that burns with a blue flame and has a spiritous odor might possibly be ethyl alcohol, but it would be foolish to try to prepare an alcohol derivative on the basis of that evidence alone.

A qualitative analysis problem can provide an exciting and challenging experience for those who apply all their skill and ingenuity to its solution. Besides testing your mastery of the operations learned previously in the laboratory, it furnishes many practical applications of the concepts learned during the lecture course in organic chemistry. Throughout the analysis of an unknown compound you may apply your knowledge of functional group chemistry, nomenclature, acid-base equilibria, structure-property relationships, spectral analysis, organic synthesis, and many other areas of organic chemistry.

Methodology

The general approach to organic qualitative analysis described in this section combines the classical "wet chemistry" scheme with limited use of instruments. It is assumed that the unknowns will be limited to the following ten classes: *alcohols, aldehydes, ketones, amides, amines, carboxylic acids, esters, halides, aromatic hydrocarbons,* and *phenols.* An unknown may contain an additional subsidiary functional group (such as the nitro group in *p*-nitrophenol or the ether group in *p*-anisic acid), but it will be classified and derivatized as a member of one of the above families. The analysis scheme is divided into three parts, as outlined in the margin. The *preliminary work* includes a gross physical examination of the unknown, an ignition test (which can provide useful clues to its structure), and an evaluation of its purity. The *functional class determination* includes the solubility tests, spectra, and classification tests necessary to assign a compound to one of the ten chemical families. The *identification* phase requires a literature search, the preparation of derivatives, and the accumulation of additional physical and chemical evidence to establish the identity of the unknown.

Qualitative analysis scheme
1. Preliminary work
   a. Gross physical examination
   b. Ignition
   c. Estimation of purity
   d. Purification (optional)
2. Functional class determination
   a. Solubility tests
   b. Infrared spectrum
   c. Functional class tests
3. Identification
   a. Examination of literature
   b. Additional tests and data
   c. Preparation of derivatives

If the unknowns provided are all in a single family or a small group of families, part 2 can be omitted or modified.

PRELIMINARY WORK

Observation of the physical state, color, and odor of a compound should provide some clues to its identity. For instance, the fact that an unknown is a solid eliminates all organic compounds that are liquids at room temperature, an intrinsic color suggests that chromophoric groups having conjugated double bonds or rings are present, and a distinctive odor may suggest the chemical family to which the unknown belongs.

Organic compounds that are nonflammable in the ignition test may contain a high ratio of halogen to hydrogen or have a very high molecular weight. A yellow, sooty flame often indicates an aromatic compound. Aliphatic hydrocarbons burn with a clean yellow flame, whereas compounds with a high oxygen content tend to burn bluer.

The melting or boiling point of an unknown gives a rough estimate of its purity. Solids having a melting point range greater than about 3–4°C should ordinarily be puri-

Light or dull colors (light yellow, tan, brown, black, etc.) may suggest that the unknown is impure, whereas intense yellows, oranges, reds, etc. are usually intrinsic.

For example, methanol burns with a bluer flame than 1-octanol because it has a higher oxygen/hydrogen ratio.

The purity of liquid unknowns can also be determined by GLC ⟨OP–32⟩.

⟨OP–23c⟩ *tells how to choose a re-*
crystallization solvent.

fied by recrystallization. A liquid with a boiling point range
greater than a few degrees (or a nonreproducible micro boiling point) should be purified by distillation and the high or
low-boiling fractions discarded. It is very important to determine the melting or boiling point accurately, since compounds are listed in order of melting and boiling points in
the tables used to identify them.

FUNCTIONAL CLASS DETERMINATION

The traditional procedure for classifying unknowns by
functional group involves the use of preliminary screening
tests that categorize the unknowns into broad groups according to solubility or other properties, followed by functional class tests that place them in chemical families. The
use of infrared and other spectrometric methods can simplify
this process considerably; a skilled analyst can often classify
a compound by its infrared spectrum alone. However, the
classical tests still have their uses, particularly for students
without extensive experience in spectral analysis. It is recommended that the solubility tests and *at least* one func-
tional-class test be performed on each unknown (except for
amines and carboxylic acids, for which the solubility tests
may suffice) even when an infrared spectrum is obtained.

Solubility Tests

Compounds that dissolve to the extent of
about 30–35 mg per ml of water will be
considered "soluble" in these tests.

Cyclic and branched compounds are
usually more soluble than straight-
chain compounds of the same carbon
number; a phenyl group has about the
same effect on solubility as *n*-butyl.

Neutral water-soluble compounds are
placed in solubility class S_n*, whereas*
acidic and basic water-soluble compounds
are placed in classes S_a *and* S_b *respec-*
tively (see Figure 1, page 384).

Amines containing two or more aryl
groups and certain hindered amines may
be insoluble in 5% HCl.

Water solubility usually indicates the presence of at least one
oxygen or nitrogen atom and a relatively low molecular
weight. The borderline for water solubility in the case of
monofunctional oxygen or nitrogen compounds is usually
around five carbon atoms. For example, 1-butanol (with
four carbon atoms) is soluble, whereas 1-pentanol (with five
carbons) is not. Of the ten classes of compounds considered
here, water solubility can be expected for low-molecular-
weight alcohols, aldehydes, ketones, amides, amines, carboxylic acids, esters, and phenols. Most amines and carboxylic acids can be distinguished from the rest by testing their
aqueous solution with litmus paper. Phenols and aromatic
amines may be too weakly acidic or basic to turn litmus red
or blue.

Water-insoluble compounds that are soluble in 5% hydrochloric acid (Class B) contain basic functional groups,
usually incorporating nitrogen atoms. Of the families considered here, only amines fall into this category. Some

amines may form insoluble hydrochloride salts as they dissolve, so solubility behavior should be observed carefully to detect any change in the appearance of the unknown when it is shaken with the solvent.

Compounds that are insoluble in water *and* 5% HCl but soluble in 5% sodium hydroxide contain acidic functional groups giving them pK_a values of approximately 12 or less. Carboxylic acids and phenols fall into this category; the latter can be recognized by their insolubility in 5% sodium bicarbonate. Care should be taken in interpreting this test. Some compounds, such as reactive esters, may react with the solvent to form soluble reaction products; long-chain carboxylic acids may form relatively insoluble salts that yield a soapy foam when shaken.

Acids with pK_a values of about 6 or less will dissolve in 5% sodium bicarbonate. This includes most carboxylic acids but not phenols, so the solubility test with sodium bicarbonate can distinguish between these two classes of compounds.

Compounds that are insoluble in all the previous solvents but dissolve in, or react with, cold concentrated sulfuric acid (Class N) include high-molecular-weight alcohols, aldehydes, ketones, amides, and esters. Unsaturated compounds and some aromatic hydrocarbons (those containing several alkyl groups on the benzene ring) are also soluble in this reagent. Solution in sulfuric acid is accompanied by protonation of a basic (nitrogen or oxygen) atom or by some other reaction such as sulfonation, dehydration, addition to multiple bonds, or polymerization.

Compounds that are insoluble in all of the solvents (Class X) include most aromatic hydrocarbons and the halogen derivatives of aliphatic and aromatic hydrocarbons.

An outline of the solubility scheme is given in Figure 1. It is seldom necessary to test an unknown with every solvent, since if it dissolves in water, for example, it will dissolve in aqueous solutions of HCl, NaOH, and $NaHCO_3$ as well. In most cases, an unknown is not tested further once it is found to dissolve in a given solvent, since its solubility classification is established at that point. The only exception to this rule concerns compounds soluble in 5% NaOH, which are also tested with 5% sodium bicarbonate to differentiate between carboxylic acids and phenols. It should be emphasized that solubility classifications can be misleading; there are exceptions and borderline cases. For instance, some phenols are acidic enough to dissolve in 5% $NaHCO_3$, and some weakly basic amines do not dissolve in 5% HCl.

Compounds soluble in both solvents are placed in class A_1; those soluble in 5% NaOH but insoluble in 5% $NaHCO_3$ are assigned to class A_2.

Some reactions in cold concentrated sulfuric acid

$$RCHO \xrightarrow{H_2SO_4} R\overset{\oplus}{C}HOH$$

$$RCH_2OH \xrightarrow{H_2SO_4} RCH_2OSO_3H$$

$$RCH_2\underset{\underset{R''}{|}}{\overset{\overset{OH}{|}}{C}}-R' \xrightarrow{H_2SO_4} RCH=\underset{\underset{R''}{|}}{C}-R'$$

$$RCOOH \xrightarrow{H_2SO_4} R\overset{\oplus}{C}(OH)_2$$

$$RCH=CHR' \xrightarrow{H_2SO_4} RCH_2\underset{\underset{R'}{|}}{\overset{\overset{OSO_3H}{|}}{C}}H$$

Therefore the solubility tests should always be supported by other chemical or spectral evidence before a definite conclusion is drawn.

Figure 1.   Flow diagram for solubility tests

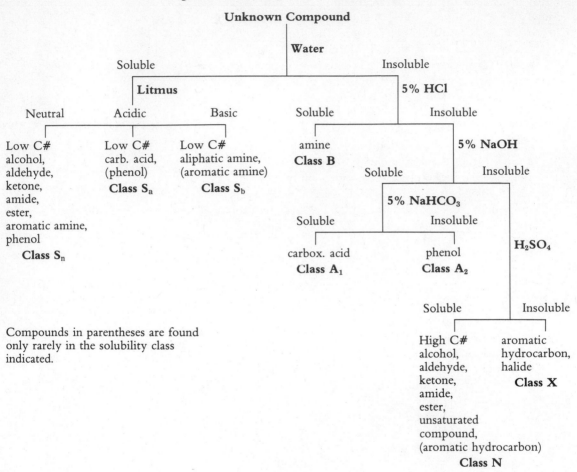

Compounds in parentheses are found only rarely in the solubility class indicated.

Infrared Spectrum

Infrared spectra of most organic molecules are very complex, since they result from the interactions of a large number of bond and structural unit vibrations. Therefore, not all absorption bands provide useful information for qualitative analysis. The following four regions are the most useful for detecting the common functional groups and should be carefully analyzed.

See Appendix VI for interpreted spectra illustrating the IR bands described here.

Region 1: 3600–3200 cm⁻¹ (2.8–3.1 μm). This region contains bands of high to medium intensity arising from O—H or N—H stretching vibrations. Care should be taken not to misinterpret any *weak* absorption band in Region 1, since carbonyl compounds (which absorb around 1700 cm⁻¹) usually display weak overtone bands at twice their carbonyl stretching frequency, and the presence of moisture or hydroxylic impurities in the unknown may result in a weak O—H band. *Alcohols* and *phenols* usually show a strong O—H stretching band between 3550 and 3200 cm⁻¹.

Amines show N—H stretching bands that are usually somewhat weaker and narrower than O—H bands. Primary amines have two bands around 3400 and 3300 cm⁻¹; these appear at higher frequencies in dilute solution. The presence of such bands, along with a medium to strong NH bending vibration in the 1650–1580 cm⁻¹ region, is good evidence for a primary amine (this should be corroborated by the solubility tests). Secondary amines display a weak to medium band around 3400–3300 cm⁻¹, whereas tertiary amines, having no NH bond, do not absorb in this region.

Amides that are not substituted on the nitrogen atom also show two N—H stretching bands in Region 1; monosubstituted amides have a single band and disubstituted amides none.

Region 2: 3100–2500 cm⁻¹ (3.2–4.0 μm). This is the C—H stretching region of the spectrum; virtually all organic compounds show a strong band between 3000 and 2850 cm⁻¹ due to the presence of sp³ hybridized C—H bonds. Any Region 2 band above 3000 or below 2800 cm⁻¹ should be examined carefully, however.

Aromatic hydrocarbons show an sp² C—H stretching band between 3000 and 3100 cm⁻¹, usually centered around 3070 cm⁻¹. The band is often not very strong, and may appear as a "shoulder" on the sp³ C—H stretching band. Other compounds also contain sp² C—H bands in this region—for instance, compounds with carbon-carbon double bonds and aromatic compounds of other chemical families, such as phenols or aromatic aldehydes. These possibilities must be eliminated before concluding that the unknown is an aromatic hydrocarbon. The presence of an aromatic ring can usually be substantiated by a distinctive pattern of weak absorption bands in the 2000–1670 cm⁻¹ region, and by one or two strong absorption bands between 900 and 690 cm⁻¹. The absence of functional groups can be established by showing that the compound is insoluble in cold concentrated

These frequencies can be converted into wavelengths (in μm) by dividing them into the number 10,000; e.g., 3550 cm⁻¹ is (10,000/3550) μm = 2.82 μm

The bands may be so close together that they look like a single two-pronged peak.

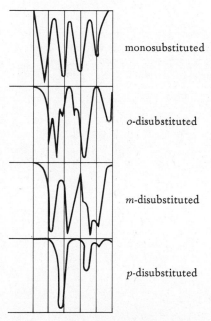

monosubstituted

o-disubstituted

m-disubstituted

p-disubstituted

Figure 2. Typical absorption patterns of substituted aromatics in the 2000-1670 cm⁻¹ region.

sulfuric acid, by performing appropriate classification tests, and by properly interpreting the other regions of the infrared spectrum.

Aldehydes exhibit distinctive C—H stretching bands arising from vibrations of the carbonyl hydrogen atom. Often there are two sharp bands of weak to medium strength located between 2830 and 2695 cm⁻¹, although some aldehydes show only one. The presence of a definite band near 2720 cm⁻¹ (along with a strong carbonyl band in Region 3) is good evidence for an aliphatic aldehyde. This C—H band is displaced to somewhat higher frequency in aromatic aldehydes.

Carboxylic acids show a very broad, intense O—H stretching band that extends between 3300 and 2500 cm⁻¹, and sometimes beyond. It is usually centered around 3000 cm⁻¹ and is superimposed over the C—H band.

Region 3: 1750–1630 cm⁻¹ (5.7–6.1 μm). This is one of the most useful regions of the spectrum; it contains the carbonyl absorption band that is characteristic of aldehydes, ketones, carboxylic acids, amides, esters, and other carbonyl compounds. This band is usually very strong and quite unmistakable. Its position is sensitive to substituent effects, so the exact frequency at which it appears can be a good indication of the specific functional group present.

Most aliphatic *aldehydes* show carbonyl absorption bands centered around 1730 cm⁻¹; in aromatic and α,β-unsaturated aldehydes the frequency is reduced to 1710–1685 cm⁻¹. Ordinary aliphatic *ketones* absorb around 1715 cm⁻¹, whereas ketones having the carbonyl group conjugated with an aromatic ring or double bond show carbonyl bands in the 1685–1665 cm⁻¹ region. In cyclic ketones, bond strain can increase the carbonyl frequency (as in cyclopentanone, which has its absorption band at 1751 cm⁻¹). Aldehydes and ketones can often be differentiated by the aldehyde C—H stretch in Region 2.

Carboxylic acids have intense carbonyl bands, which in saturated aliphatic acids occur around 1720–1705 cm⁻¹. The frequency is lowered to 1710–1680 cm⁻¹ for carboxylic acids conjugated with aromatic rings or double bonds. Aliphatic *esters* absorb at 1750–1735 cm⁻¹ in most cases, whereas formate, aromatic, and α,β-unsaturated esters have absorption bands between 1730 and 1715 cm⁻¹. The *amides* of carboxylic acids usually have carbonyl bands between 1680 and 1630 cm⁻¹. In the case of primary amides, the car-

Frequencies given are for solids or neat liquids. When spectra are run in solution, the position of a carbonyl band may change considerably.

bonyl band may overlap or obscure an N—H bending band near 1655–1620 cm^{-1}.

Region 4: 1350–1000 cm^{-1} (7.4–10.0 μm). This region, unfortunately, is usually cluttered with absorption bands from C—H bending and other vibrations, so it can be difficult to interpret accurately. Most band assignments here should be regarded as questionable until confirmed by evidence from classification tests. Nevertheless the presence of *strong* absorption bands in Region 4 may provide extremely useful information if the unknown is suspected to be an alcohol or ester.

The strong C—O stretching band of most *alcohols* occurs between 1260 and 1000 cm^{-1}; the position depends on the structure of the alcohol. Although there are exceptions to these generalizations, most saturated aliphatic primary alcohols have C—O bands between 1085 and 1050 cm^{-1}; saturated secondary alcohols absorb in the range 1125–1085 cm^{-1}; and saturated tertiary alcohols absorb around 1200–1125 cm^{-1}. The presence of aromatic rings or vinylic groups on the **carbinol carbon** can reduce the frequency by 25–50 cm^{-1}, so that α,β-unsaturated tertiary alcohols absorb in the same region as (for example) saturated secondary alcohols. Cyclic alcohols also absorb at lower frequencies—a secondary alcohol such as cyclohexanol absorbs in the same region as an aliphatic primary alcohol. *Phenols* usually show a strong C—O stretching band in the 1260–1180 cm^{-1} range, accompanied by a weaker band at 1390–1330 cm^{-1}.

Carboxylic acids display a C—O stretching band centered near 1300 cm^{-1}; this sometimes appears as a doublet (a weaker band can be found near 1400 cm^{-1}). The *esters* of carboxylic acids also show two C—O stretching bands, the most important of which is the acyl-oxygen stretch band. For most saturated esters of aliphatic acids, this band occurs around 1210–1160 cm^{-1} (except for acetates, which absorb around 1240 cm^{-1}); for esters of aromatic acids, it occurs between 1310 and 1250 cm^{-1}. The acyl-oxygen stretch band is often broader and stronger than the carbonyl band of the ester and can be found without much difficulty. The second band (alkyl-oxygen stretch) occurs in roughly the same area as the corresponding band of a comparable alcohol—for example, near 1100 cm^{-1} for esters of secondary alcohols.

Miscellaneous IR Bands. *Alkyl* and *aryl halides* are difficult to detect with certainty from their IR spectra, but a

carbinol carbon: the carbon atom to which the OH group is bonded.

These "C—O" absorption bands actually involve coupled vibrations in which some adjacent atoms are involved.

See the IR correlation chart on the back endpaper for positions of bands not described here.

C—Cl stretching band between 550 and 850 cm^{-1} is observed for alkyl chlorides and a C—Cl band around 1090 cm^{-1} is characteristic of aryl chlorides. Subsidiary functional groups such as alkoxy and nitro groups can also be detected by IR.

Besides indicating the presence of various functional groups, the infrared spectrum of an unknown can reveal the presence of aromatic rings, unsaturation, and other structural features. The discussion of aromatic hydrocarbons (see Region 1 above) describes some of the spectral features displayed by aromatic compounds. Carbon-carbon double bonds can often be detected by C—H out-of-plane bending absorptions at the low-frequency end of the spectrum, combined with sp^2 C—H stretching bands above 3000 cm^{-1}.

Table 1.   Summary of infrared band correlations

Region	Frequency Range (cm^{-1})	Bond Type	Family	Comments
1	3500–3300	N—H	amine, amide	Weak-medium. 1°: 2 bands; 2°: 1 band; 3°: no bands. See 1650–1580 for N—H band, Region 3 for C=O band.
	3550–3200	O—H	alcohol, phenol	Broad, strong. Sharp band around 3600 cm^{-1} in dilute solution. See Region 4.
2	3300–2500	O—H	carboxylic acid	Broad, strong, centered around 3000. See Regions 3, 4.
	3100–3000	C—H	aromatic hydrocarbon	May be shoulder on stronger sp^3 C—H band. See 2000–1670 and 900–690 regions.
	2850–2700	$\overset{\overset{\text{(O)}}{\parallel}}{\text{C}}$—H	aldehyde	Weak-medium, usually sharp. See Region 3.
3	1750–1630	C=O	aldehyde, ketone, carboxylic acid, ester, amide	Strong. Position depends on nature of functional group and other structural features. See Region 2 for aldehydes; Regions 2, 4 for acids; Region 4 for esters; Region 1 for amides.
4	1350–1280	C—O	carboxylic acid	Medium-strong. Usually a doublet for long-chain acids.
	1260–1180	C—O	phenol	Strong. Check 1390–1330 for weaker C—O stretch.
	1260–1000	C—O	alcohol	Strong. Frequency in order 3° > 2° > 1°.
	1210–1160 or 1310–1250	C—O	ester	Strong. Aromatic esters in the higher frequency range. Should be accompanied by weaker band in the same region as the alcohol C—O stretch.

In interpreting an infrared spectrum, it is suggested that you first observe the four regions described above to obtain evidence for specific functional groups, using Table 1 as a guide. Then, if necessary, you can derive additional structural information from other regions of the spectrum. For example, the spectrum of Figure 3 has no bands in Region 1, which eliminates compounds having O—H and N—H bonds (alcohols, phenols, carboxylic acids, and most amines and amides). Since there is a weak absorption band above 3000 cm^{-1} in Region 2, the compound may contain sp^2 C—H bonds; this observation, combined with the strong band at 700 cm^{-1}, suggests an aromatic structure. Region 3 shows a strong carbonyl band near 1720 cm^{-1}. Region 4 contains a strong C—O band at 1280 cm^{-1} that, along with the carbonyl band and a weaker C—O band at 1100 cm^{-1}, provides convincing evidence that the compound is an ester. This tentative conclusion could be confirmed by performing one or more functional class tests.

It is always a good idea to study reference spectra while interpreting your infrared spectrum. Although you may learn the approximate position and intensity of a band from this and other textbooks, only authentic spectra can show you the appearance of a specific kind of absorption band and show the interfering bands that might lead you to false conclusions. See Section F in the Bibliography for sources of published spectra and additional information on infrared spectral interpretation.

Figure 3.   Infrared spectrum of ethyl benzoate

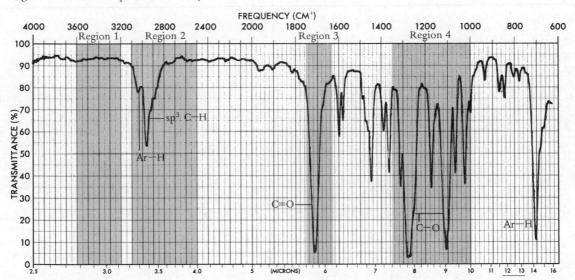

For additional practice in the analysis of infrared spectra, see Appendix VI.

Functional Class Tests

If the solubility tests and infrared analysis are performed skillfully, a classification test may be needed only to confirm the presence of the suspected functional group. Frequently, however, the results of the foregoing analyses will be inconclusive or ambiguous, and classification tests will be required to identify the functional group. In either case, the range of possibilities should be narrowed down so that only a few different tests need to be performed. To avoid wasting time, the tests should be chosen carefully to provide the desired information in a minimum number of steps.

The tests are listed under each family in approximate order of simplicity and utility. However, you must use your own judgment in deciding which tests are appropriate for your unknown.

The tests from Table 2 are most useful for detecting or confirming the presence of a given functional group; Table 3 (page 392) lists tests that provide additional structural information. Some, like the Lucas test, can function in both categories. Procedures for carrying out the tests are given in alphabetical order beginning on page 396.

Table 2.   Functional class tests

Family	Number	Test	Comments
Alcohols	C–9	Chromic acid	Negative for 3° alcohols
	C–17	Lucas's test	Negative for 1° and high M.W. alcohols
	C–1	Acetyl chloride	Useful if other tests inconclusive
Aldehydes	C–9	Chromic acid	Alcohols also react
	C–23	Tollens's test	
Aldehydes and Ketones	C–11	2,4-Dinitrophenyl-hydrazine	Ketones react with C–11, not with C–9 or C–23
Amides	C–2	Alkaline hydrolysis	Best for amides of ammonia or low M.W. amines
	C–12	Elemental analysis	Detects N; used with solubility test results
Amines		Solubility tests	Soluble in 5% HCl
Aromatic Hydrocarbons	C–3	Aluminum chloride-chloroform	Should be substantiated by tests indicating absence of functional groups (such as halogens)
Carboxylic acids		Solubility tests	Soluble in 5% NaOH and 5% NaHCO$_3$
Esters	C–14	Ferric hydroxamate	
	C–2	Alkaline hydrolysis	Can be used to prepare derivatives
Halogenated Hydrocarbons	C–5	Beilstein's test	Simple, not always reliable
	C–21	Silver nitrate/ethanol	Negative for vinyl and aryl halides
	C–10	Density test	Usually negative for monochloroalkanes
	C–12	Elemental analysis	Can distinguish Cl, Br, and I
Phenols	C–13	Ferric chloride	Positive for most (but not all) phenols
	C–8	Bromine water	Aromatic amines also react

IDENTIFICATION

After you have identified the main functional group present in your unknown, you are ready to identify the compound itself by comparing its properties with those of other compounds listed in the literature and obtaining sufficient evidence to eliminate all of the possibilities but one.

Examination of the Literature

At your instructor's option, you may either use the tables of selected organic compounds provided in Appendix VII or consult a more complete listing such as the *CRC Handbook of Tables for Organic Compound Identification* (Bib–A10).

Most literature tabulations separate the compounds into solids and liquids, listing the solids in order of increasing melting point and the liquids in order of increasing boiling point. Since their derivatives are different, tertiary amines are often listed separately from primary and secondary amines; the various classes of halides may also be listed separately. If your compound has a narrow melting or boiling point range (suggesting that it is relatively pure) around a temperature less than 200°, you should be safe in considering only those compounds having melting or boiling points within about ± 5° of your observed value. However, if your boiling or melting point is 200° or higher, particularly if you are using an uncorrected thermometer, this range should be extended to as much as ± 10°. A table listing each compound within the chosen range should then be prepared. It should include the boiling (or melting) point, the refractive index, any other appropriate physical constants, and the melting points of useful derivatives. The chemical formulas for the compounds should be drawn, with the aid of the *CRC Handbook of Chemistry and Physics, The Merck Index,* the *Dictionary of Organic Compounds,* or other appropriate sources (see Appendix VIII) when needed. At this point you should be able to tell what physical properties and chemical tests will be most useful for narrowing down the list further, or which compounds can be eliminated on the basis of spectral data and tests already performed. In some cases, it may be possible to immediately prepare a derivative. However, if the list is a long one, you will probably need to reduce its length before you can decide which is the most suitable derivative.

The dividing line between liquids and solids is usually near a melting point of 25°. Some compounds with melting points over 25° (such as t-butyl alcohol) may appear to be liquids at room temperature, particularly when impure. When in doubt, try freezing the compound in an ice bath and estimating its melting point.

Your instructor may wish to check your melting or boiling point value and tell you whether it is within the ± 5° range.

Melting points for some derivatives may not be listed, either because the compound does not form the derivative or because its melting point has not yet been reported in the literature.

Additional Tests and Data

The refractive index of a liquid can be extremely valuable in its identification. With an accurate refractometer measuring to the fourth decimal place, for example, it may be possible

This assumes that the unknown is pure and that a thermostatted refractometer is used.

to eliminate compounds having refractive index values deviating by more than ± 0.001 or so from the observed value.

Density cannot be measured as accurately as refractive index, but it can be useful in distinguishing between compounds having different structural features. Aromatic compounds are usually more dense than comparable aliphatics, for example, and density varies among halogen compounds in the order $RI > RBr > RCl$.

It will often be helpful to go back and re-interpret the infrared spectrum at this point, since it may provide information that can eliminate some compounds having (or lacking) subsidiary functional groups and certain structural features. Ultraviolet-visible spectra can be of some value in identifying compounds having chromophoric groups if tables of λ_{max} values or published spectra are available. NMR spectra can be extremely valuable if interpreted skillfully; a correct identification can often be made with very little additional evidence.

See, for example, Bib–A4 and Bib–F26.

The tests listed in Table 3 can be used to provide structural information about the unknown. Procedures are listed alphabetically beginning on page 396.

Table 3.   Chemical tests providing structural information

Family or Structural Feature	Number	Test	Use
Alcohols	C–17	Lucas's test	To classify alcohols as 1°, 2°, 3°, etc.
	C–16	Iodoform test	To detect —CH(OH)CH$_3$ groupings
Aldehydes	C–6	Benedict's test	To distinguish aliphatic from aromatic aldehydes
Aldehydes and Ketones	C–16	Iodoform test	To detect —COCH$_3$ groupings
Amines	C–15	Hinsberg's test	To classify amines as 1°, 2°, or 3°
	C–4	Basicity test	To distinguish alkyl from aryl amines
	C–20	Quinhydrone	To distinguish aliphatic, aromatic, 1°, 2°, and 3° amines
Carboxylic acids	C–18	Neutralization equivalent	To determine the equivalent weight of an acid
Halogenated Hydrocarbons	C–10	Density test	To distinguish aliphatic and aromatic Cl, Br, and I compounds
	C–22	Sodium iodide/acetone	To classify halides as 1°, 2°, and 3°, etc.
	C–21	Silver nitrate/ethanol	Complements C–22

Table 3.   Chemical tests providing structural information (*continued*)

Family or Structural Feature	Number	Test	Use
Unsaturation	C–7	Bromine/carbon tetra-chloride	To detect C=C and C≡C bonds
	C–19	Potassium permanganate	Complements C–7

Preparation of Derivatives

The preparation of at least one derivative is required to confirm the identity of your unknown. In some cases, it may be necessary to prepare two or more derivatives to be certain of your identification. Because a considerable amount of time can be wasted in preparing the wrong derivative—one that doesn't distinguish sufficiently between the possibilities—it is important to select your derivatives very carefully. For example, suppose your compound is a ketone boiling at 168° and you prepare the list of possibilities shown in Table 4.

Table 4.   Derivatives of selected ketones

Compound	b.p.	Oxime	Semicarbazone	2,4-Dinitro-phenylhydrazone
2-methylcyclohexanone	163	43	195	137
2,6-dimethyl-4-heptanone	168	210	121	92
3-methylcyclohexanone	169	· · ·	180	155
4-methylcyclohexanone	169	37	199	130
2-octanone	172	· · ·	122	58

The oxime would obviously not be a suitable derivative, since melting points have not been reported for two of the possibilities. Also, two of the oximes melt below 50°, which is undesirable since low-melting solids are usually difficult to recrystallize. Semicarbazones are reported for all the unknowns, but those for 2,6-dimethyl-4-heptanone and 2-octanone melt within a degree of each other. The 2-octanone might be eliminated by an iodoform test, but that would still leave 2-methyl- and 4-methylcyclohexanone, whose semicarbazones melt within 4° of each other—a little too close to allow any certainty about the outcome. (Your derivative might melt at 197°, for example.) The 2,4-dinitrophenylhydrazones, on the other hand, are well spread out: only the 2-methyl- and 4-methylcyclohexanone derivatives melt anywhere close to each other. This would be the derivative of choice.

A good derivative should have a melting point above 50° and below 250°. Its melting point should be significantly different from the melting point of the unknown itself.

It should be pointed out that (since even professional chemists occasionally make mistakes) there are some conflicts in the literature with regard to physical constants, including the melting points of derivatives. If you suspect that a literature value may be incorrect, it is wise to check it against the values reported for the same derivative in some of the other sources listed in the Bibliography.

See, for example, Bib–A7, Bib–A8, Bib–A10, Bib–B4, Bib–G1, Bib–G3.

Procedures for preparing derivatives are listed by chemical family beginning on page 410. Melting points of the derivatives of selected organic compounds are listed in Appendix VII.

PROCEDURES

Preliminary Work

Observe and describe the physical state and color of the unknown. Note and describe its odor (if any) by holding the

---

WARNING *Consider all unknowns to be flammable and toxic, and use great care in handling them. Do not inhale appreciable amounts of vapor; avoid contact of liquids or solids with eyes, skin, and clothing.*

---

sample well away from your nose and gently wafting the vapors toward you with your hand. Carry out an ignition test by placing a drop of a liquid or about 25 mg of a solid in a small evaporating dish and igniting it with a burning wood splint. If it burns, observe the color of the flame and whether it is clean or sooty.

A thermometer correction should be applied for temperatures over 200° (see ⟨OP–28⟩).

Obtain the melting point of a solid unknown ⟨OP–28⟩ or the boiling point of a liquid ⟨OP–29a⟩ and purify it, if necessary, by recrystallization ⟨OP–23a⟩, distillation ⟨OP–25a⟩, or another suitable method. If the compound boils at a temperature higher than 200°, it may be advisable to use vacuum distillation ⟨OP–26a⟩ for purification.

Solubility Tests

If the quantity of unknown is limited, the amounts of solvent and solute can be scaled down appropriately.

If the unknown is a liquid, measure the number of drops in 1.0 milliliter, using a medicine dropper and a small graduated cylinder. Divide this number by 5 to get the number of drops in 0.2 ml. Measure 0.2 ml of the liquid (by drops) into a 10-cm test tube and add 3.0 ml of water, then shake vigorously for a minute or so, or until the liquid appears to dissolve. If the liquid is soluble, the mixture should be homogeneous with no separate layer or suspended droplets of the unknown evident. If it is insoluble or only partly soluble, the mixture should look cloudy or show suspended droplets when shaken.

> *Always stopper the test tube before shaking it (don't use your thumb for a stopper!).*
>
> *Do not mistake an effervescence (which may occur on shaking) for suspended droplets.*

If the unknown is a solid, accurately weigh out 0.10 g of the solid, and estimate all subsequent 0.1 gram portions with reference to this quantity. Grind the solid to a fine powder and mix it with 3.0 ml of water in a small test tube, shaking or stirring it for a few minutes, or until it dissolves. Grinding the solid against the sides of the tube with a stirring rod may accelerate the process. Occasionally it may be advisable to heat the mixture gently and, if the solid dissolves, cool it back to room temperature (stirring and scratching the sides of the test tube to prevent supersaturation) to see whether it reprecipitates.

> *A small mortar and pestle or a flat-bottomed stirring rod and watch glass can be used for grinding. Be certain they are very clean.*

If the unknown is water-soluble, test it with red and blue litmus paper. If there is no reaction to litmus and you suspect that the unknown is an aromatic amine or a phenol, dissolve a little of it in 5% HCl (for an amine) or 5% NaOH (for a phenol) and see if its odor disappears.

> *Salts of amines and phenols are generally odorless.*

If the unknown is insoluble in water, test its solubility in 3.0 ml of 5% hydrochloric acid by the same procedure as described above. If it does not dissolve completely in the HCl, separate it from the solvent by filtering it, removing it with a capillary pipet, or decanting the solvent; then carefully neutralize the solvent to red litmus with dilute sodium hydroxide solution. The formation of a precipitate, a separate liquid phase, or even a cloudy solution upon neutralization classifies the unknown as "soluble" in HCl and places it in solubility class B (see Figure 1, page 384).

> *Do not heat the mixtures containing HCl or any of the following solvents, as this might cause hydrolysis or condensation reactions which could lead you to an erroneous conclusion.*

If the unknown is insoluble in 5% HCl, test it with 5% sodium hydroxide and with 5% sodium bicarbonate by the same procedure as before. This time neutralize the separated solvent to blue litmus with dilute HCl if the unknown does not dissolve completely. If it is insoluble in all of the foregoing, *carefully* test its solubility in 3.0 ml of cold (room temperature or below) concentrated sulfuric acid. Shake this mixture vigorously and look for any evidence of reaction, such as the generation of heat, a distinct change in color, the formation of a precipitate, or the evolution of a gas. Solubility or a definite reaction is considered a positive test. The nature of any reaction that takes place may provide some clues to the identity of the unknown.

> *When testing with NaOH, notice whether the mixture foams on shaking. This could indicate a long-chain carboxylic acid.*
>
> *A slight color change may be caused by impurities.*

Infrared Spectrum

Obtain an infrared spectrum of the unknown ⟨OP–33⟩ using the neat liquid or (if the sample is a solid) a mull or a KBr pellet. If the spectrum must be obtained when the sample is in solution, some of the important bands may be shifted by the solvent and it may be necessary to consult the literature for interpretation. Attempt to determine the major functional group present and choose one or more classification tests to confirm it.

Classification Tests

Unless otherwise indicated, solutions for the classification tests are aqueous.

Procedures for all classification tests, both functional class tests and those providing additional structural information, are given below. Always read any Hazard notes before starting a test. When each test is performed for the first time, it is advisable to run a *control* and (sometimes) a *blank* at the same time. Only this way will you know exactly what to look for in deciding whether the test with the unknown is positive or negative. A control is a known compound that is expected to give a predictable result with the test reagent. A blank is run by combining all reagents as in the actual test but omitting the unknown.

Always use different droppers for the reagent and the unknown, and clean droppers thoroughly after use. Do not insert your medicine droppers into reagent bottles—you may contaminate the reagents.

The volumes of most liquid reactants are given by drops in the procedures; it may be convenient to dispense classification-test reagents in dropper bottles. Droppers should be of the same kind so that they will deliver roughly the same volume per drop. If you prefer to use a graduated pipet or cylinder rather than a dropper, assume that 10 drops equals about $\frac{1}{2}$ ml.

C–1. *Acetyl Chloride.* **(a)** Carefully add about 10 drops of acetyl chloride (*Hazard!*) to 10 drops (or 0.4 g) of the unknown in a test tube. Observe any evolution of heat; carefully exhale over the mouth of the test tube to see if a cloud of HCl gas is revealed by the moisture in your breath. After a minute or two, pour the mixture into about 2 ml of water, shake it, and notice any phase separation. Carefully smell the mixture for evidence of an ester aroma, which is usually pleasant and fruit-like.

Reaction

$$CH_3COCl + ROH$$
$$\longrightarrow CH_3COOR + HCl$$

1-butanol is a suitable control for this test.

HAZARD   *Acetyl chloride is corrosive and toxic. Use gloves and a hood; do not breathe vapors.*

Positive test: Evidence of reaction (heat, HCl), especially if accompanied by phase separation or an ester-like odor. Amines and phenols also react, but the former would not yield pleasant odors. If the test is inconclusive or if you suspect a tertiary alcohol, carry out variation **(b)**.

(b) Mix 5 drops of acetyl chloride with 10 drops of *N,N*-dimethylaniline and carefully add 5 drops (or 0.2 g) of the unknown. Allow the solution to stand for a few minutes and look for evidence of a reaction (no HCl will be evolved since it reacts with the base). If heat is not evolved, warm the mixture on a 50° water bath for 15 minutes. Cool the mixture, add 1 g of ice and 1 ml of concentrated ammonia, mix it, and let it stand. If a layer separates, remove it with a dropper or pipet and test it for ester using the ferric hydroxamate test (C–14).

Tertiary alcohols will not form esters by procedure **(a)** *but should in the presence of the base used in* **(b)**.

acetyl chloride   *N,N*-dimethylaniline

C–2. Alkaline Hydrolysis. (a) Amides.

Place 0.1 g of the unknown (2 drops, if liquid) in a test tube containing 4 ml of 6M sodium hydroxide. Secure a small piece of filter paper over the top of the tube and moisten it with 2 drops of 10% cupric sulfate. Boil the mixture for a minute or two and observe any color change on the filter paper. Remove the paper and note the odor of the vapors while the solution is boiling. Acidify the solution with 6M HCl; if a carboxylic acid precipitates, save it for use as a derivative.

Positive test: Paper turning blue, ammonia or amine-like odor. Amides of higher amines that do not turn the paper blue may nevertheless give an amine-like odor. Some amides will yield a precipitate or a separate liquid phase (the carboxylic acid) when the hydrolysis mixture is acidified. The characteristic odor of a carboxylic acid may also be observed.

If the test is inconclusive, try the following:

1. Increasing the reaction time.

2. Repeating the hydrolysis at 200° using 20% KOH in glycerine.

3. Distilling off the amine and characterizing it using one of the amine classification tests.

Reactions

$$RCONR_2' + NaOH$$
$$RCOONa + R_2'NH$$

$$RCOONa + H^+ \longrightarrow$$
$$RCOOH + Na^+$$
$$R' = H, \text{ alkyl, or aryl}$$

Controls: Compare the results with benzamide and acetanilide.

sodium hydroxide

Distill a volatile amine into a receiver containing dilute HCl, then neutralize the distillate with base before testing.

Reactions

$$RCOOR' + NaOH \longrightarrow$$
$$RCOONa + R'OH$$
$$RCOONa + H^+ \longrightarrow$$
$$RCOOH + Na^+$$

(b) Esters. Mix 1 ml of the unknown (1 g, if solid) with 10 ml of 6M sodium hydroxide and reflux the mixture for $\frac{1}{2}$ hour, or until the solution is homogeneous. Note whether the organic layer has disappeared (if the unknown was water insoluble) and whether the odor of the unknown is gone. If a separate organic layer or residue remains, reflux the mixture

*Most esters boiling under 110° will hy-
drolyze in ½ hour, while higher boiling
esters may take up to 2 hours, or longer.*

*If a carboxylic acid precipitates on acidifi-
cation, save it for use as a derivative.*

Controls: Ethyl benzoate, butyl acetate.

Reaction

$$\text{ArH} \xrightarrow{\text{CHCl}_3,\ \text{AlCl}_3} \text{Ar}_3\text{C}^+\text{AlCl}_4^-$$

(and other carbonium ion species)
Controls: toluene, biphenyl

aluminum   chloroform
chloride

Reaction (in buffer)

$$\text{RNH}_2 + \text{H}^+ \xrightarrow{\text{pH 5.5}} \text{RNH}_3^+$$

(and similar reactions for 2° and 3°
amines)

Controls: aniline and *n*-butylamine;
p-toluidine and dibutylamine.

longer until it disappears or until it is apparent that no reaction is taking place. Cool the mixture, remove any organic
layer if necessary, and acidify the aqueous solution with 6M
sulfuric acid.

Positive test: Disappearance of the organic layer (for a
water-insoluble ester) and the odor (usually pleasant) of the
unknown; the formation of a precipitate (if the carboxylic acid
is a solid) or the appearance of the odor of a carboxylic
acid upon acidification. If evidence for the formation of an
acid is inconclusive, make the solution basic again with sodium hydroxide and saturate it with potassium carbonate to
see if an organic layer (the alcohol) separates. Note the odor
of this layer, which should be different from that of the original ester. Esters with boiling points higher than 200° are
usually unreactive in aqueous NaOH and may be hydrolyzed by KOH in diethylene glycol; see the reference by
Shriner *et al.* (Bib–G3) for a procedure.

C–3. *Aluminum Chloride and Chloroform.* Make up a solution containing 2–3 drops (or 0.1 g) of a Solubility Class X
unknown in 2 ml of dry chloroform (*Hazard:* Avoid contact, vapors.). Place about 0.2 g of anhydrous aluminum
chloride (*Hazard:* Avoid contact.) in a dry test tube and heat
it over a flame, angling the test tube so the AlCl₃ sublimes
onto the inner wall of the tube for a few centimeters above
the bottom. Allow the tube to cool until it can be held com-
fortably in the hand, then add a few drops of the solution
down the side of the tube so that it contacts the aluminum
chloride. Notice any color change at the point of contact.

Positive test: Bright yellow-orange, red, blue, green, or
purple color, depending on the type of aromatic compound.
A light yellow color is inconclusive or negative.

C–4. *Basicity Test.* If the unknown is water soluble, dissolve 4 drops (0.10 g of a solid) in 3 ml of water and measure
its pH using pH paper or a universal indicator. If the unknown is insoluble in water, shake about 4 drops (or 0.10 g)
with 3 ml of an acetate–acetic acid buffer (pH 5.5).

Positive test: Most aliphatic amines will dissolve in the
buffer or, if water soluble, will give pH values above 11.
Most aromatic amines will not dissolve in the buffer or, if
water soluble, will give pH values below 10. Test C–8 can
also be used to test for aromatic amines.

C–5. *Beilstein's Test.* Make a small loop in the end of a
length of copper wire (10 cm or longer), and heat the loop to

redness in a flame. Place a small amount of unknown on the loop and heat it in the nonluminous (blue) flame of a burner, near the lower edge.

Controls: chlorobenzene, 1–bromobutane.

Positive test: A distinct green or blue-green flame, due to the presence of copper halide.

C–6. *Benedict's Test.*
Add 5 drops or 0.2 g of the unknown to 5 ml of water and mix in 5 ml of Benedict's reagent. Heat the mixture to boiling and observe whether or not a precipitate forms, and if one forms, what its color is.

Controls: butyraldehyde (+), benzaldehyde (−).

Positive test: Aliphatic aldehydes generally produce a yellow to orange suspension or precipitate; this may appear green in the blue solution. Some other compounds, such as α-hydroxyketones and reducing sugars, also react. Most ketones and aromatic aldehydes do not react.

Benedict's reagent contains cupric sulfate (toxic!), sodium citrate, and sodium carbonate.

C–7. *Bromine in Carbon Tetrachloride.*
Dissolve 3 drops or 0.1 g of the unknown in 1 ml of carbon tetrachloride (*Hazard!*) and add, dropwise with shaking, a solution of 0.2M bromine in carbon tetrachloride (*Hazard!*) until the bromine color persists for at least a minute. Immediately after the addition, exhale moist air over the mouth of the test tube and observe whether a cloud of HBr gas is apparent.

Reaction

$$-\overset{|}{\underset{|}{C}}=\overset{|}{\underset{|}{C}}- + Br_2 \xrightarrow{CCl_4} -\overset{Br}{\underset{|}{\underset{|}{C}}}-\overset{Br}{\underset{|}{\underset{|}{C}}}-$$

Controls: cyclohexene, ethyl acetoacetate (reacts with HBr formation).

Bromine is corrosive and toxic, CCl₄ is toxic and possibly carcinogenic; avoid contact, do not breathe vapors.

HAZARDS

Positive test: Decolorization of more than 2 drops of the bromine/CCl₄ solution, *without* the evolution of HBr, is characteristic of olefinic unsaturation. Aldehydes, ketones, amines, phenols, and many other compounds react by substitution to evolve HBr.

bromine   carbon tetrachloride

C–8. *Bromine Water.*
Dissolve 3 drops or 0.1 g of the unknown in 10 ml of water and check the pH of the solution with pH paper. Add saturated bromine water (dropwise) until the bromine color persists; watch for evidence of a precipitate.

If the unknown is insoluble in the water, try adding just enough ethanol to bring it into solution.

Controls: Phenol, aniline.

Positive test: Decolorization of the bromine, accompanied by simultaneous formation of a white (or nearly white) precipitate. The pH of the initial solution should be less than 7 if the unknown is a phenol. Aromatic amines also react.

Reaction (and similar reactions for other phenols)

$$\text{C}_6\text{H}_5\text{OH} + 3Br_2 \longrightarrow \text{(2,4,6-tribromophenol)} + 3HBr$$

C–9. Chromic Acid. Dissolve 1 drop of a liquid or 15–30 mg of a solid unknown in 1 ml of reagent-grade acetone. If there is any doubt about the purity of the acetone, test it with a drop of the reagent beforehand. Add 1 drop of the chromic anhydride–sulfuric acid reagent and swirl, noting the time required for a color change.

Controls: 1-butanol (+), *t*-butyl alcohol (−), butyraldehyde, benzaldehyde.

HAZARDS   *The reagent is corrosive and CrO_3 may be a carcinogen; avoid contact.*

Reactions

$$2CrO_3 + H_2O \xrightarrow{H_2SO_4} H_2Cr_2O_7$$

$$3RCH_2OH + 2H_2Cr_2O_7 + 6H_2SO_4 \longrightarrow 3RCOOH + 2Cr_2(SO_4)_3 + 11H_2O$$

$$3R_2CHOH + H_2Cr_2O_7 + 3H_2SO_4 \longrightarrow 3R_2CO + Cr_2(SO_4)_3 + 7H_2O$$

$$3RCHO + H_2Cr_2O_7 + 3H_2SO_4 \longrightarrow 3RCOOH + Cr_2(SO_4)_3 + 4H_2O$$

Positive test: Formation of an opaque blue-green suspension or emulsion within 2 seconds for a primary or secondary alcohol. With aliphatic aldehydes, the solution generally

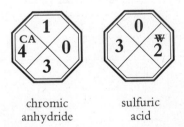

chromic   sulfuric
anhydride   acid

turns cloudy in 5 seconds and a green precipitate forms within 30 seconds, whereas aromatic aldehydes require 30 to 120 seconds or longer. The generation of some other dark color, particularly with the color of the liquid remaining orange, should be considered a negative test.

C–10. Density Test. Add a few drops of a Solubility Class X unknown to 1 ml of deionized water, stir gently, and notice whether the unknown floats or sinks.

Interpretation: Of the compounds that are insoluble in cold, concentrated sulfuric acid (see Solubility Tests), most aromatic hydrocarbons and monochloroalkanes will float,

Controls: toluene, chlorobenzene, 1-chlorobutane, 1-bromobutane.

whereas aryl chlorides and all bromides and iodides, along with polychloroalkanes, will sink. For further differentiation, measure the density of the unknown by carefully pipetting exactly one ml of the liquid (use a volumetric pipet) into a vial and weighing the liquid on an accurate balance to at least two decimal places. Density ranges for most common halides are shown in the margin.

Approximate density ranges for halogenated hydrocarbons:

alkyl chloride (mono)	0.85–1.0
alkyl bromide (mono)	1.1–1.5
alkyl iodide (mono)	over 1.4
alkyl chloride (poly)	1.1–1.7
alkyl bromide (poly)	1.5–3.0
aryl chloride	1.1–1.3
aryl bromide	1.3–2.0
aryl iodide	over 1.8

C–11. 2,4-Dinitrophenylhydrazine. Dissolve 1 drop or about 40 mg of the unknown in 1 ml (more, if necessary) of 95% ethanol and add the solution to 2 ml of the 2,4-dinitrophenylhydrazine-sulfuric acid reagent. Shake and let the

Controls: cyclohexanone, benzaldehyde.

2,4-dinitrophenylhydrazine is toxic and the acid is corrosive; avoid contact with the reagent.

HAZARD

mixture stand for 15 minutes, or until a precipitate forms. If no precipitate has formed at the end of 15 minutes, scratch the inside of the test tube or try very gentle heating.

Reaction

R, R′ = alkyl, aryl, or H

Positive test: Formation of a crystalline yellow or orange-red precipitate within 15 minutes. Some carbonyl compounds initially form oils that may or may not become crystalline; a few may require gentle heating, though heating can cause oxidation of allylic or other reactive alcohols, resulting in a false positive test. Some aromatic compounds (hydrocarbons, halides, phenols, and phenyl esters) may form slightly soluble complexes with the reagent; some alcohols may be contaminated with small amounts of the corresponding aldehyde or ketone. In these cases, the amount of precipitate should be quite small, and comparison with a control will show the difference between a positive test and a doubtful one. The color of the precipitate may give a clue to the structure of the carbonyl compound, since unconjugated aliphatic aldehydes and ketones usually yield yellow 2,4-dinitrophenylhydrazones while aromatic and α,β-unsaturated aldehydes and ketones yield orange-red precipitates.

An orange-red color can also be caused by impurities, including 2,4-dinitrophenylhydrazine itself.

This procedure is adapted from an article in the *Journal of Chemical Education, 54,* 187 (1977).

The sodium-lead alloy is much less hazardous than elemental sodium; it reacts with water to give a comparatively gentle reaction with no fire hazard.

If the unknown is quite volatile, it may be advisable to place it in the test tube before adding the alloy, and to heat them together gently until charring occurs. Then heat the tube more strongly and proceed as directed.

Reactions

$$RX \xrightarrow{Na(Pb)} NaX$$

$$\text{Nitrogen Compound} \xrightarrow{Na(Pb)} NaCN$$

$$AgNO_3 + NaX \longrightarrow \mathbf{AgX} + NaNO_3$$

$$AgCl + 2NH_3 \longrightarrow Ag(NH_3)_2Cl$$

$$2Br^- + Cl_2 \xrightarrow{CCl_4} 2Cl^- + \underset{\text{(red-brown)}}{Br_2}$$

$$2I^- + Cl_2 \xrightarrow{CCl_4} 2Cl^- + \underset{\text{(purple)}}{I_2}$$

$$(X = Cl, Br, \text{ or } I)$$

Controls: acetamide (N), bromobenzene (Br).

The boiling removes (as HCN) the cyanide ion that is formed if nitrogen is present, and that would react to yield AgCN in this test.

If your instructor has told you that more than one kind of halogen may be present in your unknown, you can consult the book by Shriner et al. (Bib–G3) or a similar reference for ways to analyze halogen mixtures.

Controls: phenol, salicylic acid.

C–12. *Elemental Analysis.* Place about 0.5 g of sodium-lead alloy in a small *dry* test tube held vertically by an asbestos-lined clamp. Melt the alloy with a burner flame and continue heating until the sodium vapor rises about 1 cm up the tube, then add two drops of the unknown (about 10 mg of a solid) directly onto the molten alloy so that it does not touch the sides of the tube. Heat gently to start the reaction, remove the flame until the reaction subsides, then heat the tube for a minute or two with the bottom a dull red color. Let the tube cool to room temperature, then add 3 ml of water and heat gently for a minute or so until the excess sodium has been decomposed and gas evolution ceases. If necessary, filter the solution, wash the filter paper with 2 ml of water, and combine the wash water with the filtrate. The filtrate should be colorless or just slightly yellow; if it is darker, repeat the fusion with stronger heating or more of the alloy.

To test for *nitrogen,* add 10 drops of the fusion solution to a small test tube and add enough solid sodium bicarbonate, with stirring, to saturate it (excess solid should be present). Add 1–2 drops of this solution to a test tube containing 20 drops of PNB reagent. A purple color is a positive test for nitrogen (green indicates sulfur).

To test for the *halogens,* acidify ten drops of the fusion solution with dilute nitric acid, boil gently under the hood for a few minutes, and add a drop or two of 0.3M aqueous silver nitrate. The formation of a heavy white precipitate of silver chloride indicates the presence of chlorine, whereas bromine yields a pale yellow and iodine a yellow precipitate, respectively. If only a faint turbidity is produced, it may be caused by traces of impurities or by incomplete sodium fusion. To confirm the presence of chlorine, remove the solvent by filtration (or by centrifugation and decanting), add 2 ml of 3M aqueous ammonia, and shake. If the precipitate is silver chloride, it will dissolve. Silver bromide will be only slightly soluble and silver iodide will be insoluble. To test further for bromine and iodine, acidify 3 ml of the original stock solution with dilute sulfuric acid, boil for a few minutes, and add 1 ml of carbon tetrachloride and a drop or two of freshly prepared chlorine water. Shake and look for a color in the CCl$_4$ layer (purple for iodine, reddish-brown for bromine).

C–13. *Ferric Chloride.* Dissolve 1 drop or about 40 mg of the unknown in 2 ml of water (or in a water-alcohol mixture if it does not dissolve in water) and add 1–3 drops of 2.5% ferric chloride solution.

Positive test: Formation of an intense red, green, blue, or purple color suggests a phenol or an easily enolizable compound. Some phenols do not react under these conditions. Many aromatic carboxylic acids form tan precipitates; aliphatic hydroxy acids yield yellow solutions.

Compare the test color with the color resulting when ferric chloride is added to 2 ml of pure water.

C–14. *Ferric Hydroxamate Test.* Before carrying out the ferric hydroxamate test, dissolve 1 drop or about 40 mg of the unknown in 1 ml of 95% ethanol, add 1 ml of 1M hydrochloric acid, then add 2 drops of 2.5% ferric chloride. If a definite color other than yellow results, the ferric hydroxamate test cannot be used. Save the solution for comparison with the following test.

The preliminary test eliminates those phenols and enols that give colors with ferric chloride in acidic solution, and that would therefore give a false positive result in the ferric hydroxamate test.

The reagent is toxic and can cause a form of anemia; avoid contact.   **HAZARD**

Mix 1 ml of 0.5M ethanolic hydroxylamine hydrochloride with 0.2 ml of 6M sodium hydroxide, add 1 drop (or about 40 mg) of the unknown, and heat the solution to boiling. Allow it to cool slightly and add 2 ml of 1M hydrochloric acid. If the solution is cloudy at this point, add enough 95% ethanol to clarify it. Add 2 drops of 2.5% ferric chloride solution and observe any color produced. If the color does not persist, continue to add the ferric chloride solution until the color becomes permanent.

hydroxylamine hydrochloride

Control: butyl acetate.

Reactions

$$\underset{\displaystyle \|}{R\overset{O}{C}}-OR' + H_2NOH \longrightarrow \underset{\displaystyle \|}{R\overset{O}{C}}-NHOH + R'OH$$

$$3RCONHOH + FeCl_3 \longrightarrow (RCONHO)_3Fe + 3HCl$$
$$\text{ferric hydroxamate}$$

Positive test: A burgundy or magenta color that should be distinctly different from the color obtained in the preliminary test.

C–15. *Hinsberg's Test.* In a test tube mix together 0.1 ml (or 0.1 g) of the unknown, 5 ml of 3M sodium hydroxide, and 0.3 g of *p*-toluenesulfonyl chloride (*Hazard!*). Stopper

Benzensulfonyl chloride may be used instead of p-toluenesulfonyl chloride, but it has a greater tendency to form oils.

HAZARD   *Arenesulfonyl chlorides are toxic and corrosive; avoid contact, do not breathe vapors.*

Some N,N-dialkylanilines form a purple dye if the reaction mixture gets too hot. If this occurs, repeat the reaction in a 15-20° water bath.

Reactions

$1°$ $RNH_2 + ArSO_2Cl +$
      $2NaOH \rightarrow ArSO_2NR^-Na^+ +$
                  (soluble)
      $NaCl + 2H_2O$
      $ArSO_2NR^-Na^+ + HCl \rightarrow$
      $ArSO_2NHR + NaCl$
        (insoluble)
$2°$ $R_2NH + ArSO_2Cl +$
      $NaOH \rightarrow ArSO_2NR_2 + NaCl$
      (insoluble)
$3°$ $R_3N + ArSO_2Cl \rightarrow$ no reaction
      $R_3N + HCl \rightarrow R_3NH^+Cl^-$
                  (soluble)

Controls: aniline, N-methylaniline, N,N-dimethylaniline.

If a pure amine yields a solid residue after the original reaction and forms a precipitate upon acidification of the filtrate, it is probably a primary amine that has produced some of the disulfonyl derivative.

Use 3 ml or more of dioxane if the unknown is not soluble in water.

Controls: isopropyl alcohol, 2-butanone.

the tube and shake it intermittently for 3–5 minutes, then remove the stopper and warm the solution over a steam bath for a minute. The solution should be basic at this point (if not, add more NaOH). If there is a liquid or solid residue in the tube, separate it from the solution (by filtration, use of a pipet, or some other means) and test its solubility in 5 ml of water and in dilute hydrochloric acid. Acidify the original solution with 6M hydrochloric acid and, if no precipitate forms immediately, scratch the sides of the test tube and cool.

Interpretation: Primary amines usually yield no appreciable amount of liquid or solid residue after the initial reaction; a p-toluenesulfonamide should precipitate when the solution is acidified. Most *secondary amines* yield a solid p-toluenesulfonamide that does not dissolve in water or dilute HCl. *Tertiary amines* should not react; the residue will be the original liquid or solid amine, which should dissolve in dilute HCl. If the test yields an oil, it should be possible to tell whether it is the original (liquid) amine or a p-toluenesulfonamide that has failed to crystallize—the latter are more dense than water.

Some high-molecular-weight and cyclic primary alkylamines form sodium salts that are insoluble in the alkaline solution. These salts should dissolve in the 5 ml of water used to test the solubility of the precipitate; if not, the test will falsely suggest a secondary amine. Other primary amines may yield disulfonyl derivatives that are also insoluble in the original solution, so it is wise to reserve judgment on any test suggesting a secondary amine unless it can be confirmed independently. Water-soluble tertiary amines should yield a clear solution that does not form a separate phase on acidification.

C–16. *Iodoform Test.* Dissolve 3 drops (or 0.1 g) of the unknown in 1 ml of water and add 1 ml of 3M sodium hydroxide solution. Add the 0.5M iodine–potassium iodide reagent dropwise until the iodine color persists after shaking. Place the test tube in a 60° water bath and add more of the I_2–KI solution until the brown color persists for 2 minutes; then add 3M NaOH dropwise until the color just disappears (a light yellow color may remain). Remove the test tube from the water bath and add 10 ml of cold water. If a precipitate does not form immediately, let it stand for 15 minutes. If there is any doubt about the identity of the precipitate, obtain its melting point, which should be about 119–121°.

iodine

Reactions

1. Methyl carbinols

$$\underset{\text{OH}}{\text{RCH}}-\text{CH}_3 + 4\text{I}_2 + 5\text{NaOH} \longrightarrow \underset{\text{O}}{\text{RC}}-\text{CI}_3 + 5\text{NaI} + 5\text{H}_2\text{O}$$

$$\underset{\text{O}}{\text{RC}}-\text{CI}_3 + \text{NaOH} \longrightarrow \underset{\text{O}}{\text{RC}}-\text{ONa} + \underset{\text{(iodoform)}}{\text{CHI}_3}$$

2. Methyl ketones and acetaldehyde

$$\underset{\text{O}}{\text{RC}}-\text{CH}_3 + 3\text{I}_2 + 3\text{NaOH} \longrightarrow \underset{\text{O}}{\text{RC}}-\text{CI}_3 + 3\text{NaI} + 3\text{H}_2\text{O}$$

$$\underset{\text{O}}{\text{RC}}-\text{CI}_3 + \text{NaOH} \longrightarrow \underset{\text{O}}{\text{RC}}-\text{ONa} + \text{CHI}_3$$

Positive test: Formation of a yellow precipitate of iodoform. Methyl carbinols must have at least one hydrogen atom on the carbinol carbon in order to react, since they are oxidized to methyl ketones initially. Other compounds that can also yield iodoform in this test include some conjugated aldehydes, such as acrolein and furfural, and certain 1,3-dicarbonyl or dihydroxy compounds.

C–17. Lucas's Test. Place 2 ml of the Lucas reagent (*Hazard!*) in a small test tube, add 3–4 drops of the unknown,

zinc   hydrochloric
chloride   acid

Lucas's reagent (ZnCl₂ in conc. HCl) can cause serious burns; avoid contact.

HAZARD

---

stopper the tube immediately and shake vigorously, then allow the mixture to stand for 15 minutes or more.

Interpretation: Tertiary alcohols that are soluble in the reagent should almost immediately cause turbidity and the formation of a separate layer of alkyl chloride. *Secondary alcohols* usually turn the clear solution cloudy in 3–5 minutes and form a distinct layer within 15 minutes. *Primary alcohols*

Reaction

$$\text{ROH} + \text{HCl} \xrightarrow{\text{ZnCl}_2} \text{RCl} + \text{H}_2\text{O}$$

Controls: n-butyl, sec-butyl, and *t-butyl* alcohols.

If there is any question of whether the alcohol is secondary or tertiary, repeat the test using concentrated HCl instead of Lucas's reagent. A tertiary alcohol should react within minutes, whereas a secondary alcohol should not react at all.

Polyhydroxy alcohols boiling higher than 140° are often soluble in the reagent and can be tested.

Reaction

RCOOH + NaOH $\longrightarrow$
                    RCOONa + H$_2$O

Control: adipic acid.

Use bromothymol blue as the indicator if the solvent is ethanol.

N.E. = $\dfrac{\text{mass of sample (mg)}}{\text{ml of NaOH} \times \text{molarity NaOH}}$

If alcohol is the solvent, it is best to run a blank using the pure solvent and compare the result with that for the unknown.

The reaction rate depends on the solubility of the unknown in the solvent. A solid that is only sparingly soluble should be finely powdered and the reaction mixture should be shaken vigorously.

Reaction

3—C=C— + 2KMnO$_4$ + 4H$_2$O $\longrightarrow$

$$
\underset{\substack{| \quad |}}{3-\overset{\displaystyle \overset{HO}{|}}{C}-\overset{\displaystyle \overset{OH}{|}}{C}-} + 2MnO_2 + 2KOH
$$

do not react under these conditions. Most allylic and benzylic alcohols give the same result as tertiary alcohols, except that the chloride formed from allyl alcohol is itself soluble in the reagent and separates out only upon addition of ice water. The test is not valid for alcohols that are not soluble in the reagent. This group includes most alcohols having six or more carbons and boiling points higher than 140–50°. You should interpret with caution any apparent positive result, such as immediate phase separation, that has been obtained from a water-insoluble alcohol having a boiling point above 140°.

C–18. *Neutralization Equivalent.* Accurately weigh (to 3 decimal places) about 0.2 g of the unknown carboxylic acid and dissolve it in 50–100 ml of water, ethanol, or a mixture of the two, depending on its solubility. Titrate the solution with a standardized solution of 0.1M sodium hydroxide using phenolphthalein as the indicator. Calculate the neutralization equivalent (N.E.) of the acid using the formula shown in the margin.

Interpretation: The neutralization equivalent (equivalent weight) of a carboxylic acid is equal to its molecular weight divided by the number of carboxyl groups. For instance, the N.E. of adipic acid (HOOC(CH$_2$)$_4$COOH; M.W. = 146) is 73. An acid that has an unusually low neutralization equivalent for its boiling or melting point probably contains more than one carboxyl group.

C–19. *Potassium Permanganate.* Dissolve 1 drop (or 30 mg) of the unknown in 2 ml of water or 95% ethanol and add 0.1M potassium permanganate dropwise until the purple color of the permanganate persists. Count the number of drops required. If a reaction does not take place immediately, shake the mixture and let it stand for up to 5 minutes. Disregard any decolorization that takes place after 5 minutes have elapsed.

Positive test: Decolorization of more than 1 drop of the purple permanganate solution, accompanied by the formation of a brown precipitate (or reddish-brown suspension) of manganese dioxide. The test is positive for most compounds containing double and triple bonds, including some alkenes that do not react with bromine in carbon tetrachloride. Easily oxidizable compounds such as aldehydes, aromatic amines, phenols, formic acid, and formate esters also give positive tests. Alcohols that contain oxidizable impurities may react with a small amount of permanganate, so decolorization of only the first drop should not be considered a positive test; most pure alcohols will not react in less than 5 minutes. Aromatic hydrocarbons and conjugated alkadienes do not, as a rule, give positive tests with cold, neutral potassium permanganate.

Controls: cyclohexene (+), 1-butanol (−).

potassium permanganate

C–20. *Quinhydrone.* This test should be run in conjunction with test C–4 or with an IR spectrum that shows whether the amine is aliphatic or aromatic. Shake 1 drop (or 30 mg) of the unknown amine with 6 ml of water in a 15-cm test tube. If the amine dissolves, add 6 ml more of water; otherwise add 6 ml of ethanol. Shake the mixture, add 1 drop of 2.5% quinhydrone in methanol, and let it stand for 2

Controls: butylamine, dibutylamine, tributylamine, aniline, *N*-methylaniline, *N,N*-dimethylaniline.

The components of quinhydrone (hydroquinone, benzoquinone) are toxic and irritant. Avoid contact; wash hands after using.

HAZARD

---

minutes. If no distinct color develops, add 5 more drops of the reagent, shake the mixture again, and allow it to stand for another 2 minutes.

    Interpretation: Most alkyl amines produce strong colors with one drop of the reagent; aryl amines require 6 drops. Controls should be run for comparison, since the colors produced (especially by aromatic amines) may be difficult to characterize accurately. The test is not applicable to nitro-substituted aromatic amines and phenylenediamines.

Colors of quinhydrone test:
  1° aliphatic: violet
  2° aliphatic: rose
  3° aliphatic: yellow
  1° aromatic: rose
  2° aromatic: amber
  3° aromatic: yellow

C–21. *Silver Nitrate in Ethanol.* Add 1 drop or 50 mg of the unknown (dissolved in a little ethanol, if it is a solid) to

Controls: n-butyl chloride, t-butyl chloride, n-butyl bromide, bromobenzene, ethyl iodide.

Reaction

$$RX + AgNO_3 \longrightarrow RONO_2 + AgX$$

(other organic products, such as ROH, ROEt, and alkenes, may also be formed).

2 ml of 0.1M ethanolic silver nitrate. Shake the mixture and let it stand. If no precipitate has formed after 5 minutes, heat the solution to boiling and boil it for 30 seconds. If a precipitate forms, note its color and see if it dissolves when the mixture is shaken with 2 drops of 1M nitric acid.

silver
nitrate

HAZARD   *Silver nitrate is caustic and toxic; avoid contact with skin or eyes.*

Halides that react at room temperature:

     Chlorides: 3°, allyl, benzyl.

     Bromides: 1°, 2°, 3° alkyl (except geminal di- and tribromides); allyl, benzyl, CBr_4.

     Iodides: All aliphatic and alicyclic except vinyl.

Halides that react upon heating:

     Chlorides: 1°, 2° alkyl.

     Bromides: alkyl *gem*-di- and tribromides.

     Some activated aryl halides, such as 2,4-dinitrohalobenzenes.

Halides that do not react:

     Chlorides: alkyl *gem*-di- and trichlorides, CCl_4.

     Most aryl and vinyl halides.

Reactions

$$RCl + NaI \xrightarrow{acetone} RI + NaCl$$

$$RBr + NaI \xrightarrow{acetone} RI + NaBr$$

Controls: n-butyl chloride, *t*-butyl chloride, *n*-butyl bromide, bromobenzene, 1,2-dichloroethane.

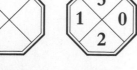

       sodium   acetone
       iodide

Interpretation: The rate of reaction of a halide with silver nitrate depends on its structure. Tertiary, allylic, and benzylic halides react fastest, followed by secondary and primary halides. Most aryl and vinyl halides are unreactive, except for aryl halides activated by 2 or more nitro groups. Alkyl halides having more than one halogen atom on the same carbon (geminal halides) are generally less reactive than the corresponding monohaloalkanes. This is not true, however, if the halogen atoms are on an allylic or benzylic carbon atom—such compounds react at room temperature. Alkyl iodides react faster than the corresponding bromides, which are more reactive than alkyl chlorides. In the margin are listed the types of compounds that react at room temperature, after boiling the solution, or not at all.

     The color of the silver salt formed may indicate the type of halide responsible. Silver chloride is white, silver bromide is pale yellow or cream-colored, and silver iodide is yellow. These salts will not dissolve when the mixture is acidified. Some carboxylic acids and alkynes yield silver salts, but these should dissolve in the acidic solution.

C–22. Sodium Iodide in Acetone. Into a small test tube, place 1 ml of the sodium iodide/acetone reagent. Add 2 drops of the halogen compound (or 0.1 g of a solid dissolved in the minimum volume of acetone). Shake the mixture and allow it to stand for three minutes, noting whether a precipitate or a red-brown color forms. If there is no reaction at the end of this time, place the test tube in a 50° water bath (replenish the acetone if necessary) and leave it there for an additional six minutes; then cool the mixture to room temperature.

Interpretation: Certain alkyl chlorides and bromides react to precipitate sodium chloride or bromide; others may also liberate elemental iodine, which is red-brown in acetone solution. Since the reaction is, in most cases, a bimolecular displacement by iodide ion, the order of reactivity is generally methyl > primary > secondary > tertiary > aryl, vinyl. Many cycloalkyl halides (cyclopentyl is an exception) react more slowly than the corresponding open-chain compounds and may give no precipitate after heating. Bromides react faster than chlorides, and the test is not applicable to iodides. Vicinal (having substituents on adjacent carbon atoms) and some geminal halides, along with triarylmethyl halides, undergo redox reactions to liberate iodine while precipitating the sodium halide. A summary of the reactivities of various halides is given in the margin.

If a precipitate forms upon mixing but does not persist, the test should be considered negative. Some halides may be contaminated with isomeric impurities that could yield a little precipitate of sodium halide, so it is important to run a control and compare the amount of precipitate obtained in each case.

C–23. Tollens's Test.

Prepare the reagent immediately before use as follows. Measure 2 ml of 0.3M aqueous silver nitrate into a thoroughly cleaned test tube and add 1 drop of 3M sodium hydroxide. Then add 2M aqueous ammonia dropwise, with shaking, until the precipitate of silver oxide just dissolves (avoid an excess of ammonia). Add 1 drop (or 30–50 mg) of the unknown to this solution, shake the mixture, and let it stand for 10 minutes. If no reaction has occurred by the end of this time, heat the mixture in a 35° water bath for 5 minutes. Immediately after the test is completed, wash the contents of the test tube down the drain with plenty of water and rinse the tube with dilute nitric acid (*Important!*).

Halides that react at room temperature (X = Br, Cl only; starred compounds liberate iodine, turning the solution red-brown.)

1° alkyl bromides; benzyl and allyl halides; α-halo ketones, esters, and amides; CBr_4★; *vic*-dihalides★; triaryl halides★.

Halides that react at 50°: most 1°, 2° alkyl chlorides; most 2°, 3° alkyl bromides; *gem*-di- and tribromides★.

Halides that do not react: 3° alkyl chlorides; aryl, vinyl halides; cyclopropyl, cyclobutyl, cyclohexyl halides; *gem*-polychloro compounds (except benzyl and allyl).

Reaction
$$RCHO + 2Ag(NH_3)_2OH \longrightarrow 2Ag + RCOONH_4 + H_2O + 3NH_3$$
Control: benzaldehyde

silver nitrate

The reagent and test solution must never be stored—explosive silver fulminate forms on standing.

HAZARD

Positive test: Formation of a silver mirror on the inside of the test tube. If the tube is not sufficiently clean, a black precipitate or a suspension of metallic silver may form instead. In addition to aldehydes of all kinds, certain aromatic amines, phenols, and α-alkoxy or dialkylaminoketones may give positive tests.

Some cyclic ketones (as cyclopentanone) may also give a positive Tollens test.

DERIVATIVES

The following procedures should be suitable for preparing derivatives of most common organic compounds in the appropriate class. In some cases, a variation of the procedure given (the use of different reaction conditions or recrystallization solvents, etc.) may be required for satisfactory results. If you do not obtain the expected derivative, consult one of the references listed in the Bibliography (Bib–B4, Bib–G1 to Bib–G3) for alternative procedures. You may also wish to prepare a derivative that is not described in this section, but that will better differentiate between the possibilities. If so, you should consult your instructor about the feasibility of preparing such derivatives (the reagents may not be available in your stockroom) and find an appropriate procedure in one of the sources listed.

The procedures specify quantities of the unknown in grams. If the unknown is a liquid, its density can usually be estimated from the tables and an equivalent volume measured out instead. In most cases, the relative quantities of unknown and reagent are not crucial, since any excess is removed during the isolation and purification of the derivative. In cases where the quantities could affect the results, the quantity of the unknown is given in millimoles so that the required mass can be calculated using an estimate of its molecular weight.

In preparing a derivative, you will be using very small quantities of reactants and reagents, so you should use small-scale apparatus whenever feasible. Reactions are usually run in large test tubes and only a few require refluxing. Filtrations, distillations, extractions, and other operations are carried out on a semimicro scale as described in ⟨OP–12a⟩, ⟨OP–25a⟩, ⟨OP–13a⟩, etc. Whenever the recrystallization solvent is specified as an alcohol-water mixture or some other mixture, the procedure for recrystallization from mixed solvents ⟨OP–23b⟩ with semimicro apparatus ⟨OP–23a⟩ should be used. The derivative will usually be dissolved in the less polar solvent and the solution saturated by adding the more polar solvent dropwise.

As always, appropriate precautions should be taken to avoid contacting or inhaling the reagents and unknown. Many of the reagents are, of necessity, highly reactive and may react violently with water or other substances. All should be considered toxic to a degree (some are highly so) and many of them are corrosive, lachrymatory (tear-producing), or have other unpleasant properties. Always read the *Hazard* notes before beginning any derivative preparation and follow the precautions indicated.

Read ⟨OP–7b⟩ for methods of carrying out small-scale reactions under reflux.

Alcohols

D–1. *3,5-Dinitrobenzoates*, p-*Nitrobenzoates*. Mix 0.2 g of pure 3,5-dinitrobenzoyl chloride *or* p-nitrobenzoyl chloride

Reaction

ArCOCl + ROH $\longrightarrow$
                ArCOOR + HCl

HAZARD

(*Hazard!*) with 0.1 g of the alcohol. Heat the mixture over a *small* flame so that it is just maintained in the liquid state (do not overheat—decomposition will result). If the alcohol boils under 160°, heat the mixture for 5 minutes; otherwise heat it for 10–15 minutes. Allow the melt to cool and solidify, break it up with a stirring rod, and stir in 4 ml of 0.2M sodium carbonate. Heat the mixture to 50–60° on a steam bath, stir it at that temperature for half a minute, then cool it and collect the precipitate by small-scale vacuum filtration. Wash the precipitate several times with cold water and recrystallize it from ethanol or an ethanol-water mixture.

If a microburner is not available, remove the barrel of your burner and light it with the gas turned down.

In general, the higher boiling (or melting) alcohols will require a higher ethanol/water ratio in the recrystallization solvent.

D–2. α-*Naphthylurethanes, Phenylurethanes.* Unless you are certain that the alcohol is anhydrous, dry it with magnesium sulfate or another suitable drying agent, and dry a small test tube in an oven or over a flame. Stopper the tube and allow it to cool, then mix 5 drops (or 0.2 g) of the alcohol with 5 drops of phenyl isocyanate *or* α-naphthyl isocyanate (*Hazard!*). If no reaction takes place immediately,

The presence of moisture in the alcohol will result in the formation of diphenyl- or di-α-naphthylurea, which melt at 241° and 297° respectively.

Tertiary alcohols do not form urethanes readily.

The isocyanates are irritants and lachrymators. Avoid contact with liquid and vapors; use a hood.

HAZARD

heat the mixture in a 60–70° water bath for 5–15 minutes. Cool the test tube in ice and scratch its sides, if necessary, to induce crystallization. Collect the precipitate by decantation or vacuum filtration, then recrystallize it from about 5 ml of high-boiling petroleum ether or commercial heptane (filter the hot solution to remove high-melting impurities).

Reaction

$$ArN{=}C{=}O + ROH \longrightarrow$$

$$ArNH\overset{O}{\underset{\|}{C}}{-}OR$$

Aldehydes and Ketones

D–3. *2,4-Dinitrophenylhydrazones.* Dissolve 0.25 g of the unknown in 10 ml of 95% ethanol, add 7½ ml of the 2,4-

The reagent contains 27 g of 2,4-dinitrophenylhydrazine per liter of solution.

2,4-Dinitrophenylhydrazine is toxic and the acid is corrosive; avoid contact with the reagent.

HAZARD

Reaction

R, R' = alkyl, aryl, or H

$$R-\underset{\underset{R'}{|}}{C}=O + H_2N-NH-\underset{O_2N}{\bigcirc}-NO_2$$

$$R-\underset{\underset{R'}{|}}{C}=N-NH-\underset{O_2N}{\bigcirc}-NO_2 + H_2O$$

If the unknown is water-soluble, omit the ethanol.

$$-\underset{|}{\overset{O}{\overset{\|}{C}}} + H_2N-NHCNH_2$$

$$-\underset{|}{C}=N-NHCNH_2 + H_2O$$

Reaction

$$-\underset{|}{\overset{O}{\overset{\|}{C}}} + H_2N-OH \longrightarrow$$

$$-\underset{|}{C}=N-OH + H_2O$$

dinitrophenylhydrazine–sulfuric acid reagent, and allow the solution to stand at room temperature until crystallization is complete. If necessary, warm the solution gently for a minute on a steam bath. If no precipitate appears after 15 minutes (or if precipitation does not seem complete), add water dropwise to the warm solution until it is cloudy, heat to clarify, and cool. Recrystallize the derivative by dissolving it in up to 15 ml of 95% ethanol and adding water (to a maximum of 5 ml) dropwise. If the derivative does not dissolve in 15 ml of boiling ethanol, add ethyl acetate dropwise until it goes into solution. Complete crystallization may require several hours, but you can usually collect enough derivative for a melting point after 15–30 minutes.

D–4. *Semicarbazones.* Mix together 0.2 g of semicarbazide hydrochloride, 0.3 g of sodium acetate, 2 ml of water, and 2 ml of 95% ethanol in a test tube. Add 0.2 g of the aldehyde or ketone and stir. (If the solution is cloudy, add more ethanol until it clears up.) Shake the mixture for a minute or two and let it stand, cooling it in ice if necessary to induce crystallization. If no crystals form, place the test tube in a boiling water bath for a few minutes and allow it to cool. Collect the crystals, wash them with cold water, and recrystallize the product from alcohol or an alcohol-water mixture.

D–5. *Oximes.* Oximes are suitable derivatives for most ketones and for some (though not all) aldehydes. They can be prepared by the method given for semicarbazones (D–4), using hydroxylamine hydrochloride in place of semicarbazide hydrochloride. It is usually necessary to warm the reactants on a steam bath for 10 minutes or more. Adding a few milliliters of cold water to the reaction mixture may hasten precipitation.

HAZARD   *The reagent is toxic and can cause a form of anemia. Avoid contact.*

Amides

If the alkaline hydrolysis classification test (C–2) suggests that the amide is difficult to hydrolyze, use 6M NaOH or a longer reaction time, or both.

D–6. *Hydrolysis Products.* The acid and amine portions of an amide can be obtained by hydrolysis and one or both of them characterized. Reflux 0.6 g of the amide with 10 ml of 3M sodium hydroxide for 15 minutes or more, then reas-

semble the apparatus for semimicro simple distillation, placing 4 ml of 3M hydrochloric acid in the receiver (a 25-ml
round-bottomed flask is recommended). Since the amine
may be a gas, the receiving flask should be attached directly
to the vacuum-adapter outlet and the vacuum side arm
should be connected to a gas trap containing dilute HCl.
Distill the reaction mixture until about 6 ml of distillate has
been collected.

   To characterize the amine, add 10 ml of 3M sodium hydroxide to the distillate, then add 0.4 ml of benzenesulfonyl
chloride or 0.6 g of *p*-toluenesulfonyl chloride (*Hazard!*)
and proceed according to the directions given in procedure
D–9 for preparing arenesulfonamides.

   To characterize the carboxylic acid, carefully acidify the
residue in the boiling flask with 6M hydrochloric acid. If a
precipitate forms, collect it by vacuum filtration, wash it, recrystallize it (from water, alcohol-water, or another suitable
solvent), and obtain its melting point. If no precipitate
forms, prepare a *p*-nitrobenzyl derivative according to the
directions in procedure D–15.

D–7. *N-Xanthylamides.*   This procedure is suitable only for
N-unsubstituted amides. Dissolve 0.4 g of xanthydrol in
5 ml of glacial acetic acid (*Hood!*) and shake until the solid
dissolves. Decant the solution from any undissolved residue,
add 0.2 g of the amide (or as close to 1.5 mmoles as you can
estimate), and warm the mixture in an 85° water bath for
15–20 minutes. If the amide is not soluble in acetic acid, dissolve it in a minimum volume of ethanol before adding it to
the reaction mixture, then add 1 ml of water after the reaction is complete (just before cooling). Let the solution cool
until precipitation is complete, collect the solid by vacuum
filtration, and recrystallize the derivative from 70% aqueous
dioxane or from an ethanol-water mixture.

*If the amide is known to be unsubstituted
on nitrogen, omit this procedure.*

*Hazard: Arenesulfonyl chlorides are
toxic and corrosive; avoid contact, do
not breathe vapors.*

Reaction

acetic acid

---

*Acetic acid can cause burns and its vapors are very irritating. Use
under hood; avoid contact.*   **HAZARD**

Amines (Primary and Secondary)

D–8. *Benzamides.*   Combine 0.2 g of the primary or secondary amine (or as close to 2 mmoles as you can estimate)
with 2 ml of 3M sodium hydroxide, then add 0.8 ml of ben-

Reaction

$$\underset{\text{(R' can be alkyl, aryl, or H.)}}{PhC-Cl + R-NH \longrightarrow PhC-N-R + HCl}$$

$$\overset{O}{\underset{\|}{PhC}}-Cl + R-\underset{\underset{R'}{|}}{N}H \longrightarrow \overset{O}{\underset{\|}{PhC}}-\underset{\underset{R'}{|}}{N}-R + HCl$$

(R' can be alkyl, aryl, or H.)

benzoyl chloride

zoyl chloride (*Lachrymator!*) dropwise, with vigorous shaking. Continue to shake the stoppered test tube for about 5 minutes, then carefully neutralize the solution to a pH of 8 with 3M HCl (use pH paper). Break up the solid mass with a stirring rod, if necessary, and collect the derivative by vacuum filtration. After washing the product thoroughly with cold water, recrystallize it from an alcohol-water mixture.

HAZARDS   *Benzoyl chloride is a lachrymator and strong irritant. Avoid contact with liquid or vapors; use under hood.*

Reactions:

$$RNH_2 + ArSO_2Cl \xrightarrow{\text{NaOH}} \xrightarrow{\text{HCl}}$$
$$ArSO_2NHR$$

$$R_2NH + ArSO_2Cl \xrightarrow{\text{NaOH}}$$
$$ArSO_2NR_2$$

D–9. *Benzenesulfonamides and* p-*Toluenesulfonamides.* Combine in a test tube 0.2 g (or as close to 2 mmoles as you can estimate) of the primary or secondary amine, 0.4 ml of benzenesulfonyl chloride *or* 0.6 g of p-toluenesulfonyl chloride (*Hazard!*), and 10 ml of 3M sodium hydroxide. Stopper the tube and shake it frequently over a period of 3–5 minutes.

HAZARD   *Arenesulfonyl chlorides are toxic and corrosive. Avoid contact; do not breathe vapors.*

The precipitate formed in the Hinsberg test (C–15) can be used as a derivative after it has been purified as described here.

Remove the stopper and warm the tube gently over a steam bath for a minute or two. Let the solution cool and (*a*) if a precipitate forms, collect it by vacuum filtration, wash it with water, and recrystallize it from an ethanol-water mixture. (*b*) If no precipitate forms on cooling, acidify the solution to pH 6 with 6M hydrochloric acid, cool the mixture to complete crystallization, and collect and purify the product as described above.

D–10. α-*Naphthylthioureas and Phenylthioureas.* Dissolve 0.2 g (or as close to 2 mmoles as you can estimate) of the primary or secondary amine in 2 ml of ethanol and add 0.2 ml of phenylisothiocyanate (*Hazard!*) *or* 0.3 g of α-naphthyliso-thiocyanate (*Hazard!*). Reflux the mixture for 5 minutes (using a water or steam bath) so that the alcohol boils gently. Cool the mixture and scratch the sides of the test tube to in-duce crystallization. If no crystals form, reflux the mixture for another 20 minutes or so. Add 1 ml each of ethanol and water to facilitate transfer, collect the derivative by vacuum filtration, wash it with dilute ethanol (one washing with 50% and one with 95% ethanol are recommended), and re-crystallize the product from ethanol or from an alcohol-water mixture. If the derivative appears impure, it may be necessary to extract the impurities by boiling the solid with 4 ml of petroleum ether before recrystallizing it, or to carry out several recrystallization steps.

Adding water dropwise until the solution becomes cloudy may facilitate crystalliza-tion. If an oil forms, scratching the sides of the tube will usually cause it to crystallize.

Reaction

$$ArN{=}C{=}S + \underset{R'}{R}NH \longrightarrow Ar\underset{R'}{N}H\overset{S}{\overset{\|}{C}}NR$$

(R′ may be alkyl, aryl, or H.)

The isothiocyanates are toxic and irritant. Avoid contact; do not inhale vapors. Use gloves and hood.

HAZARD

Amines (Tertiary)

D-11. Methiodides. Mix 0.2 g of the tertiary amine with 0.2 ml of methyl iodide (*Hazard!*) in a test tube and warm the mixture in a 50–60° water bath for five minutes. Cool the mixture in ice, collect the product by vacuum filtration, and recrystallize it from alcohol, ethyl acetate, or a mixture of the two.

Reaction

$$R_3N + CH_3I \longrightarrow R_3NCH_3{}^+I^-$$

(The R groups may be alkyl, aryl, or a combination of the two.)

Methyl iodide is poisonous and can cause burns after prolonged skin contact; it is also suspected of causing cancer. Avoid contact; do not breathe vapors.

HAZARD

D-12. Picrates. Dissolve 0.2 g of the unknown in 5 ml of 95% ethanol and add 5 ml of a saturated solution of picric

Reaction

acid in ethanol. Heat the solution to boiling over a steam bath and let it cool to room temperature. Collect the crystals by vacuum filtration and wash them with cold ethanol or methanol. (The product may be recrystallized from alcohol if necessary.)

Picrates of primary amines, secondary amines, and aromatic hydrocarbons can be prepared by essentially the same method. The picrates of many aromatic hydrocarbons dissociate when heated and cannot be recrystallized. Some picrates explode when heated, so use caution when obtaining the melting point.

picric acid

HAZARD   *In solid form, picric acid is unstable and can explode when subjected to heat or shock. It is also toxic and irritant. Avoid contact with solutions containing picric acid.*

Carboxylic Acids

Reactions

$$RCOOH + SOCl_2 \longrightarrow$$
$$RCOCl + HCl + SO_2$$
$$RCOCl + NH_3 \longrightarrow$$
$$RCONH_2 + HCl$$

The reaction apparatus must be dry.

If you do not intend to prepare an anilide or p-toluidide, add the entire reaction mixture to 15 ml of ammonia.

D–13. *Amides.*   This derivative is suitable for aromatic acids and for most aliphatic acids having six or more carbon atoms. Since procedures D–13 and D–14 both involve preparation of the acid chloride, it is convenient to prepare one of the D–14 amides with the same solution.

Reflux 1 g of the carboxylic acid with 5 ml of thionyl chloride (*Hazard!*) under a fume hood for 20–30 minutes using a steam bath or a hot water bath. Cool the mixture in an ice bath and (if desired) save part of it in a stoppered container to prepare an anilide or *p*-toluidide. Still under the hood, *slowly* add 2 ml of this solution to a beaker containing 5 ml of ice-cold concentrated aqueous ammonia (*Hazard!*), with constant stirring. Let the solution stand for five minutes, separate the derivative by vacuum filtration, and recrystallize it from water or from an alcohol-water mixture.

HAZARDS   *Thionyl chloride and ammonia can both irritate or cause burns to skin and eyes; their vapors are highly irritating and toxic. Use rubber gloves and a hood.*

D–14. *Anilides and* p-*Toluidides.*   Prepare the acid chloride as directed in D–13, or use the reaction mixture saved from the amide preparation. Dilute about 2 ml of the acid chloride reaction mixture with 2 ml of ethyl ether and add portion-wise (shaking after each addition) a solution containing 0.8 g of p-toluidine *or* 0.7 ml of aniline (see *Hazards*) dissolved in 10 ml of ethyl ether. Continue the addition until the odor of the acid chloride has disappeared. If the odor persists after all the arylamine solution has been added, heat the mixture gently on a steam bath for a minute or two. Wash the ether solution in a large test tube with 5 ml of 1.5M hydrochloric acid followed by 5 ml of water, separating the aqueous layers from the ether layer using a capillary pipet. Evaporate the solvent under aspirator vacuum. Recrystallize the derivative from water or from an alcohol-water mixture.

Reaction

$$RCOOH + SOCl_2 \longrightarrow RCOCl + HCl + SO_2$$

$$RCOCl + ArNH_2 \longrightarrow RCONHAr + HCl$$

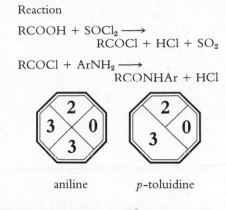

aniline   p-toluidine

The aromatic amines are toxic by inhalation, ingestion, and skin absorption. Avoid contact with skin, eyes, and clothing; do not breathe vapors. Avoid contact with the acid chloride; do not breathe vapors.

HAZARDS

D–15. p-*Nitrobenzyl Esters.* Preparation of this derivative is recommended only when the previous derivatives are not suitable, or when the carboxylic acid is obtained in aqueous

Reaction

$$RCOONa + O_2N-\!\!\!\bigcirc\!\!\!-CH_2Cl \longrightarrow RCOOCH_2-\!\!\!\bigcirc\!\!\!-NO_2 + NaCl$$

solution (for instance, after the hydrolysis of an ester or amide). If the acid is obtained by hydrolysis, the alcohol or amine portion should be identified first so that the molecular weight of the acid portion can be estimated from a list of possibilities. Then you can estimate the amount of acid in the hydrolysis solution and measure out enough of it to provide 2 mmoles of the acid.

Mix an estimated 2 mmol of the carboxylic acid with 3 ml of water (or use the measured solution from amide or ester hydrolysis) and add a drop of phenolphthalein solution. Add 3M sodium hydroxide dropwise until the solution turns pink. Heat (if necessary) to put the acid salt into solution, then add ~2 drops of 1.5M hydrochloric acid to discharge the pink color. Add 0.3 g (1.75 mmoles) of p-nitrobenzyl chloride (*Hazard!*) dissolved in 10 ml of 95% ethanol and reflux the mixture gently for about 1½ hours. Allow the solution to cool to room temperature and, if precipitation

It is important that p-nitrobenzyl chloride not be present in excess, since it is difficult to remove from the product.

Di- and triprotic acids should be refluxed for 2–3 hours.

has not occurred, add 1 ml of water and scratch the sides of the reaction vessel. When crystallization is complete (it may take 20 minutes or so), collect the product by vacuum filtration and wash it twice with small portions of 5% sodium carbonate, then once with water. Recrystallize the derivative from alcohol or an alcohol-water mixture. If the melting point of the product is close to 71° (the melting point of p-nitrobenzyl chloride), the reaction may not be complete and a longer reflux time may be required.

HAZARDS   *p-Nitrobenzyl chloride is lachrymatory (tear-producing) and can cause blisters. Avoid skin and eye contact with the reagent or its vapors. Avoid contact of product with the skin—blistering may result.*

Esters

Reactions

$$RCOOR' + NaOH \longrightarrow$$
$$RCOONa + R'OH$$

$$RCOONa + H^+ \longrightarrow$$
$$RCOOH + Na^+$$

The acidified residue can also be used to prepare a p-nitrobenzyl ester directly.

It may be necessary to purify the acid or alcohol before attempting to prepare a derivative.

D–16. *Hydrolysis Products.*   Carry out the hydrolysis of the ester according to the procedure in test C–2, but use 5 ml of the ester and 40 ml of 6M sodium hydroxide. If a solid carboxylic acid precipitates upon acidification of the solution (use concentrated HCl, with care) it can be recrystallized from water or an alcohol-water mixture and used as a derivative itself. It can also, if necessary, be converted to one of the carboxylic acid derivatives. If no precipitate forms, make the solution alkaline with dilute sodium hydroxide and distill it until about 5 ml of distillate has been collected. Acidify the residue in the boiling flask and extract it with ether, then evaporate the ether to isolate the carboxylic acid for a derivative preparation. The alcohol portion of the ester often separates out when the distillate is saturated with potassium carbonate. It can then be isolated with a separatory funnel or (if necessary) by extraction with ether and used to prepare a derivative as well.

Reaction

$$RCOOR' + CH_3OH \xrightarrow{CH_3ONa}$$
$$RCOOCH_3 + R'OH$$

Procedure D–17 forms a derivative of the carboxylic acid portion of the ester; procedure D–18 forms a derivative of the alcohol portion.

D–17. *N-Benzylamides.*   This procedure works well only with methyl and ethyl eters. To prepare the N-benzylamide of an ester of a higher alcohol, reflux the ester (about 1 g) with 0.25 g of sodium methoxide in 5 ml of anhydrous methanol for about 30 minutes. Evaporate the methanol and use the resulting methyl ester in this procedure.

Combine 1 g of the ester with 3 ml of benzylamine (*Hazard!*) and 0.1 g of powdered ammonium chloride. Reflux the mixture for one hour, then cool the reaction mixture and wash it with a few milliliters of water, using a capillary pipet to remove the water. If no precipitate forms, add a drop or two of 3M hydrochloric acid and scratch the sides of the test tube. If a precipitate still does not form, transfer the mixture to an evaporating dish (use 1 ml of water in the transfer) and boil it for a few minutes to remove the excess ester. Collect the derivative by vacuum filtration, wash it with ligroin, and recrystallize it from an alcohol–water mixture or ethyl acetate.

Reaction

$$RCOOR' + PhCH_2NH_2 \xrightarrow{NH_4Cl}$$
$$RCONHCH_2Ph + R'OH$$

(R′ should be ethyl or methyl.)

Benzylamine is highly irritating to skin, eyes, and mucous membranes. Avoid contact; do not breathe vapors.

HAZARD

D–18. *3,5-Dinitrobenzoates.* Combine 1 g of the ester with 0.8 g of 3,5-dinitrobenzoic acid (*Hazard!*) and add a drop of concentrated sulfuric acid (*Hazard!*). Reflux the mixture until the 3,5-dinitrobenzoic acid dissolves, then continue refluxing for an additional 30 minutes. Dissolve the cooled

This procedure is not satisfactory for esters of 3° and some unsaturated alcohols.

Reaction

$$O_2N-C_6H_3(O_2N)-COOH + RCOOR' \longrightarrow O_2N-C_6H_3(O_2N)-COOR' + RCOOH$$

Prevent contact with sulfuric acid and 3,5-dinitrobenzoic acid.

HAZARDS

reaction mixture in 20 ml of anhydrous ethyl ether and wash the ether solution twice with 10-ml portions of 0.5M sodium carbonate (*Caution:* foaming), then with water. Evaporate the ether and dissolve the residue (often an oil) in 2–3 ml of boiling ethanol. Filter the hot solution, add water until the mixture becomes cloudy, and cool the solution to crystallize the product. Recrystallize the derivative from an alcohol–water mixture, if necessary.

Alkyl Halides

In addition to the following derivative, density values (test C–10) and refractive index values are often very useful in characterizing alkyl and aryl halides.

Tertiary alkyl halides will not form this derivative. If the unknown has been shown to be an alkyl chloride, use the alternate procedure with ethylene glycol.

D–19. *S-Alkylthiuronium Picrates.* Mix together 0.2 g of the alkyl halide, 0.2 g of powdered thiourea, and 2 ml of 95% ethanol. Boil the mixture on a steam bath for a few

---

HAZARDS *Picric acid solutions are toxic and corrosive (see D–12). Thiourea is toxic by ingestion or skin absorption. Avoid contact with both reagents.*

---

minutes, add 2 ml of a saturated solution of picric acid in ethanol, and boil the mixture for a few minutes longer. Allow the solution to cool, collect the derivative by vacuum filtration, and recrystallize it from ethanol or from an ethanol-water mixture.

Reaction

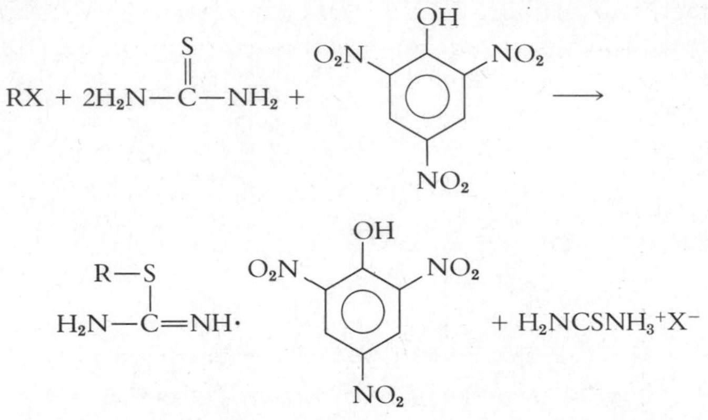

$$RX + 2H_2N-\overset{\overset{\displaystyle S}{\|}}{C}-NH_2 + $$

$$\longrightarrow$$

$$+ H_2NCSNH_3{}^+X^-$$

$$X = Cl, Br, I \quad R = alkyl$$

If the derivative fails to form under these conditions (alkyl chlorides are often unreactive), follow the same procedure using 6 ml of ethylene glycol as the solvent. Heat the solution in an oil bath at 120° for 30 minutes before adding the picric acid solution, then continue heating it for 15 minutes. Add 6 ml of water and cool the mixture in an ice bath to assist crystallization.

Aryl Halides

In addition to the following derivative, aryl halides having alkyl side chains can be characterized by oxidizing the side chain (procedure D–21). Density values (test C–10) and refractive index values are also very useful.

D–20. *Nitro Compounds.* **(a)** Mix 0.5 g of the unknown with 2 ml of concentrated sulfuric acid (*Hazard!*). Cautiously add 2 ml of concentrated nitric acid (*Hazard!*) dropwise with shaking or stirring, then heat the mixture on a 60° water bath for 10 minutes, shaking frequently. Pour the product onto 15 ml of cracked ice, with stirring. After the ice has melted, collect the precipitate by vacuum filtration, wash it with water, and recrystallize it from an ethanol-water mixture. Additional recrystallizations may be necessary if the product melts at a temperature lower than expected or over a wide range.

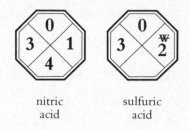

nitric
acid

sulfuric
acid

Sulfuric acid, nitric acid, and (especially) fuming nitric acid are toxic, reactive, and very corrosive. Avoid contact; do not inhale any brown fumes of nitrogen dioxide that may be generated. Some compounds react violently; use a hood and rubber gloves.

HAZARDS

(b) Follow procedure **(a)**, but use 2 ml of fuming nitric acid (*Hazard!*) instead of concentrated nitric acid, and heat the solution for 10 minutes on a steam bath. Add the acid slowly to minimize the generation of brown (NO₂) fumes. Procedure **(a)** will yield mononitro derivatives of most comparatively reactive aryl halides and hydrocarbons, whereas procedure **(b)** should produce dinitro or trinitro derivatives of reactive compounds and mononitro derivatives of unreactive ones. Di- and trialkylbenzenes (such as the xylenes and mesitylene) usually yield trinitro derivatives by procedure **(b)**, whereas monoalkylbenzenes such as toluene yield dinitro compounds. Since the di- and trinitro derivatives are often easier to purify than mononitro derivatives, procedure **(b)** is usually preferred.

Reaction

$$ArH + HNO_3 \xrightarrow{H_2SO_4} ArNO_2 + H_2O$$

(and di- or trinitro derivatives in some cases)

If the product separates as an oil, it is probably a mixture of compounds with different numbers of nitro groups. See the reference by Cheronis, Entrikin, and Hodnett (Bib–G1) for a more detailed discussion of nitration, as well as additional procedures.

Aromatic Hydrocarbons

In addition to the following derivative, aromatic hydrocarbons can be characterized by the preparation of picrates (D–12) and nitro compounds (D–20).

D–21. *Aromatic Carboxylic Acids.* In a small boiling flask equipped with a condenser, combine **(a)** 0.5 g of an aromatic hydrocarbon or halide having one alkyl side chain with **(b)** 1.5 g of potassium permanganate dissolved in 25 ml of water and **(c)** 0.5 ml of 6M sodium hydroxide so-

Reaction

$$Ar{-}R \xrightarrow{KMnO_4} ArCOOH$$

HAZARDS *Avoid contact with the caustic NaOH solution and with potassium permanganate. Keep the latter away from oxidizable materials.*

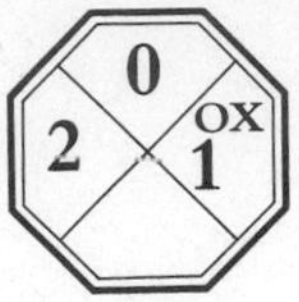

potassium permanganate

lution. Reflux the mixture for 1 hour or until the permanganate color has disappeared. Cool the reaction mixture to room temperature, acidify it with 6M sulfuric acid, and boil it for a few minutes. If necessary, stir a little solid sodium bisulfite into the hot solution to destroy any excess manganese dioxide, keeping the solution acidic during the treatment. Cool the mixture, collect the product by vacuum filtration, and recrystallize the acid from water or from an ethanol-water mixture.

If the unknown is suspected to have two or three alkyl side chains, double or triple the amounts of all reactants except the unknown.

Phenols

Reaction

$$ArOH + ClCH_2COOH \longrightarrow ArOCH_2COOH + HCl$$

Congo Red solution may be used.

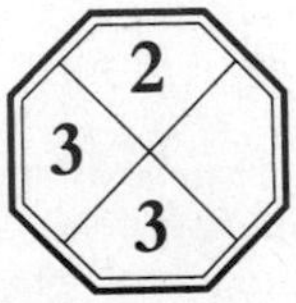

chloroacetic acid

D–22. *Aryloxyacetic Acids.* Combine 0.5 g of the unknown with 0.7 g of chloroacetic acid (*Hazard!*) and 3 ml of 8M sodium hydroxide. Heat the mixture on a steam bath for an hour, then cool it to room temperature and add 6 ml of water. Acidify the solution to pH 3 with 6M hydrochloric acid, extract it with 20 ml of ethyl ether, and wash the ether extract with 5 ml of cold water. Extract the derivative from the ether using about 10 ml of 0.5M sodium carbonate, then acidify the sodium carbonate solution with 6M hydrochloric acid (*Care:* foaming). Collect the precipitate by vacuum filtration and recrystallize it from water.

HAZARDS *Chloroacetic acid is corrosive and toxic. Avoid contact with skin, eyes, and clothing; do not breathe vapors. Avoid contact with the highly caustic NaOH solution.*

The equivalent weight of the aryloxyacetic acid (and thus of the phenol as well) can be determined by obtaining its neutralization equivalent (test C–18).

D–23. *Bromo Derivatives.* Prepare a brominating solution by dissolving 4.5 g of potassium bromide in 30 ml of water and *carefully* adding 1 ml (about 3 g) of pure bromine (*Hazard!*). Dissolve 0.3 g of the phenol in 6 ml of 50% ethanol

Reaction

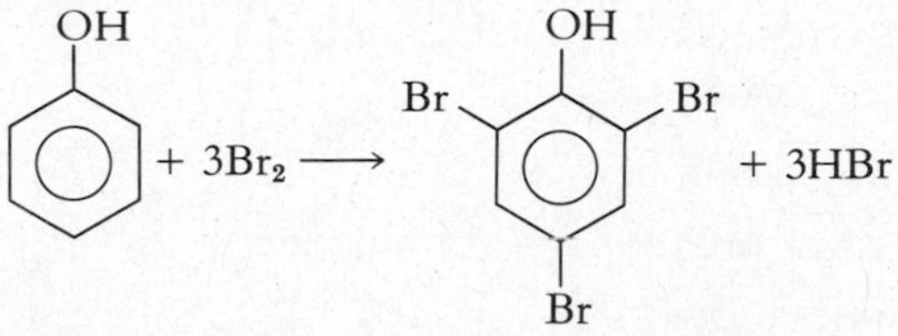

$$\text{OH} + 3\text{Br}_2 \longrightarrow \text{(Br-substituted OH)} + 3\text{HBr}$$

(and similar reactions for substituted phenols)

(try 95% ethanol, acetone, or dioxane if it doesn't dissolve) and add the brominating solution dropwise, with stirring, until the bromine color persists after shaking. Add 15 ml of water and shake the mixture, then collect the derivative by vacuum filtration. Wash the product with 1M sodium bisulfite to remove excess bromine and recrystallize it from ethanol or from an alcohol-water mixture.

bromine

Bromine is very poisonous and causes severe burns. Handle with rubber gloves under the hood; do not breathe vapors.   **HAZARD**

D–24. *α-Naphthylurethanes.* Carry out the general procedure for preparing α-naphthylurethanes of alcohols (D–2),

Reaction

$$\text{ArN}{=}\text{C}{=}\text{O} + \text{Ar}'\text{OH} \longrightarrow \text{ArNH}\overset{\text{O}}{\overset{\|}{\text{C}}}{-}\text{OAr}'$$

but add a drop of pyridine to catalyze the reaction. For particularly unreactive phenols, it may be helpful to add 1 ml of pyridine and a drop of 10% triethylamine in petroleum ether to the reaction mixture, and to heat the mixture at 70° for about 30 minutes. If the urethane does not precipitate on cooling, add 1 ml of 0.5M sulfuric acid.

pyridine

Pyridine is toxic and irritant and has an unpleasant odor. α-Naphthyl isocyanate is lachrymatory and a strong irritant. Use a hood.   **HAZARDS**

REPORT

Your report should include information under each of the following categories:

Unknown number

Results of preliminary examination

Physical constants determined

Results of solubility tests

Spectrometric results

Results of functional class tests

Interpretation of spectra and tests

Results of literature examination

Additional tests and data

Probable compounds

Derivatives prepared

Discussion and conclusion

Where convenient, you should tabulate your data and results so that they are presented clearly and concisely. Some of the categories may be combined—for example, the interpretation of the functional class tests might be included in a tabulation describing your observations. Insofar as possible, the report should reveal the thought processes leading to your conclusion as well as the physical and chemical data supporting it.

Topics for Report

1. Write balanced equations for the chemical reactions undergone by your specific unknown, including those involved in classification tests and the preparation of derivatives.

2. Specify the functional class (chemical family) of each of the following unknowns and give any additional structural information suggested by the tests. Explain the reasoning behind your deductions.

(a) Unknown #22 is insoluble in all of the solubility test solvents except cold, concentrated sulfuric acid. It gives a blue-green suspension with the chromic acid reagent, yields no precipitate with Benedict's solution, and immediately forms a separate organic layer when shaken with concentrated HCl and $ZnCl_2$ (Lucas's test). The C—H stretching band in the infrared spectrum extends from about 2750 to 3050 cm^{-1}, but there is no strong absorption at a higher frequency.

(**b**)   Unknown #7B is insoluble in water but soluble in dilute sodium hydroxide. It decolorizes potassium permanganate and bromine in carbon tetrachloride solutions. Addition of a saturated solution of bromine in water causes decolorization of the bromine and yields a white precipitate.

(**c**)   Unknown #128 is in solubility class B and yields an infrared spectrum with strong absorption centered around 3400 cm^{-1}. The unknown is insoluble in a pH 5.5 buffer, and forms an insoluble precipitate when shaken with p-toluenesulfonyl chloride in aqueous sodium hydroxide.

(**d**)   Unknown #33C has a strong infrared band in the $1750-1650 \text{ cm}^{-1}$ region, but it forms no precipitate with 2,4-dinitrophenylhydrazine reagent. When refluxed with 6M sodium hydroxide, a vapor is generated that turns red litmus paper blue. Acidification of the alkaline hydrolysis solution yields a white precipitate.

(**e**)   Unknown #2B is insoluble in cold concentrated sulfuric acid, and a drop of the unknown in water sinks to the bottom. The unknown gives an orange-red color with aluminum chloride and chloroform. It reacts immediately to form precipitates with both ethanolic silver nitrate and sodium iodide in acetone; the colors of the precipitates are pale yellow and white, respectively.

(**f**)   Unknown #44 is water soluble and reacts with acetyl chloride to yield a liquid that smells like nail polish. It dissolves in the Lucas reagent but gives no separate phase or cloudiness within an hour, even when ice water is added. Treating the unknown with iodine in sodium hydroxide forms a yellow precipitate with a medicinal odor.

3.   Indicate the solubility class to which each of the following should belong.

(**a**)   propanoic acid
(**b**)   toluene
(**c**) p-anisidine
(**d**) p-cresol
(**e**)   3-methylpentanal
(**f**) p-toluic acid
(**g**)   2-chlorobutane
(**h**)   2-aminobutane
(**i**)   ethylene glycol
(**j**)   acetophenone

4.   Give reasonable mechanisms for the reactions involved in each of the following classification tests or derivative preparations.

(*a*)   The alkaline hydrolysis of *N*-ethylbenzamide.

(*b*)   The reaction of *p*-cresol with bromine water.

(*c*)   The preparation of the 2,4-dinitrophenylhydrazone of 2-butanone.

(*d*)   The reaction of ethyl acetate with hydroxylamine in alkaline solution.

(*e*)   The reaction of diethylamine with *p*-toluenesulfonyl chloride in aqueous sodium hydroxide.

(*f*)   The iodoform reaction of 2-butanone (give mechanisms for each reaction step involved).

(*g*)   The Lucas reaction of *t*-butyl alcohol.

(*h*)   The reaction of 2-chloro-2-methylbutane with ethanolic silver nitrate.

(*i*)   The reaction of 1-bromobutane with sodium iodide in acetone.

(*j*)   The formation of the 3,5-dinitrobenzoate of 2-butanol.

(*k*)   The formation of the methiodide of triethylamine.

(*l*)   The formation of the aryloxyacetic acid of *p*-cresol.

PART IV

Advanced Projects

The following research-type projects are intended for advanced or honors students and may require equipment or laboratory skills not possessed by the average student of organic chemistry. Many of the projects are open-ended and can lead to further investigations not suggested in the outlines. The student may have to make extensive use of the library; he or she should also exercise independent judgment in planning the best course of action. Although some of the projects will not lead directly to new or unique contributions to the sum of chemical knowledge, others could (at least potentially) yield hitherto unknown facts and new discoveries.

Each project is outlined only briefly to give the student considerable latitude in developing and pursuing a course of action. Each is provided with one or more key references—additional references can be located by using Appendix VIII. A record of all laboratory work should be kept in a bound notebook, and each project should be written up in a form acceptable for publication in a professional journal such as the *Journal of Organic Chemistry*.

Isolation and Testing of a Natural Growth Inhibitor

PROJECT 1

juglone

Key References:
J. Chem. Educ. 49, 436 (1972). Synthesis of juglone.
J. Chem. Educ. 50, 782 (1973). General information, isolation of juglone.
J. Chem. Educ. 54, 156 (1977). Isolation of juglone from walnut hulls.
K. Paech and M. V. Tracey, *Modern Methods of Plant Analysis,* Vol. 3 (Berlin: Springer-Verlag, 1955), pp. 565–625. Cultivation and biological testing of plants.

Juglone is a natural substance occurring in walnut trees that is known to "poison" some kinds of plants and inhibit their development. It has been speculated that juglone acts as a kind of chemical defense agent for walnut seedlings, favoring their growth at the expense of competing species.

Your major objectives are:

1. To isolate and purify a quantity of juglone from black-walnut hulls.

2. To prepare synthetic juglone and establish the identity of the natural with the synthetic compound by spectrometric methods.

3. To test the inhibitory activity of purified juglone against the growth of various species or varieties of plants.

Juglone is an effective growth inhibitor for many plants of the heath family (*Ericaceae*), so both wild and domesticated species of such plants could be tested. The use of wild plants may require special cultivation methods. Different varieties of tomatoes or other domestic plants could be tested to determine whether certain varieties are more resistant than others. Biological testing could be carried out on a cross-section of typical plants from other families to determine what characteristics the juglone-sensitive plants may have in common. The juglone molecule could also be modified chemically to see what effect, if any, the modifications have on its inhibitory properties. Possible modifications include introducing substituents on the aromatic ring and converting the hydroxyl group to an ester linkage.

Isolation and Analysis of an Essential Oil

PROJECT 2

The leaves, stems, and other parts of a plant can usually be steam distilled to yield appreciable quantities of an essential oil, which contains the plant's more volatile components and is largely responsible for its odor. The number and relative abundance of the components of such an oil can be determined by GLC. The major components can be isolated by fractional distillation, column chromatography, and other methods. Individual components can often be identified by spectrometric methods (using collections of standard spectra, when available) along with traditional qualitative analysis procedures such as the determination of physical properties and the preparation of derivatives.

Vacuum fractional distillation is recommended for oils that may decompose at atmospheric pressure.

Your major objectives are:

1. To identify and collect specimens of a plant growing wild in your geographical area.

2. To obtain an essential oil from the leaves and stems (or other aromatic parts) of the plant by steam distillation.

3. To analyze the essential oil by GLC.

4. To isolate and identify at least one (preferably more) component of the essential oil.

In selecting a plant species to analyze, keep in mind, first, that the plant must be abundant enough to provide the

Key References:
E. Guenther, *Essential Oils* (3 Volumes) (New York: Van Nostrand, 1948). Isolation methods, sources, and properties of essential oils.

K. Paech and M. V. Tracey, *Modern Methods of Plant Analysis* (Berlin: Springer-Verlag, 1955). Methods for the isolation and analysis of plant constituents.

M. Grieve, *A Modern Herbal* (New York: Dover, 1971). Species-by-species description of medicinal plants and their uses.

B. Angier, *Field Guide to Medicinal Wild Plants* (Harrisburg, PA: Stackpole Books, 1978) Guide to medicinal plants in the U.S. (see also the field guides listed for Project 3).

large amounts of plant material that may be needed, and, second, that plants containing appreciable quantities of essential oil will ordinarily have a pronounced odor. It should be especially interesting to select a plant whose volatile oil is known to possess medicinal properties and to attempt to identify the active component or components of the oil. Many such plants are described in the references by Grieve and Angier. If the essential oil is too complex or does not have any very abundant constituents, it may be extremely difficult to separate and identify its components unless special analytical tools (GC/IR, GC/MS, HPLC, etc.) are available. In any event, it may be wise to screen a number of different species by gas chromatography before selecting one that has a comparatively simple mixture of constituents.

Most of the commonly encountered essential oils are described in the reference by Guenther; however, this comprehensive work is over thirty years old and some of its information is out of date. You should plan on searching through *Chemical Abstracts* to determine whether the chemical makeup of any given plant has already been reported in the chemical literature. It is clearly more interesting and challenging to analyze a new plant than to repeat a previous analysis, but your project can also be designed to verify the presence of previously characterized constituents in a species under study.

Isolation and Identification of Pigments from Wild Plants

PROJECT 3

Key References:
T. W. Goodwin, *Chemistry and Biochemistry of Plant Pigments* (London: Academic Press, 2nd ed., 1976). Isolation and analysis methods.

E. Gibbons, *Stalking the Healthful Herbs* (New York: David McKay, 1966). Nutritional aspects of wild plant usage.

E. Gibbons and G. Tucker, *Euell Gibbons' Handbook of Edible Wild Plants* (Virginia Beach, VA: Donning, 1979). Comprehensive listing of edible wild plants of the U.S.

Carotenoids and other natural plant pigments can be isolated from plant material by solvent extraction and column chromatography and identified by spectrometric methods. Although the pigments in tomatoes, carrots, spinach leaves, and many other domesticated plants have been analyzed thoroughly, the same cannot be said for many common wild plants. Carotene content, in particular, has nutritional significance since the carotenes are Vitamin A precursors. Some edible wild plants such as curled dock (*Rumex crispa*) are said to contain more "Vitamin A" than carrots.

Your major objectives are:

1.   To identify and collect specimens of one or more edible wild plants.

2.   To isolate and identify as many as possible of the pigments present in the leaves or other edible parts of the plant.

3.   To determine the amount of β-carotene and other Vitamin A precursors as a fraction of the dry weight of the plant matter, and convert this to equivalent Vitamin A content.

General methods for isolating and analyzing plant pigments are described in the Goodwin reference. Carotenoid pigments are always found in the leaves of plants; they may also occur in brightly colored fleshy fruits or in the roots of some plants. For example, the very common "Queen Anne's lace" (*Daucus carota*) is simply a wild version of the domestic carrot. It might be of interest to compare the carotene content of its root with that of a domestic carrot at the same stage of growth.

Many other possibilities should suggest themselves after you have scanned one or both of the Gibbons references. The field guides by Peterson and Kirk will help you identify the relevant species in your area.

Lee Peterson, *A Field Guide to Edible Wild Plants of Eastern and Central North America* (Boston: Houghton Mifflin, 1978). Guide to the identification and uses of wild plants.

D. R. Kirk, *Wild Edible Plants of the Western United States* (Healdsburg, CA: Naturegraph Publishers, 1970). Field guide for plants of the Western U.S.

The Effect of
Molecular Modification on Odor   PROJECT 4

Slight changes in molecular structure can have drastic (and sometimes unpredictable) effects on the odor of a substance. *p*-Hydroxybenzaldehyde is described in *The Merck Index* as having a "slight, agreeable, aromatic odor." Its two functional groups and benzene ring make possible a number of interesting molecular modifications. Aldehyde functions can be oxidized to yield carboxylic acids, reduced to form alcohols, condensed with active hydrogen compounds, treated with Grignard reagents, etc.; and the resulting functional groups can, in turn, be altered to form esters and many other derivatives. Phenolic hydroxyl groups can be converted to ester and ether linkages. The benzene ring itself is subject to a wide variety of electrophilic substitution reactions. For example, nitration of the ring followed by reduction and

p-hydroxybenzaldehyde

Key References:
R. W. Moncrieff, *The Chemical Senses* (London: Leonard Hill, 1951). Treatise on the senses of taste, smell, and vision.

R. B. Wagner and H. D. Zook,
Synthetic Organic Chemistry (New
York: Wiley, 1953). Guide to tradi-
tional synthetic methods.

diazotization should allow the synthesis of a number of
1,2,4-trisubstituted benzene derivatives.

Your major objectives are:

1.   To prepare, starting with *p*-hydroxybenzaldehyde,
as many related compounds as you have time for, using the
reaction types indicated above or any other reactions that
suggest themselves.

2.   To obtain sufficient analytical data, including spectra and GLC chromatograms where feasible, to establish the
identity and purity of each product.

3.   To observe and characterize the odor of each product prepared, preferably with the help of a panel of students
or other volunteers.

Bear in mind that the presence of two functional groups
may cause complications in certain reactions unless one of
them is "masked" or otherwise altered. In a sodium borohydride reduction of the aldehyde functional group, for example, the hydroxyl group will decompose the reducing agent
unless $-OH$ is converted to $-O^-Na^+$ by running the reaction in basic solution. Oxidation of the carbonyl group will
also cause oxidation of the hydroxyl group unless the latter
is first converted to an ester or ether functional group. Accordingly, you must plan each synthesis carefully in order to
optimize the yield of the desired product—or, in some
cases, to get any product at all. General synthetic procedures
can be found in many of the references on organic synthesis
listed in the Bibliography. A search of the chemical literature, using Wagner and Zook as a guide, may yield procedures for specific conversions as well. The Moncrieff reference provides considerable information, with specific
examples, about the effect of molecular structure on odor.
     You may wish to prepare a "perfume" from one or
more of your products by making up dilute solutions in ethanol and adding a fixative. Since some of your products may
be irritants or have other undesirable properties, it may be
inadvisable to test them on the skin. Such testing, in any
event, should only be carried out using very small quantities
of the chemical.

See Appendix VIII and section B of
the Bibliography.

Stereochemistry of a Hydride Reduction of 4-*t*-Butylcyclohexanone

Reductions and Grignard reactions of 4-*t*-butylcyclo-hexanone have been extensively studied to help elucidate the stereochemistry of carbonyl addition reactions. Since the bulky *t*-butyl group effectively locks the molecule into one conformation (that in which *t*-butyl is equatorial), the conformation of the hydroxyl group in the product should be the same as its developing conformation in the transition state leading to the product. Both "steric approach control" and "product development control" concepts have been used to rationalize the results of such reactions as lithium aluminum hydride reduction, which yields mostly *equatorial* alcohol, and Grignard additions, which result in predominantly *axial* OH groups. A comparatively new reducing agent, sodium

Key References:
E. J. Goller, "Stereochemistry of Carbonyl Addition Reactions," *J. Chem. Educ. 51*, 182 (1974).

A. S. Kushner and T. Vaccariello, "Use of a New Chemical Reducing Agent in the Undergraduate Organic Laboratory," *J. Chem. Educ. 50*, 154 (1973).

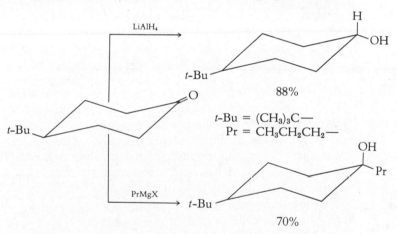

t-Bu = $(CH_3)_3C$—
Pr = $CH_3CH_2CH_2$—

88%

70%

bis (2-methoxyethoxy) aluminum hydride, is considerably safer and easier to handle than lithium aluminum hydride. It also contains two organic side chains that could conceivably affect the stereochemical outcome of a reduction reaction. Other new hydride reducing agents are discussed in source B–20 in the Bibliography.

$NaAlH_2(OCH_2CH_2OCH_3)_2$

sodium bis(2-methoxyethoxy)aluminum hydride

[also called Vitride, Red-al, and sodium dihydrido-bis(2-methoxyethoxy)aluminate]

Your major objectives are:

1. To carry out the reduction of 4-*t*-butylcyclohexanone using sodium bis(2-methoxyethoxy)aluminum hydride or some other "new" hydride reducing agent (not LiAlH$_4$ or NaBH$_4$) and analyze the product mixture by GLC.

2. To separate the axial and equatorial products by an appropriate separation method and identify each product by a spectroscopic method, so that the stereoisomer responsible for each GLC peak can be determined.

3. To compare the results with those obtained in other carbonyl addition reactions and attempt to explain them on the basis of the concepts discussed in the Goller article.

A study of the literature should reveal the most productive methods of separation and analysis. If time permits, additional reducing agents (or some Grignard reagents not mentioned in the literature) may be used as well.

Comparison of the Migratory Aptitudes of Aryl Groups

PROJECT 6

$$\begin{array}{c} \text{OH} \quad \text{R} \\ | \quad \quad | \\ \text{R}'-\text{C}-\text{C}-\text{R}' \\ \backslash \; / \\ \text{R}^+ \end{array}$$

$$2\text{Ph}-\overset{\text{O}}{\overset{\|}{\text{C}}}-\text{Ar} \longrightarrow \text{Ph}-\overset{\text{HO}}{\underset{\text{Ar}}{\overset{|}{\text{C}}}}-\overset{\text{OH}}{\underset{\text{Ar}}{\overset{|}{\text{C}}}}-\text{Ph}$$

Key Reference:
N. M. Zaczek, et al., "Migratory Aptitudes," *J. Chem. Educ. 48,* 257 (1971).

The pinacol rearrangement is a well-known reaction that involves the migration of an alkyl or aryl group, presumably by means of a bridged intermediate of the kind illustrated in the margin. Since the intermediate contains a positive charge distributed over the migrating group, migration rates should be enhanced by substituents that stabilize such a charge and retarded by substituents that destabilize it. By preparing different pinacols from aryl phenyl ketones and analyzing the products of their rearrangement, it should be possible to compare the migratory aptitudes of different aryl groups with that of an unsubstituted benzene ring.

Your major objectives are:

1. To prepare different aryl phenyl ketones by Friedel-Crafts reactions and convert them to pinacols.

2. To induce a pinacol rearrangement of each pinacol and analyze the product mixture by NMR or by cleavage and GLC (or both).

3. To compare the migratory aptitudes of the aryl groups in terms of some appropriate quantitative measure.

The key reference outlines the procedures used with *p*-chlorophenyl and *p*-toluyl migrating groups. You should extend the scope of that study by testing a wider range of substituents, perhaps using *ortho*- and *meta*-substituted benzene rings, different *para*-substituents, polynuclear aryl groups such as naphthyl, etc.

Investigation of the Relative Stability of Endocyclic and Exocyclic Double Bonds

1-Benzylcycloalkanols can be dehydrated to yield alkenes containing either an *endocyclic* or an *exocyclic* double bond. This is illustrated below by the dehydration of 1-benzylcyclopentanol, which yields about 75% of the *endo* product. Since the reaction is believed to involve thermodynamic control of the product mixture, it is evident that the alkene containing a double bond inside the ring is more stable, even though the *exo* double bond is stabilized by a phenyl group. Changing the ring size can alter the relative amounts of *endo* and *exo* products, but it does not appear to increase the

A reaction exhibits thermodynamic control when the composition of the product mixture is determined by the relative stabilities of the products, rather than by their relative rates of formation.

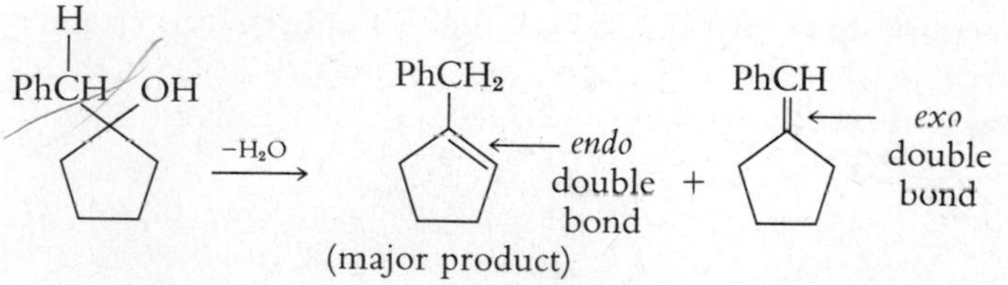

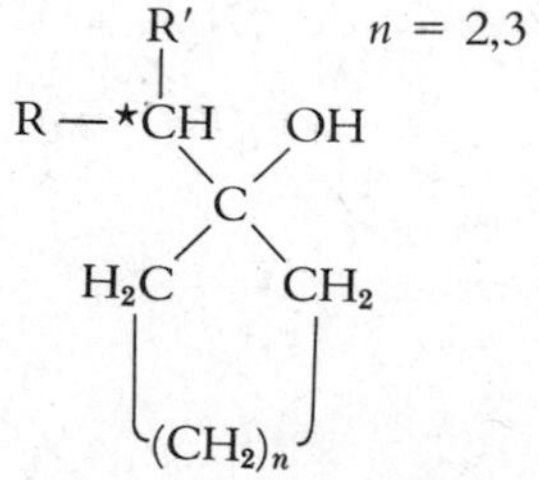

(major product)

amount of *exo* alkene to much over 25%. The use of different substituents on the starred carbon atom in structure **1** will presumably alter the *exo/endo* ratio. Some of them might even make the *exo* product the predominant one.

1

Your major objectives are:

1.   To prepare several 1-alkylcyclopentanols and 1-alkylcyclohexanols from the appropriate cyclic ketones and Grignard reagents.

2.   To dehydrate each of the cyclic alcohols using the same reaction conditions.

Key Reference:
G. R. Newkome, J. W. Allen, and
G. M. Anderson, "The Preparation and
Dehydration of 1-Benzylcycloal-
kanols," *J. Chem. Educ. 50*, 372
(1973).

*R and R' can be alkyl, aryl, or hydro-
gen. They can also contain functional
groups.*

3. To determine the percentage of *endo* and *exo* products in each reaction mixture using one or more appropriate analytical techniques.

4. To attempt to explain the observed variations in the *exo/endo* ratio with ring size and substituents.

In choosing the groups R and R' for incorporation into your cycloalkanols, consider the factors involved in stabilizing double bonds, and also the availability of starting materials for the corresponding Grignard reagents. The method used for analyzing the product mixture should establish the identity as well as the quantity of each component, so a GLC chromatogram alone is not sufficient unless the identity of the alkene peaks can be established by an independent method. The study can be extended by starting with small- or large-ring cyclic ketones other than cyclopentanone and cyclohexanone.

PROJECT 8

Synthesis and Structure Proof of an Insect Pheromone

Key Reference:
Naturwissenschaften 59, 469 (1972).
Article on isolation of the pheromone, in English.

The sex pheromone of the Douglas-fir beetle was recently found to be 3-methyl-2-cyclohexene-1-ol. One way of establishing the structure of a natural product with certainty is to prepare the substance by an independent synthesis and compare the properties of natural and synthetic chemicals.

Your major objectives are:

1. To prepare an authentic sample of 3-methyl-2-cyclohexene-1-ol starting with cyclohexanol.

2. To establish the structure and purity of your synthetic product by GLC and spectral analysis, along with any appropriate chemical tests.

3. To accurately measure the physical properties of your preparation and compare them with literature values.

Your synthetic pathway should be carefully planned to afford a reasonable yield of the final product. You must consider the possibility that one or more synthetic steps may yield a mixture of products requiring careful separation of the desired compound. If you can capture some Douglas-fir beetles, you may wish to observe the effect of your synthetic pheromone on their behavior.

Preparation of a
New Organic Compound

When Baeyer proved the structure of α-pinene, he started by oxidizing it to pinonic acid with potassium permanganate. Recently this reaction has been shown to proceed in very high yield with "purple benzene," a solution of potassium permanganate in benzene containing a crown ether. Pinonic acid is a relatively uncommon compound containing two reactive functional groups. It should therefore be possible, using it as the starting material, to prepare a totally new organic compound—one that has never been reported in the chemical literature.

pinonic acid

Crown ethers are toxic and have been reported to cause testicular atrophy in males.

WARNING

Your major objectives are:

1. To synthesize pinonic acid from α-pinene using "purple toluene."

2. To plan the synthesis of a new organic compound from pinonic acid, and conduct a literature search to establish that the compound has not been reported in the literature.

3. To carry out the synthesis of the compound and establish its structure by spectrometric and chemical methods.

4. To characterize the new compound as thoroughly as possible by measuring its physical properties, obtaining spectra, observing its chemical properties, etc.

Your literature search should encompass both *Beilstein's Handbuch* and *Chemical Abstracts*. Since nomenclature may vary considerably from year to year, the formula indexes of these references must be consulted. Since crown ethers are quite expensive, you should try to make do with

The oxidation can also be carried out in aqueous solution without the crown ether, or by using a phase-transfer catalyst as in Experiment 29.

The new compound should not be a simple functional group derivative (such as a phenylhydrazone or ester), of pinonic acid itself. Its preparation should require two or more synthetic steps and should be approved by your instructor beforehand.

Key Reference:
D. J. Sam and H. F. Simmons, "Crown Polyether Chemistry. Potassium Permanganate Oxidations in Benzene," *J. Am. Chem. Soc. 94,* 4024 (1974).

See Appendix VIII and section B of the Bibliography for sources of synthetic methods.

only a few grams of pinonic acid. This means that every reaction step in your synthesis must be carefully planned to provide the maximum yield of pure product. Keep in mind that some reagents may react with both functional groups and may even cause ring cleavage or other unanticipated reactions. It may be necessary to use a protective group for one function (such as ethylene ketal for the ketone function) while carrying out a reaction on the other. The final product should be obtained in as pure a form as possible, so that the physical and spectral data obtained would be suitable for submission to a professional journal.

PART V

The Operations

Elementary Operations

Cleaning and Drying Laboratory Glassware

Cleaning Glassware

Always clean any dirty glassware at (or before) the end of each laboratory period.

See ⟨OP–2⟩ for methods for dealing with frozen joints.

Used glassware that has been scratched or etched may not wet evenly.

A nylon scrubbing cloth is useful for cleaning spatulas, stirring rods, large beakers, and the outer surfaces of other glassware. Pipe cleaners are ideal for cleaning narrow funnel stems, eyedroppers, etc.

Use organic solvents sparingly and recycle them, as they are much more costly than water. Never use reagent grade acetone for washing; keep wash acetone separate from other grades.

Clean glassware is essential for satisfactory results in the organic chemistry laboratory, since small amounts of impurities can inhibit chemical reactions, catalyze undesirable side-reactions, and invalidate the results of chemical tests or rate studies. It is important to clean glassware as soon as possible after its use. Otherwise, residues may harden and become more intractable to cleaning agents; they may also attack the glass itself, weakening it and making future cleaning more difficult. It is particularly important to wash out strong bases like sodium hydroxide or sodium methoxide quickly, since these substances can etch the glass permanently and cause "frozen" joints after prolonged contact. When glassware has been thoroughly cleaned, water applied to its inner surface should wet the whole surface and not form droplets or leave dry patches.

Most glassware can be cleaned adequately by vigorous scrubbing with hot water and soap (or a good laboratory detergent), using a brush of appropriate size and shape to reach otherwise inaccessible spots. A plastic trough or another suitable container can serve as a "dishpan." Organic residues that do not come clean with detergent and water will often dissolve in organic solvents such as acetone or methylene chloride, followed by detergent and water. An acetone rinse can be used to remove excess methylene chloride (or another chlorinated solvent) before scrubbing with soap and water. After washing, always rinse the glassware thoroughly with water (a final distilled-water rinse is a good idea) and check it to make sure that the water wets its surface evenly. If it doesn't pass the "wettability" test, scrub it some more or use a cleaning solution.

Very troublesome residues can usually be removed with a chromic acid cleaning solution (*Hazard!*), but only after as much organic matter as possible has been removed by scrubbing or scraping. Put a few drops of the solution

Commercial cleaning solutions that do not contain chromic acid are available.

HAZARD

Chromic acid cleaning solutions cause severe burns; chromic acid may be carcinogenic. The solutions may react violently with some organic residues. Wear rubber gloves; avoid contact with skin, eyes, or clothing; wash up spills immediately.

into the *dry* piece of glassware to see that there is no violent reaction, then add a few milliliters and rotate, shake, or otherwise manipulate the glass object to distribute the solution evenly. After allowing the object to stand a short while, pour out the excess chromic acid (it can be recycled for as long as the original amber-brown color remains) and rinse the glassware with a little water (*Care:* heat will be generated). Then scrub it with soap and water and rinse thoroughly with water.

Drying Glassware

The easiest (and cheapest) way to dry glassware is to let it stand overnight in a position that allows easy drainage. The outer surfaces of glassware may be dried with a soft cloth or chamois, if necessary. However, the surfaces that will be in contact with chemicals should *not* be dried in this way, because of the possibility of contamination. If a piece of glassware is needed shortly after washing, it can be drained of excess water and rinsed with one or two *small* portions of technical grade acetone, then dried in a gentle stream of *dry* air (see ⟨OP–22⟩) or with a length of glass tubing connected to an aspirator.

The acetone should then be recycled for washing.

Compressed air from an air line often contains pump oil and moisture.

Glassware that must be very dry for reactions run under anhydrous conditions can be dried in an oven, or by playing a "cool" bunsen burner flame over the outer surface of the assembled apparatus. When using a flame, be very careful to keep it away from flammable solvents, and avoid overheating any part of the apparatus. Protect the glassware from atmospheric moisture while it is cooling (use a drying tube ⟨OP–22*a*⟩ on the assembled apparatus) to keep water from condensing on its inner surfaces.

The burner oxygen control should be opened just far enough that the flame does not deposit soot on the glassware.

Using Standard Taper Glassware

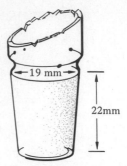

Figure 1.   19/22 Standard taper joint

Transparent joints of the clear-seal type need not be greased.

Each joint should be examined to see that the grease extends completely around the joint with no breaks.

Equipment and supplies:
Organic chemistry labkit (or equivalent), stopcock grease, ringstands, clamps; rings and wire-gauze squares as needed.

Most ground-joint glassware used in organic chemistry is of the straight, standard taper type with rigid joints. The size of a tapered joint is designated by two numbers (e.g., 19/22), in which the first number is the diameter, in millimeters, of the large end of the joint, and the second the length of the taper. Straight adapters are available that make it possible to use different joint sizes in the same assembly. Organic labkits are commercially available that can be assembled to perform many different operations, from refluxing to fractional distillation. The setups for these operations are usually illustrated in the instruction booklet that comes with each kit.

Lubricating Joints

A thin layer of stopcock grease should be applied to the joints of most standard-taper glassware to provide an airtight seal and prevent freezing. Hydrocarbon greases can be easily removed by acetone and other solvents, but because they are soluble in many organic liquids they may introduce undesired contaminants during such operations as distilling or refluxing. Silicone lubricants are suitable for most operations, particularly under conditions of high vacuum or high temperature.

Stopcock grease is applied only to the inner (male) joint of each component, by spreading a very thin coating on the top half of the joint. When the components are assembled, the joints are pressed firmly together with a slight twist (if necessary) to form a seal around the entire joint. In no case should excess grease extend beyond the joint inside the apparatus.

Assembling Glassware

Because standard-taper joints are rigid, apparatus must be assembled carefully to avoid excess strain that can result in breakage. First the necessary clamps and rings are placed at appropriate locations on the ringstands. The apparatus is then assembled *from the bottom up, starting at the heat source.* If a heating bath or mantle is to be used, it should be placed at a location from which it can be lowered and removed when the heating period is over. The reaction flask or boiling flask is then clamped securely at the proper distance from the heat source (see ⟨OP–7⟩). As other components are added they are clamped to the ringstands, but the clamp jaws are not tightened completely until all of the components are in place and aligned properly. When the clamps have been positioned so that all glass joints slide together without applying excessive force, the joints are seated and the clamps tightened carefully. Each joint is then examined for possible leaks or loose connections.

One should use as many clamps as are necessary to provide adequate support for all parts of the apparatus. A vertical setup such as the one for addition and reflux ⟨OP–10⟩ requires at least two clamps for security, because if the setup is bumped the clamp holding the reaction flask may rotate and deposit your glassware on the lab bench—usually with very expensive consequences. Some vertical components, such as Claisen adapters, need not be clamped if they are adequately supported by the component below; nonvertical components such as distilling condensers should generally be clamped, since they may be left unsupported and fall if the ringstands are accidentally moved apart.

For clamping condensers and other components at an angle to a ringstand, an adjustable clamp with a wing nut on the shaft is required. This wing nut must be tightened after the apparatus is aligned. Occasionally a strong rubber band can substitute for a clamp when there will be little force on a joint; the vacuum adapter, for example, can be fastened to a condenser by passing a rubber band around the tubulation on both. This adapter should never be allowed to hang unsupported, since stopcock grease cannot be relied on to hold joints together against the force of gravity.

Figure 2 on page 444 illustrates the steps followed in assembling a typical ground-glass apparatus.

Disassembling and Cleaning Glassware

Ground-joint glassware should be disassembled promptly after use, since joints left coupled for extended periods of time may freeze together and become difficult or impossible to separate without breakage. Frozen joints can sometimes be loosened by tapping the outer joint gently with the wooden end of a spatula or by applying steam to the outer joint while rotating it slowly. They are then separated by pulling the components apart with a twisting motion. If these procedures do not work, the instructor should be consulted.

If it will be necessary to lower a receiving flask (or other container) to add or remove substances, it can be supported by a ring and wire gauze or by a lab jack, blocks of wood, etc.

Spring clamps can also be used to hold glass joints together.

Alkalis can cause ground-glass joints to fuse together permanently.

After use, each joint should be wiped free of excess grease and cleaned with methylene chloride or another appropriate solvent. The glassware should always be cleaned thoroughly ⟨OP–1⟩ after use, since residues may otherwise dry and harden, becoming more difficult to remove and sometimes attacking the surface of the glass.

Steps

1. Position clamps, rings.

2. Position heat source.

3. Secure boiling flask (clamp tightly).

4, 5. Add Claisen, 3-way adapters.

6. Clamp condenser in place.

7. Attach vacuum adapter with rubber band or spring clamp.

8. Attach receiving flask, hold in place with ring and wire gauze.

9. Readjust all clamps to align.

10. Press joints together tightly.

11. Tighten clamps.

12. Add stopper.

13. Add thermometer adapter, position thermometer.

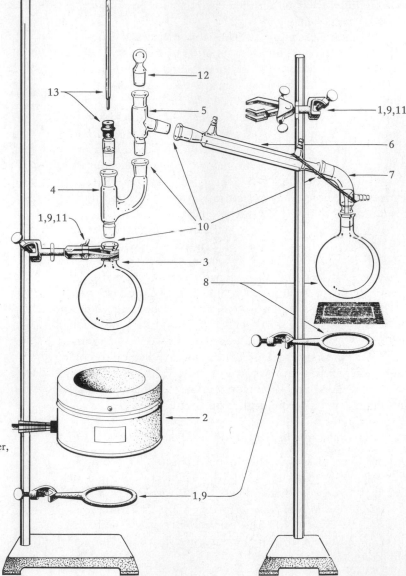

Figure 2.   Assembly of ground-glass apparatus

When clean and reasonably dry, each component should be returned to its proper location in the labkit or to the stockroom, as directed by the instructor.

Using Corks and Rubber Stoppers

Corks and rubber stoppers have been largely replaced by ground-glass joints for many operations such as refluxing and distillation, but they still have some important applications in organic chemistry. Rubber stoppers sometimes swell in contact with organic liquids and vapors and may contaminate organic solvents, but they provide a tighter seal than corks and should be used for vacuum filtration ⟨OP–12⟩ and other reduced-pressure operations. Corks should generally be used to stopper test tubes or flasks and in other applications not requiring airtight seals. Ground-glass stoppers should be used to stopper standard-taper glassware; materials such as Parafilm or aluminum foil are often preferred over corks for sealing flasks and other containers when contamination may be a problem.

Boring Corks

Before boring a hole in a cork, it is desirable to soften it either with a special cork softener or by rolling it between two hard, flat objects. The cork borer should be sharp (if it is not, the instructor will show you how to use a cork-borer sharpener) and should be slightly smaller in diameter than the object to be inserted in the cork. The borer shold *not* be pushed forcefully through the cork, but should be allowed to *cut* its way through by using a rotating motion and only a small amount of pressure. The borer and cork should be rotated in opposite directions and the alignment should be checked frequently to insure that the borer is going in straight. To avoid breaking off large chunks of cork, the borer should be removed when it is about halfway through and applied to the other end until the two holes meet.

Figure 1.   Cork borer

OH OH   OH
| | |
CH₂CH — CH₂

glycerol
(glycerine)

Boring Rubber Stoppers

Rubber stoppers are bored in about the same manner as corks, but the borer must be *very* sharp and a lubricant such as glycerol should be applied to the cutting edge to reduce friction. It is particularly important to avoid using excessive force in boring rubber stoppers, since trying to "punch" out a hole rather than cutting it out will result in a hole that is much smaller than the diameter of the cork borer. It is difficult to enlarge a hole already cut in a stopper, so if a large hole is required it should be made in a solid stopper and not in a stopper that already has one or more holes.

Inserting Glass Tubing and Thermometers

Glass tubing and thermometers can generally be inserted through corks and rubber stoppers without difficulty if a lubricant such as glycerol is first applied *generously* to the glass. The tubing is then grasped *close to the stopper* and *twisted* through the hole with firm, steady pressure. Holding the tube too far from the stopper is dangerous, since the glass may break and lacerate or puncture the hand. Holding the

WARNING *Improper insertion of glass tubing is one of the most frequent causes of laboratory accidents. The resulting cuts and puncture wounds can be very severe, requiring medical treatment and sometimes causing the victim to go into shock. Thermometers are particularly easy to break (especially at the 76-mm immersion line) and are expensive to replace.*

glass tubing in a towel or other piece of cloth offers partial protection to the hands. It is important that force be applied with gentle but firm pressure *directly along the axis* of the tubing; any sideways force may cause the tube to break. Forcing a tube through a hole too small for it can also cause breakage. After the tube or thermometer is correctly positioned in the stopper, the glycerol should be rinsed off thoroughly with water.

Removing Tubing

In most cases, a tube or thermometer can be removed by twisting it out of the stopper with a firm, continuous motion. Hold the tubing close to the stopper and avoid applying any sideways force that could cause it to break. A little glycerol applied to the part of the glass that will pass through the stoper can facilitate the process. If the glass tubing or thermometer cannot be removed easily by this method, the stopper can be cut off with a single-edged razor blade. Alternatively, a cork borer having a slightly larger diameter than the tube can be placed around it and twisted gently through the stopper until the glass can be dislodged.

Basic Glass Working

Glass connecting tubes and other simple glass items are required for many operations in organic chemistry. Soft-glass tubing, which is suitable for most applications of this type, can be worked easily with a Bunsen burner. Pyrex tubing may require the hotter flame provided by an oxygen torch or by some large burners.

Equipment and supplies:
Burner, flame spreader, glass tubing or glass rod.

Cutting Glass Rod and Tubing

Most glass rods and tubes can be "cut" by scoring them at the desired location and snapping them in two. The rod or tube is scored by drawing a sharp triangular file across the surface at a right angle to the axis of the tubing. Often only a single stroke is needed to make a deep scratch in the surface. The file should not be used like a saw, and the scratch should be on only one side of the glass. The scratch is moistened with water or saliva, then the rod or tube is broken by placing the thumbs about 1 cm apart on the side opposite the scratch and, while holding the glass firmly in both hands, pushing out quickly with the thumbs and pulling the hands apart simultaneously. A towel placed over the glass helps protect the hands.

Special glass-scoring knives are also available.

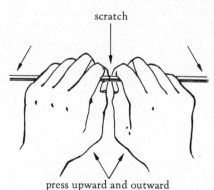

press upward and outward
Figure 1.   Breaking glass tubing

Working Glass Rod

The ends of a glass rod can be rounded by rotating them at an angle in a burner flame. An end can be flattened by rotating it in the flame until it is incandescent and very soft, then pressing the rod straight down onto a hard surface such as the base of a ringstand.

See ⟨OP–7⟩ for directions on the use of a burner.

WARNING   *Do not burn yourself on the hot end of the glass rod or tube you are working with. Do not lay hot glass directly on the desk top or onto combustible materials.*

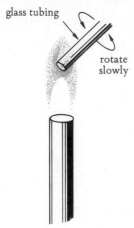

glass tubing

rotate slowly

Figure 2.   Fire polishing

CAUTION: *Acetone is flammable.*

A micro burner (or a Bunsen burner with the barrel removed) should be used to seal small-diameter tubing.

Experiment 24 utilizes sealed-tube reactions.

The flame will turn yellow as the glass begins to soften.

Fire Polishing and Sealing Glass Tubing

Cut glass tubing (and glass rod) should always be *fire polished* to remove sharp edges and prevent accidental cuts. The cut end of the tube is rotated slowly in a burner flame at about a 45° angle to the flame until the edge becomes rounded and smooth.

The end of a soft-glass tube can be closed by rotating it in a flame as for fire polishing until the soft edges come together and eventually merge. To obtain a sealed end of uniform thickness, remove the tube from the flame as soon as it is closed and blow gently into the open end. Occasionally a tiny hole will remain in the closed end. This can be detected by allowing the tube to cool to room temperature, attaching the open end to an aspirator, and placing the closed end in a test tube containing a small amount of acetone or methylene chloride. If the tube is not properly sealed, the liquid will leak into it when you apply suction.

Another way of sealing a glass tube is to heat it about 5 cm from one end, rotating it constantly, and draw the ends 3-5 cm apart when the glass is soft. The narrow neck remaining between them is then heated until the tubing separates, leaving a sealed tip on the end. The following method is often used to prepare completely sealed tubes for certain chemical reactions. A soft-glass tube is cut to length with about 5 cm to spare, then closed at one end. The reactants are placed inside and the tube is sealed in a flame by heating it about 5 cm from the open end and drawing the ends apart, as described above. If the reactants are volatile, the closed end should be immersed in an ice bath while the open end is being sealed.

Bending Glass Tubing

Unless a large-barreled burner (Meker burner, etc.) is available, a flame spreader should be used for all glass-bending operations. The tubing is held over the burner flame parallel to the long axis of the flame spreader and rotated constantly at a slow, even rate until it is soft enough to bend easily. (At this time, it should be nearly, but not quite, soft enough to

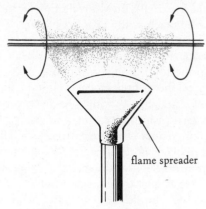

Figure 3.   Bending tubing

bend under its own weight). The hot tubing is then removed from the flame and immediately bent to the desired shape with a firm, even motion. If the glass is soft enough, it should not be necessary to use much force to bend it, and the bend should follow a smooth curve with no constrictions.

Bend the glass in the vertical plane with the ends up and the bend at the bottom.

Weighing

Many different kinds of balances are used in chemistry laboratories, including (*a*) triple-beam balances, which require manual adjustment of weights on each of three balance beams; (*b*) top-loading automatic balances; and (*c*) analytical balances, which are accurate to closer than 1 mg. Strictly speaking, such balances measure mass rather than weight; however, the word "weight" is commonly used for "mass" when this will not cause error or confusion. Your instructor will explain or demonstrate the operation of the balances used in your laboratory.

Mass is a measure of the quantity of matter in an object, whereas weight is a measure of the earth's gravitational attraction for the object.

Weighing Solids

Most solids can be weighed in glass containers (such as vials or beakers) or on special glossy weighing papers. Filter paper and other absorbent papers should not be used for accurate weighings, since a few particles will always remain in

Hygroscopic solids (those that absorb water from the atmosphere) should be weighed in closed containers.

the fibers of the paper. If you are weighing an indeterminate amount of a solid (such as the product of a reaction), its container or the weighing paper should first be weighed separately and the mass recorded (unless the balance has a taring system). Then the solid is added, the total mass is recorded, and the mass of the container or weighing paper is subtracted. If the balance has a taring system, the mass readout scale is set to zero (using the taring control) while the container or weighing paper is on the balance pan. Then the solid is added, and its mass is read directly from the scale.

If you are measuring out a specified quantity of a solid (such as a solid reactant), the expected total mass of the solid plus its container should be calculated and the balance should be set to approximately that value. Then the solid is added in small portions, using a clean scoop or spatula, until the desired reading is attained. If the balance has a taring system, its readout scale is zeroed with the container on the pan and is then set to the desired mass of the solid.

Most products obtained from a preparation are transferred to vials or other small containers, which should be weighed empty and then reweighed after the product has been added. As a general practice, the container should be weighed with its cap and label on and this **tare** weight recorded. Then the mass of the contents at any given time can be obtained by subtracting the tare weight from the total mass of container and contents.

See *J. Chem. Educ. 55,* 455 (1978) for a description of a convenient solids dispenser.

Allow a moistened label time to dry before weighing a labeled container.

tare: allowance for the weight of a container.

Weighing Liquids

The mass of an indeterminate quantity of a liquid is measured as described above for a solid. A tared container should be used, and it should be kept stoppered during the weighing to avoid loss by evaporation. To weigh out a specified quantity of a liquid from a reagent bottle, you should first measure out the approximate quantity of the liquid *by volume* and then weigh that quantity accurately in a closed container. For example, if 9.0 g of chloroform ($d \approx 1.5$ g/ml) is required for a synthesis, a little more than 6 ml (9.0 g $\div$ 1.5 g/ml) of the liquid is measured into a graduated cylinder and transferred to a tared vial. If the measured weight is not close enough to that required, liquid can be added or removed with a clean dropper. If the liquid is provided in dropper bottles, it can be transferred directly from the bottle to the container on the balance pan (*Care:* Do not get any on the balance) without pre-measuring.

Never withdraw liquid directly from a reagent bottle with your dropper or pipet— you may contaminate the liquid.

Calculate the volume from

$$V = \frac{\text{mass}}{\text{density}}$$

Clean up any spills in the vicinity of the balance immediately.

A given volume of a liquid can be measured using either a graduated cylinder, a pipet, or a syringe, depending upon the quantity and accuracy required. Burets and volumetric flasks are also used to measure liquid volumes accurately. Their use is discussed in most general and analytical chemistry laboratory manuals.

Graduated Cylinders

Graduated cylinders are not highly accurate, but they are adequate for measuring specified quantities of solvents and wash liquids as well as liquid reactants that are present in excess. The liquid volume should always be read from the bottom of the liquid meniscus.

Pipets

Graduated or volumetric pipets can be used to accurately measure relatively small quantities of a liquid. Suction is required to draw the liquid into a pipet; however, using mouth suction is unwise because of the danger of drawing

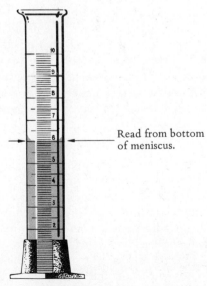

Read from bottom of meniscus.

The volume contained is 6.0 ml.
Figure 1.   Reading a graduated cylinder

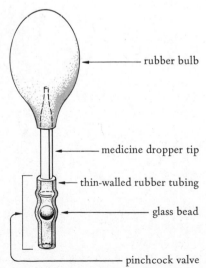

rubber bulb

medicine dropper tip

thin-walled rubber tubing

glass bead

pinchcock valve

Figure 2.   Pipetting bulb [*J. Chem. Educ. 51,* 467 (1974)].

See *J. Chem. Educ. 54*, 639 (1977) and *J. Chem. Educ. 55*, 38 (1978) for other types of pipetting bulbs.

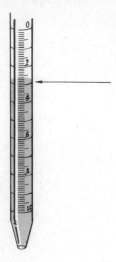

The pipet has delivered
0.30 ml of liquid.

Figure 3.   Graduated pipet

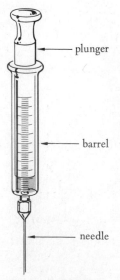

Figure 4.   Syringe

toxic or corrosive liquids into the mouth. An ordinary ear syringe works quite well as a pipetting bulb. Another convenient pipetting-bulb assembly (illustrated in Figure 2) is operated as follows:

1.   The top end of the pipet is inserted into the pinchcock valve.

2.   The pinchcock is opened by pinching it at the glass bead and the bulb squeezed to eject the air.

3.   The pipet tip is placed in the liquid and the pinchcock is squeezed open to fill the pipet to just above the calibration mark.

4.   The bulb is removed from the narrow end of the dropper and the pinchcock is carefully opened until the liquid falls to the calibration mark.

5.   The liquid is then delivered into another container by opening or detaching the pinchcock valve.

Most volumetric pipets are calibrated "to deliver" (TD) a given volume, meaning that the measured liquid is allowed to drain out by gravity, leaving a small amount of liquid in the bottom of the pipet. This liquid is not removed, since it is accounted for in the calibration. Graduated pipets are generally filled to the top (zero) calibration mark and then drained into a separate container until the calibration mark for the desired volume is reached. The remaining liquid is either discarded or returned to its original container. The maximum indicated capacity of some graduated pipets is delivered by draining to a given calibration mark and of others by draining completely. It is important not to confuse the two, since draining the first type completely will deliver a greater volume than the indicated capacity of the pipet.

Syringes

Syringes are most often used for the precise measurement and delivery of very small volumes of liquid, as in gas-chromatographic analysis ⟨OP–32⟩. A syringe is filled by placing the needle in the liquid and slowly pulling out the plunger until the barrel contains a little more than the required volume of liquid. Then the syringe is held with the needle pointed up and the plunger is pushed in to eject the excess sample. Excess liquid is wiped off the needle with a tissue.

Syringes should be cleaned immediately after use by rinsing them several times with a volatile solvent, then re-

vacuum release valve
pinchclamp

syringe barrel

medicine dropper
bulb

glass tube

filter flask

to aspirator

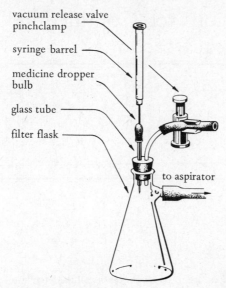

Figure 5.   Apparatus for drying microsyringes

moving the plunger and letting the barrel dry. Microsyringes can be dried rapidly by aspiration using the apparatus illustrated in Figure 5. The needle is inserted carefully
through the dropper bulb and the aspirator is turned on for a
minute or so. The pinchclamp is then opened to release the
vacuum, the aspirator is turned off, and the syringe is removed.

Operations for Conducting Chemical Reactions

Heating

Heat Sources

Many kinds of heat sources are available for such applications as heating reaction mixtures, evaporating volatile solvents, and carrying out distillations. The choice of a heat source for a particular application depends on such factors as the temperature required, the flammability of a liquid being heated, the need for simultaneous stirring ⟨OP–9⟩, and the cost and convenience of the heating device.

Burners. Bunsen burners and similar heat sources are simple and convenient to operate and can bring a liquid rapidly to its boiling point; however, they have some serious disadvantages that limit their use. It is difficult to control the temperature of a burner flame precisely, although rough adjustments can be made by varying its distance from the vessel being heated and its air/gas mix. The temperature at different parts of the vessel can vary widely; generally it is hottest at the bottom. The resultant local superheating can cause reaction mixtures to decompose or undergo unwanted side reactions.

Because of the high risk of a fire, burners should never be used to heat flammable liquids in open containers. Some flammable liquids with relatively high boiling points can be safely heated under reflux or in distillation setups; however, all joints must be tight, water condensers must be working properly, and the flame must be extinguished when the liquid is being introduced into, or removed from, the apparatus. The experimenter must also be certain that other students will not be handling flammable solvents nearby while the burner is in use.

Never leave a burner flame unattended; it may change in intensity, or even go out and possibly cause an explosion.

When refluxing or distilling liquid mixtures, a burner should be used with an asbestos-centered wire gauze to spread the flame and promote even heating. The gauze is placed on a ring about 2–3 cm from the barrel of the burner and the reaction flask is clamped to the ringstand with its bottom *just* above the center of the gauze. The gas valve (usually on the bottom of the burner) is opened if necessary and the burner ignited, then the flame is adjusted by rotating the barrel counter-clockwise for a hotter, more oxygen-rich flame or clockwise for a cooler flame. The size of the flame can be controlled by readjusting the gas valve and oxygen control in sequence.

Heating a flask directly can cause strains that result in breakage.

Local superheating will be reduced if the flask is not actually touching the gauze.

Steam Baths. A steam bath is a safe, convenient heat source that is limited by the fact that it has only one operating temperature, 100°C. It is particularly useful for heating recrystallization mixtures, evaporating volatile solvents, and refluxing low-boiling liquids. The condensation of steam in the vicinity of a steam bath may be a nuisance, but this can be reduced by using enough rings to bridge any gaps between a flask and the steam bath, and by maintaining a slow rate of steam flow. Beyond a certain point there is no advantage to increasing the steam-flow rate, since the temperature is constant. If the flask is placed correctly, heating is comparatively even and efficient, and the low operating temperature helps prevent the decomposition of heat-sensitive substances.

When used to heat a reaction mixture under reflux, the steam bath should be positioned on a lab jack, a set of wood blocks, or some other support so that it can be lowered and removed when the heating period is over. Rubber tubing is used to connect the steam line to the steam inlet and to extend from the water outlet to a sink for drainage. The boiling flask and its attached apparatus should be securely clamped to a ringstand. The concentric rings on top of the bath are removed until the flask fits quite snugly in the bath with its liquid level below the level of the rings. The steam valve is first opened fully to purge the steam line of water, then adjusted to provide the desired rate of heating.

Flat-bottomed containers like beakers and Erlenmeyer flasks can be heated on top of the steam bath by leaving just enough rings to support them safely. If a low rate of heating is desired, more rings can be left on the steam bath to reduce the area of direct contact between the steam and the container. If a higher heating rate is needed, enough rings can be re-

steam inlet

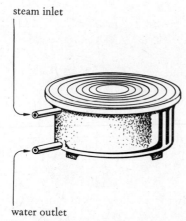

water outlet

Figure 1. Steam bath

See *J. Chem. Educ.* 52, 348 (1975) for construction of an inexpensive lab jack.

The steam bath should be drained of water before and after use.

moved so that the flask or beaker can be clamped inside the steam bath, with its bottom below the level of the rings.

Heating Mantles.   A heating mantle is perhaps the most satisfactory laboratory device yet developed for heating over a wide temperature range. Unlike a Bunsen burner, a heating mantle can be used with magnetic stirrers and its heat output can be controlled precisely. Unfortunately, most heating mantles are costly, take a long time to warm up, and accommodate only a narrow range of flask sizes. It is also difficult to monitor the operating temperature of a heating mantle. Despite these disadvantages, heating mantles are highly recommended for most heating operations.

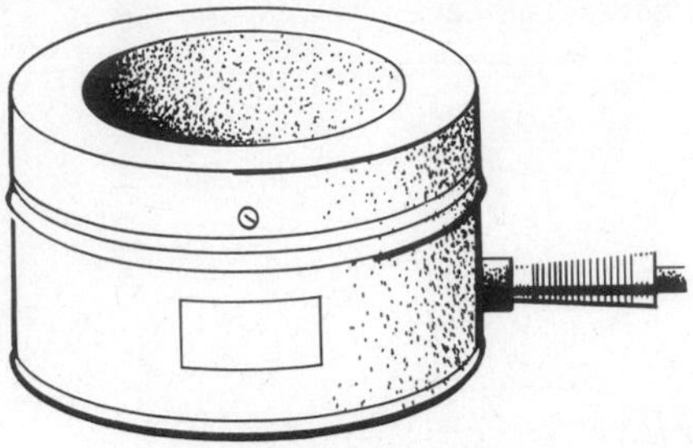

A heating mantle must be used in conjunction with a variable transformer or a time-cycling heat control to regulate the heat output. Since the temperature of the mantle itself can be measured only by a thermocouple, it is difficult or inconvenient to set a heating mantle for operation at a particular temperature. A mantle will generally not be at thermal equilibrium with the contents of a flask, so if the flask is not filled to a level near the top of the mantle, the part of the flask above the liquid level will be hotter than its contents and can cause decomposition of materials splashed onto it. When possible, the mantle should have a well of nearly the same diameter as the flask being heated. Some kinds of all-purpose mantles are intended for operation with a range of flask sizes; however, heating efficiency is reduced and the chance of superheating is increased when a small flask is heated in a large mantle.

The mantle is mounted on a lab jack, ring, set of wood blocks, or some other support so that it can be lowered and removed quickly if the rate of heating becomes too rapid. The flask is clamped in place so that it is in direct contact with the heating well, and the heating control dial is adjusted until the desired rate of heating is attained. Because a heating mantle responds slowly to changes in the control setting, it is easy to overshoot the desired temperature by turning the control too high at the start. If this occurs, the mantle should be lowered so that it is no longer in contact with the flask. The voltage input should then be reduced and the mantle allowed to cool down. Further adjustment may be required to maintain heating at the desired rate.

Oil Baths.   A good oil bath is capable of uniform heating and precise temperature control. Unlike the preceding heat

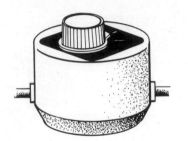

Figure 2.   Heating mantle

Figure 3.   Heat control for mantle

A 250-ml Briskeat beaker heater with integral heat control works passably for 250-ml boiling flasks and quite well for 100-ml and smaller flasks. Its heating rate can be increased by lowering the flask in the well and by using glass wool for insulation.

Never heat an empty flask; this may burn out the heating element.

Hydrocarbon oil fires are most easily extinguished by dry-chemical fire extinguishers or powdered sodium bicarbonate.

sources, an oil bath can operate at thermal equilibrium with the contents of a reaction flask. Accordingly, decomposition and side reactions caused by local overheating are less likely, and the reaction temperature can be determined by suspending a thermometer in the bath liquid.

The main disadvantages of most oil baths are that they are messy to work with, difficult to clean, and can cause dangerous fires or severe burns. It is best not to heat an oil bath above the *flash point* of the heating oil (about 190° for ordinary mineral oil), since above this point the oil can suddenly burst into flames with a spark. Old, dark oil is more likely to flash than new, and most bath oils will start to smoke and decompose at elevated temperatures. Any hot bath oil can cause severe injury if accidentally spilled on the skin—the oil, being difficult to remove and slow to cool, remains in contact with the skin long enough to produce deep, painful burns. Water should be kept away from hot oil baths since it causes dangerous splattering; oil that contains water should not be used until the water is removed. An oil bath that is dark and contains gummy residues should be replaced.

Some commonly used bath liquids are listed in the margin. Mineral oil is useful for most applications, although it presents a potential fire hazard and is hard to clean up. Polyethylene glycol bath oils such as Carbowax 600 are water-soluble (which makes cleanup much easier) and can be used at relatively high temperatures without appreciable decomposition. Silicone oils can be used at even higher temperatures.

Oil baths can be heated by a coil of resistance wire, a power resistor, a Calrod heating element, or some other device that can be safely immersed in the bath liquid. External heat sources, such as burners and hot plates, are less satisfactory—particularly burners, because of the fire hazard. The output of the heating element is controlled by a variable transformer and the temperature of the bath is measured ⟨OP–7c⟩ with a thermometer suspended in the liquid. A porcelain casserole dish makes a convenient bath container, since it is less easily broken than a glass container and has a handle for convenient placement and removal.

The oil bath must be positioned so that it can easily be lowered and removed from the vessel being heated. A lab jack or a set of wood blocks is better than a ring support, since hot oil may spill when the ring is raised or lowered if this is not done with great care. The flask to be heated is placed in the bath so that the liquid level inside it is just below the oil level. The heating control unit is turned up

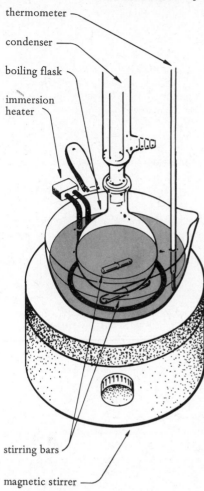

thermometer
condenser
boiling flask
immersion heater
stirring bars
magnetic stirrer

Figure 4.   Oil bath assembly

Useful heating bath liquids:
  Mineral oil
    b.p. 360°
    flash pt. 193°
    Fire hazard, smokes at about 150°
    and above.
  Glycerol
    b.p. 290°
    flash pt. 160°
    Water soluble, nontoxic, viscous.
  Dibutyl phthalate
    b.p. 340°
    flash pt. 157°
    Viscous at low temperatures.

Triethylene glycol
  b.p. 276°
  flash pt. 177°
  Water soluble.
Polyethylene glycols (Carbowaxes)
  b.p. and flash points vary, depending on M. W. range. Some are solid at room temperature.
  Water soluble.
Dow Corning silicone oils
  DC 330: flash pt. 290°
  DC 550: flash pt. 310°
  Expensive. Decomposition products are very toxic.

The bath should be set directly on a magnetic stirring unit when the bath liquid or reaction mixture must be stirred (see ⟨OP–9⟩).

until the desired temperature is obtained, then adjusted to maintain that temperature. Very uniform heating is possible if the oil bath is stirred magnetically; a reaction mixture can be stirred simultaneously if desired, as illustrated in Figure 4.

Water Baths.   Water baths are useful for evaporating volatile solvents ⟨OP–14⟩ and in other applications requiring gentle, even heating. When precise temperature control is not essential, a water bath can simply be a large beaker or porcelain casserole filled with preheated water (from a hot water tap or another source), with its temperature adjusted by adding hot or cold water. As the bath cools, some of the bath water can be siphoned off and replaced by fresh hot water. When there is no significant fire hazard, a water bath may be heated with a burner or hot plate. Like an oil bath, a water bath can be stirred for more uniform heating and its temperature can be monitored by suspending a thermometer in the water. Commercial water baths with removable metal rings are available.

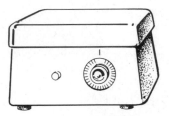

Figure 6.   Hot plate   Figure 5.   Metal water bath

Hot Plates.   A hot plate can be used to heat Erlenmeyer flasks or beakers, unless they contain low-boiling flammable liquids that could splatter on the hot surface and ignite. Hot plates can also, in some cases, be used to heat oil baths or water baths for reactions or distillations. A hot plate, however, should never be used to heat a round-bottomed flask directly, since local overheating and breakage may result. Hot plates with built-in magnetic stirrers are available.

Asbestos gloves or beaker tongs should be kept handy so that a vessel can be quickly removed from a hot plate if necessary.

Miscellaneous Heat Sources.   Other heating devices such as infrared heat lamps, electric forced-air heaters (heat guns), and electrically heated "air baths" can be used in some heating applications. A heat lamp plugged into a variable transformer provides a safe and convenient way to heat comparatively low-boiling liquids. The boiling flask is usually fitted

Figure 7.   Heat lamp

with an aluminum-foil heat shield to concentrate the heat on the reaction mixture.

Smooth Boiling Devices

When liquids are heated to their boiling points or above, they may erupt violently as large bubbles of superheated vapor are discharged from the solution. A porous object such as a boiling chip or boiling stick can prevent this "bumping" by emitting a steady stream of small bubbles that breaks up the large vapor bubbles. Boiling chips (or boiling stones) are generally made from pieces of pumice, carborundum, marble, or glass. Acid- or base-resistant boiling chips (carborundum, for example) should be used when heating strongly acidic or alkaline mixtures, since ordinary boiling chips may break down in such mixtures.

J. Chem. Educ. 53, 50 (1976) and *J. Chem. Educ. 54,* 611 (1977) describe the construction and use of a "boiling tube" to prevent bumping.

Microporous boiling chips, made from a special grade of anthracite, can be used when liquids are distilled under reduced pressure ⟨OP–26⟩. Wooden applicator sticks can be broken in two and the broken ends used to promote smooth boiling in nonreactive solvents; they should not be used in reaction mixtures because of the possibility of contamination.

It is important to add boiling chips *before* heating begins, since a liquid may froth violently and boil over if they are added when the liquid is hot. When a liquid is cooled below its boiling point and then reheated, a new boiling chip or two should be added before heating is resumed.

When boiling stops, liquid is drawn into the pores of a boiling chip, making it less efficient.

Stirring causes turbulence that breaks up large bubbles, so boiling chips are not needed when a liquid is stirred constantly at the boiling point. Unless you are instructed differently, you should *always* add boiling chips to an unstirred reaction mixture or a liquid being distilled.

Refluxing

Heating accelerates the rate of a chemical reaction by increasing the average kinetic energy of the molecules, resulting in a larger fraction of molecules that have sufficient energy to react at a given instant. The temperature of a reaction mixture can be controlled in several ways, the simplest and most convenient being to use a reaction solvent that has a boiling point within the optimum temperature range for the reaction. The reaction is then conducted at the boiling point of the solvent, using a water-cooled condenser to return the solvent vapors to the reaction vessel and prevent their escape. This process of boiling a reaction mixture and condensing the solvent vapors back into the reaction vessel is known as *refluxing* or heating under reflux. It is the most widely used technique for carrying out organic reactions at elevated temperatures.

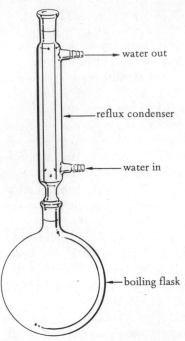

Figure 1.   Simple reflux assembly

Operational Procedure

Equipment and supplies:
Round-bottomed flask, reflux condenser, boiling chips, two lengths of rubber tubing (about 1 m each), heat source.

See ⟨OP–9⟩ if the reaction mixture must be stirred while refluxing, and ⟨OP–22a⟩ or ⟨OP–22b⟩ if a dry atmosphere or a gas trap is required.

Position the heating device ⟨OP–7⟩ at the proper location on or near a ringstand so that it can be quickly removed if the flask should break or the reaction become too vigorous. Select a round-bottomed flask of the right size (the reactants should fill it about half full, or a little less) and clamp it onto the ringstand at the proper location in relation to the heat source (see ⟨OP–7⟩). Solids should be added to the flask through a powder funnel or a makeshift "funnel" constructed from a piece of glazed paper; liquids should be added through a stemmed funnel of some kind. (This keeps grease on the ground joint from contaminating the reactants.) Mix the reactants by swirling or stirring and add a few boiling chips, then insert a reflux condenser into the flask and clamp it to the ringstand.

A clamp on the condenser keeps the apparatus from toppling over if jarred. Be sure the ground-glass joints don't separate when the clamp is tightened.

Never begin heating before the condenser water is turned on; solvent vapors may escape and cause a fire or health hazard.

WARNING

Connect the water inlet (the lower connector) on the condenser jacket to a cold-water tap with a length of rubber tubing and run another length of tubing from the water outlet (the upper connector) to a sink, making sure that it is long enough to prevent splashing when the water is turned on. Turn on the water carefully so that the condenser jacket slowly fills with water from the bottom up, and adjust the water pressure so that a narrow stream flows from the outlet. The flow rate should be just great enough to **(a)** maintain a continuous flow of water in spite of pressure changes in the water line, and **(b)** keep the condenser at the temperature of the tap water during the reaction.

If the rubber tubing slips off when pulled with moderate force, replace it by smaller diameter tubing or secure it with wire.

High water pressure may force the tubing off the condenser or cause an accidental "flood."

Turn on the heat source and adjust it to maintain gentle boiling of the solvent. A continuous stream of bubbles should emerge from the liquid, but bumping and excessive foaming should be prevented. The vapors passing into the condenser will form a *reflux ring* of condensate that should be clearly visible. Below this point, solvent will be seen flowing back into the flask; above it, the condenser should be dry. If the reflux ring rises more than halfway up the condenser, the heating rate should be reduced to prevent the escape of solvent vapors.

If the reflux ring is too high, it may be necessary to change to a more efficient condenser or use two condensers in tandem.

At the end of the reaction period, turn off the heat source and remove it from contact with the flask. Let the apparatus cool, then turn off the condenser water and pour the reactants into a container suitable for the next operation. Clean the reaction flask ⟨OP–1⟩ as soon as possible so that residues do not dry on the glass.

Cooling can be accelerated by passing an air stream over the flask.

Summary

1. Position heat source.

2. Clamp flask at proper location in relation to heat source.

3. Add reactants and solvents to flask through funnel, and mix.

4.   Add boiling chips.

5.   Insert reflux condenser, clamp in place, and attach rubber tubing.

6.   Turn on and adjust water flow.

7.   Commence heating; adjust heat until reaction mixture boils gently.

8.   Check position of reflux ring; readjust water flow or heating rate if necessary.

9.   At end of reflux period, remove heat source, allow flask to cool, transfer reactants, disassemble and clean apparatus.

OPERATION 7 *b*

Semimicro Refluxing

Small amounts of reactants can be heated under reflux using a *cold finger condenser* inserted into a test tube or small flask. When water is passed through the condenser, it cools the surrounding area enough to condense the rising vapors. To prevent pressure buildup in the container, a stopper with a groove in one side must be used. The reflux ring that appears on the sides of the container should be well below the top of the test tube or flask to prevent the escape of vapors. Some microcondensers are available that resemble an ordinary reflux condenser (except in size) and that are used in the same way.

The operational procedure for small-scale refluxing is essentially the same as that for ordinary refluxing ⟨OP–7a⟩, except that the cooling water goes into the *upper* connector of a cold-finger condenser and comes out the lower one.

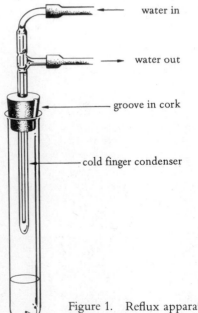

water in

water out

groove in cork

cold finger condenser

Figure 1.   Reflux apparatus with cold finger condenser

OPERATION 7 *c*

Temperature Monitoring

The temperature of a liquid in a reaction flask, oil bath, or other container can be monitored with a thermometer positioned so that its bulb is *entirely immersed* in the liquid. If the thermometer is used in an open container, it can be held in place by a three-fingered clamp or inserted into a stopper that is held by an ordinary utility clamp. If it is used in a container that must be closed off (as in a distillation assembly), the thermometer is usually inserted into a rubber stopper or (preferably) a thermometer adapter.

Some reactions are conducted at temperatures below the boiling point of the solvent, either because decomposition or unwanted side reactions occur at the boiling point, or because the reaction proceeds at a more convenient rate at a lower temperature. When water or other high-boiling solvents are used, such a reaction might be run in an Erlenmeyer flask into which a thermometer has been inserted to monitor the reaction temperature. (The thermometer should not be used as a stirring rod, since the bulb is fragile and breaks easily.) However, if the solvent has a high vapor pressure at the specified reaction temperature—and particularly if it may present a health or fire hazard—a condenser will be required. Magnetic or mechanical stirring ⟨OP–9⟩ is usually necessary, since the absence of boiling action prevents thorough mixing. An apparatus for heating, stirring, and temperature monitoring is shown in Figure 1.

If mechanical stirring ⟨OP–9⟩ or addition of reagents ⟨OP–10⟩ is also required during the reaction period, a three-necked flask can be used. One neck carries the reflux condenser, a second holds a thermometer, and a third receives an addition funnel or stirrer sleeve. Some flasks are provided with a thermometer well, freeing one neck of the flask for other purposes.

If possible, the thermometer should be immersed down to an inscribed immersion mark (usually 76 mm).

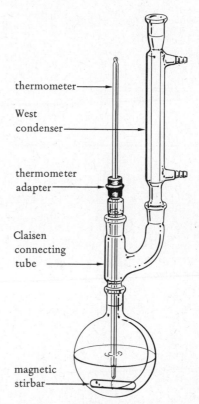

thermometer

West condenser

thermometer adapter

Claisen connecting tube

magnetic stirbar

Figure 1.   Assembly for heating with temperature monitoring

Cooling

Occasionally a reaction proceeds too violently at room temperature or involves reactants or products that decompose at room temperature. In such cases, it may be necessary to cool the reaction mixture by using some kind of cold bath, which can be anything from a beaker filled with cold water to an electrically refrigerated bath.

A mixture of ice and water in a suitable container can be used for cooling down to 0°C; a freezing mixture consisting of 3 parts of finely crushed ice or snow with one part sodium

Equipment and supplies:
bath container (beaker, crystallization dish, Dewar flask etc.), ice water or other cooling mixture, thermometer.

The ice should be finely divided; snow is ideal. Enough water should be present to form a thick slurry, since ice alone is not an efficient heat-transfer medium.

chloride is good for temperatures down to about −20°; and mixtures of $CaCl_2 \cdot 6H_2O$ containing up to 1.4 g of the calcium salt per gram of ice or snow can offer temperatures down to −55°C. In practice, the minimum values may be difficult to attain; the actual temperature of an ice bath depends on the fineness of the ice and salt, the rate of stirring, and the insulating ability of the container.

Ice-salt combinations must be mixed thoroughly and stirred frequently.

CAUTION   *Dry ice should not be handled with bare hands.*

Temperatures below −40°C cannot be measured using a mercury thermometer; mercury freezes at that temperature.

Even lower temperatures can be reached by mixing small chunks of dry ice (solid carbon dioxide) with a suitable solvent (for example, acetone, chloroform, or ethanol) in a Dewar flask or other insulated container. Such a bath can reach minimum temperatures around −75°C. Specific temperatures between about −26° and −72°C can be attained by using dry ice in mixtures of *ortho* and *meta*-xylene of varying composition [*J. Chem. Educ. 45,* 664 (1968)]. Temperatures below those attainable with dry ice are possible using liquid nitrogen (b.p. −196°C), either alone or with an appropriate solvent.

A beaker, crystallization dish, or metal water bath is usually satisfactory for a cold bath; the latter two have a more convenient shape. The useful life of a cold bath can be extended by wrapping the bath container with glass wool and placing it inside a larger container. If a cold bath is needed for a long period of time or at a particularly low temperature, an insulated container such as a Dewar flask or Thermos jar should be used. The temperature of the bath and of the reaction mixture, if desired, can be monitored ⟨OP −7c⟩. For the most efficient cooling, the flask (or other container) should be immersed deeply in the cooling bath and its contents swirled or stirred frequently.

OPERATION 9   # Mixing

Reaction mixtures are frequently stirred, shaken, or otherwise agitated to promote efficient heat transfer, improve contact between the components of a heterogeneous mix-

---

ture, or mix in a reactant that is being added during the course of a reaction. If the reaction is being carried out in an Erlenmeyer flask, this can be accomplished by manual shaking and swirling, or by using a stirring rod. If the apparatus is not too unwieldy and the reaction time is comparatively short, ground glass assemblies can sometimes be manually shaken for adequate mixing. This is most easily done by clamping the assembly *securely* to the ringstand and carefully sliding the base of the ringstand back and forth. But when more efficient and convenient mixing is required, particularly over a long period of time, it is necessary to use some kind of magnetic or mechanical stirring device.

Magnetic Stirring. A magnetic stirrer consists of an enclosed unit containing a motor attached to a magnet, underneath a platform. As the magnet inside the unit rotates, it can in turn rotate a teflon- or glass-covered stirring bar inside a container placed on (or above) the platform. The rate of stirring is controlled by a dial on the stirring unit. Since no moving parts extend outside this container, a magnetically-stirred reaction assembly can be completely enclosed if necessary.

Magnetic stirrers can be used with heating mantles or heating baths that are not constructed of ferrous metal. They work particularly well with oil baths, since they can be used to stir the oil and a reaction mixture simultaneously. The reaction flask must be positioned close enough to the bottom of the oil bath to allow sufficient transfer of magnetic torque from motor to stirring bar. When a copper or aluminum steam bath is used for heating, the flask should be clamped inside the rings, close to the bottom of the steam bath. Some hot plates have an integral magnetic stirrer, and these units can be used (with a heating bath) to simultaneously heat and stir a reaction mixture.

When magnetic stirring is used during a reaction, the heat source is set directly on the stirring unit and a stir bar is placed in the reaction flask in place of boiling chips. The stirring motor should be started and cooling water for the reflux condenser turned on (if applicable) before heating is begun.

Mechanical Stirring. Mechanical stirring utilizes a stirring motor connected to a paddle or agitator by means of a shaft extending through the neck of the reaction vessel. A glass sleeve or bearing is used to align the shaft, which is ordinarily made of glass to reduce the likelihood of contamination.

A motion combining shaking with swirling is more effective than swirling alone.

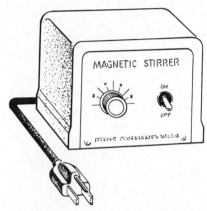

Figure 1.   Magnetic stirring unit

See *J. Chem. Educ. 54,* 229 (1977) for a description of an easily constructed magnetic stirrer.

The stirring bar in the bath should be larger than the one in the reaction mixture.

Boiling chips are not used with magnetic or mechanical stirring, since the stirring action prevents bumping.

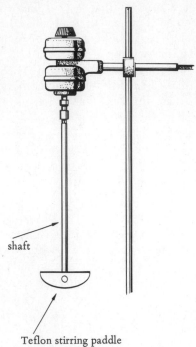

shaft

Teflon stirring paddle
Figure 2.   Mechanical stirrer

Mechanical stirrers can exert more torque than magnetic stirrers, and are preferred when viscous liquids or large quantities of suspended solids must be stirred. A variety of stirring paddles made of teflon, glass, and chemically resistant wire are available.

OPERATION 10

Addition

In many organic syntheses, the reactants are not all mixed together at the start of the reaction; instead, one or more of them is added over a period of time. This is necessary when the reaction is strongly exothermic or when one of the reactants must be kept in excess to prevent side reactions. Cylindrical addition funnels incorporating a pressure release tube are specially constructed for this purpose. However, an ordinary separatory funnel can be used to perform the same function.

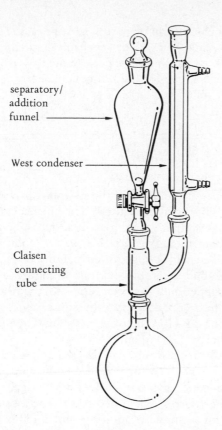

separatory/
addition
funnel

West condenser

Claisen
connecting
tube

Figure 1.   Apparatus for addition and
reflux

If a reaction is run in an open container such as an Erlenmeyer flask, the funnel can simply be clamped to a ringstand above the flask, which is shaken and swirled during
the addition. More frequently, a reaction will be conducted
with *addition under reflux,* using an apparatus such as that illustrated in Figure 1. The addition funnel should be placed
on the straight arm of the Claisen connecting tube, directly
over the reaction flask, and the reflux condenser should be
on the bent arm. Addition of the reactant can be carried out
portionwise or dropwise. In portionwise addition small portions of the reactant are added at regular intervals, with
shaking or stirring after each addition. To allow pressure
equalization between the addition funnel and the reaction
flask, a strip of filter paper can be placed between the stopper
and the top of the addition funnel. In dropwise (continuous)
addition, the stopcock is kept open just far enough to bring
about the desired rate of addition, until all of the reactant has
been added. The reaction mixture can be stirred magnetic-
ally⟨OP–9⟩ or shaken and swirled manually to provide adequate mixing during the addition.

Equipment and supplies:
Boiling flask, Claisen connecting
tube, separatory/addition funnel, reflux condenser, stopper.

*Alternatively, the stopper can be removed
just before each addition and replaced after
it.*

*The stopper can be omitted if the liquid
being added is quite involatile (e.g., an
aqueous solution).*

The apparatus can be modified to provide for temperature monitoring ⟨OP–7c⟩ or mechanical stirring if a three-necked flask is available. One neck of the flask is used for the addition funnel, one for the reflux condenser, and the third for another function.

OPERATION 10 a

Semimicro Addition

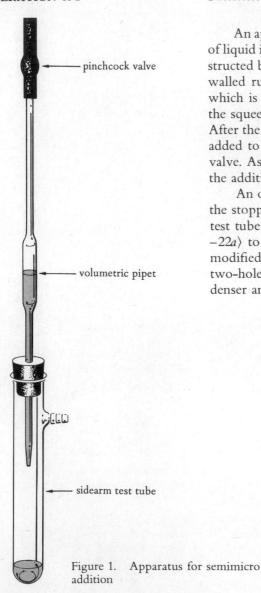

pinchcock valve

volumetric pipet

sidearm test tube

Figure 1.   Apparatus for semimicro addition

An apparatus suitable for the addition of small quantities of liquid is illustrated in Figure 1. The pinchcock valve (constructed by inserting a solid glass bead into a length of thin-walled rubber tubing) is attached to the volumetric pipet, which is filled with the specified volume of reactant using the squeeze-bulb assembly pictured in Figure 2 of ⟨OP–6⟩. After the rubber bulb has been removed, the reactant can be added to the reaction mixture by squeezing the pinchcock valve. As in ⟨OP–10⟩, the reactants should be mixed during the addition by shaking and swirling.

An ordinary test tube can be used as a reaction vessel if the stopper is notched for pressure release, but the sidearm test tube allows a gas trap ⟨OP–22b⟩ or drying tube ⟨OP–22a⟩ to be attached when necessary. The assembly can be modified for semimicro addition under reflux by using a two-holed rubber stopper provided with a cold-finger condenser and an addition pipet.

Separation Operations

Gravity Filtration

Filtration is used for two main purposes in organic chemistry: to remove solid impurities from a liquid or solution, and to isolate an organic solid from a reaction mixture or a crystallization solvent. In most instances, gravity filtration is preferred for the first operation and vacuum filtration for the second. Gravity filtration is often used to remove drying agents such as magnesium sulfate from dried organic liquids or solutions, and to remove solid impurities from hot recrystallization solutions.

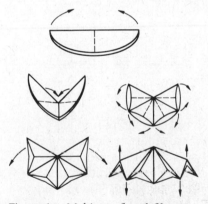

Figure 1.   Making a fluted filter paper

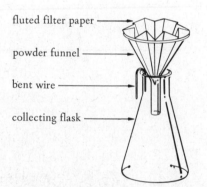

Figure 2.   Apparatus for gravity filtration.

Gravity filtration of most organic liquids is best performed using a funnel with a short, wide stem (such as a powder or filling funnel) and a relatively fast, fluted filter paper. Ordinary filter paper can be folded or "fluted" for this application, but good commercial fluted filter papers are available at a reasonable price from most chemical supply houses. Aqueous solutions are sometimes filtered using a long-stemmed funnel and an ordinary folded filter cone.

Operational Procedure

If you are filtering the liquid into a narrow-necked container like an Erlenmeyer flask, support the funnel on the neck of the flask, with a bent wire (like a paper clip) between them to provide space for pressure equalization (see Figure 2). Otherwise support it on a ring or a funnel support directly over the collecting container. Place the fluted filter paper inside the funnel and add the liquid fast enough to keep it well-filled throughout the filtration. If the mixture to be filtered contains an appreciable amount of finely divided solid, allow it to settle and decant the liquid carefully so that most of the solid will remain behind until the end of the filtration. (Otherwise, the pores of the filter paper may become clogged and retard the filtration.) It is a good practice to stir any solid remaining on the filter paper with a small amount of fresh solvent and drain this wash solvent into the **filtrate.** This will reduce losses due to adsorption of organic materials on the solid.

Allowing a hot recrystallization solution to stand for too long before filtering can cause premature crystallization.

Be careful not to tear the filter paper while stirring.

filtrate: the liquid that has been filtered.

OPERATION 12

Vacuum Filtration

Vacuum filtration (also called suction filtration) offers a fast, convenient method for isolating a solid from a solid–liquid mixture or for removing impurities from large quantities of a liquid. A circle of filter paper is laid flat on a perforated plate inside a *Buchner funnel* which is supported on a *filter flask* (see Figure 1). When a partial vacuum is created in the flask by a water aspirator, liquid is rapidly forced through the filter paper by the unbalanced external pressure. (See *J. Chem. Educ. 53,* 45 (1976) for the construction of a unitized filtration stand.) A filter trap interposed between the trap and the aspirator keeps water from "backing up" into the flask when the water pressure changes. Because of the pressure on the mixture being filtered, solid particles are more likely to pass through the filter paper than in the case of gravity filtration, so a finer grained ("slower") filter paper should be used. An all-purpose paper like Whatman #1 is adequate for filtration of most solids.

When filtering finely divided solids from a liquid, it is sometimes necessary to use a *filtering aid* (such as Celite) to keep the solid from plugging the pores in the filter paper. The filtering aid is mixed with a solvent to form a thin paste or slurry, which is poured into the filter under suction until a bed about 2–3 mm thick has been deposited. The solvent is then removed from the filter flask before continuing with the filtration. This technique should *not* be used when the solid is to be saved, since it would be contaminated with the filtering aid.

Filtering aids are made of diatomaceous earth, which consists of the microscopic shells of deceased diatoms.

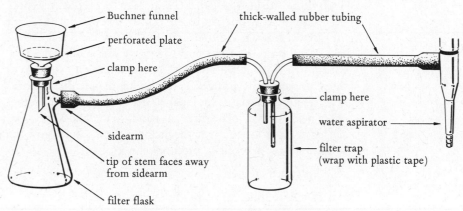

Figure 1.   Apparatus for vacuum filtration

Operational Procedure

Clamp the filter flask and filter trap to a ringstand, connect them as illustrated in Figure 1, and connect the trap to a vacuum line or water aspirator. Use thick-walled rubber tubing for all connections. Insert a Buchner funnel with a snug-fitting rubber stopper into the filter flask and place a circle of filter paper inside the funnel. The diameter of the filter paper should be slightly less than that of the perforated plate, so that the paper covers all of its holes but does not fit too snugly. Moisten the paper with a few drops of solvent (preferably that present in the filtration mixture), then apply suction by turning on the aspirator. Add the filtration mixture in portions, keeping the funnel nearly full throughout. Stir and swirl the filtration mixture near the end of the filtration to get most of the remaining solid onto the filter. Transfer any remaining solid to the filter paper with a flat-bladed spatula, using a little of the filtrate or some additional pure solvent to facilitate the transfer. Unless the trap has a pressure release valve (see Figure 1, page 545), disconnect the rubber tubing at the aspirator before turning off the aspirator; this will keep water from backing up into the system.

Equipment and supplies:
Buchner funnel, filter flask, water trap, filter paper, thick-walled rubber tubing, flat-bottomed stirring rod, flat-bladed spatula, washing solvent.

The filter paper should not extend up the sides of the funnel.

If the solid is finely divided, it may be advisable to carefully decant the liquid after the solid has settled and wait until the end of the filtration to transfer the bulk of the solid.

Any fresh solvent used in the transfer can be used for the first washing as well.

1-2 ml of wash solvent per gram of solid is usually sufficient.

Sometimes the filter cake is compressed with the top of a clean cork, etc., to remove excess water.

Any particles remaining in the funnel should be scraped out.

Unless you are told otherwise, *wash* the solid on the filter paper by the following method. Use the fresh solvent or another suggested solvent; if the solid was filtered from a solvent mixture, use the solvent in which it is least soluble. With the aspirator turned off, add enough of the *cold* washing solvent to just cover the solid and mix them intimately with a spatula or a flat-bottomed stirring rod. Be careful not to disturb the filter paper. Apply suction to drain off the wash liquid, then turn off the aspirator and repeat the process with at least one more portion of fresh solvent. After the last washing, leave the aspirator on a few minutes to *air dry* the solid on the filter and make it easier to handle. Run the tip of a flat-bladed spatula around the circumference of the filter paper to dislodge the filter cake, then invert the funnel carefully over a large filter paper or a watch glass (or another container) to remove the filter cake and paper. The filter paper can be scraped clean with the help of the spatula, and the solid dried by one of the methods described in ⟨OP–21⟩.

Summary

1. Assemble apparatus for vacuum filtration.

2. Position and moisten filter paper, turn on aspirator.

3. Add filtration mixture (in portions).

4. Transfer remaining solid to funnel, using additional solvent if necessary.

5. Disconnect and turn off aspirator.

6. Wash solid on filter with cold solvent, turn on aspirator to drain solvent. Repeat as necessary.

7. Air-dry solid on filter, remove filter cake.

8. Remove solvent from filter flask, disassemble and clean apparatus.

OPERATION 12 *a*

Semimicro Vacuum Filtration

Small quantities of solids can be filtered by the same general method as described in ⟨OP–12⟩, using a Hirsch funnel and a small filter flask or a sidearm test tube. Since the perforated plate of a Hirsch funnel is small in diameter, filter paper of the right size is not generally available. Small circles of filter paper can be cut from an ordinary filter paper using a sharp cork borer on a flat cutting surface, such as the bottom of a large cork.

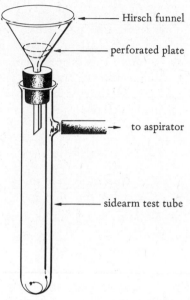

- Hirsch funnel
- perforated plate
- to aspirator
- sidearm test tube

Figure 1.   Apparatus for semimicro vacuum filtration

Extraction

Principles and Applications

Extraction is a convenient method for separating an organic substance from a mixture, such as an aqueous reaction mixture or a steam distillate. The *extraction solvent* is usually a volatile organic liquid that can be removed by evaporation ⟨OP–14⟩ after the desired component has been extracted. [The related process of *washing* liquids, in which impurities are extracted from an organic liquid, is described in ⟨OP –19⟩.]

The extraction technique is based on the fact that, if a substance is soluble to some extent in two immiscible liquids, it can be transferred from one liquid to the other by shaking together the solute and the two liquids. For example, acetanilide is partly soluble in both water and ethyl ether. If a solution of acetanilide in water is shaken with a

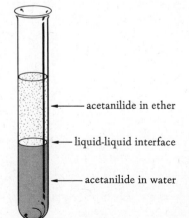

- acetanilide in ether
- liquid-liquid interface
- acetanilide in water

Figure 1.   Distribution of a solute between two liquids

portion of ethyl ether (which is immiscible with water) some of the acetanilide will be transferred to the ether layer. The ether layer, being less dense than water, separates above the water layer and can be removed and replaced with another portion of ether. When this is in turn shaken with the aqueous solution, more acetanilide passes into the new ether layer. This new layer can then be removed and combined with the first. By repeating this process enough times, virtually all of the acetanilide can be transferred from the water to the ether.

The ability of an extraction solvent, S_2, to remove a solute A from another solvent, S_1, depends on the distribution coefficient (K) of solute A in the two solvents, as defined in Equation 1:

$$K = \frac{\text{Concentration of A in } S_2}{\text{Concentration of A in } S_1} \qquad \textbf{(1)}$$

A very rough estimate of K can be obtained by using the ratio of the solubilities of the solute in the two solvents, that is:

$$K \approx \frac{\text{Solubility of A in } S_2}{\text{Solubility of A in } S_1}$$

This approximate relationship can be helpful in choosing a suitable extraction solvent.

In the example of acetanilide in water–ethyl ether, the distribution coefficient is given by $K = [\text{acetanilide}]_{\text{eth}}/[\text{acetanilide}]_{\text{w}}$. The larger the value of K, the more solute will be transferred to the ether with each extraction, and the fewer portions of ether will be required for essentially complete removal of the solute.

Extraction Solvents

If a solvent is to be used to extract an organic compound from an aqueous mixture or solution, it must be virtually insoluble in water, and it should have a low boiling point so that the solvent can be evaporated after the extraction ⟨OP –14⟩. The solute should also be more soluble in the extraction solvent than in water, since otherwise too many extraction steps will be required to remove all of the solute.

Ethyl ether is the most commonly used extraction solvent. It has a very low boiling point (34.5°C) and can dissolve a large number of organic compounds, both polar and nonpolar. However, ethyl ether must be used with great care, since it is extremely flammable and tends to form explosive peroxides on standing. Methylene chloride (dichloromethane) has most of the advantages of ethyl ether; in addition, it is nonflammable and more dense than water. However, it has a tendency to form emulsions, which can make it difficult to separate the layers cleanly. Other useful solvents and their properties are listed in Table 1. Various grades of petroleum ether (a mixture of low-boiling hydrocarbons) can be used in place of pentane and hexane.

If the extraction solvent is less dense than water, the aqueous layer must be drained out after each extraction and returned to the separatory funnel for the next extraction. This is not necessary if the solvent is more dense than water.

Table 1. Properties of commonly used extraction solvents

Solvent	b.p.	d.	Comments
ethyl acetate	77°	0.90	Absorbs much water.
ethyl ether	34.5°	0.71	Good general solvent. Absorbs some water; easy to remove. Very flammable. Vapors should not be inhaled.
methylene chloride	40°	1.34	Good general solvent; easy to dry and remove. Possible health hazard.
chloroform	62°	1.48	Can form emulsions; easy to dry and remove. Health hazard; suspected carcinogen.
benzene	80°	0.88	Can form emulsions. Hazardous to health; may cause leukemia.
pentane	36°	0.63	Easy to dry and remove. Very flammable.
hexane	69°	0.66	Easily dried.

Hazards

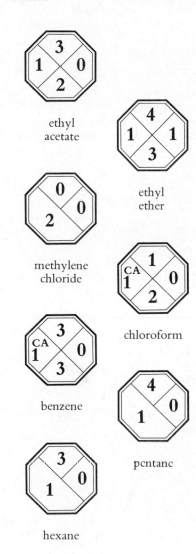

ethyl acetate

ethyl ether

methylene chloride

chloroform

benzene

pentane

hexane

(See page 13 for explanation of Hazard symbols.)

Potential hazards must always be considered in selecting and using an extraction solvent. Because of its proven carcinogenic potential, benzene is not recommended as an extraction solvent unless appropriate precautions are taken, such as the use of an efficient fume hood and rubber gloves. Flames must not be allowed in the laboratory when highly flammable solvents, such as ethyl ether, pentane, and petroleum ether, are in use. Precautions must be taken with all solvents to minimize skin or eye contact, inhalation of vapors, and exposure to possible ignition sources.

Experimental Considerations

Separatory funnels are very expensive and break easily. Never prop the funnel on its stem, but support it on a ring, a funnel support, or some other stable support. Glass stopcocks should be lubricated by applying thin bands of stopcock grease on both sides, leaving the center (where the drain hole is located) free of grease to prevent contamination (see Figure 2C). A teflon stopcock should not be treated with stopcock grease, nor should the glass stopper.

Frequently the volume of extraction solvent and the number of extraction steps are specified in an experimental

Rule of thumb: total volume of extraction solvent ≈ volume of liquid being extracted. (There are many exceptions to this "rule.")

procedure. If not, it is usually sufficient to use a volume of extraction solvent about equal to the volume of liquid being extracted, divided into at least two portions. For example, 60 ml of an aqueous solution can be extracted with two 30 ml (or three 20 ml) portions of ethyl ether. It is more efficient to use several small portions of extraction solvent than one large portion of the same total volume

A liquid emulsion contains microscopic droplets of one liquid suspended in another.

Avoid using too much salt; excess solid can clog the stopcock drain hole.

When certain kinds of solutes are present, an emulsion may form at the interface between the two liquids, making them impossible to separate sharply. Emulsions can sometimes be broken up by gently swirling or stirring the liquids and allowing the funnel to stand open and undisturbed for a time, or by dissolving a little sodium chloride in the aqueous layer with stirring. Small amounts of insoluble material that sometimes form at the interface can be removed by filtering the mixture through a loose pad of glass wool. After each extraction, the extraction solvent should be transferred to a collecting flask, which is kept stoppered to prevent evaporation and reduce any fire hazard.

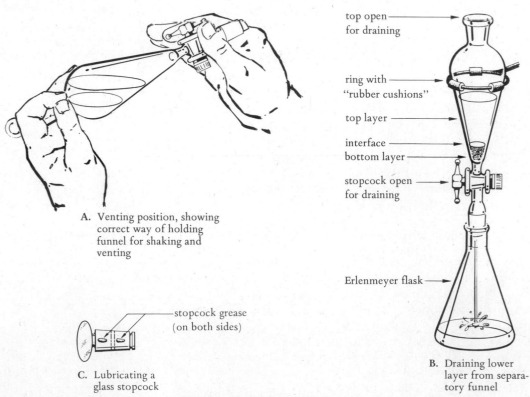

A. Venting position, showing correct way of holding funnel for shaking and venting

top open for draining

ring with "rubber cushions"

top layer

interface
bottom layer

stopcock open for draining

Erlenmeyer flask

B. Draining lower layer from separatory funnel

stopcock grease (on both sides)

C. Lubricating a glass stopcock

Figure 2.   Extraction techniques

Operational Procedure

Support a separatory funnel on a ring of suitable diameter, equipped with some lengths of split rubber tubing (see Figure 2B) to cushion the funnel and prevent damage. (A special funnel support can also be used.) Close the stopcock by turning the handle to a horizontal position (*Important!*) and add the liquid to be extracted through the top of the funnel. Measure the required volume of extraction solvent using a graduated cylinder (or some other graduated container—the exact volume is not crucial) and pour it into the funnel. The total volume of both liquids should not exceed ¾ of the funnel's capacity; if it does, obtain a larger funnel or carry out the extraction in two or more steps, using a fraction of the liquid in each step. Insert the stopper (lubricated with a little water or other solvent), then pick up the funnel in both hands and *invert* it with the right hand holding the stopcock (left hand for a southpaw) and the first two fingers of the left hand holding the stopper in place (see Figure 2A). Vent the funnel by slowly opening the stopcock to release any pressure buildup. Close the stopcock, shake or swirl the funnel gently for a few seconds, then vent it again (be sure the funnel is inverted). Shake the funnel more vigorously, with occasional venting, for 2–3 minutes. Venting should not be necessary after there is no longer an audible hiss of escaping vapors when the stopcock is opened.

Replace the funnel on its support, remove the stopper, and allow the funnel to stand until there is a sharp demarcation line between the two layers. Drain the bottom layer into an Erlenmeyer flask (flask A) by opening the stopcock fully; turn it to slow the drainage rate as the interface approaches the bottom of the funnel. When the interface *just* reaches the outlet, quickly close the stopcock to separate the layers cleanly.

If the extraction solvent is *more* dense than the aqueous solution being extracted, the aqueous layer will remain in the separatory funnel after each extraction and can be extracted with a fresh portion of solvent, which is then combined with the first extract in flask A. Additional extractions with fresh solvent can be performed as needed. After the last extraction is finished, pour the aqueous layer out of the *top* of the funnel into a separate container (flask B).

If the extraction solvent is *less* dense than the aqueous layer, it will remain in the separatory funnel after the bottom (aqueous) layer has been drained into flask A. Pour it out of

Equipment and supplies:
Separatory funnel, 2 flasks, extraction solvent, graduated cylinder, ring support, ringstand, closure for solvent flask.

The liquid in the funnel must be at room temperature or below if the extraction solvent is very volatile.

Be sure that the stem of the funnel is pointed away from you and your neighbors when you are venting it.

A combined shaking and swirling motion is more efficient than swirling alone. Avoid vigorous shaking if the system is prone to emulsions.

Careful stirring with a stirring rod or wooden applicator stick can sometimes accelerate the consolidation of layers.

Most aqueous solutions will have densities close to 1 g/ml or a little higher.

If there is any doubt about which layer should be discarded, save both *until the correct one is identified. Mixing a drop or so of water with a little of each layer will establish which is the aqueous layer.*

the *top* of the funnel into flask B after each extraction. Return the aqueous solution to the funnel from flask A and extract it with another portion of solvent, then drain it into flask A and again pour the solvent into flask B. Repeat the process, as necessary, with fresh extraction solvent.

Summary

1.   Place separatory funnel on support, add liquid to be extracted.

2.   Add extraction solvent, stopper funnel, invert and vent funnel.

3.   Shake and swirl funnel, with occasional venting, to extract solute into extraction solvent.

4.   Remove stopper, drain lower layer into flask A.
IF extraction solvent is the *lower* layer,
GO TO 5.
IF extraction solvent is the *upper* layer,
GO TO 6.

Usually 2 or 3 extraction steps should be performed.

5.   Stopper flask A.
IF another extraction step is needed,
GO TO 2.
IF extraction is complete, empty and clean funnel,
STOP.

6.   Pour upper layer into flask B and stopper it.
IF another extraction step is needed, return contents of flask A to separatory funnel,
GO TO 2.
IF extraction is complete, clean funnel,
STOP.

OPERATION 13 *a*

Semimicro Extraction

See ⟨OP–6⟩ on the use of volumetric pipets.

Small–scale extractions can be carried out by shaking the aqueous solution with the extraction solvent in a stoppered test tube (with occasional venting) and separating the layers using a capillary pipet, a volumetric pipet, or an "aspirator pipet." An aspirator assembly using a capillary (Pasteur-type) pipet is illustrated in Figure 1. In using the aspirator pipet, the aspirator is turned on, the pipet is lowered into the appropriate liquid layer, and tube A is closed with a finger to

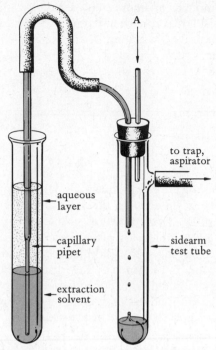

A

to trap,
aspirator

aqueous
layer

capillary
pipet

extraction
solvent

sidearm
test tube

Figure 1.   Aspirator pipet assembly for semimicro extraction (Illustrated for an extraction solvent that is more dense than water)

provide suction until the layer is transferred to the sidearm test tube. Operating the pipet effectively requires some care and practice, and it may be necessary to remove the last few drops manually using an ordinary Pasteur pipet with a dropper bulb. The basic extraction procedures resemble those described in ⟨OP–13⟩, except that *either* the top or the bottom layer—depending on the density of the extraction solvent —can be removed in order to leave the liquid being extracted in the test tube after each extraction step.

Separation of Liquids

Sometimes a distillate or the product of a reaction consists of two immiscible liquids that must be separated for subsequent operations. This can be accomplished by transferring the liquid mixture to a separatory funnel and allowing it to stand until a sharp demarcation line forms between the liquids. The lower layer is then drained into one container and the upper layer poured out the top into another.

For small amounts of liquids, the separation method described in ⟨OP–13a⟩ can be utilized. See ⟨OP–13⟩ for information on dealing with emulsions, identifying the aqueous layer, etc.

OPERATION 13 c

Salting Out

Adding an inorganic salt (such as sodium chloride or potassium carbonate) to an aqueous solution containing an organic solute often reduces the solubility of the organic compound in the water, and thus promotes its separation. This *salting out* technique is often used in extractions and liquid-liquid separations to maximize the transfer of an organic solute from the aqueous to the organic layer or to separate an organic liquid from its aqueous solution. Usually enough of the salt is added to saturate the aqueous solution, which is stirred or shaken to dissolve the salt, filtered by gravity ⟨OP–11⟩ if necessary, and transferred to a separatory funnel for separation ⟨OP–13b⟩ or extraction ⟨OP–13⟩.

Filtering the solution through a loose plug of glass wool may be satisfactory.

OPERATION 13 d

Solid Extraction

Some substances, particularly natural products, contain components that can be separated from a solid residue by solvent extraction. The simplest technique for solid extraction is to mix the solid intimately with an appropriate solvent using a flat-bottomed stirring rod in a beaker, filter off the solid by gravity ⟨OP–11⟩ or vacuum filtration ⟨OP–12⟩, and repeat the process as many times as necessary. The mixing should be as thorough as possible, using the flat end of the stirring rod to press down or crush the solid in contact with the solvent, in order to extract more of the desired component. After each filtration, the solid is returned to the beaker for the next extraction. The liquid extracts are combined in the same collecting flask.

More thorough solid extraction can be accomplished by using the special apparatus called a Soxhlet extractor.

A mortar and pestle can also be used.

This process is called trituration.

Evaporation

Evaporation refers to the vaporization of a liquid at or below its boiling point. It can be used to remove a volatile solvent from a comparatively involatile residue. Partial solvent removal (*concentration*) is frequently used to bring a recrystallization solution to its saturation point (see ⟨OP –23⟩). Complete solvent removal is utilized to isolate an organic solute after such operations as extraction ⟨OP–13⟩ or column chromatography ⟨OP–16⟩.

Small quantities of solvent can be evaporated simply by leaving the solution under a fume hood in an evaporating dish, or by passing a slow, dry stream (see ⟨OP–22⟩) of air or nitrogen over it (Figure 1). The solution can be protected from airborne particles by supporting a filter paper some distance above it. Relatively large quantities of solvent can be removed by simple distillation ⟨OP–25⟩ or vacuum distillation ⟨OP–26⟩ when convenient. This procedure is most useful when the residue is to be distilled immediately after solvent removal. Commercial flash evaporators are available for fast removal of solvents under reduced pressure, but these are expensive and not generally available for use in undergraduate laboratories.

Experimental Considerations

For most purposes, moderate quantities of solvent can be evaporated efficiently with one of the setups pictured in Figure 2. Evaporation is accelerated by using reduced pressure and gentle heating from a water bath or steam bath ⟨OP–7⟩. Gently swirling the solution during evaporation can also speed up the process and prevent bumping. Although bumping is not a problem with some low-boiling solvents, the use of boiling sticks or boiling chips may be advisable in other cases. A solvent trap similar to that pictured in Figure 1, ⟨OP–12⟩ should always be interposed between the evaporation vessel and the aspirator to collect the solvent, which can then be returned to a used-solvent container for recycling.

Because of possible health and fire hazards, it is not advisable to evaporate solvents by heating open containers in the laboratory. Even when one uses the method described below, the hazards associated with each solvent should be known and allowed for.

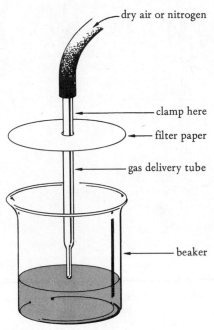

Figure 1.   Apparatus for small-scale evaporation

Cooling the trap in an ice bath ⟨OP–8⟩ will retard loss of the solvent by evaporation.

Hazard signs for many of the solvents commonly encountered are given in ⟨OP–13⟩, page 475, and ⟨OP–23c⟩, page 526.

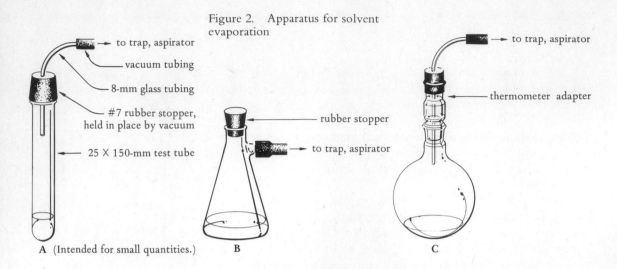

Figure 2.   Apparatus for solvent evaporation

A (Intended for small quantities.)   B   C

Equipment and supplies:
Evaporation vessel (Figure 2), rubber tubing, solvent trap, aspirator, water bath or steam bath.

It should be possible to tell from the odor and volume of the residue whether all of the solvent has been removed.

Some residues may solidify only after cooling.

Ethyl ether is often used for this purpose.

Operational Procedure

Assemble one of the setups pictured in Figure 2, parts A, B, or C. Add the liquid to be evaporated and connect the apparatus to a solvent trap and aspirator. Turn on the aspirator and hold (or clamp) the evaporation vessel in a hot water bath (or over a steam bath). Be ready to disconnect the aspirator or remove the heat source if excessive foaming occurs. Adjust the temperature of the water bath (or the rate of heating on a steam bath) to attain a satisfactory rate of evaporation. The liquid should boil, but must not foam over into the trap. Swirling the liquid will increase the evaporation rate and may reduce foaming. When sufficient solvent has evaporated, or when evaporation has ceased entirely, break the vacuum by removing the hose from the aspirator (or, in the case of apparatus A, by sliding the stopper off the mouth of the test tube). Turn off the aspirator. If the product is a solid, it can be removed with a flat-bladed spatula. The last traces of residue are sometimes removed by rinsing the container with a little volatile solvent and allowing the solvent to evaporate under a hood or in a dry air stream. Place the solvent from the trap in a waste-solvent container or dispose of it as recommended by your instructor.

Summary

1.   Assemble apparatus for solvent evaporation.

2.   Add liquid, stopper evaporation vessel, turn on aspirator.

3. Apply heat and swirl to maintain rapid evaporation.

4. Discontinue heating, disconnect aspirator vacuum, recover residue and solvent.

Codistillation

Principles and Applications

When a *homogeneous solution* of two liquids is distilled, the vapor pressure of each liquid is lowered by an amount roughly proportional to the mole fraction of the other liquid present. This usually results in a solution boiling point somewhere between the boiling points of the components. When a *heterogeneous mixture* of two immiscible liquids is distilled, however, each liquid exerts its vapor pressure more-or-less independently of the other, so that the total vapor pressure over the mixture roughly equals the sum of the vapor pressures that would be exerted by the separate pure liquids at the same temperature. This has several important consequences: **(1)** the vapor pressure of a mixture of immiscible components must be *higher* than that of the most volatile component; therefore **(2)** the boiling point of the mixture must be *lower* than that of the lowest-boiling component; and **(3)** the boiling point will remain constant during distillation of such a mixture, as long as each component is present in significant quantities.

A mixture of toluene and water at 1 atmosphere boils at 85°C, which is considerably lower than the normal boiling point of either toluene (111°) or water. The vapor pressure of water at 85° is (from the *CRC Handbook of Chemistry and Physics*) 434 torr and, by subtraction from 760, the vapor pressure of toluene must be about 326 torr. According to Avogadro's law, the number of molecules of each component of an ideal-gas mixture is proportional to its partial pressure in the mixture (see Equation 1). Thus the ratio of moles (toluene) to moles (water) is 326/434, and the *mole fraction* of toluene in the vapor is 326/760 or 0.43. However, since toluene has a higher molecular weight than water, the *weight fraction* of toluene is considerably higher than this. Using Equation 2, the weight ratio of toluene to water in the vapor is found to be about 3.84/1, and thus the weight fraction

Distillation of a mixture of two or more immiscible liquids is called codistillation.

Vapor pressure of a mixture of two immiscible components

$$P \approx P_A^\circ + P_B^\circ$$

(P° refers to the vapor pressures of the pure liquids A and B)

Equations for calculating mole and weight ratios

$$\frac{n_A}{n_B} = \frac{P_A}{P_B} \qquad (1)$$

$$n = \text{no. moles} = \frac{\text{mass}}{\text{mol. wt.}} = \frac{w}{M}$$

so

$$\frac{w_A}{w_B} = \frac{P_A M_A}{P_B M_B} \qquad (2)$$

Weight fraction of toluene

$$\frac{w_{\text{tol}}}{w_w} = \frac{326 \times 92}{434 \times 18} = 3.84$$

$$\text{wt. fraction} = \frac{w_{\text{tol}}}{w_{\text{tol}} + w_w}$$

$$= \frac{3.84}{3.84 + 1.00} = 0.79$$

484

Part V   The Operations

Calculated values may deviate from the experimental values because most liquids are at least slightly miscible with water and their vapors are not ideal.

of toluene in the distillate is 0.79, which agrees well with an experimental value of 80% toluene.

Table 1.   Boiling points and compositions of heterogeneous mixtures with water

Component A	b.p. of A	b.p. of mixture	weight % of A
toluene	111°	85°	80%
chlorobenzene	132°	90°	71%
bromobenzene	156°	95°	62%
iodobenzene	188°	98°	43%
quinoline	237°	99.6°	10%

Note: Component B is water.

Because of its relatively low molecular weight and its immiscibility with many organic compounds, water is nearly always one of the components used in a codistillation. In general, the higher the boiling point of the organic component in a mixture containing water, the lower will be its proportion in the vapor (and therefore in the distillate) and the closer the mixture boiling point will be to 100° (see Table 1). If the vapor pressure of the organic component at 100° is much below 5 mm, codistillation becomes impractical since it would require the distillation of a large amount of water to recover a small quantity of the organic component. Besides having an appreciable vapor pressure at that temperature, the organic component in a codistillation must be insoluble enough in water to form a separate phase, and must not react with (or be decomposed by) hot water or steam.

Codistillation of a mixture containing water is sometimes called direct steam distillation.

Codistillation is of little use for the final purification of a liquid, since it cannot effectively separate components having similar boiling points.

In this book, the term *codistillation* will be used to describe the distillation of any mixture of immiscible components *without* using externally generated steam (that process will be referred to as *steam distillation* ⟨OP–15a⟩). Both codistillation and steam distillation are useful for isolating organic components from reaction mixtures or natural products, by separating them from high-boiling residues such as tars, inorganic salts, or other relatively involatile components. Because distillation occurs well below the normal boiling point of the organic component, decomposition by excessive heat is minimized.

Experimental Considerations

Refer to ⟨OP–25⟩ for experimental details.

A mixture of water with an organic liquid can be distilled by essentially the same method as used for simple distillation

⟨OP–25⟩, except that additional water can be added during the distillation if necessary. If water is to be added, a setup like that illustrated in Figure 1 should be used; otherwise an ordinary simple distillation apparatus is adequate. The thermometer is not essential, but it can sometimes indicate the end of the distillation, since the temperature should rise to 100° at that point.

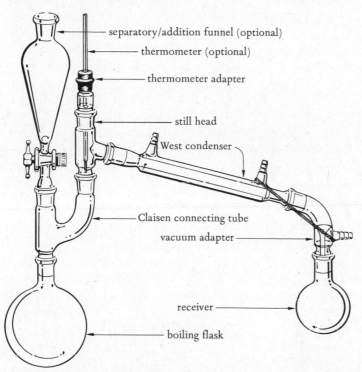

Figure 1.   Apparatus for codistillation of immiscible liquids

Operational Procedure

Set up the apparatus illustrated either in Figure 1 or in Figure 2 of ⟨OP–25⟩. Add the organic material, boiling chips, and enough water to fill the boiling flask about ⅓ to ½ full (unless enough water is already present). Heat the flask ⟨OP–7⟩ to maintain a rapid rate of distillation. Do not allow vapors to escape at the receiver (it may be necessary to cool the receiver in an ice bath to prevent this). Add water, as needed, to replace that lost during the distillation. Discontinue heating when the distillate contains no more of the organic component *and* the distillation temperature is about 100°C.

Supplies and equipment:
Boiling flask, Claisen head, three-way connecting tube, West condenser, distillation adapter, receiving flask, separatory/addition funnel, thermometer, thermometer adapter, heat source, condenser tubing.

The distillate should no longer be cloudy or contain droplets of organic liquid at this point.

OPERATION 15 *a*   *Steam Distillation*

Experimental Considerations

The external steam helps prevent bumping caused by solids or tars.

The use of live steam is preferred for many distillations, especially those involving solids or comparatively involatile liquids. The steam can be obtained from a steam line or by boiling water in a steam generator like the one pictured in Figure 1. The safety tube is needed to allow pressure release in case there is an obstruction in the apparatus, such as a solid plugging up the steam-delivery tube. It also sounds a warning when the water in the generator is getting low.

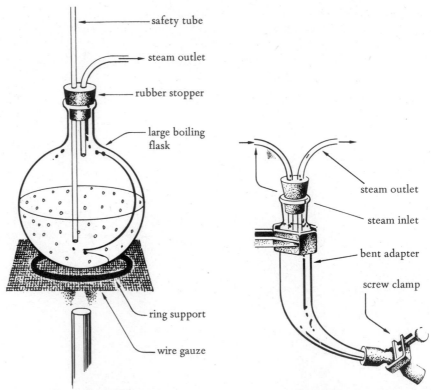

Figure 1.   Steam generator   Figure 2.   Steam trap

When steam is obtained from a steam line, a steam trap must be used to remove condensed water and foreign matter such as grease or rust. An ordinary water trap can be used in some cases, but the trap illustrated in Figure 2 is preferable, since it includes a valve for draining off excess water. A steam trap is also desirable (if not essential) when a steam generator is used.

The boiling flask should be large enough that the total volume of water and organic material in the flask never fills it more than half full. Some steam will condense in the flask during the distillation and can be removed by external heating, if necessary, during extended distillations. The Claisen head is used to prevent mechanical transfer of materials in the spray expelled from the liquid surface. The thermometer can indicate the end of the distillation when the organic distillate is quite volatile. For example, in a toluene–water mixture the temperature will rise rather rapidly from 85° to about 100° when the toluene is nearly gone. With less volatile materials, the temperature may be close to 100° throughout the distillation, so a thermometer will be of little use. The distillation should be carried out rapidly to reduce condensation in the distilling flask, and to compensate for the large volume of water-laden distillate that may have to be collected to yield much of the organic component. Due to the rapid distillation rate and the high heat content of steam, efficient condensing is essential, so the cooling water should

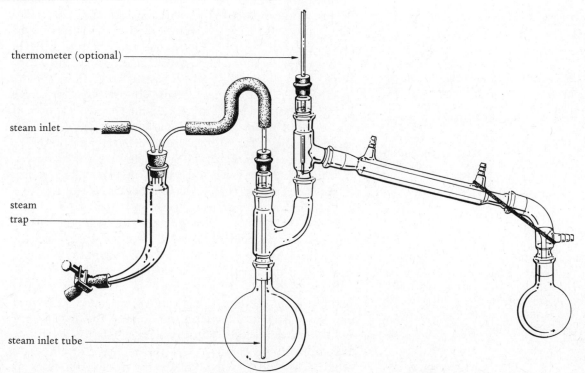

Figure 3. Apparatus for steam distillation with external generation of steam

The distillation adapter should be cool to the touch throughout the distillation; no vapor should escape from the outlet.

Position the boiling flask high enough that heat can be applied if necessary. In some cases, it may be advisable to apply heat from the beginning of the distillation.

The flask should be no more than ⅓ full at the start.

The steam should be turned on for a while with the steam-trap valve open to remove condensed water from the line.

Check the connection between the condenser and 3-way connecting tube frequently to make sure no vapor is escaping; these joints sometimes separate because of the violent action of the steam.

As long as there is some organic material in the distillate, it should appear cloudy or inhomogeneous. When all of the organic component has distilled, the distillate should be clear water.

run faster than for ordinary distillations. Other ways of increasing the cooling rate are using two condensers in tandem or placing the receiving flask in an ice bath.

Equipment and Supplies: Boiling flask, Claisen head, 3-way connecting tube, condenser, distillation adapter, receiving flask, thermometer adapter, thermometer (optional), steam-delivery tube, condenser tubing, steam source. Trap: Bent adapter, screw clamp, 2-hole rubber stopper, bent glass tubes, rubber tubing.

Operational Procedure

Assemble the apparatus pictured in Figure 3, using a large (250–500 ml or larger) boiling flask and, as the steam-delivery tube, a 6 mm O.D. glass tube extending to within about 0.5 cm of the bottom. Connect this tube through a trap to the steam line or to a steam generator. Add the organic mixture and a small amount of water (unless the mixture already contains water) to the distilling flask. Make sure that the condenser water is running rapidly enough to condense the distillate efficiently. Turn on the steam and, after distillation begins, adjust the steam flow to maintain a rapid rate of distillation. Do not distill so rapidly that vapor escapes at the receiver; use an ice bath, if necessary, to prevent this. Drain the trap periodically to remove condensed water. If the flask begins to fill up excessively (it should not be much more than half full) heat it with a steam bath or other heat source to reduce the condensation and keep the water level down. If you must interrupt the distillation for any reason, open the steam-trap valve (or raise the steam-delivery tube out of the liquid) before turning off the steam. If a solid is being distilled, check the receiver outlet frequently to make sure that it does not become obstructed. If it begins to plug up, turn the condenser water off *momentarily* and drain the condenser jacket so that the hot vapors will melt the solid and carry it into the receiver; then immediately turn the water on again.

When the distillate is clear (no oily droplets or cloudiness in the condensate) and the temperature is about 100°, collect and examine a few drops of the distillate on a watch glass to see if it contains any immiscible liquid. If it does, continue distilling and collect and examine the distillate at 5 or 10 minute intervals. When you are certain the distillation is complete, open the steam-trap valve fully, *then* turn off the steam. This will keep the liquid in the pot from backing up into the steam line.

The organic liquid can be obtained by separation in a

separatory funnel ⟨OP–13*b*⟩ or by extraction with ether or another suitable solvent ⟨OP–13⟩. Extraction is preferred if the volume of the organic liquid is small compared to that of the water.

If the aqueous layer appears cloudy, dissolve some NaCl or other salt in it ⟨OP–13c⟩ to improve the separation.

Summary

1. Assemble apparatus for steam distillation (Figure 3).
IF steam line is available,
GO TO 3.
IF no steam line is available,
GO TO 2.

2. Assemble steam generator (Figure 1).

3. Assemble steam trap (Figure 2); connect trap to steam line (or steam generator) and to steam-delivery tube.

4. Add water to boiling flask (if necessary), turn on condenser water, turn on steam line (or boil water in steam generator).

5. Distill rapidly, using external heat if necessary, until distillate is clear. Drain trap periodically.

6. Open steam-trap valve (or raise steam-inlet tube out of liquid), turn off steam.

7. Separate or extract organic liquid from distillate, disassemble and clean apparatus.

Column Chromatography

Principles and Applications

Column chromatography is a simple, efficient method for separating the components of a mixture. It operates on the principle that different substances are adsorbed on the surface of a solid adsorbent (such as alumina) to an extent that depends on their polarity and other structural features. Since some compounds are more strongly adsorbed than others, they will be washed down a column of adsorbent at a slower rate and thus become separated from those less strongly adsorbed. The method discussed in this operation is classified as *liquid-solid adsorption chromatography;* other separation principles are also used in column chromatography.

Approximate strength of adsorption of different functional groups in order of increasing adsorption power

—Cl, —Br, —I	—CHO
—C=C—	—SH
—OR	—NH$_2$
—COOR	—OH
—C=O	—COOH

Chromatography terms:
stationary phase: in column chromatography, a solid or liquid that remains immobile in the column and separates components by adsorption, partition, etc.

mobile phase: a fluid that passes through the column and carries the components with it at varying rates.

adsorbent: a substance that retains molecules of other substances by surface attraction.

limonene   carvone

Common chromatography adsorbents

Alumina (Al$_2$O$_3$)   strong

Charcoal (C)

Florisil (MgO/SiO$_2$)

Silica (SiO$_2$)

Magnesia (MgO)   weak

(in approximate order of adsorbent strength)

Liquid-solid adsorption chromatography involves the use of a solid **stationary phase** (such as alumina or silica) and a liquid **mobile phase** (such as hexane or chloroform). The stationary phase—usually called the **adsorbent** in adsorption chromatography—is packed firmly into a glass tube called the *column*. The sample, consisting of two or more components in a neat liquid or solution, is placed on top of the adsorbent in a narrow band and washed down the column (eluted) by a suitable mobile phase, also called the *eluent*. As the eluent passes down the column, the components of the sample spread out to form separate bands of solute, some passing down the column rapidly with the solvent, others lagging behind.

Consider, for example, a separation of limonene and carvone on a silica adsorbent. At any given time, a molecule of one component will either be adsorbed on the silica (stationary phase) or dissolved in the mobile phase. While it is adsorbed, the molecule will stay put; while dissolved, it will move down the column with the eluent. A relatively polar molecule of carvone is strongly attracted to the polar adsorbent and spends more time adsorbed on the silica than dissolved in a nonpolar eluent like petroleum ether. It will therefore pass down the column very slowly with this solvent. On the other hand, a nonpolar molecule of limonene is very soluble in petroleum ether and only weakly attracted to the adsorbent; it will therefore spend less time sitting still and more time moving than a carvone molecule. As a result, the limonene passes down the column rapidly and will soon separate from the slow-moving carvone. If a more polar solvent such as methylene chloride is then added, the carvone will spend a greater fraction of its time in solution and be washed down the column in turn. The kind of separation attained by column chromatography thus depends on a number of factors, including the quantity and kind of adsorbent used, the polarity of the mobile phase, and the nature of the components in the mixture.

Experimental Considerations

Adsorbents. A number of different adsorbents are used for column chromatography, but alumina and silica are the most popular. Adsorbents are available in a wide variety of activity grades and particle-size ranges; alumina can be obtained in acidic, basic, or neutral forms as well. The *activity* of an adsorbent is a measure of its attraction for solute molecules, the most active grade of a given adsorbent being one from which all water has been removed. The most active grade is not always the best for a given application, since too active an adsorbent may catalyze a reaction or cause solute bands to move too slowly. Alumina is *deactivated* by mixing in 3–15% water (see Table 1), whereas silica is generally deactivated with 10–20% water.

Some mixtures should not be separated on certain kinds of adsorbents. For example, basic alumina would be a poor choice to separate a mixture containing aldehydes or ketones, which might undergo aldol condensation reactions on the column; it would also be unsuitable for carboxylic acids, which bond so strongly to alumina that they cannot be easily desorbed. Deactivated silica, although less active than alumina, is a good all-purpose adsorbent that can be used with most kinds of functional groups.

The amount of adsorbent required for a given application depends on the sample size and the difficulty of the separation. If the components of a mixture differ greatly in polarity, a long column of adsorbent should not be necessary, since the separation will be easy. The more difficult the separation, the more adsorbent will be needed. About 20–50 grams of adsorbent per gram of sample is sufficient for most separations, but ratios of 200:1 or higher are occasionally required for problem cases.

Eluents. In a typical elution process, the eluent acts primarily as a solvent to differentially remove molecules of solute from the surface of the adsorbent. In some cases, polar solvent molecules will also *displace* solute molecules from the adsorbent by becoming adsorbed themselves. If the solvent is too strongly adsorbed, the components of a mixture will remain largely in the mobile phase and will not separate efficiently. For this reason, it is generally best to start with a solvent of low polarity and then (if necessary) increase the polarity gradually to elute the more strongly adsorbed components. Table 2 lists a series of common chromatographic solvents in order of increasing eluting power from alumina and silica. Such a listing is called an *eluotropic series.* The eluotropic series for a nonpolar adsorbent like charcoal is nearly the reverse of the one for alumina; less polar solvents are the more effective eluents in this case.

Elution Techniques. Many chromatographic separations cannot be performed efficiently with a single solvent, so several solvents or solvent mixtures are used in sequence, start-

Table 1. Alumina activity grades

Grade	wt% water
I	0
II	3
III	6
IV	10
V	15

Since all adsorbents are deactivated by water, it is important to keep their containers tightly closed and to minimize their exposure to atmospheric moisture.

Table 2. Eluotropic series for alumina and silica

alumina	silica
Pentane	Cyclohexane
Petroleum ether (light)	Petroleum ether
Hexane	Pentane
Cyclohexane	Carbon tetrachloride
Carbon tetrachloride	Benzene
Benzene	Chloroform
Ethyl ether	Ethyl ether
Chloroform	Ethyl acetate
Methylene chloride	Ethanol
Ethyl acetate	Water
2-Propanol	Acetone
Ethanol	Acetic acid
Methanol	Methanol
Acetic acid	

Note: For additional solvents, see *The Chemist's Companion*, p. 375 (Bib-A21).

Elution solvents should be as pure and water-free as possible.

See *J. Chem. Educ.* **50**, 401 (1973) for a description of an alternative elution technique called dry-column chromatography.

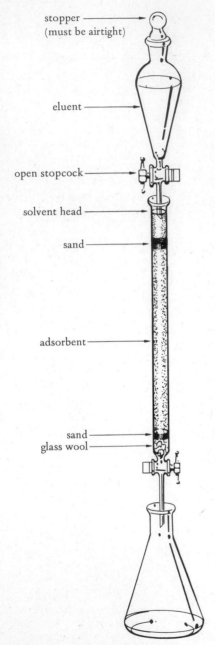

stopper
(must be airtight)

eluent

open stopcock

solvent head

sand

adsorbent

sand
glass wool

Figure 1.   Packed column with continuous-feed reservoir

ing with the weaker eluents near the top of the eluotropic series. These will wash down the most weakly adsorbed components while strongly adsorbed solutes remain near the top of the column. By adding more powerful eluents, the remaining solute bands can then be washed off the column one by one.

In practice, it is best to change eluents gradually by using solvent mixtures rather than to change directly from one solvent to another. In *stepwise elution* the strength of the eluting solvent is changed in small stages by adding small portions of a stronger eluent to the weaker one. Because subsequent portions of the stronger eluent have less effect on elution power than the first one, the proportion of that eluent is increased more-or-less exponentially. For example, 5% methylene chloride in hexane may be followed by 15% and 50% mixtures of these solvents. One "rule of thumb" suggests that the eluent composition should be changed after three column volumes of the previous eluent have passed through; for example, if the packed volume of the adsorbent is 15 ml, then the eluent composition should be changed with every 45 ml or so of eluent.

Columns.   In choosing a column for a particular chromatographic separation, an experimenter must first consider the amount of adsorbent needed for a given amount of sample and then choose a column that will completely contain the adsorbent with about 10–15 cm to spare. Ordinarily, the height of the column packing should be at least 10 times its diameter.

There are many different kinds of chromatography columns, from a simple glass tube with a constriction at one end to an elaborate column with a porous plate to support the packing and a detachable base to facilitate extrusion of the adsorbent. A buret will suffice for many purposes, particularly if it has a Teflon stopcock. If the column does not have a stopcock, the tip can be closed with a piece of flexible tubing equipped with a screw clamp. Unless the tubing is resistant to the eluents (polyethylene and Teflon will not contaminate most solvents) it should be removed before elution begins. If the column contains a porous plate to support the packing, no additional support will be necessary; otherwise the column packing should be supported on a layer of glass wool and clean sand.

Flow Rate.   The rate of solvent flow through the column should be slow enough that the solute can attain equilibrium, but not so slow that the solute bands will broaden appreciably by diffusion. For most purposes, a flow rate of between five and fifty drops per minute should be suitable—difficult separations require the slower rates. The flow rate can be reduced, if necessary, by partly closing the stopcock on the column or by reducing the *solvent head* (the height of the solvent above the adsorbent). The flow rate can be increased by increasing the solvent head or by applying uniform air pressure to the top of the column through a large ballast volume (for example, with a rubber bulb connected to a carboy).

It is not advisable to try increasing the flow rate by applying suction to the bottom of the column; this procedure can cause the column adsorbent to separate.

Packing the Column

To achieve good separation with a chromatographic column, it must be packed properly. The packing must be uniform, without air bubbles or channels, and its surface must be even and horizontal. Columns using alumina are generally packed with the dry adsorbent, while those using silica gel are packed with a *slurry* containing the adsorbent in a solvent.

Equipment and supplies:
Buret (or other column), stopper, powder funnel, Pasteur pipet, tapper, adsorbent, eluent, clean sand, glass wool, filter paper.

To pack a column with *silica gel,* first fill it about half full (close the outlet!) with the least powerful eluting solvent to be used in the separation (often petroleum ether or hexane). Push a plug of glass wool to the bottom of the column with a glass rod and tamp it down enough to form a level surface and press out any air bubbles. At this time, be sure the column is absolutely vertical and clamp it tightly so that it remains that way. Carefully pour clean sand (enough to form a 1 cm layer) through a funnel into the column. As it filters down through the solvent, tap the column gently and continuously with a "tapper" made of (for example) a rubber stopper on the end of a pencil, so that the sand forms a level, uniform deposit at the bottom of the column. Mix the measured amount of silica gel (or another slurry-packed adsorbent) thoroughly with enough solvent to make a fairly thick, but pourable, slurry. Pour a little of this slurry into the column, with tapping, through a powder funnel, so that the adsorbent gently filters down to form a layer 2 cm thick at the bottom. Then open the column outlet to let the solvent drain into a flask while you slowly add the rest of the slurry, tapping constantly to help settle and pack the adsorbent. When all of the adsorbent has been added, close the outlet. The surface of the adsorbent should be as level as possible, so continue tapping until it has completely settled. (Stirring the top of the solvent layer while the adsorbent is settling can help form a level surface.) Rinse down any adsorbent ad-

The column should be tapped gently near the middle (where it is clamped) to avoid displacing it from the vertical.

Keep the adsorbent container tightly capped!

If the slurry becomes too thick to pour, add more solvent to it.

There should be enough solvent in the column so that the solvent level is always well above the adsorbent level; if not, add more solvent.

hering to the sides of the column using an eyedropper or Pasteur pipet filled with solvent. Place a piece of filter paper cut to size on top of the adsorbent column. If the column will not be used immediately, carefully fill it with solvent and stopper it tightly.

To pack a column with *alumina,* fill it about 2/3 full with the appropriate solvent and add glass wool and a layer of sand as described above. With the stopcock closed, pour enough *dry* alumina through a dry powder funnel, with tapping, to form a 2 cm layer at the bottom of the column. Then open the outlet and add the rest of the dry alumina while the solvent drains, tapping constantly so that it settles uniformly. Finish preparing the column as described above for a silica column.

Operational Procedure

If the sample is a solid, dissolve it in a minimum amount of a suitable nonpolar solvent; otherwise, use the neat liquid. Open the column outlet until the solvent level comes down to the top of the adsorbent (no lower). Apply the sample carefully around the circumference of the adsorbent using a capillary pipet, so that it spreads evenly over the surface of the adsorbent, and open the outlet until the liquid level again falls to the top of the adsorbent. Pipet a little of the initial eluent around the inside of the column to rinse down any adherent sample and again open the outlet to bring the liquid level to the top of the adsorbent. Carefully add 3–5 ml of eluent and sprinkle enough clean sand through the liquid to provide a uniform protective layer about 0.5 cm thick.

Clamp a separatory/addition funnel over the column and measure the initial eluent into it, then add enough eluent to cover the adsorbent with 10–15 cm or more of liquid. Place a collecting vessel at the column outlet, open the outlet, and continue adding solvent (or use a continuous-feed reservoir as shown in Figure 1) to keep the liquid level nearly constant throughout the elution. When the eluent is being changed, allow the previous solvent to drain to the level of the sand before adding the next eluent from the reservoir.

If the components are colored or can be observed on the column by some visualization method, change collectors each time a new band of solute begins to come off the column and when it has about disappeared from the column. If two or more bands overlap, collect the overlapping regions in separate containers to avoid contaminating the pure fractions. When all of the desired components have been eluted from the column, evaporate the solvent ⟨OP–14⟩ from the pure fractions unless the Procedure indicates otherwise.

If the components are not visible and the procedure does not specify the volume of eluent to be collected, collect equal fractions (usually 50–100 ml each) in tared collecting vessels, evaporate the solvent from each fraction, weigh the collectors, and plot the mass of each residue versus the fraction number to obtain an elution curve such as that illustrated in Figure 2. From the elution curve it should be possible to identify separate components and decide which fractions can be safely combined.

J. Chem. Educ. 50, 401 (1973) describes a specially-constructed solvent evaporator for column chromatography.

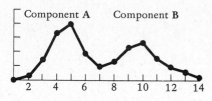

Figure 2.   Elution curve

Summary

1.   Pack column using designated adsorbent and solvent.

2.   Drain column to top of adsorbent, add sample, drain, rinse down sample, drain again.

Each time, drain only until the liquid level reaches the top of the adsorbent.

3.   Add a little solvent and 0.5 cm of sand, drain to top of sand.

4.   Add eluent from reservoir, put collector in place, open column outlet, keep eluent level in column nearly constant.

5.   To change elution solvents, drain previous solvent to top of sand layer, add next eluent from reservoir.
IF components are visible,
GO TO 6.
IF components are not visible,
GO TO 7.

6.   Change collectors when new band starts or ends, and also for overlapping areas between bands.
GO TO 8.

7.   Change collectors after a predetermined volume has been collected.

8.   When appropriate, evaporate and weigh desired fractions; combine fractions containing the same component.

Thin-Layer Chromatography

Principles and Applications

As its name suggests, "thin-layer" chromatography (TLC) is carried out on a very thin layer of chromatographically active material dispersed on the surface of an inert support. In TLC, as in column chromatography ⟨OP–16⟩, a liquid mobile phase moves along a solid stationary phase, carrying with it the components of a mixture. In the process, the components are separated by differential partitioning between the solid and liquid phases. In its usual form, the stationary phase is an adsorbent such as alumina or silica gel, so separation is effected by the same basic mechanism as in adsorption chromatography on a column. Although TLC is not applicable to large-scale preparative separations, it is much faster than column chromatography, and can be carried out with very small amounts of material, so that little is wasted. It also offers greater speed and resolution than the related technique of paper chromatography ⟨OP–18⟩, and can be applied to a broader range of organic compounds.

The principles of liquid-solid adsorption chromatography are discussed in ⟨OP–16⟩.

TLC is often used by organic chemists to help identify unknown compounds, analyze reaction mixtures, determine the purity of products, and monitor various processes. For example, by running thin-layer chromatograms of a reaction mixture at regular intervals, an investigator can determine the optimum reaction time by comparing the relative amounts of product, reactants and by-products appearing on each chromatogram. If a product mixture is to be separated by column chromatography, TLC can help determine the best solvent for the separation; minute amounts of the mixture are spotted on TLC plates having the same adsorbent as the column and different solvents are applied to each spot to see which gives the best separation. The fractions eluted from a column can also be analyzed by TLC to determine which component is in each fraction, so that the fractions containing the same component can be combined and evaporated. Finally, if a product is further purified by recrystallization, TLC analysis can quickly show whether the "pure" product contains appreciable amounts of impurities.

The procedure is described in "Choosing a Developing Solvent" below.

Experimental Considerations

Adsorbents.   In contrast to column chromatography, most TLC separations are carried out with the mobile phase moving *up* the thin layer of adsorbent by capillary action, rather than *down* through the adsorbent by gravity. The adsorbent is more finely divided than that used in a column and is usually provided with a *binder* to make it stick to the support. The most commonly used binder for silica gel and alumina is plaster of Paris ($CaSO_4 \cdot \frac{1}{2}H_2O$), which hardens when combined with water or other hydroxylic solvents by forming gypsum ($CaSO_4 \cdot 2H_2O$) or other solvated forms of calcium sulfate. The most popular backing for "do-it-yourself" TLC plates is glass, in the form of either microscope slides or glass plates measuring (typically) 20×20 cm. Commercial plates on glass, plastic, and aluminum backings are available in a variety of sizes and compositions.

TLC Plates. Microscope-slide TLC plates are usually made by dipping two slides at a time into a slurry of the adsorbent in water or an organic solvent. Although water gives a more durable layer, organic solvents such as chloroform or a chloroform-methanol mixture are satisfactory and considerably easier to work with (see Table 1 for the compositions of several of these slurries). With some practice, the dipping technique can produce a reasonably uniform layer of adsorbent on microscope slides; however, it cannot be used conveniently for large plates, which are coated by spreading, pouring, or spraying techniques. After the layer has been applied, a plate should be inspected by holding it up to a strong light to see that there are no streaks, thin spots, lumps, or loose particles. If the coating is not uniform or is otherwise imperfect, the plate should be scraped clean, washed, and coated again.

If a plate has been prepared with water or another hydroxylic solvent, the adsorbent must be *activated* by heating it for a period of time, usually at 110°, to remove solvent and increase its adsorptive power. The activated slides should be used immediately or stored in a dessicator over an efficient drying agent (blue silica gel is recommended), since they lose their activity rapidly in a moist atmosphere.

Spotting. The activated plate is prepared for development by *spotting* it with a solution of the mixture to be analyzed. Since incorrectly-placed spots may run into each other or onto the edge of the adsorbent layer, it is important to position them accurately. This is most conveniently done with a template, although a ruler (preferably a transparent plastic one) will work if it can be supported just above the surface of the plate so that it does not touch the adsorbent. On a mi-

TLC-grade adsorbents containing $CaSO_4 \cdot \frac{1}{2}H_2O$ are sold under trade names like Silica Gel G (for "gypsum") and Alumina G. It is much simpler to use the premixed adsorbent than to prepare it yourself.

Table 1. Composition of nonaqueous TLC slurries for dipping

Adsorbent	grams adsorbent	ml CHCl₃	ml CH₃OH
Silica Gel G	35	67	33
Alumina G	60	70	30
Cellulose★	50	50	50

★ No binder required

Aqueous slurries of adsorbents containing a binder will harden in a few minutes; chloroform slurries can be kept for several days.

When preparing and handling TLC plates, it is very important to avoid touching the surface of the glass or adsorbent and to protect the plate from foreign materials.

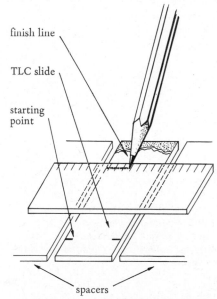

Figure 1. Scoring a TLC slide

This procedure prevents accidental overde-velopment of the chromatogram, since the solvent will stop moving when it reaches the scored finish line.

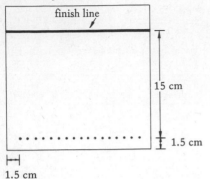

Figure 2.   Spotted TLC plate (20 × 20 cm)

The solvent should, if possible, be quite nonpolar and have a boiling point of 50–100°. Chloroform is a good choice if the solute is sufficiently soluble in it.

Commercial TLC sheets usually have thinner coatings and require less sample than homemade plates.

croscope-slide TLC plate, the starting point is marked on both edges about 1 cm from the bottom (see Figure 1). The finish line, if desired, is inscribed across the width of the slide about 0.5 cm from the end of the absorbent layer. Two spots can be placed about 1 cm apart between the starting marks and equidistant from the edges of the plate. If three spots are to be developed, one is placed in the center and the other two about 8 mm from the edges.

A 20 × 20 cm TLC plate can accommodate up to 18 spots spaced 1 cm apart along the starting line, with a distance of 1.5 cm between the end spots and the edges of the plate. The starting line is 1.5 cm from the bottom of the plate; the finish line is usually 10 or 15 cm from the starting line. It can, however, be farther if additional separation is required.

The substance to be analyzed is dissolved in a suitable solvent to make an approximately 1% solution. Column chromatography fractions and other solutions can often be used as is, if the solute is present at a concentration of about 0.1–2%. The spots are best applied with a syringe or micro-pipet, delivering up to 5 μl of liquid to form a spot 1–3 mm in diameter. Larger volumes of the solution should be delivered in successive applications on the same spot, letting the solvent dry after each application.

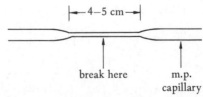

Figure 3.   Drawing a capillary micropipet

Capillary micropipets can be reused a few times if rinsed by solvent between applica-tions; however, it is best to use a new pipet for each different mixture.

It is important not to dig a hole in the ad-sorbent surface while spotting; this will obstruct solvent flow and distort the chro-matogram.

A pair of capillary micropipets can be prepared by heat-ing an open-ended melting-point capillary in the middle over a microburner and drawing it out to form a very fine capillary about 4–5 cm long. The tube is then allowed to cool, scored, and snapped apart in the middle. To spot a plate, the narrow capillary tip is dipped in the solution, then very gently applied to the surface of the TLC plate, at the proper location, for only an instant. With a little practice, it should be possible to deliver just the right volume of solu-tion. Too much solute can result in "tailing" (a diffuse solute zone following the spot), "bearding" (ditto, preceding the spot), overlap of components, or unreliable R_f values; too little solute will make it difficult to detect some of the components. It is often worthwhile to try 1, 2, and 3 applications at three points on the TLC plate to determine which quantity gives the best results.

Choosing a Developing Solvent. Solvents that are suitable eluents for column chromatography are equally suitable as TLC developers; the eluotropic series of Table 2 〈OP–16〉 should be of help in choosing a solvent for a particular application. It is best to choose the least polar solvent that will give a satisfactory separation, and to aim for a separation in which the major components have traveled one-third to one-half the distance to the finish line by the end of the development. A quick way to find the best solvent is to spot a TLC plate with as many spots as you have solvents to test (you can use a grid pattern with the spots 1.5–2 cm apart), then to apply enough solvent directly to each spot to form a circle of solvent 1–2 cm in diameter. After visualization (see below), the best solvent should show an outer belt of solute about midway from the center. If none of the solvents gives this kind of separation, choose two mutually miscible solvents that bracket the midpoint (one with too much solute migration, the other with too little) and test them in varying proportions.

Development. TLC plates are *developed* by placing them in a *developing chamber* containing the developing solvent and a paper liner, which helps saturate the atmosphere with the solvent vapors. A suitable chamber can be a jar with either a screw-cap or ground-glass lid (Figure 4), a beaker covered by aluminum foil or plastic (polyethylene) wrap, or any of a variety of cylindrical or rectangular containers with tight-fitting closures. The container should be as small as possible so that it will quickly fill with solvent vapors. Development should be carried out in a place away from direct sunlight or drafts, to prevent temperature gradients. When the solvent reaches the finish line, the TLC plate is removed and allowed to dry.

Visualization. If the spots are colored, they can be observed immediately; otherwise, they must be *visualized* by some method that will allow them to be distinguished from the background. A good general visualization procedure is to place the dry plate in a closed chamber such as a wide-mouthed jar with a screw-cap lid, add a few crystals of io-

You can practice your spotting technique on a used plate or on the portion of a plate above the finish line.

Solvents should be chromatography grade or redistilled for good results.

Shorter distances allow insufficient separation; longer ones can result in spreading of the spots by diffusion.

Mark the circumference of each circle before the solvent dries.

Hexane, benzene, chloroform, and methanol or ethanol (alone or in binary combinations) can effect most of the commonly encountered separations.

A 250-ml Erlenmeyer flask works well for 2 × 10 cm strips cut from commercial TLC plates.

A solvent advance of 10–15 cm may take 15–45 minutes; a microscope-slide plate can often be developed in 5–10 minutes.

The TLC plate should not be left in the solvent after development is complete, since the spots will spread by diffusion.

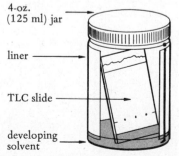

4-oz. (125 ml) jar

liner

TLC slide

developing solvent

Figure 4. Developing chamber for TLC slides

The iodine color fades in time, so the spots should be marked promptly. Saturated hydrocarbons and alkyl halides may not form spots with iodine.

dine, and gently heat the chamber on a steam bath so that the iodine vapors sublime onto the adsorbent. Most organic compounds will form brown spots; unsaturated compounds will often show up as white spots against the dark background. Another simple method is to hold the plate under an ultraviolet light (about 254 nm), which will reveal fluorescent compounds on an ordinary adsorbent or compounds that quench fluorescence if the adsorbent contains an added phosphor.

Spots can often be located by spraying a special visualizing reagent onto the TLC plate. A general-purpose visualizing reagent is 10% potassium dichromate in 50% sulfuric acid, which reveals all components as black spots when the plate, after spraying, is heated at 150° for several minutes. Visualizing reagents for specific classes of compounds include ninhydrin (for amino acids) and 2,4-dinitrophenylhydrazine (for aldehydes and ketones). All spraying should be done under a hood or in a spraying chamber. An aerosol or special spraying bottle is used, and a thin spray is applied from about 2 feet away. Large plates are sprayed by crisscrossing them with horizontal and vertical passes.

$$R_f = \frac{\text{Distance traveled by spot}}{\text{Distance traveled by solvent}}$$

The R_f value of a given component can depend on the nature, thickness, and activity of the adsorbent; the identity and purity of the solvent; the size of the sample; and the temperature of the developing chamber.

Analysis.   To identify the components on a developed chromatogram, the R_f values of the spots must be determined. This is done by measuring the perpendicular distance from the starting line to the center of each spot (or its "center of gravity" if it is irregular) and dividing this by the distance traveled by the solvent. Although R_f values are reported in the literature for certain compounds, the absolute values can rarely be used to prove the identity of a substance, since they depend on a number of factors that are difficult to standardize. They are, however, a useful guide to the relative migra-

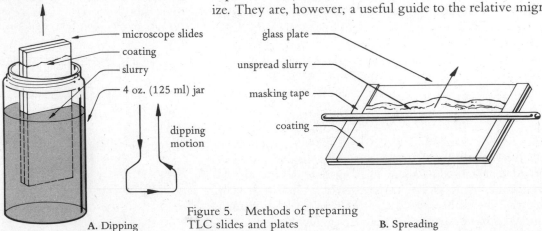

Figure 5.   Methods of preparing TLC slides and plates

A. Dipping — microscope slides, coating, slurry, 4 oz. (125 ml) jar, dipping motion

B. Spreading — glass plate, unspread slurry, masking tape, coating

tion distances of various compounds, since with the same solvent and adsorbent, components should migrate in the same sequence. The only way to be reasonably sure that an unknown is identical to a known compound is to run the known beside it on the same TLC plate and compare the distances traveled by the two. Even then, its identity must often be confirmed by an independent method.

Preparation of TLC Plates

TLC plates can be prepared by one of the following methods:

A. *Dipping.* Combine 33 ml of methanol and 67 ml of chloroform in a 4 oz. (125 ml) screw-cap jar, stir in 35 g of Silica Gel G, and shake the capped jar vigorously for about a minute. Stack two *clean* microscope slides back-to-back, holding them together at the top. Without delay, dip them into the slurry for about 2 seconds, using a smooth, unhurried, paddle-like motion (see Figure 5**A**) to coat them uniformly with the adsorbent. (Dip the slides shortly after shaking, so that the adsorbent does not have time to settle, and immerse them deeply enough so that only the top 1 cm or so remains uncoated.) Touch the bottom of the stacked slides to the jar to drain off the excess slurry, let them air dry a minute or so to evaporate the solvent, then separate them and wipe the excess adsorbent off the edges with a tissue paper. Activate the slides by heating them in a 110° oven for 15 minutes, or by placing them in a covered beaker heated to that temperature.

Several students should work together so they can use the same slurry. Add more slurry (or dilute it to replace evaporated solvent) as necessary.

Handle the clean slides by the edges or at the very top, or use forceps. The slides can be cleaned with detergent and water, then rinsed with distilled water and 50% aqueous methanol before dipping.

Slides can be stored in a microscope-slide box inside a dessicator.

B. *Spreading.* Put down some newspapers or other paper on your bench top. Place two strips of masking tape along two parallel edges of a *clean* glass plate (20 × 20 cm is standard) so that each strip of tape covers a 0.5-cm strip along the edge of the plate. Mix about 10 g of Silica Gel G with 20 ml of water (stirring and shaking well to get out any lumps) and pour some of it along one of the untaped ends. Spread it out by drawing a heavy glass rod across the plate (see Figure 5–B) using one firm, continuous motion. Let the plate air dry for ten minutes or more, remove the tape, and set the plate vertically in a well-ventilated 110° oven for at least 30 minutes to activate the adsorbent.

Additional layers of tape can be used for thicker coatings.

The exact amount of slurry will have to be determined by trial and error.

Since the binder sets rapidly, the mixing and spreading must be completed in 2–3 minutes. Making more than one pass with the rod is usually not desirable.

Operational Procedure

Equipment and supplies:
TLC plate(s), developing chamber
with lid, filter-paper liner, developing
solvent, micropipet or syringe, pencil,
ruler, unknown solution, standard so-
lution(s), visualizing agent (spraying
bottle, iodine chamber, UV lamp,
etc.).

Do not actually draw a starting line
across the plate—just mark it at the
edges.

Prepare a covered developing chamber containing a liner
made of filter paper (or chromatography paper) which extends at least halfway around the circumference. Pour in
enough developing solvent to form a 3–5 mm layer on the
bottom. Tip the chamber and slosh the solvent around to
soak the liner and saturate the atmosphere with solvent
vapors. Keeping the apparatus saturated with solvent vapors
reduces the development time and results in better shaped
spots.

Mark the starting point and finishing line on a TLC
plate and spot it **(a)** with the solution to be chromatographed and **(b)** with any solutions of known standards (if
used). When the spots are dry, place the slide or plate in the
chamber, spotted end down, propping the back (uncoated)
side against the liner. Cover the chamber immediately.

When the solvent reaches the finish line, carefully remove the plate and mark the centers of any visible spots
with a pencil. Let the plate dry thoroughly and visualize the
spots (if necessary) by one of the methods described above.
Measure the distance from the center of each spot to the
starting line, and calculate the R_f values. If requested, make a
permanent record of the chromatogram **(a)** by tracing it on
semitransparent paper, **(b)** by photographing or photocopy-
ing it, or **(c)** by pressing strips of transparent tape onto the
surface and stripping off the adsorbent layer.

Summary

1.   Obtain or prepare TLC plate(s).

2.   Mark starting point and finishing line on each plate.

3.   Spot TLC plate with unknown solution(s) and
with standard(s).

4.   Place TLC plate in developing chamber containing
solvent and liner; cover chamber.

5.   Remove plate when solvent reaches finish line,
allow it to dry.

6.   Visualize spots, if necessary, and mark their centers.

7.   Measure R_f values; copy or otherwise preserve the
chromatogram.

Paper Chromatography

Principles and Applications

Paper chromatography is similar to thin-layer chromatography ⟨OP–17⟩ in practice, but quite different in principle. Although paper consists mainly of cellulose, the stationary phase in a typical paper chromatography experiment is not cellulose, but the water that is adsorbed or chemically bound to it. Development is carried out by passing a comparatively nonpolar mobile phase through the cellulose fibers, partitioning the solutes between the bound water and the mobile phase. Paper chromatography thus operates by a liquid-liquid partitioning process rather than by adsorption on the surface of a solid.

Since only polar compounds are appreciably soluble in water, ordinary paper chromatography is limited to the separation of polar substances, especially polyfunctional compounds like sugars and amino acids. Manufactured chromatography paper is quite uniform and doesn't have the activation requirements of TLC adsorbents, so R_f values obtained by paper chromatography are generally more reproducible than those from TLC. However, the resolution of spots is often poorer and the development times are usually much longer. Nevertheless, paper chromatography is a valuable analytical technique that, in its many forms, can be applied to a large variety of separations.

Experimental Considerations

Paper. Ordinary filter paper can be used for paper chromatography; however, the same paper is manufactured in convenient rectangular sheets, strips, and many other shapes for chromatographic use. For good results, the paper surface must be kept clean. Sheets of paper or plastic should be laid on the bench top to prevent contamination, and the paper should be handled only by the edges or along the top, beyond the anticipated finish line. Forceps are helpful; it is sometimes a good idea to leave a "handle" (which can be cut off just before development) on one (or both) ends of the paper. For much qualitative work, the paper can be 10–15 cm high (in the direction of development) and wide enough to accommodate the desired number of spots, spaced 1.5–2.0 cm apart. Narrow strips accommodating two or more spots can be developed in test tubes, bottles,

Cellulose chromatography paper contains up to 22% water, corresponding to two water molecules per $C_6H_{10}O_5$ unit.

The principle is the same as that of liquid-liquid extraction ⟨OP–13⟩.

Paper can be specially treated to separate less polar compounds by impregnating it with other stationary phases, such as formamide or paraffin oil.

These disadvantages can be minimized by using powdered cellulose as the coating on a TLC plate.

Whatman #1 and S&S 4043b are standard chromatography papers.

The spots tend to spread out more than with TLC.

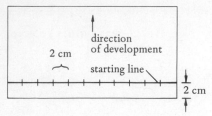

Figure 1.   11 × 22-cm paper marked for spotting

Acetone, ethyl ether, and chloroform are frequently used.

During the spotting, it is best that nothing touch the underside of the paper at the point of application.

Practice your spotting technique on a piece of filter paper before attempting to spot the chromatogram. Keep in mind that too little is usually better than too much.

The water is needed to maintain the composition of the aqueous stationary phase on the cellulose.

cylinders, or Erlenmeyer flasks. A wider sheet can be rolled into a cylinder and developed in a beaker. The paper should be cut so that its grain is parallel to the direction of development, and the starting line should be marked in pencil about 2 cm above the bottom edge. The positions of the spots can also be marked lightly with pencil, but the finish line is not drawn until development is complete.

Spotting.   As in TLC, the substance to be analyzed should be dissolved in a suitable solvent, usually at a concentration of about 1%. The spotting can be performed by the same techniques as for TLC (see ⟨OP–17⟩) using a micropipet or syringe. However, for many purposes a platinum wire loop 2–3 mm in diameter—or even a round toothpick saturated with the solution—is quite adequate. Each spot can be made with about 10 μl of solution and should be 2–5 mm (no more) in diameter. Larger volumes of solution should be applied stepwise, drying the spot after each application.

Developing Solvents.   The developing solvents used for paper chromatography are nearly always mixtures containing an organic solvent, water, and often a third component to increase the solubility of the water in the solvent or to provide an acidic or basic medium. The solvent mixtures are frequently prepared by saturating the organic solvent(s) with water in a separatory funnel, separating the two phases, and using the organic phase for development. Alternatively, monophase mixtures having about the same composition as the organic phase of the saturated mixtures can be used; these yield about the same results. Some typical monophase solvent mixtures are listed in Table 1. Most solvent mixtures should be made up fresh each time they are used and not kept more than a day or two.

Table 1.   Monophase solvent mixtures for paper chromatography

Solvents	Composition	Equivalent 2-Phase System
2-propanol–ammonia–water	9:1:2 (by volume)	none
1-butanol–acetic acid–water	12:3:5 (by volume)	4:1:5
phenol–water	500 g PhOH, 125 ml H_2O	saturated solution
ethyl acetate–1-propanol–water	14:2:4 (by volume)	6:1:3

Development.   Paper chromatograms can be developed by ascending, descending, horizontal, and radial techniques. An ascending development is carried out in much the same way as for TLC. However, the solvent mixture is usually left longer in the chamber before development to saturate the atmosphere with solvent, and the paper liner can be omitted.

Visualization. Visualization of spots by ultraviolet light is quite useful in paper chromatography, since paper fluoresces dimly in a dark room and many organic compounds will quench that fluorescence. Paper chromatograms can also be visualized by spraying them or by dipping them into a solution of a suitable visualizing reagent.

J. Chem. Educ. 49, 20 (1972) describes "an easily constructed aerosol sprayer."

Analysis. Although R_f values in paper chromatography are somewhat more reproducible than those obtained from TLC, it is usually necessary to run a standard along with the unknown for qualitative analysis. If possible, the solvent and concentration should be the same for the standard as for the unknown. Spot migration distances are customarily measured from the starting line to the *front* of each spot, rather than to the center as for TLC.

The distance from the starting line to each spot should be measured along a line passing through the spot perpendicular to the starting line.

(cover omitted for clarity)

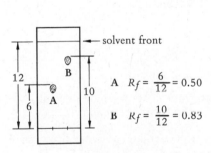

$A \quad R_f = \dfrac{6}{12} = 0.50$

$B \quad R_f = \dfrac{10}{12} = 0.83$

$$R_f = \frac{\text{Distance to leading edge of spot}}{\text{Distance to solvent front}}$$

Figure 2. Measuring R_f values in paper chromatography

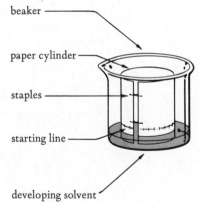

Figure 3. Apparatus for paper chromatography

Operational Procedure

Obtain chromatography paper and cut it to form an 11 × 22 rectangle (or another appropriate size). Without touching the surface of the paper, draw a starting line 2 cm from one long edge (the "bottom" edge) and lightly mark the posi-

Alternatively, narrow strips can be suspended in jars, test tubes, or other appropriate containers.

Equipment and supplies:
Chromatography paper, pencil, ruler, capillary micropipet(s) (or toothpicks, etc.), solutions to be spotted, developing chamber, developing solvent, visualizing reagent and equipment.

The paper must not touch the sides of the beaker.

Development may take an hour or more, depending on the solvent and the distance traveled.

Smaller chromatograms can be pre-punched and hung on a wooden applicator stick (clamped in a horizontal position) to air dry.

tions for spots, spacing them about 2 cm from each side and 1.5–2.0 cm apart. Spot the paper with the solutions and standards to be chromatographed, then roll it into a cylinder with the starting line at the bottom and staple the ends together, leaving a small gap between them. Add enough developing solvent to a 600 ml beaker to give a liquid depth of about 1 cm on the bottom, cover the beaker tightly with aluminum foil or polyethylene film, and slosh the solvent around in it for a short time. After allowing about 10 minutes for the solvent to saturate the chamber, quickly uncover it, place the paper cylinder inside (spotted end down) and cover it tightly again. Put the beaker in an area away from sunlight and strong drafts.

When the solvent front is a centimeter or less from the top of the paper, carefully remove the cylinder and separate the edges, then draw a line along the *entire* solvent front with a pencil. If any spots are visible at this time, outline them with a pencil (carefully, to avoid tearing the wet paper), as they may fade in time. Again roll the paper into a cylinder and stand it on edge to air dry. When it is completely dry, visualize the spots by an applicable method. Measure the appropriate spot and solvent migration distances and calculate the R_f values.

Summary

1. Obtain chromatography paper and cut it to size.

2. Mark starting line, apply spots.

3. Develop chromatogram in appropriate solvent and chamber.

4. Remove chromatogram, mark solvent front and visible spots, dry chromatogram.

5. Visualize spots if necessary.

6. Measure R_f values.

Washing and Drying Operations

Washing Liquids

Organic liquids or solutions often contain impurities that can be removed by extracting them into a suitable wash liquid. The organic substance being washed may be either a *neat* liquid (without solvent) or a compound dissolved in a solvent. In either case, it must not dissolve appreciably in the wash liquid or it will be extracted along with the impurities.

Water and saturated aqueous sodium chloride are used to remove water-soluble impurities such as salts and low-boiling polar organic compounds. Saturated sodium chloride is frequently used for the last washing before a liquid is dried ⟨OP–20⟩ because it removes excess water from the organic liquid by the salting-out effect. It is preferred to water in some other cases because it helps prevent emulsions at the interface between the liquids. (Emulsions can also be broken up or prevented by the methods described in ⟨OP–13⟩.)

Some wash liquids contain chemically reactive solutes that are used to convert water-insoluble impurities into soluble salts. Aqueous solutions of sodium hydroxide, sodium carbonate, and sodium bicarbonate remove acidic impurities by converting them into soluble sodium salts. Dilute aqueous solutions of hydrochloric acid or sodium hydrogen sulfate are useful for removing alkaline impurities such as amines. Aqueous sodium hydrogen sulfite can remove certain aldehyde and ketone impurities by forming soluble bisulfite addition compounds. When a chemically reactive wash liquid is used, one should perform a preliminary washing with water or aqueous sodium chloride to remove most of the water-soluble impurities. This may prevent a potentially violent reaction between the reactive wash liquid and the impurities.

Reactions of washing solutions

$$NaHCO_3 + HA \text{ (acidic impurity)}$$

$$Na^+A^- + H_2O + CO_2$$

$$Na_2CO_3 + 2HA$$

$$2Na^+A^- + H_2O + CO_2$$

$$HCl + RNH_2 \text{ (amine)}$$

$$RNH_3^+Cl^- \text{ (amine salt)}$$

$$NaHSO_3 + RCHO \text{ (aldehyde)}$$

$$\underset{|}{OH}$$

$$RCHSO_3^-Na^+$$

(bisulfite addition compound)

Operational Procedure

The procedure for washing liquids is essentially the same as that for solvent extraction ⟨OP–13⟩, except that the extrac-

If the washing reaction generates a gas like CO₂, the layers should be stirred or swirled until gas evolution subsides before the separatory funnel is stoppered and shaken.

See ⟨OP–13⟩ for experimental details and a summary.

tion solvent (wash liquid) is discarded while the liquid being washed is retained. For example, if an ethereal solution of benzaldehyde is being washed with aqueous sodium hydroxide to remove benzoic acid impurities, the less dense ether layer will remain in the separatory funnel after the first portion of wash liquid has been drained out. It can thus be washed repeatedly with additional aqueous portions until all of the benzoic acid has been removed, then poured out the top of the funnel. If the liquid being washed is *more* dense than the washing solvent, it must be returned to the separatory funnel after every washing step except for the final one.

OPERATION 20

Drying Liquids

Organic liquids and solutions that have been separated from a reaction mixture or isolated from a natural product often retain traces of water. This water must often be removed before subsequent operations are performed. Liquids can be dried conveniently by allowing them to stand in contact with a *drying agent,* which is then removed by decanting or filtration.

Drying Agents

Hydration of Magnesium Sulfate

$$MgSO_4 + 7H_2O \leftrightharpoons MgSO_4 \cdot 7H_2O$$

Most drying agents are inorganic salts that form hydrates by combining chemically with water. A drying agent is evaluated according to the amount of water it can absorb per unit mass (capacity), the degree of dryness it can bring about (intensity), and how fast the drying takes place (speed). An ideal drying agent should have a high capacity, high intensity, and short drying time; it also must not react with or dissolve in the liquid being dried. Anhydrous magnesium sulfate probably comes closest to meeting all these criteria, although other drying agents have advantages for specific applications. For example, sodium sulfate can be used to predry very wet solutions because it has a high water capacity. The final drying is then accomplished with a high-intensity drying agent such as calcium sulfate. Table 1 summarizes the properties of the most common drying agents.

Table 1. Properties of commonly-used drying agents

Drying Agent	Speed	Maximum Capacity	Intensity	Comments
magnesium sulfate	fast	105%	medium	Excellent general drying agent, suitable for nearly all organic liquids.
calcium sulfate (Drierite)	very fast	7%	high	Very fast and efficient, but low capacity.
sodium sulfate	slow	126%	low	Inefficient and very slow, good for pre-drying. Loses water above 32.4°C.
calcium chloride	slow to fast	97%	medium	Removes traces of water quickly, large amounts slowly. Reacts with many oxygen and nitrogen-containing compounds.
potassium carbonate	fast	26%	medium	Cannot be used to dry acidic compounds.
potassium hydroxide	fast	very high	high	Used mainly to dry amines; reacts with many other compounds. Causes severe burns.

Note: Capacity is given as the percentage of its weight that a drying agent can (in theory) absorb.

Experimental Considerations

Ethyl ether, ethyl acetate, and other comparatively polar liquids absorb appreciable quantities of water and should be washed with saturated sodium chloride ⟨OP–19⟩ to remove excess water just before drying. They should then be dried with a high-capacity drying agent such as magnesium sulfate. Less polar liquids, such as petroleum ether and chlorinated hydrocarbons, absorb little water and can be dried by calcium sulfate or calcium chloride.

The *quantity* of drying agent needed depends on the capacity and particle size of the drying agent and on the amount of water present. As a rough rule of thumb, about 1 gram of drying agent should be used for each 25 ml of liquid; more should be added if the drying agent has a low capacity or a large particle size and if the liquid has a high water capacity. It is best to start with a small amount of drying agent and then add more if necessary, since starting with too much results in excessive losses by adsorption. The appearance of the drying agent after the drying time is up often suggests whether more is required. Magnesium sulfate clumps together in large crystals as it absorbs water, calcium chloride takes on a glassy surface appearance, and indicating Drierite changes color from blue to pink. If it appears that most of

Small amounts of liquid can be dissolved in ether or another volatile solvent before drying to reduce adsorption losses. The solvent is subsequently removed by evaporation ⟨OP–14⟩.

Spent drying agent contains hydrates which reduce drying efficiency.

The color in indicating Drierite can be leached out by polar liquids such as alcohols. The color reaction is caused by hydration of cobalt (II) chloride.

the drying agent is exhausted, it is best to remove it by decanting or filtering before adding fresh drying agent.

The *time* required for drying depends on the speed of the drying agent and the amount of water present. Calcium sulfate and magnesium sulfate attain most of their drying potential in a few minutes, whereas sodium sulfate may require a half hour or more. A rough test for excessive moisture is to add a granule of indicating Drierite to the liquid. If any pink color is observable, the drying time should be lengthened or more drying agent used. All drying agent should be removed before subsequent operations, since heating a drying agent causes it to give up some of the water it absorbed.

Equipment and Supplies: Erlenmeyer flask, stopper, drying agent, funnel, fluted filter paper.

Operational Procedure

If a second (aqueous) liquid phase forms, remove it with a pipet and add more drying agent.

Select an Erlenmeyer flask that will hold the liquid with plenty of room to spare, and add the liquid and the estimated quantity of a suitable drying agent (keep its container tightly closed!). Stopper and shake the flask, then set it aside for 3–5 minutes, swirling and shaking it frequently throughout the drying period. (If the drying agent is calcium chloride or sodium sulfate, let the flask stand 15-30 minutes or longer.) Examine the drying agent carefully; if most of it is exhausted, remove it by gravity filtration (wash the residue as described below) and replace it by fresh drying agent. If the drying agent is not exhausted but drying appears incomplete, extend the drying time, add more drying agent, or replace it by a more efficient drying agent. When drying appears complete, filter the mixture by gravity ⟨OP–11⟩ using a coarse, fluted filter paper. Wash the drying agent with a small volume of fresh solvent (if applicable) to remove adsorbed solute, and combine the wash liquid with the filtrate.

Coarse-grained drying agents can sometimes be removed by decanting the liquid or filtering the mixture through glass wool.

Summary

1. Choose drying agent, estimate quantity required (if necessary).

2. Mix drying agent with liquid in flask, stopper and shake, set aside.

3. Swirl occasionally until drying time is up.
IF drying agent is exhausted *or* aqueous phase separates,
GO TO 4.
IF not,
GO TO 5.

4. Remove aqueous phase or used-up drying agent,
add fresh drying agent,
GO TO 3.

5. Check for completeness of drying.
IF liquid is not sufficiently dry,
GO TO 6.
IF liquid is sufficiently dry,
GO TO 7.

6. Extend drying time *or* add more drying agent *or*
filter and add a more efficient drying agent.
GO TO 3.

7. Filter or decant to remove drying agent.

8. Wash drying agent with fresh solvent, combine
wash solvent with filtrate.

*Step 8 applies to solutions rather than
neat liquids.*

Drying Solids

Solids that have been separated from a reaction mixture
or isolated from other sources usually retain traces of water
or other solvents used in the separation. Solvents can be removed by a number of drying methods, depending on the
kind of solvent being removed, the amount of material to be
dried, and the melting point and thermal stability of the
solid compound.

folded filter paper

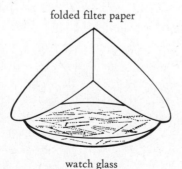

watch glass

Figure 1. Covered watch glass for
drying crystals

Experimental Considerations

Solids that have been collected by vacuum filtration ⟨OP–12⟩ are usually dried on the filter by leaving the vacuum turned on for a few minutes after filtration is complete. Unless the solvent is very volatile, further drying is then required. Comparatively volatile solvents can be removed by simply spreading the solid out on a watch glass (cover it to keep out airborne particles) and placing it in a location with good air circulation (such as a hood) for a sufficient period of time. If faster drying is necessary and the compound is not heat-sensitive, it can be placed in an evaporating dish and warmed over a steam bath, using a low rate of steam flow to prevent condensation inside the container. Clamping an inverted funnel over the evaporating dish and passing a gentle, dry ⟨OP–22⟩ air stream through it will accelerate the drying rate. Fast drying is also possible in a laboratory oven set at 110°C or another suitable temperature. A shallow tray constructed of heavy aluminum foil promotes fast drying, although glass or porcelain containers can also be used.

A good general method for removing water from all kinds of solids requires the use of a *desiccator* containing a suitable desiccant (drying agent) such as calcium chloride or calcium sulfate. Indicating Drierite, which is calcium sulfate containing cobalt(II) chloride, is very convenient since it changes from blue to pink when the desiccant is exhausted. Although commercial desiccators are readily available, a simple desiccator jar for efficient drying of small amounts of solid can be constructed from an 8-ounce (about 250-ml) wide-mouthed jar with a screw cap (Figure 2). The solid desiccant is added to a depth of 1 cm or so, and a piece of wire screen is cut and bent to fit on top of it, providing a stable, horizontal surface for the sample. The jar is kept tightly closed until ready for use.

The compound must not melt, decompose, or sublime under 100°C if a steam bath is used, or under the oven temperature if a drying oven is used.

Care: Don't blow the crystals away!

The oven temperature should normally be at least 20–30°C below the melting point of the solid.

Containers should be labeled to prevent mixups.

Sulfuric acid and phosphorus pentoxide are also widely used, but these agents are somewhat hazardous for general use in undergraduate laboratories.

Vacuum dessicators combine the use of a dessicant with low air pressure for very fast, efficient drying.

Operational Procedure

Transfer the air-dried solid to a large circle of filter paper on a clean surface and blot it with another filter paper to remove excess water. Break up the solid with a large flat-bladed spatula and rub it against the filter paper with the blade of the spatula until it is finely divided and friable. You may have to blot it with fresh filter papers or pulverize it with a flat-bottomed stirring rod. Spread out the solid in a low container, such as a tray constructed of heavy aluminum

This procedure can be used to remove water from solids when quick drying is not essential. If faster drying is necessary, use a steam bath, a drying oven, or a vacuum dessicator.

foil or a small evaporating dish, and place it in a desiccator charged with fresh desiccant. Leave the solid overnight or longer in the tightly closed desiccator, then remove and weigh it. Leave it in the desiccator for another hour or so (add fresh desiccant if necessary) and then reweigh it to see if drying is complete. Drying should be sufficient for most purposes when two successive weighings differ by less than 1%.

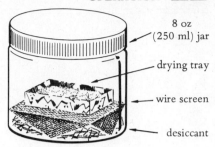

Figure 2.   Desiccator jar

8 oz (250 ml) jar

drying tray

wire screen

desiccant

Drying Gases

The gas most often dried in organic chemistry experiments is air. For instance, it may be necessary to exclude moisture ⟨OP–22a⟩ from a reaction mixture or to evaporate a solvent in a stream of dry air. Other gases are available in cylinders and can be purchased in a form that is dry enough for most applications. In some cases, however, a cylinder gas or a gas generated in the laboratory may also require drying before use.

Air from an air line is sometimes dried by passing it slowly through an ordinary calcium chloride drying tube or a U-tube filled with a drying agent. The gas drying trap in Figure 1 is more efficient for this purpose; it removes oil, rust, and other impurities from the air line as well. About 2–3 cm of glass wool is loosely placed in the bottom of the sidearm test tube, then the glass tube is inserted far enough to contact it. The tube is held there to keep the end clear while the drying agent is added. Another layer of cotton or glass wool is added to absorb residual impurities and keep the sidearm free of desiccant particles. The stopper is then inserted tightly and, if necessary, wired in place. Since most drying agents require a few seconds of contact time for adequate drying, air should be passed through the trap in a *slow* stream.

Gases other than air can be dried by the same general method. Some gases react with certain drying agents, so it is important to make the right choice. Obviously, an acidic gas such as HCl should not be dried with a basic substance like silica or alumina, and gaseous ammonia should not be dried with sulfuric acid. Care must be taken when drying flammable gases like methane or hydrogen to avoid fires and explosions.

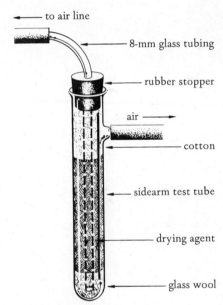

to air line

8-mm glass tubing

rubber stopper

air

cotton

sidearm test tube

drying agent

glass wool

Figure 1.   Trap for drying air and other gases

Indicating silica gel and granular alumina are preferred drying agents; indicating Drierite is also satisfactory.

For very efficient drying, a drying train containing one or more drying towers and gas scrubbing bottles is used. Sulfuric acid is the preferred dessicant for gas scrubbing bottles, which must always be preceded by a trap.

Excluding Moisture

If it is necessary to exclude moisture from an experimental setup during an operation such as refluxing or distillation, a *drying tube* containing a suitable drying agent should be attached to any part of the apparatus that is open to the atmosphere. Granular alumina, silica gel, calcium sulfate

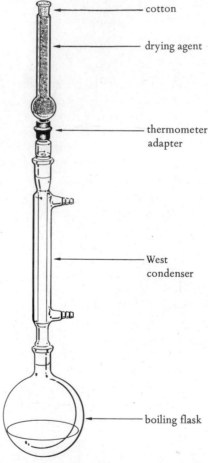

— cotton

— drying agent

— thermometer adapter

— West condenser

— boiling flask

Figure 1.   Assembly for refluxing in a dry atmosphere

(Drierite), and calcium chloride are commonly used for this purpose. Calcium chloride is the least efficient of the four, but it is still adequate in many cases. The drying tube is filled by tamping a plug of dry cotton or glass wool into the bottom opening, adding the dessicant, and inserting another plug of cotton or glass wool to prevent spillage. It must never be stoppered, as this would result in a closed system that might explode or fly apart when heated. The drying tube can be inserted in the top of a reflux condenser through a rubber stopper or thermometer adapter, or attached to the sidearm of a vacuum adapter and similar devices with a short length of rubber tubing. When a very dry atmosphere is required, the apparatus should be swept out with dry nitrogen or another suitable gas before the operation is begun.

The nitrogen is introduced through a gas delivery tube, preferably one with a fritted glass outlet.

Trapping Gases

Some reactions release toxic or corrosive gases that must be *trapped* to keep them out of the atmosphere. An efficient *gas trap* can be constructed by inverting a narrow-stemmed

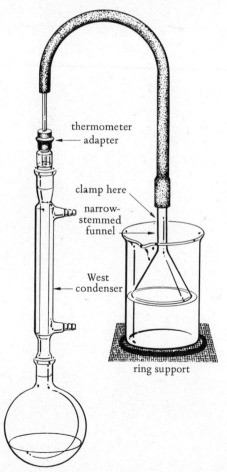

thermometer adapter

clamp here

narrow-stemmed funnel

West condenser

ring support

Figure 1.   Apparatus for trapping gases during reflux

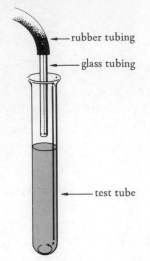

rubber tubing

glass tubing

test tube

Figure 2.   Semimicro gas trap

funnel inside a beaker containing a suitable gas-absorbing liquid, so that the rim of the funnel extends just below the liquid surface. Water alone is suitable for absorbing many gases, but dilute aqueous sodium hydroxide (1–2M) is generally used to absorb acidic gases such as hydrogen bromide or hydrogen chloride. For small amounts of gases, a straight glass tube inserted in a large test tube is often adequate. In this case, however, the outlet of the tube must be a millimeter or so *above* the surface of the liquid to keep the liquid from backing up into the system when gas evolution ceases or heating is discontinued. A gas trap is connected to the reaction apparatus at any point that is open to the atmosphere (usually the top of the reflux condenser) by means of a length of tubing and a stopper or thermometer adapter.

Purification Operations

Recrystallization

Principles and Applications

Recrystallization is the most frequently used operation for purifying organic solids. This technique is based on the fact that the solubility of an organic compound in a given solvent will often increase greatly as the solvent is heated to its boiling point. When an impure organic solid is first heated in such a solvent until it dissolves, then cooled to room temperature or below, it will usually **recrystallize** from solution in a much purer form. Most of the impurities will either fail to dissolve in the hot solution (from which they can be filtered) or remain dissolved in the cooled solution (from which the pure crystals are filtered).

Salicyclic acid dissolves in water to the extent of 0.20 g per 100 ml at 20°C and 7.5 g/100 ml at 100°C, whereas acetanilide has a solubility in water of 0.50 g/100 ml at 20°. Suppose a 5.00 g sample of salicylic acid contaminated by 0.25 g of acetanilide is dissolved in and recrystallized from boiling (100°) water. The amount of water required to just dissolve all of the salicylic acid at the boiling point is 67 ml (see calculation); all of the acetanilide impurity will also dissolve in the hot solution. When the solution is cooled down to 20°, about 0.13 g of salicylic acid will remain dissolved; the other 4.87 g will crystallize out if sufficient time is allowed. *All of* the acetanilide should remain in solution, since up to 0.35 g of acetanilide can dissolve in 67 ml of water at 20°. Therefore one should obtain, under ideal conditions, about a 97% recovery of essentially pure salicylic acid (see calculation).

This is an oversimplified description of a complex process; a number of factors may bring about results different from those predicted.

1. Very slow crystallization yields large crystals that may trap or *occlude* impurities within the crystal lattice, whereas fast crystallization results in small crystals that readily adsorb impurities on their surfaces and are difficult to wash.

crystallize: to obtain crystals of a substance.

recrystallize: to *again* crystallize substances that were initially in crystalline form.

Calculations:

$$\frac{5.0 \text{ g}}{7.5 \text{ g}} \times 100 \text{ ml} = 67 \text{ ml water}$$

$$.20 \text{ g} \times \frac{67 \text{ ml}}{100 \text{ ml}} = \begin{array}{l} .13 \text{ g salicylic} \\ \text{acid} \end{array}$$

$$.50 \text{ g} \times \frac{67 \text{ ml}}{100 \text{ ml}} = .35 \text{ g acetanilide}$$

$$\frac{4.87 \text{ g}}{5.00 \text{ g}} \times 100 = 97\% \text{ recovery}$$

Medium-sized crystals are usually the purest.

2. The solubility of one solute in a saturated solution of another solute is not usually the same as its solubility in pure solvent.

3. Adding only enough solvent to dissolve a solid may result in premature crystallization from the saturated solution and consequent losses. In practice, an excess of solvent is often used to prevent this.

Experimental Considerations

If the solid is unusually heat sensitive or prone to oiling out, heating can be carried out below the boiling point of the solvent.

J. Chem. Educ. 54, 639 (1977) describes a novel "gauze bandage method" for recrystallization.

In its simplest form, the recrystallization of a solid is carried out by dissolving the solid in the hot (usually boiling) recrystallization solvent, cooling the resulting solution to room temperature or below, and collecting the crystals by vacuum filtration ⟨OP–12⟩. In practice, experimental factors will often require additional steps such as filtering or decolorizing the hot solution.

The size of the crystals formed depends on the rate of cooling; rapid cooling yields small crystals and slow cooling on a thermally-nonconducting surface yields large ones. Large crystals are not necessarily the purest, but slow cooling is recommended when the solid is initially quite impure. If further purification is required it should be carried out with rapid cooling to form small crystals. In some instances, it may be desirable to collect a second or third crop of crystals by concentrating ⟨OP–14⟩ the **mother liquor** from the first crop. These crystals will contain more impurities than the first crop and may require recrystallization from fresh solvent. A melting point ⟨OP–28⟩ or TLC ⟨OP–17⟩ can be used to detemine whether a solid is sufficiently pure after recrystallization.

mother liquor: the liquid from which the crystals are filtered after crystallization.

Filtering the Hot Solution. Some impurities in a substance being crystallized may be insoluble in the boiling solvent and should be removed by filtration after the organic compound has dissolved. It is important not to mistake such impurities for the compound being purified and add too much solvent in an attempt to dissolve them. If, after most of the crude solid has dissolved, the addition of another portion of hot solvent does not appreciably reduce the amount of solid in the flask, that solid is probably an impurity — particularly if it is different in appearance from the remainder of the solid.

Dust, laboratory debris, filter-paper fibers, and other organic or inorganic impurities may be among the insoluble contaminants.

If too much solvent is used, the product yield may be greatly reduced, or the compound may not crystallize at all. Excess solvent can be removed by evaporation ⟨OP –14⟩.

The hot solution is filtered by the usual procedure for gravity filtration ⟨OP–11⟩, carried out as quickly as possible so that the solution does not have time to cool appreciably. Vacuum filtration ⟨OP–12⟩ can be used when large quantities of solvent are involved. However, reduced pressure cools the solvent by evaporation, which can cause premature crystallization and make it necessary to reheat the filtrate or add more solvent to redissolve the product after filtration. The funnel and collecting flask should be kept warm during filtration to prevent premature crystallization; it is important to preheat the filtering apparatus on a steam bath (or in an oven) until it is ready to use. Any cloudiness or precipitate that forms in the collecting flask during the filtering operation should be redissolved by heating before the flask is set aside to cool.

The hot solution should be returned to the heat source after each addition to the funnel.

Crystals forming in the filter or funnel stem can be dissolved with small portions of the hot solvent, which are combined with the filtrate.

The flask that was used to preheat the pure solvent can be used as a collecting flask.

Removing Colored Impurities. If a crude sample of a compound known to be white yields a colored recrystallization solution, *activated carbon* (Norit) can often be used to remove the colored impurity. The hot solution is removed from the heat to cool down a few degrees below the boiling point and a small quantity of activated carbon (about 0.2 g per 100 ml solution) is stirred in, along with an equal weight of a filtering aid such as Celite. The solution is then heated

Using too much carbon results in adsorption of the product as well as the impurities.

WARNING

Never add the carbon to a solution near the boiling point —it may boil over.

back to the boiling point with stirring and filtered by gravity, using a filter paper retentive enough to keep the carbon from passing through. If the solution is still colored, the treatment can be repeated with fresh carbon.

Inducing Crystallization. If no crystals form after the hot solution is cooled to room temperature, it is likely that the solution is supersaturated and that crystallization can be induced by one of the following methods. The tip of a glass rod is rubbed against the side of the flask with an up-and-down motion just over the liquid surface, so that it touches the liquid on the downstroke. If several minutes of scratching does not affect crystallization, then a few seed crystals (saved from the crude solid, or taken from a pure sample, if available) are dropped into the solution with cooling ⟨OP–8⟩ and stirring. Applying a piece of dry ice to the side of the flask can also help induce crystallization in difficult cases. If

Some compounds have been known to remain in supersaturated solutions for years before crystallizing!

crystals do not form after these procedures and after the solution has been allowed to stand overnight or longer, it is likely that the recrystallization solvent is unsuitable or that too much was used. In the first case, the solvent must be removed by evaporation ⟨OP–14⟩ and a new solvent tried; in the second, the solution must be concentrated by evaporation ⟨OP–14⟩ until it reaches the saturation point, then cooled to effect crystallization.

Dealing with Oils and Colloidal Suspensions.   When a compound being recrystallized is quite impure or has a low melting point, it may separate as a second liquid phase (an "oil") on cooling. Oils are undesirable, since even if they solidify on cooling the solid retains most of the original impurities. The following general approach is recommended unless the cause of the oiling is known and it can be prevented by another method. Heat the solution until the oil dissolves completely, then cool it slowly with constant stirring, adding a seed crystal or two (if available) at the approximate temperature where oiling occurred previously. If this is not successful, add about 25% more solvent and repeat the process. If oiling still occurs, cool the solution ⟨OP–8⟩ until the oil crystallizes (seeding and rubbing the oil with a stirring rod may help), then collect it by vacuum filtration and recrystallize it from the same solvent or a more suitable one, using the above techniques as necessary to prevent further oiling.

If a compound separates from solution as a fine colloidal suspension, the colloid can often be coagulated to form normal crystals by extended heating on a water bath or by adding an electrolyte such as sodium sulfate. Colloids can sometimes be prevented by adding activated carbon to the hot solution, or by cooling it very slowly.

Equipment and Supplies: Two Erlenmeyer flasks, recrystallization solvent, graduated cylinder, heat source, boiling sticks, flat-bottomed stirring rod, funnel, fluted filter paper, small watch glass, vacuum filtration apparatus ⟨OP–12⟩, cold washing solvent. *Optional:* decolorizing carbon, filtering aid.

Occasionally crystallization can be effected by adding a compatible solvent in which the compound is less soluble, then proceeding as described in ⟨OP–23b⟩.

Some compounds, such as acetanilide in water, separate as a second liquid phase when a certain saturation level is exceeded. This can be prevented by using additional solvent.

Add more solvent, if necessary, to dissolve the oil.

Crystallization may also be induced by removing the solvent, adding a seed crystal to the oil, and letting it stand in a refrigerator or freezer.

Colloids cannot be filtered—they will pass through the filter paper.
Electrolytes are useful only in solvents polar enough to dissolve them.

HAZARDS   *See ⟨OP–23c⟩ for hazards associated with common solvents. Do not heat flammable solvents with a burner; extinguish all nearby flames before filtering the hot solution.*

Operational Procedure

Measure the estimated volume of solvent needed for recrystallization (plus a little more to allow for error and evaporation) into an Erlenmeyer flask (the *solvent flask*). Add a boiling stick or a few boiling chips and heat the solvent to boiling on an appropriate heat source ⟨OP–7⟩. Place the solid to be purified in a second Erlenmeyer flask (the *boiling flask*) and add about half to three-quarters of the estimated volume of hot solvent, or 2–3 ml per gram of crude solid if the amount of solvent is not known. Heat the mixture at the boiling point with constant stirring, breaking up any large particles with a flat-bottomed stirring rod, until it appears that no more solid will go into solution with further heating and stirring. Add more hot solvent in *small* portions (5–10% of the total volume) and boil the mixture (with stirring) after each addition until the compound dissolves. Leave a wide-necked funnel and fluted filter paper on the solvent flask between additions so that they are warmed by the hot solvent vapors, and pour out any unused solvent just before filtration. Unless there are no solid impurities remaining, filter the hot solution ⟨OP–11⟩ into the preheated solvent flask (as described above) and wash the residue with a little hot solvent. If colored impurities must be removed, add decolorizing carbon before the filtration.

Set the flask on a surface (such as wood or cork) that is a poor heat conductor, cover it with a small watch glass or a piece of filter paper to keep out airborne particles, and let it stand undisturbed until crystallization is complete. If desired, the crystals can be cooled further in an ice bath after the mixture reaches room temperature. This should increase the yield, but it may decrease the purity of the crystals slightly. Once crystals have begun to form, 15–30 minutes of standing at room temperature or below is usually sufficient to insure complete crystallization. (If no crystals have formed by the time the solution reaches room temperature, use one of the methods described above to induce crystallization.) Collect the crystals by vacuum filtration ⟨OP–12⟩ and wash them on the filter with small portions of the ice-cold recrystallization solvent (or another suggested solvent). Do not allow the recrystallization solvent to dry on the filter before adding the first portion of wash solvent, since this will deposit impurities on the crystals. Air dry the crystals on the filter, then dry them further ⟨OP–21⟩ if necessary.

If crystals form in the filtrate during filtration, crystallization was incomplete. The filtrate should be allowed to cool longer, then it should be refiltered. If a second crop of

If the amount of solvent is not known, start with 5–10 ml per gram of solid and use more if needed.

A steam bath is preferred for volatile organic solvents; water can be boiled using a burner or hot plate. Some hazardous solvents should be heated only under reflux or in a fume hood.

If the solvent is quite volatile or extended heating is required, place a small watch glass on the flask or insert a cold-finger condenser (see ⟨OP–7b⟩) to reduce evaporation.

If small crystals are desired, swirl the flask under a cold water tap until crystallization begins, then set it aside to complete the process.

Use a little of this ice-cold solvent for the transfer of residual crystals to the filter, and for the first washing.

crystals is to be collected, retain the mother liquor in the filter flask and concentrate it by evaporation ⟨OP–14⟩ to the saturation point. Heat the solution (or add a little solvent) to discharge any turbidity, then cool it to induce crystallization and proceed as before.

Summary

1. Measure estimated volume of solvent (plus a small excess) into solvent flask and heat to boiling.

2. Add part of solvent to solid in boiling flask, boil with stirring.

3. Add more hot solvent portionwise (if necessary) until organic solid dissolves.
IF solution contains colored impurities,
GO TO 4.
IF not,
GO TO 5.

4. Cool below boiling point, stir in decolorizing carbon and filtering aid (optional), heat to boiling.

5. IF solution contains undissolved impurities,
GO TO 6.
IF not,
GO TO 7.

6. Filter hot solution by gravity.

See Experimental Considerations if crystals do not form, or if an oil or colloid forms.

7. Cover flask and set aside to cool until crystallization is complete.

8. Collect crystals by vacuum filtration, wash and air dry on filter.

OPERATION 23 *a*

Semimicro Recrystallization

Recrystallization of quantities from a few tenths of a gram up to a gram or two should be carried out using special small-scale apparatus to avoid excessive losses. Test tubes of the appropriate size can be used in place of Erlenmeyer flasks for dissolving and recrystallizing the solute, and small-scale filtration equipment can be used to filter the hot solution and the crystals. The following procedure is suggested for semimicro recrystallization, although other kinds of apparatus and techniques may be equally effective.

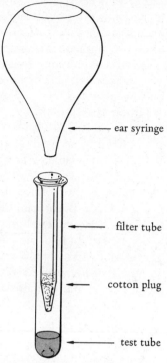

ear syringe

filter tube

cotton plug

test tube

Figure 1.   Use of filter tube

Operational Procedure

Dissolve the solid in a test tube by adding small portions of hot solvent with stirring and heating *gently* so as to avoid excessive bumping or evaporation. (A semimicro refluxing setup ⟨OP–7b⟩ can be used if necessary.) If the solution contains solid impurities, filter it through a special filter tube fitted with a small cotton plug and rinse the cotton with a small portion of hot solvent. Combine the rinse liquid with the filtrate. Concentrate the solution ⟨OP–14⟩ if necessary, then set it aside until crystallization is complete. Collect the crystals by semimicro vacuum filtration ⟨OP–12a⟩ and wash them with very small quantities of ice-cold solvent. Air-dry them on the filter, and dry them further ⟨OP–21⟩, if necessary.

The test tube and filter tube should be kept warm to prevent premature crystallization.

A conical funnel provided with a small fluted filter paper can also be used.

Excessively rapid cooling can be prevented by setting the test tube in a warm water bath and allowing the tube and bath to cool to room temperature together.

Recrystallization from Mixed Solvents

Sometimes no single solvent is suitable for recrystallizing a particular compound. In such a case, it is necessary to use a

Table 1.   Some compatible solvent
pairs

ethanol—water

methanol—water

acetic acid—water

acetone—water

ethyl ether—methanol

ethanol—acetone

ethanol—petroleum ether

ethyl acetate—cyclohexane

chloroform—petroleum ether

benzene—ligroin

benzene—cyclohexane

See *J. Chem. Educ. 51,* 602 (1974) for
an alternative mixed-solvent recrystallization method.

*See ⟨OP-23c⟩ for hazards associated
with recrystallization solvents.*

*It may be necessary to use an excess of
solvent A to prevent premature crystallization.*

*Keep the solution boiling during the
addition of solvent B.*

*The recovered crystals should generally be
washed (on the filter) with solvent B.*

mixture consisting of two compatible solvents—one (solvent A) in which the compound is quite soluble and another
(solvent B) in which it is comparatively insoluble. It may be
feasible, in some cases, to prepare a mixture of the solvents
beforehand and purify the compound by the usual method.
But it is often more practical to recrystallize the compound
by dissolving it in solvent A and then adding enough of solvent B to form a saturated solution.

If the compound is very soluble in solvent A, the total
volume of solvent may be quite small compared to that of
the crystals, which may then separate as a dense slurry. In
such a case, it is desirable to use more of solvent A than is
needed to just dissolve the compound, and to add corre-
spondingly more of solvent B to bring about saturation.
(Recrystallization from a very small volume of solvent is not
advisable.) One must be careful to avoid adding so much of
solvent A that *no* amount of solvent B will result in saturation; if that occurs, it will be necessary to concentrate
⟨OP–14⟩ the solution or to remove all of the solvent and
start over.

Operational Procedure

Place the crude solid in a flask and add solvent A (previously
heated to boiling) portionwise, with heating and stirring,
until the solid is dissolved. If the hot solution contains colored impurities, decolorize it with carbon; if it contains solid
impurities, remove them by gravity filtration at this point.
Add solvent B in small portions to the hot solution until a
persistent cloudiness forms or crystals appear. Add just
enough of hot solvent A dropwise to clarify the solution (or
dissolve any precipitate), set the mixture aside to cool to
room temperature, and proceed as for ordinary recrystallization ⟨OP–23⟩.

OPERATION 23 *c*

Choosing a
Recrystallization Solvent

Several characteristics should be considered when
choosing a recrystallization solvent for an organic compound.

1.   The compound being purified must dissolve to a
substantially greater extent in the boiling solvent than in the
cold solvent.

2.   The boiling point of the solvent should be high

enough to take advantage of the compound's temperature coefficient of solubility, but not so high that it cannot be easily removed from the crystals or concentrated by evaporation. Most good recrystallization solvents boil in the 50–85° range (water is a notable exception).

3. Ideally, the impurities should either be insoluble or more soluble in the recrystallization solvent than the compound being purified. If this is not the case, recrystallization can still be effective unless the amount of impurity is quite large.

4. The solvent must not react with the compound being purified and should not be excessively hazardous to work with. When possible, solvents like benzene and carbon tetrachloride should be avoided in favor of other, less toxic solvents.

5. The high freezing points of solvents like acetic acid (17°) and benzene (5.5°C) can be disadvantageous, since they limit the minimum temperature attainable on cooling.

Some common recrystallization solvents and their properties are listed in Table 1.

In choosing a suitable solvent, information available in reference books may be helpful. For example, the entry under "crystalline form" for naphthalene in the *CRC Handbook* reads "mcl pl (al)" meaning that naphthalene crystallizes from alcohol (methanol or ethanol) in the form of monoclinic plates. If no recrystallization solvent is listed, solubility data should be useful. A solvent in which the compound is designated as *sparingly soluble* or *insoluble* when cold and *very soluble* or *soluble* when hot might be suitable for recrystallization. Alternatively, a mixture of two compatible solvents may work if the compound is insoluble or sparingly soluble in one and soluble or very soluble in the other. If solubility data are not available, a solvent may have to be chosen by trial and error from a selection like that in Table 1. Ordinarily the recrystallization solvent should be somewhat *more* or *less* polar than the solute, since a solvent of similar polarity may dissolve too much of it. Once a possible solvent or solvent mixture is chosen, it should be tested by the following procedure.

The *CRC Handbook of Chemistry and Physics, Lange's Handbook of Chemistry, The Merck Index, Dictionary of Organic Compounds, Beilstein's Handbuch,* and others list recrystallization solvents and solubility data for many organic compounds. See Bibliography, part A.

If possible, the compound should have a solubility of about 5–25 g/100 ml in the hot solvent and less than 2 g/100 ml in the cold solvent, with at least a 5:1 ratio between the two values.

A table in *The Chemist's Companion,* pp. 442–43, (Bib–A21) gives the classes of compounds that can be recrystallized from different solvents.

Table 1. Properties of common recrystallization solvents

Solvent	b.p.	f.p.	Comments
water	100°	0°	Good solvent for polar compounds; crystals dry slowly.

Table 1. Properties of common recrystallization solvents (*continued*)

Solvent	b.p.	f.p.	Comments
acetic acid	118°	17°	Hard to remove, unpleasant to work with, corrosive.
methanol	64.5°	−98°	Toxic.
ethanol	78°	−116°	Good general solvent; 95% ethanol is commonly used.
acetone	56°	−95°	Limited by low b.p.; 2-butanone may be preferable in some cases.
ethyl acetate	77°	−84°	
chloroform	62°	−64°	Good general solvent, easily removed. Suspected carcinogen.
benzene	80°	5.5°	Good solvent for aromatics. Major health hazard, carcinogenic.
petroleum ether	~30–60°	low	Boiling range varies; ligroin is similar. Very flammable.
hexane	69°	−94°	Good solvent for less polar compounds.

Note: Solvents are listed in approximate order of decreasing polarity.

Hazards

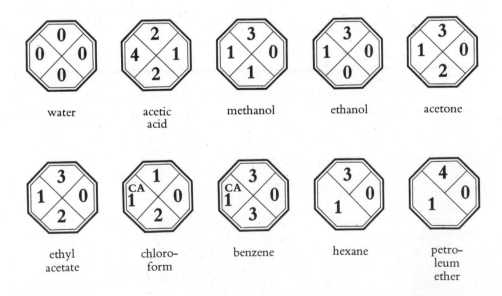

water   acetic acid   methanol   ethanol   acetone

ethyl acetate   chloroform   benzene   hexane   petroleum ether

The quantities can be scaled down if you can spare only a little solid for testing.

A flat-bottomed stirring rod and watch glass can be used to grind the solid.

Testing Potential Recrystallization Solvents. Weigh about 0.3 g of the finely divided solid into a test tube, add 3 ml of the solvent, and shake and stir the mixture to see if the solid will dissolve. If it does not, warm the mixture very gently, with stirring and shaking. If the solid dissolves in the cold solvent or with gentle warming, the solvent is unsuitable. Now carefully heat the mixture to boiling, with stirring. If the solid does not dissolve readily, add more solvent, in 1 ml portions (gently boiling and stirring the mixture after each addition), until it dissolves *or* until the total volume of added solvent is about 10 ml. If the solid does not dissolve (or nearly so) by this time, the solvent is probably not suitable; if it does dissolve (except for insoluble impurities), cool the solution with scratching (see Inducing Crystallization, p. 519) to see if crystallization occurs. If crystals separate, examine them visually for apparent yield and evidence of purity (absence of extraneous color, good crystal structure). If necessary, repeat the process with other solvents until a suitable one is identified. If no single solvent is satisfactory, choose one solvent in which the compound is quite soluble and another in which it is comparatively insoluble (make sure that the two are miscible!). Dissolve about 0.1 g of the solid in the first solvent with heating and stirring, then add the second hot solvent dropwise until saturation occurs. Cool the resulting mixture to induce crystallization. If crystals form, examine them as before; if saturation cannot be attained or no crystals form, try another solvent pair.

Use a water bath or steam bath if the solvent is flammable. Heat gently to reduce solvent evaporation.

Use an ice or ice-salt bath if necessary ⟨OP–8⟩.

It is more efficient to test several solvents at one time and choose the best among them.

See Table 1 ⟨OP–23b⟩ for a list of compatible solvents.

Sublimation

Principles and Applications

Sublimation is a phase change in which a solid passes directly into the vapor phase without going through an intermediate liquid phase. Many solids having appreciable vapor pressures below their melting points can be purified by sublimation, either at atmospheric pressure or under vacuum. Such purification is most effective if the impurities have either low or very high vapor pressures at the sublimation temperature, so that they either do not sublime appreciably or fail to condense after vaporization. Although not as selective as recrystallization or chromatography, sublimation offers advantages in that no solvent is required, losses in transfer can be kept very low, and the process is rapid with the right apparatus and conditions. Sublimation can be accelerated by performing it under reduced pressure (vacuum sublimation, ⟨OP–24a⟩) or in an air stream (entrainer sublimation).

Most compounds that sublime readily have highly symmetrical molecules, which results in a comparatively high melting point and low boiling point.

Experimental Considerations

sublimand: the solid before it has sublimed.

sublimate: the solid after it has sublimed and condensed.

Good beaker combinations are 250 ml/400 ml and 400 ml/600 ml.

entrain: to draw up and transport by the flow of air or another fluid.

Sublimation is usually carried out by heating the **sublimand** on an oil bath or other uniform, easily controllable heat source and collecting the **sublimate** on a cool surface. For best results, the sublimand should be finely divided and free of solvent. A very simple sublimator consists of two nested beakers having about a 1–2-cm space between them at the bottom. The sublimand is spread out inside the larger beaker and the condensing (inner) beaker is partially filled with ice water (ice being added periodically to maintain the temperature). As the outer beaker is heated, crystals of sublimate collect on the bottom of the condensing beaker. Figure 1 illustrates a sublimator operating on the same principle, except that an Erlenmeyer flask is used as a condenser and the temperature is controlled by flowing water.

A sublimation apparatus based on the **entrainment** principle is illustrated in Figure 2. Air drawn in through the lip of the evaporating dish flows over the sublimand, sweeping the vapors up through the asbestos sheet so that they condense on the funnel or glass wool. The air flow must be gentle enough that the sublimate will not pass through the glass wool and condense inside the funnel stem or rubber tubing.

Figure 1.   Sublimation apparatus

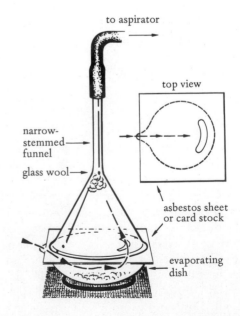

Figure 2.   Apparatus for entrainer sublimation

Equipment and Supplies: sublimation apparatus, heat source, cooling fluid, flat-bladed spatula.

Operational Procedure

Construct one of the sublimation setups described above. Powder the dry sublimand finely (if necessary) and spread it in a thin, even layer over the bottom of the sublimand container (outer beaker, evaporating dish, etc.). Assemble the apparatus and turn on the aspirator or cooling water (or fill the condenser with ice water). Heat the sublimand with an appropriate heat source until sublimate begins to collect on the condenser, then adjust the temperature to attain a suitable rate of sublimation without melting or charring the sublimand.

A burner with a wire gauze is sometimes suitable, but tends to cause charring. An oil bath is much better.

If the condenser becomes overloaded with sublimate, cool and disassemble the apparatus to remove the sublimate, then reassemble it and resume heating. When all of the compound has sublimed or only a nonvolatile residue remains, remove the apparatus from the heat source and let it cool. *Carefully* remove the condenser (avoid dislodging any sublimate) and scrape the crystals into a suitable container using a flat-bladed spatula.

Melting the sublimand will decrease the sublimation rate by reducing surface area; also, molten sublimand may splatter onto the condenser.

The last traces of sublimate can be removed by rinsing the collector with a volatile solvent and evaporating the solvent ⟨OP–14⟩.

Summary

1. Construct sublimation apparatus, spread solid out in sublimand container.

2. Add cooling fluid to condenser *or* turn on aspirator or water tap.

3. Heat until sublimation begins and control heating to maintain sublimation rate.

4. When sublimation is complete (or collector reaches its capacity), discontinue heating, cool and disassemble apparatus, remove sublimate.

Vacuum Sublimation

Many solids that do not sublime rapidly at atmospheric pressure will do so under reduced pressure. The sublimation assemblies pictured in Figure 1 can be evacuated for that purpose. Apparatus **A** requires a sleeve, which can be a rubber stopper bored out with a large cork borer, or a length of

J. Chem. Educ. *52,* 720 (1975) describes the construction of a vacuum sublimator for student use.

rubber tubing having a suitable diameter. For larger quantities of solid, a filter flask can be used in place of the sidearm test tube. Sublimators similar to apparatus **B** are available commercially. A vacuum sublimation is carried out as in ⟨OP–24⟩, except that the vacuum is turned on before heating is begun, and turned off after sublimation is complete *and* the heat source has been removed. If aspirator vacuum is used, a trap must be placed between the sublimation apparatus and the aspirator. The trap should have a pressure-release valve (see Figure 1, ⟨OP–26⟩) so that the vacuum can be broken slowly; otherwise air entering the sublimator may disturb the crystals of sublimate.

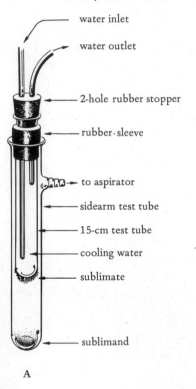
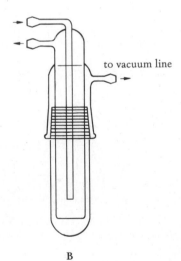

Figure 1.   Apparatus for vacuum sublimation

water inlet
water outlet
2-hole rubber stopper
rubber-sleeve
to aspirator
sidearm test tube
15-cm test tube
cooling water
sublimate
sublimand

A

to vacuum line

B

OPERATION 25

Simple Distillation

Principles and Applications

When a mixture of several liquids (or of a liquid with solid impurities) is heated in a boiling flask to its boiling point, the resulting vapor ordinarily has a composition different from that of the liquid itself. If this vapor is condensed into a separate vessel, the resulting liquid will contain a high proportion of the most volatile component present in the boiling flask. This process of vaporizing a liquid in one vessel and condensing the vapor into another is called *distillation;* it is an important method for purifying organic liquids.

Simple distillation involves only a single vaporization-condensation cycle. It is useful for purifying a liquid that contains either (*a*) involatile impurities or (*b*) small amounts of higher- or lower-boiling liquids. *Fractional distillation* ⟨OP–27⟩ is a more efficient way of separating miscible liquids having comparable volatility and boiling points, since it allows several cycles of vaporization and condensation to occur in a single operation. *Vacuum distillation* ⟨OP–26⟩ is used to purify high-boiling liquids and liquids that decompose when distilled at atmospheric pressure.

If a pure liquid (or a liquid containing involatile impurities) is heated at standard pressure in a distillation apparatus such as that pictured in Figure 2 (p. 535), it will begin to boil when its vapor pressure equals 1 atmosphere, at a temperature defined as its *normal boiling point*. Its vapors will then remain at that temperature throughout its distillation. When a mixture of liquids, each exerting an appreciable vapor pressure at the boiling temperature, is distilled, the process is more complicated. To simplify the explanation somewhat we shall consider a mixture of ideal liquids that have ideal vapors, which obey both Raoult's Law (Equation 1) and Dalton's Law (Equation 2). Unfortunately there are no *real* liquids that obey these laws perfectly, so we shall invent some imaginary hydrocarbons for that purpose. *Entane* ($C_{5.5}H_{13}$) is a pseudoalkane having a normal boiling point of 50°C, whereas *orctane* ($C_{7.5}H_{17}$) boils at 100°C at 1 atmosphere. The equilibrium vapor pressures of entane and orctane at several temperatures between their boiling points are listed in Table 1. As expected, the vapor pressure of entane

Boiling occurs when the vapor pressure over a liquid equals the external pressure (usually about 1 atmosphere in simple distillations).

Condensation usually takes place in a water-cooled condenser like a reflux condenser.

$$P_A = X_A P_A^0 \qquad (1)$$

$$P_A = Y_A P \qquad (2)$$

P_A = partial pressure of component A in a mixture.

P_A^0 = vapor pressure of pure A at the same temperature.

P = total pressure over the mixture.

X_A = mole fraction of A in the liquid.

Y_A = mole fraction of A in the vapor.

(The same relationships hold for any other component in the mixture, such as B)

Entane and orctane exist only in Middle-earth, where they are used for fuel by Ents and Orcs, respectively (cf. J. R. R. Tolkien, *The Lord of the Rings*).

Table 1.   Equilibrium vapor pressures and mole fractions of *entane* and *orctane* at different temperatures

T, °C	Entane			Orctane		
	P^0, torr	X	Y	P^0, torr	X	Y
50°	760	1.00	1.00	160	0.00	0.00
60°	1030	0.67	0.90	227	0.33	0.10
70°	1370	0.42	0.76	315	0.58	0.24
80°	1790	0.24	0.57	430	0.76	0.43
90°	2300	0.11	0.32	576	0.89	0.68

Table 1.   Equilibrium vapor pressures and mole fractions of *entane* and *orctane* at different temperatures (*continued*)

T, °C	Entane			Orctane		
	P^0, torr	X	Y	P^0, torr	X	Y
100°	2930	0.00	0.00	760	1.00	1.00

P^0 = equilibrium vapor pressure of the pure liquid; X = mole fraction in liquid mixture; Y = mole fraction in vapor. Vapor pressures were calculated using the Clausius-Clapeyron equation and Trouton's rule.

$$X_A = \frac{P - P_B^0}{P_A^0 - P_B^0}$$ (3)

$$Y_A = \frac{P_A^0}{P} \cdot X_A$$ (4)

at its normal boiling point of 50° is 760 torr (1 atm); the less volatile orctane has a much lower vapor pressure at that temperature. As the temperature increases, the vapor pressure of orctane also rises until it attains a value of 760 torr at 100°C.

If a mixture of entane and orctane is heated at normal atmospheric pressure, it will begin to boil at a given temperature, producing vapor of a given composition. The boiling temperature and the composition of the vapor phase will both be determined by the composition of the liquid mixture. For example, a mixture containing an equal number of moles of entane and orctane will boil at $66\frac{1}{2}$°C and the vapor will contain over 4 moles of entane for every mole of orctane (81.4 mol% entane). The liquid and vapor composition at any temperature can be calculated using Equations 3 and 4, which are derived from Dalton's Law and Raoult's Law; compositions at several temperatures are given in Table 1. It is apparent that the *vapor* over such a mixture of liquids is richer in the more volatile component (entane) than the liquid mixture itself; this is the key to the purification effected by distillation.

For a better understanding of the process, refer to Figure 1 in which the composition of liquid and vapor are plotted against the boiling temperature of the mixture. Suppose we wish to purify a mixture containing 2 moles of entane for every mole of orctane (67 mole percent entane). From the graph (and Table 1) it can be seen that such a mixture will boil at 60° (point **A**) and that its vapor will contain 90 mol% entane (point **A'**). Thus the *liquid* that is condensed from this vapor (point **A"**) will be much purer in entane than was the original mixture in the boiling flask. As the distillation continues, however, the more volatile component will boil away faster and the boiling flask will contain progressively less entane. The vapor will also contain less entane and the

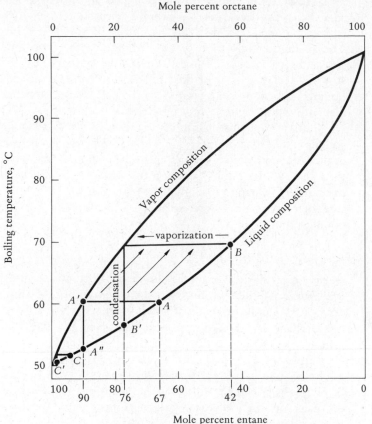

Mole percent orctane

Boiling temperature, °C

Mole percent entane

Figure 1.   Temperature-composition
diagram for entane-orctane mixtures

boiling temperature will rise. When the mole fraction of entane has fallen to 0.42 (point **B**), the boiling temperature will
have risen to 70° and the **distillate** will contain only 76
mol% entane (point **B′**). Only when nearly all of the entane
has been distilled will the distillate be richer in the less volatile component; at 90°, for example, over $\frac{2}{3}$ of the distillate
will be orctane.

It can be seen that the purification effected by the simple
distillation of such a mixture of volatile liquids is very imperfect. In the example given, the distillate never contains
more than 90 mol% entane, and it may be considerably less
pure than that, depending on the temperature range over
which it is collected. However, if we had started with a
mixture containing only 5 mole percent orctane (point **C**),
considerably greater purity could have been achieved—the

distillate: the liquid that has been
distilled and has condensed into the
receiving flask.

initial distillate would be 99 mol% entane (point C') at 51° and, if the distillation were continued until the temperature rose to 55°, the final distillate would be 95 mol% entane. The average composition of the distillate would lie somewhere between those values; so, if the distillation were stopped at 55°C, most of the orctane would remain in the boiling flask along with some unrecovered entane, and the distillate would be relatively pure entane.

Thus, simple distillation can be used to purify a liquid containing *small* amounts of volatile impurities if **(a)** the impurities have boiling points appreciably higher or lower than that of the liquid, and **(b)** the distillate is collected over a narrow range (usually 4–6°) starting at a temperature that is within a few degrees of the liquid's normal boiling point.

Experimental Considerations

Apparatus.   A typical setup for simple distillation is pictured in Figure 2. The size of the components should be consistent with the volume of the **distilland**—otherwise, excessive losses will occur. For example, 250 ml of vapor (about 11 mmoles) will remain in a 250-ml boiling flask after distillation is completed. This loss corresponds to about $1\frac{1}{2}$ ml of a liquid like butyl acetate. An additional 1–3 ml of liquid may be lost as vapor in the still head and as liquid adsorbed on the glass surface inside the apparatus. Thus it is not practical to use a large boiling flask for a small volume of distilland or to use the apparatus in Figure 2 to distill less than about 20 ml of a liquid.

The boiling flask should be about one-third to one-half full of the distilland and should be supplied with several boiling chips or other smooth boiling devices to prevent bumping (see ⟨OP–7a⟩). The thermometer is inserted into the *still head* (a 3-way connecting tube) through a stopper or thermometer adapter. It should be well centered in the still head (not closer to one wall than another) with the entire bulb below a line extending from the bottom of the sidearm. In other words, the *top* of the bulb should be aligned with the *bottom* of the sidearm, as illustrated by Figure 3. In this way, the entire bulb will become moistened by condensing vapors of the distillate. The placement of the thermometer is extremely important, since otherwise the observed distillation temperature will be erroneous (usually too low) and the distillate will be collected over the wrong temperature range.

distilland: the liquid in the boiling flask, which is to be purified by distillation.

See ⟨OP–25a⟩ *for apparatus used to distill smaller quantities of a liquid.*

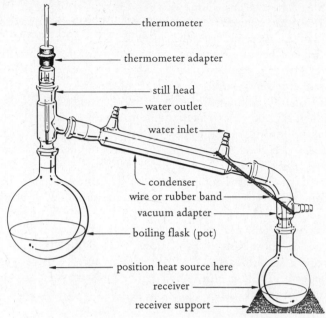

Figure 2.   Apparatus for simple distillation

- thermometer
- thermometer adapter
- still head
- water outlet
- water inlet
- condenser
- wire or rubber band
- vacuum adapter
- boiling flask (pot)
- position heat source here
- receiver
- receiver support

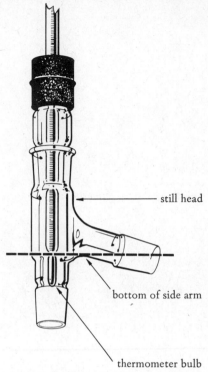

- still head
- bottom of side arm
- thermometer bulb

Figure 3.   Thermometer placement

Condenser water should flow in the lower end of the condenser and out the upper end.

See ⟨OP–22a⟩ if moisture must be excluded from the system during distillation.

Alternatively, a special collecting adapter can be used. See *J. Chem. Educ. 53,* 39 (1976).

The condenser should have a straight inner section of comparatively small diameter. A typical example is the Lie-big-West condenser (also used for refluxing) found in a typical organic labkit. (Do not mistakenly use a jacketed distillation column as a condenser; its larger inner diameter results in less efficient condensation.) The vacuum adapter should be secured to the condenser by a rubber band or a spring clamp. If the distillate is quite volatile, or if it is flammable and a burner is used for heating, the receiver should be a ground–joint flask fitting snugly on the vacuum adapter. If not, another container such as an Erlenmeyer flask or a tared vial can be used. If a ground–joint receiver is used, the vacuum-adapter sidearm must be open to the atmosphere—otherwise, heating the system will build up pressure that

Never heat a closed system! It may explode or fly apart.

WARNING

could result in an explosion and severe injury from flying glass.

Heat Sources.   Almost any of the heat sources described in ⟨OP–7⟩ can be used for simple distillation. Heating mantles and oil baths are preferred because they are less likely to start a fire and because they can provide constant, even heating over a wide temperature range. Steam baths can be used to distill some low-boiling liquids; a burner provided with a wire gauze is adequate for higher boiling, less flammable liquids. It is important to keep the heating rate constant—temperature drops can cause the vapor level to fall so that the observed boiling temperature will fluctuate and distillation may cease. An excessive heating rate can reduce the efficiency of the distillation and cause decomposition or mechanical carryover of liquid to the receiver. If a burner is used, it should be positioned close enough to the wire gauze that drafts will not cause appreciable temperature fluctuations. In most simple distillations, a distilling rate of 1–3 drops a second (about 3–10 ml per minute) will provide adequate distilling efficiency; higher rates may reduce the purity of the distillate.

Some burners are provided with a "chimney" to keep out drafts. It is not a good idea to hold the burner in your hand and try to control the heating rate by moving it around.

In order to maintain a good distilling rate as the more volatile components distill over, it may be necessary to increase the heating rate.

Boiling Range.   The boiling range for a distillation is often specified in the experimental procedure. For example, in Experiment 6 the boiling range for the impure butyl acetate is given as 120–126°, which extends from 5° below its normal boiling point (125°) to 1° above it. This kind of behavior is often observed for a liquid that is contaminated by a lower-boiling impurity—in this case *n*-butyl alcohol, having a boiling point of 117°. Liquids contaminated by higher-boiling impurities might be collected (for example) over a range extending from a degree or two below their normal boiling point to 4–5° above it.

Liquids that form azeotropes may exhibit different behavior, however.

If the boiling range of a liquid is *not* specified in a procedure, you can sometimes estimate the range by analogy. For example, since the isoamyl acetate prepared in Experiment 6 is, like butyl acetate, contaminated by a lower-boiling alcohol, its boiling range should also be from about $T - 5°$ to $T + 1°$, where T is the normal boiling point of the liquid. If you are unable to estimate a boiling range accurately, plan to collect the distillate over a range of 6° or less, beginning no more than 5° below the normal boiling point and ending no more than 5° above it. If, after you have done this, the boiling temperature is still reasonably close to T and an appreciable amount of distilland remains in the pot, it may be best to keep distilling to $T + 5°$ or higher and then to redistill the entire distillate, collecting it over a narrower range the second time.

Equipment and Supplies: Boiling flask, 3-way connecting tube (still head), thermometer adapter, thermometer, condenser, vacuum adapter, receiving vessel, boiling chips, condenser tubing, heat source, utility clamp, condenser clamp, 2 ringstands, rubber band or spring clamp(s), support for receiver, and heat source.

Operational Procedure

Assemble the apparatus pictured in Figure 2, using a boiling flask of appropriate size and taking great care to position the thermometer correctly. Add the distilland through a stemmed funnel, drop in a few boiling chips, replace the thermometer, and turn on the condenser water to provide a slow but steady stream of cooling water. Apply heat at such a rate that the liquid begins to boil gently and the reflux ring of condensing vapors rises slowly into the still head. Shortly after the reflux ring reaches the thermometer, the temperature reading should rise sharply and vapors will begin passing through the sidearm into the condenser, coalescing into droplets that run into the receiving flask. As the first few droplets come over, the thermometer reading should rise to an equilibrium value. (At this time, the entire thermometer bulb should be bathed in condensing liquid that drips off the end of the bulb into the pot.) Record the temperature where the reading stabilizes and observe it frequently throughout the distillation. Distill the liquid at a rate of about 1–3 drops a second.

    The initial thermometer reading may be substantially below the expected boiling point of the liquid being purified. If this happens, carry out the distillation until the *lower* end of its expected boiling range is attained, then quickly replace the receiver by another one. If necessary, heating can be discontinued during the switch; but it is usually possible to maintain a slow distillation rate and to make the change "between drops." The **forerun** in the first receiver contains volatile impurities and is ordinarily discarded. If the initial thermometer reading is within the expected boiling range, there is no need to change receivers.

    Collect the distillate until the *upper* end of the expected boiling range is reached or until only a small volume of liquid remains in the boiling flask. Remove the heat source and (if necessary) transfer the distillate to a tared vial or another suitable container. If another fraction is to be collected, change receivers at the specified temperature without interrupting the distillation and collect that fraction over its ex-

Review ⟨OP–2⟩ for directions on the proper assembly of ground glass apparatus.

To insure equilibrium, do not allow distillation to begin until at least a minute after vapors first enter the still head.

The temperature will usually stabilize after 5–10 drops of distillate have been collected. Check the thermometer placement if the initial boiling temperature is lower than expected.

See ⟨OP–14⟩ if a low-boiling solvent must be removed before the main component is distilled.

forerun: a volatile liquid fraction that distills before the main fraction.

It is important to stop heating when the bottom of the pot is still moist. Heating a dry flask might decompose the residue and cause tar formation or even an explosion; it might also break the flask.

pected boiling range. Disassemble and clean the apparatus as soon as possible after the distillation is completed.

Summary

1. Assemble distillation apparatus.

2. Add liquid to pot through funnel, add boiling chips.

3. Turn on and adjust condenser water flow.

4. Commence heating, adjust heat so that vapors rise slowly into still head.

5. Record temperature after distillation begins and thermometer reading stabilizes.
IF initial temperature is below expected range,
GO TO 6.
IF initial temperature is within expected range,
GO TO 8.

6. Check thermometer placement, adjust if necessary.

7. Collect distillate until temperature rises to lower end of expected range, change receivers.

8. Distill liquid until temperature rises to upper end of expected range *or* until $\frac{1}{2}$ ml or less of liquid remains in boiling flask.
IF another fraction is to be collected,
GO TO 9.
IF not,
GO TO 10.

9. Change receivers, increase heating rate if necessary, GO TO 5.

10. Remove heat source, retain distillate(s) (except forerun), disassemble and clean apparatus.

OPERATION 25 *a*

Semimicro Distillation

The vacuum adapter should be secured by a wire or spring clamp—a rubber band exposed to the heat of the vapors may break.

The distillation of small quantities of liquids is best carried out in a special semimicro apparatus utilizing a microcondenser or cold-finger condenser. However, the apparatus pictured in Figure 1 can be constructed from ordinary organic labkit parts and works reasonably well. The water-cooled condenser of an ordinary simple distillation apparatus is replaced by a cooling bath surrounding the receiver, which can be insulated (if necessary) by a second beaker containing glass wool. If a heating mantle or other bulky heat source is used, it may be necessary to slant the vacuum adapter and receiver to allow room for the heat source. If the distilland boils over 150° and the receiver is well insulated from the heat source, it may be feasible to omit the cooling bath. Tap water is a suitable coolant for liquids boiling near 100° or above; ice water can be used for liquids distilling above 50°. Below this temperature an ice-salt bath (see ⟨OP–8⟩) should be used. Whenever the cold bath is much below room temperature, the vacuum-adapter sidearm should be connected to a drying tube ⟨OP–22*a*⟩ so that moisture will not condense inside the receiver. If a burner is used for distillation and the distillate is flammable, a length of rubber tubing should be run from the vacuum-adapter sidearm to a sink to conduct flammable vapors away from the burner.

The receiver can be attached to the vacuum adapter by a length of wire or clamped in place.

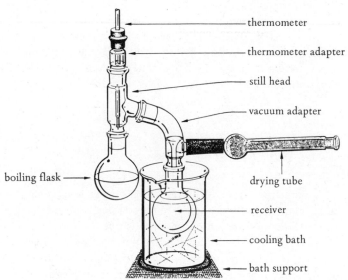

thermometer

thermometer adapter

still head

vacuum adapter

boiling flask

drying tube

receiver

cooling bath

bath support

Figure 1.   Apparatus for semimicro distillation

Distillation is carried out as described in the ⟨OP–25⟩ procedure, except that the distillation rate can be reduced to 1 drop/second or slower since less liquid is being distilled. With a cooling bath, it is generally necessary to remove the heat source to stop the distillation while changing receivers.

Distillation of Solids

Comparatively low-melting solids can be distilled using the apparatus pictured in Figure 1 or with a special apparatus designed for that purpose. No condenser is necessary, although the condensation of low-*boiling* solids must sometimes be assisted by passing an air or water stream over the receiver. It may be necessary to heat the vacuum adapter with a burner or heat lamp to keep solids from plugging up the vacuum-adapter outlet. It is *very* important to keep this outlet free from solid, since if it is plugged the resulting pressure buildup could result in an explosion. When distillation is complete, the distillate is usually melted (if necessary) by heating the receiver in a water or steam bath so that the liquefied solid can be transferred to a suitable container. The last traces of solid can be transferred with a small amount of ethyl ether or another volatile solvent, which is then evaporated under a hood.

The apparatus pictured in Figure 1 can be used to distill viscous or high-boiling liquids as well as solids; it can also be adapted for the vacuum distillation of low-melting solids, if a suitable smooth boiling device is used. Because the narrow vacuum-adapter outlet is particularly easy to plug up, this apparatus should *not* be used to distill high-melting solids.

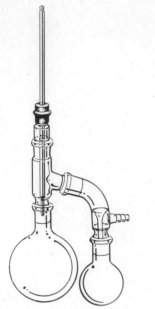

Figure 1.   Apparatus for distillation of solids

Water Separation

Water may have to be removed during the course of a reaction to increase the yield of a reversible reaction or to prevent the decomposition of water-sensitive compounds. This can be done by simple distillation, but it is often more convenient to use a device called a *water separator*. Commercial water separators such as the Dean–Stark trap (Figure 1) can be used, but they are generally not available to undergraduates. A "homemade" Dean–Stark trap can easily be constructed from a large test tube and a length of glass tubing, as described in *J. Chem. Educ. 40,* 349 (1963). If neither of these devices is available, the parts from an organic labkit can be assembled as shown in Figure 2 so that the 25 ml flask with adapters function as a water separator. This assembly works better if a "broken" vacuum adapter, with the drip tube removed, is used; otherwise this tube will fill up with liquid if the heating rate is too rapid.

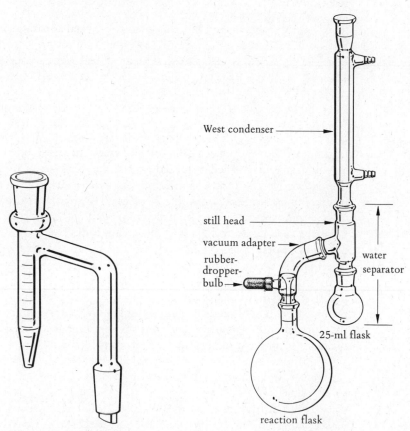

West condenser

still head

vacuum adapter

rubber-
dropper-
bulb

water
separator

25-ml flask

reaction flask

Figure 1.   Dean-Stark trap   Figure 2.   Apparatus for water separation

The reaction is generally run in a solvent (such as benzene or toluene) that is less dense than water, and the trap is filled with that solvent beforehand. As water forms during the reaction, its vapors and those of the solvent pass into the reflux condenser, where they condense and drip down into the trap. The organic condensate overflows through the sidearm and returns to the reaction flask, while the water collects at the bottom of the trap. The volume of water expected from the reaction can be estimated beforehand, so that by measuring the amount of water in the trap at any given time, one can tell whether the reaction has attained completion.

Alternatively, a water separator can be used (as in Experiment 19) to slowly remove a volatile product during a reaction *without* continuously returning distillate to the reaction flask. In this case, the trap is left empty at the beginning of the reaction.

OPERATION 26

Vacuum Distillation

Principles and Applications

Table 1.   Approximate boiling points of liquids at 25 torr

normal b.p. 760 torr	b.p. 25 torr
150°	60°
200°	100°
250°	140°
300°	180°

These chemical transformations may include condensation, dehydration, isomerization, oxidation, pyrolysis, polymerization, or rearrangement reactions.

The vapor bubble that occupies a volume of ¼ ml at 760 torr will expand to nearly 10 ml at 20 torr.

A Claisen head helps prevent mechanical carryover; many devices can be used to reduce bumping.

Distillation under reduced pressure, or vacuum distillation as it is frequently called, is based on the fact that lowering the external pressure about a liquid also lowers its boiling point. For example, a liquid that boils at 200°C at a pressure of one atmosphere (760 torr) will boil near 100°C when the pressure is lowered to 25 torr, and around 40°C at 1 torr. This is important when high-boiling liquids are to be distilled, since many of them undergo chemical transformations at high temperatures (especially in the presence of impurities that may be present in a reaction mixture). As a general rule, most liquids boiling over 200°C at atmospheric pressure should be purified by distillation under reduced pressure to prevent losses due to such transformations.

Vacuum distillation has certain inherent features that can cause experimental difficulties or hazards. The volume of the vapor generated when a given amount of liquid vaporizes is much greater at low pressures; this can cause excessive bumping in the boiling flask and mechanical carryover of liquid to the receiver. The vapor velocity is also greater because there are fewer molecules around to bump into; this can cause superheating of the vapor at the still head and a pressure differential throughout the system, so that the observed distillation temperature may be too high and the pressure reading too low. These difficulties can be overcome (or at least lived with) by using the right kind of apparatus and by carrying out the distillation slowly. Even then, the separation attainable under vacuum distillation does not equal that possible at atmospheric pressure, and there is always the danger that the apparatus may shatter (implode) because of the unbalanced external pressure on the system. Accordingly, a vacuum distillation must be performed with great care and attention to detail to obtain satisfactory results and prevent accidents.

Experimental Considerations

Vacuum Pumps. The most convenient vacuum source for most undergraduate laboratories is the water aspirator, which is capable (under optimum conditions) of pulling a vacuum of 10–25 torr, depending on the water temperature. In theory, an aspirator should be able to attain a vacuum equal to the vapor pressure of the water flowing through it, which is a function of the water temperature (Table 2). In practice, the pressure is often 5–10 torr higher because of insufficient water pressure, leaks in the system, or deficiencies in the aspirator itself.

An aspirator pump must be provided with a water trap to prevent backup of water into the receiving flask (due to changes of water pressure) and to reduce pressure fluctuations throughout the system. A thick-walled filter flask of the largest convenient size makes a suitable trap, if it is provided with a pressure release valve and hooked up to the aspirator and distillation apparatus as illustrated in Figure 1, p. 545.

For lower pressures, an oil diffusion pump can be used. Most oil pumps will routinely pull a 1 torr vacuum if they are in good condition, and 2–10 torr otherwise. Since the oil is subject to dilution or decomposition from vapors passing out of the distillation apparatus, an oil pump must always be used with one or more cold traps; it should also be provided with a manostat for adjusting the pressure.

Pressure Measurement. In order to know when the desired product is distilling and how pure it is, you must know the pressure inside the system and the approximate boiling point at that pressure. The pressure can be measured by a *manometer* such as that illustrated in Figure 1. The pressure (in torr) is equal to the vertical distance between the two columns of mercury, in millimeters, as measured by means of a movable scale or a sheet of metric graph paper. The mercury in the manometer can present a hazard when the vacuum is broken suddenly—inrushing air can push the mercury column forcefully to the closed end of the tube, breaking it and spraying poisonous mercury all over the laboratory. For this and other reasons, the vacuum must always be released *slowly*.

Boiling Points at Reduced Pressure. If the boiling point of a substance under reduced pressure is not known, it can be es-

Table 2. Vapor pressure of water below 30°C

T, °C	P, torr	T, °C	P, torr
30°	31.8	18°	15.5
28°	28.3	16°	13.6
26°	25.2	14°	12.0
24°	22.4	12°	10.5
22°	19.8	10°	9.2
20°	17.5	8°	8.0

If for any reason a pressure above the value attainable by an aspirator is desired, the screw clamp can be replaced by a bleed valve consisting of the bottom part of a bunsen burner. The pressure is then varied by adjusting the needle valve of the burner.

A cold trap consists of a chamber cooled by dry ice in isopropyl alcohol (or some other solvent) that condenses the vapors and keeps them out of the pump.

When reporting a boiling range from a vacuum distillation, always report the pressure at which the boiling range was recorded.

Vapors from the distillation may also migrate to the closed end of the manometer, invalidating the pressure readings. Therefore, it is advisable to open the manometer to the system only when a pressure reading is being made.

$$T^p \approx \frac{5.46 \cdot T^{760}}{8.34 - \log p} \qquad \textbf{(1)}$$

(T is in kelvins, p in torr)

Example:

$$T^{10} = \frac{5.46 \times (211 + 273) \text{ K}}{8.34 - \log 10}$$

$$= 360 \text{ K } (87°C) \text{ for } p\text{-nitrotoluene}$$

$$\log p = -\frac{0.05223A}{T} + B \qquad \textbf{(2)}$$

(T is in kelvins, p in torr)

Never use thin-walled glassware, such as Erlenmeyer flasks, in an evacuated system.

Receiving flasks can be tared to facilitate weighing the distillate.

Foaming can be dealt with by using a larger boiling flask, distilling slowly, or using an antifoaming agent.

timated from Equation 1 (with fair accuracy) for liquids that are not associated by hydrogen bonding. Associated liquids, such as alcohols and carboxylic acids, often have boiling point reductions about 10% less than the calculated decrease. For example, the boiling point of p-nitrotoluene is 211°C at 760 torr. Using Equation 1, its boiling point at 10 torr is calculated to be 87°C, compared to the literature value of 85°. However, the 10 torr boiling point of 1-octanol (normal b.p. = 195°C) is calculated to be 75°C, compared to an experimental value of 88°C; thus, the actual reduction is 13° (or 11%) less than the calculated reduction. Reduced-pressure boiling points can also be estimated by using the vapor pressure-temperature nomograph on the front endpaper of this book. More precise estimates can be made by referring to tables such as those in R. R. Dreisbach, *Pressure-Volume-Temperature Relationships of Organic Compounds* (New York: McGraw-Hill, 1952), by applying Equation 2 using A and B values given in older editions of *The CRC Handbook of Chemistry and Physics* (Bib–A1) (e.g., 50th edition, 1969, p. 167), or by using various empirical relationships.

Heat Sources. The heat source should be capable of providing constant, uniform heating to prevent bumping and superheating and to maintain a constant distillation rate. An oil bath ⟨OP–7⟩ is best, especially if it can be electrically heated and stirred. Heating mantles and heat lamps are satisfactory, and a burner can be used in a pinch.

Assembling Apparatus. The glassware used in assembling the apparatus *must* be inspected very closely to make certain that it is free from cracks and star fractures. Joints should be well lubricated with silicone vacuum grease to prevent leaks, and all rubber tubing should be stretched or bent to make sure it is pliable and free from cracks. All connecting tubing in the vacuum line should be (**a**) heavy-walled to prevent collapse, and (**b**) as short as possible to reduce the pressure differential between vacuum pump and distilling flask. Connections between rubber and glass, as well as ground-joint connections, must be secure and air-tight.

Smooth Boiling Devices. Numerous techniques are used to prevent bumping in vacuum distillations. One simple method is to use a magnetic stirbar in the distilling flask (possibly coupled with a larger one in the heating bath). Ordinary boiling chips are of little use since they are soon exhausted under vacuum, but special microporous boiling chips are quite suitable for most vacuum distillations; wooden applicators (boiling sticks) have also been used.

The classical method of vacuum distillation utilizes a flexible capillary bubbler about as fine as a cat's whisker, drawn from a length of thick-walled capillary tubing. This kind of bubbler is inserted into the boiling flask through a thermometer adapter (or a rubber stopper if necessary) so that its fine tip extends to within a millimeter or two of the bottom. Under vacuum, the capillary delivers a very fine stream of air bubbles that prevents the development of the large bubbles that cause bumping. Aside from the difficulty of constructing one, the capillary bubbler method has the following disadvantages: the air entering the system raises the pressure slightly and decreases the accuracy of the manometer reading; air may oxidize the product at high temperatures (this can be prevented by using nitrogen or another inert gas); and the capillary may plug up when the vacuum is broken. Nevertheless, a well-constructed capillary bubbler may work better than any of the other devices mentioned.

The microporous chips distributed by the Todd Scientific Company are made of specially purified anthracite coal.

Other fine-tipped tubes can be used, if provided with a screw-clamp valve to regulate the rate of ebullition.

Capillaries drawn from ordinary glass tubing tend to be inflexible and can break easily unless they are prepared with special care.

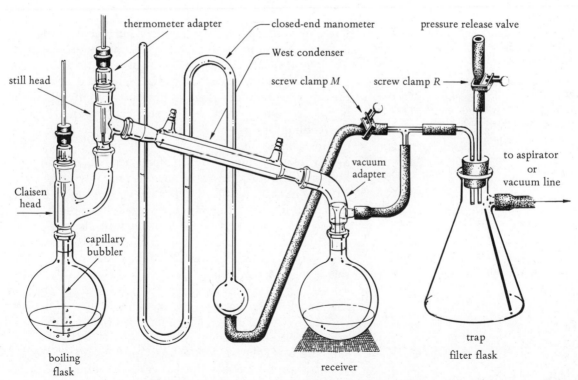

Figure 1.   Apparatus for distillation at reduced pressure

Equipment and Supplies: Boiling flask, Claisen head, still head (3-way connecting tube), stopper (optional), thermometer, thermometer adapter, condenser, vacuum adapter, receiving flask(s), microporous boiling chips (or capillary bubbler, stirbar, etc.), rubber band or spring clamp(s), utility clamp, condenser clamp, heat source, supports for receiver and heat source, condenser tubing, heavy-walled rubber tubing, glass tee, screw clamp, manometer, aspirator or oil pump. *Trap:* Large filter flask, utility clamp, two-hole rubber stopper, 2 short glass tubes (8 mm O.D.), screw clamp, short rubber tube.

With an oil pump, add one or more cold traps, a manostat, and a pressure release valve such as a 3-way stopcock. Omit the filter flask trap assembly.

Operational Procedure

Inspect your glassware and rubber tubing for imperfections, then assemble the apparatus illustrated in Figure 1 (substi-

---

WARNING   *Because of the danger of an implosion, safety glasses MUST be worn during a vacuum distillation.*

It may be necessary to insulate the still head with glass wool or aluminum foil for high-boiling liquids.

If the liquid may contain a volatile solvent, remove it first by distilling at atmospheric pressure, then let the apparatus cool before proceeding.

If the pressure is within 10 mm of that estimated from Table 2, the apparatus should be satisfactory.

tute a glass stopper for the capillary bubbler assembly if a bubbler will not be used). See that all joints and connections are tight, then add the material to be distilled through a funnel. Drop in a few microporous boiling chips or use one of the other methods described to prevent bumping. Raise the heat source into position, but do not begin heating. Turn on the cooling water for the condenser, open both screw clamps, and turn on the aspirator to its maximum flow rate. *Slowly* close clamp **R** on the pressure-release valve. (If bumping and foaming occur, there is probably some residual solvent in the distilland. Open the pressure-release valve, then close it down to a point where the solvent can be evaporated without excessive bumping.) With clamp **R** completely closed, wait a minute or two until the pressure equilibrates, then read the manometer. If the pressure is higher than expected, check the system for leaks and, if necessary, release the vacuum by opening clamp **R** and repair them. If the pressure is satisfactory, use the nomograph on the front endpaper (or one of the other methods described above) to estimate the boiling range at that pressure. Commence heating (and magnetic stirring, if used) until distillation begins. Adjust the heat so that a distillation rate of about 1 drop/second is attained, then record the temperature and pressure readings. Clamp **M** should then be closed to protect the manometer from vapors, and opened slowly when another pressure reading is made. Pressure fluctuations may occur as a result of changes in the water flow rate; this will require heating adjustments to reestablish the distillation rate.

If possible, only a few students should use the aspirators at the same time, to maintain pressure and minimize fluctuations.

WARNING

A rapid pressure increase, accompanied by a thick fog in the distilling flask, indicates decomposition of the distilland. If this occurs, remove the heat source immediately and get away (warning others to do so also) until the flask cools. Then try to determine the cause.

If the initial vapor temperature is lower than the estimated boiling point of the product at the pressure used, distill the volatile forerun until the temperature stabilizes close to the expected value. Then change receivers by the following procedure:

1. Lower the heat source and wait about 5 minutes to let the system cool down.

2. Open clamp **R** *slowly* until the system is at atmospheric pressure.

3. Replace the receiver with a clean one (preferably tared so the product can be weighed directly).

4. Open clamp **M**, then slowly close clamp **R** and read the manometer to be sure the vacuum is adequate.

5. Commence heating until distillation begins, then record the temperature and pressure.

6. Close clamp **M**.

It is a good idea to add another boiling chip or two each time the vacuum is broken.

Do not turn off the aspirator or oil pump while changing collectors.

Continue distilling until the upper end of the anticipated temperature range is reached, or until a significant drop in temperature indicates that the product is completely distilled. If a higher boiling fraction is to be distilled, change collectors by the same procedure as before. Stop the distillation before the boiling flask is completely dry.

When the distillation is completed, follow steps **(1)** and **(2)** above for bringing the system back to atmospheric pressure, then turn off the aspirator or oil pump. Disassemble the apparatus and clean the glassware promptly.

A temperature drop may also occur because the pressure has changed, the boiling chips (or bubbler) are not working, or the heating source is not hot enough. It may be necessary to raise the heating bath temperature at the end of the distillation to drive over the last milliliter or so of distilland.

Summary

1. Assemble apparatus, check connections and joints.

2. Add distilland and microporous boiling chips (or use another smooth-boiling device).

3. Position heat source, start cooling water, open pressure-release valve and manometer valve, turn on aspirator or oil pump.

4. Close pressure-release valve slowly, let pressure equilibrate.
IF bumping and foaming occur, adjust pressure-release valve to evaporate solvent.

5. Read manometer and, if necessary, estimate boiling range.
IF pressure is too high, check system for leaks, repair as necessary.

6. Adjust heat to attain desired distillation rate.

7. Record temperature and pressure, close manometer valve.
IF temperature is below expected boiling range,
GO TO 8
IF temperature is within expected boiling range,
GO TO 12

8. Distill until temperature reaches lower end of expected boiling range.

9. Lower heat source, let cool 5 minutes, open pressure-release valve slowly, change receivers.

10. Open manometer valve, close pressure-release valve slowly, read manometer.

11. Heat to resume distillation, record temperature and pressure, close manometer valve.

12. Distill until upper end of temperature range is attained *or* only a little distilland remains.

13. Lower heat source, let cool 5 minutes, open pressure release valve slowly.
IF last fraction has been collected,
GO TO 14.
IF additional fractions are to be collected, change receivers,
GO TO 4.

14. Turn off vacuum source, disassemble and clean apparatus.

Semimicro Vacuum Distillation

The apparatus pictured in Figure 1 ⟨OP–25a⟩ for semimicro distillation can be used under reduced pressure if the vacuum adapter sidearm is connected to a vacuum line (rather than to a drying tube) through a trap and manometer as illustrated in Figure 1, ⟨OP–26⟩. Except that a cooling bath is used instead of condenser water, the procedure is essentially the same as that used for ordinary vacuum distillation ⟨OP–26⟩. It may be necessary to insert a Claisen head between the boiling flask and the still head to prevent mechanical carryover of the distilland, or to allow for insertion of a capillary bubbler.

Fractional Distillation

Principles and Applications

Simple distillation is not a very efficient way of separating liquids whose boiling points are less than 80–100° apart. With closer-boiling liquids, the distillate will be a mixture of constantly changing composition and boiling point, containing an excess of the more volatile component at the beginning of the distillation and an excess of the less volatile component at the end (see the volume-temperature graph in Figure 1). The separation could be improved, of course, by redistilling the distillate fractions and combining fractions of similar composition, then repeating the process until a reasonably good separation is obtained. Alternatively, one might "invent" an apparatus such as that in Figure 2, in which each distillate is delivered directly to another distilling flask that redistills the condensed vapors and delivers them to another flask, and so on. This process can be understood by referring to the temperature-composition diagram in Figure 3 for our imaginary *entane-orctane* mixture (see ⟨OP–25⟩).

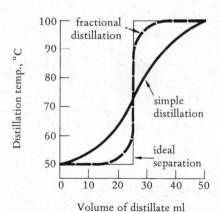

Figure 1.   Separation efficiency of simple and fractional distillations

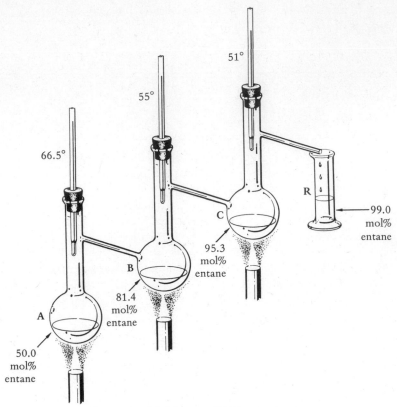

Figure 2.   Multi-stage distilling apparatus

The boiling point in flask B is lower than in flask A because the liquid in B is richer in the more volatile component.

If we boil a 50:50 mole percent mixture of entane and orctane in flask **A** (point **A**), the *vapor* over that mixture will have a composition of 81.4 mol% entane and only 18.6 mol% orctane (point **A′**). This is because the entane has a considerably higher vapor pressure at the boiling point of the mixture (66.5°) than does orctane. This vapor is then condensed into flask **B** (line **A′–B**), where it is boiled at 55° to yield a vapor containing 95.3 mol% entane (line **B–B′**), which is condensed (**B′–C**) into flask **C**. The condensed liquid in **C** boils at 51° to yield a vapor that is 99.0 mol% entane (**C–C′**), which is delivered to the receiver **R** as a liquid of the same composition (line **C′–R**).

Unfortunately, only the first few drops of distillate will attain this kind of purity; because as the more volatile entane is removed, the less volatile orctane will accumulate in the distilling flasks, reducing the proportion of entane in the vapor. The efficiency of our apparatus can be improved con-

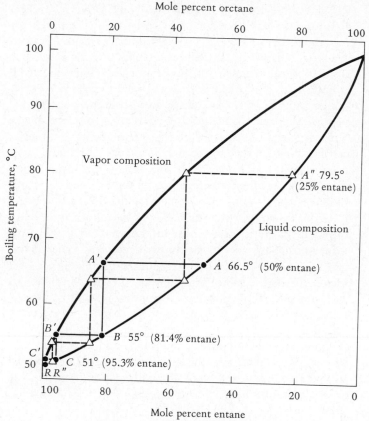

Figure 3. Temperature-composition
diagram for *entane-orctane* mixture

siderably if some of the orctane-enriched liquid in each flask
is drained back into the previous flask through an overflow
tube so that orctane will not accumulate as rapidly in the
upper stages. Even then, the purity of the distillate will decrease with time. For example, when half of the entane has
distilled out of flask **A** (point **A''**, 25 mol% entane), the
vapor condensing into the receiver (point **R''**) will be only
96.6 mol% entane, as shown by the broken lines in Figure 3.
In other words, during a distillation process one always
climbs inexorably up the temperature-composition graph,
and the composition of the distillate changes accordingly.
Nevertheless, the separation effected by several distillation
stages is considerably better than by only one, as can be seen
by comparing the curves for simple distillation and fractional
distillation in Figure 1.

Fractional distillation *is, in effect, a
multistage distillation process performed in
a single operation.*

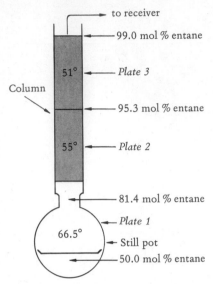

to receiver

99.0 mol % entane

51° ← Plate 3

Column

95.3 mol % entane

55° ← Plate 2

81.4 mol % entane

← Plate 1

66.5° ← Still pot

50.0 mol % entane

Figure 4.   Operation of a fractionating column

In certain multistage columns, the "theoretical" plates are real. The Bruun column consists of a series of horizontal plates stacked at intervals inside a vertical tube, and one vaporization-condensation cycle occurs on each plate.

Volatility definition

$$v_c = \frac{Y_c}{X_c}$$

v = volatility of component
Y_c = mole fraction of component in vapor
X_c = mole fraction of component in liquid, etc.

The apparatus described above would be very cumbersome and difficult to operate efficiently. A much simpler approach is to interpose a vertical *distilling column* between the boiling flask and the still head, and allow the successive distillation operations to take place on the column. Such a column is usually filled with some kind of *column packing,* which provides a large surface area from which repeated vaporization and condensation operations can take place. Suppose our 50:50 mol% mixture of entane and orctane is distilled through such a column. The vapor entering the column from the still pot should have the same composition (81.4 mol% entane) as before, but as it passes onto the column it will cool, condense on the packing surface, and begin to trickle down the column on its way back to the pot. Since the temperature is higher near the pot, part of it will vaporize on the way down, yielding a vapor still richer in entane. This vapor passes up the column until, at a higher level than before (since its boiling point is lower), it cools enough to recondense. This process of vaporization and condensation may be repeated a number of times on the way to the top of the column, so that when the vapor finally arrives at that point it is nearly pure entane. Although this is a continuous process involving the simultaneous upward flow of vapor and downward flow of liquid, the net result is the same as that produced by a discrete process of successive distillations.

Each section of the distillation apparatus that provides a separation equivalent to one cycle of vaporization and condensation (one "step" on the temperature-composition graph) is called a *theoretical plate.* Since the first cycle occurs in the pot, it provides one theoretical plate; the column illustrated in Figure 4 contributes two more. Notice that there is a continuous variation in both temperature and vapor composition as one proceeds up the column, and that the temperature is established automatically by the liquid-vapor composition. For instance, the entane-rich liquid near the top of the column boils at a lower temperature than the original mixture, so the average temperature at plate 3 is lower than that of the plates below it.

The efficiency of a particular column can be measured in terms of its theoretical plate rating. If the *volatility* (v) of liquid C is defined as the ratio of its vapor composition to its liquid composition at a given temperature, then the *relative volatility* of two liquids C and D is just the ratio of their respective volatilities at that temperature (Equation 1). By convention, the volatility of the more volatile component is put in the numerator; thus, the relative volatility (α) is a measure of the enrichment of the vapor in the more volatile component after each vaporization-condensation cycle. The enrichment after two cycles will be α^2, and after n cycles, α^n, where n is equal to the number of theoretical plates provided by the distillation apparatus. If the mole fractions of C and D in the vapor as it emerges from the top of the column are given by Z_C and Z_D, then the number of plates, n, can be expressed in terms of α as shown in Equation 2, the Fenske equation. Therefore, the theoretical plate rating of a given column can be determined by distilling a liquid mixture for which α is known, measuring the composition of the distillate and the liquid in the pot, and plugging these values into the Fenske equation.

Another useful measure of column efficiency is the *height equivalent to a theoretical plate (HETP)*, which is equal to the length of the column divided by the number of theoretical plates. For example, a 48-cm Vigreux column with 6 theoretical plates has a HETP value of 8 cm.

In practice, a number of factors limit the efficiency of a given column. The number of theoretical plates is ordinarily determined under the equilibrium condition of *total reflux,* in which all the vapors are returned to the pot. In practice, however, part of the vapor is always being drawn off and condensed into the receiver, which disturbs the equilibrium and reduces the column's efficiency. For efficient operation, the ratio of liquid returned to liquid distilled (the reflux ratio, R) should be carefully controlled. According to one rule of thumb, the reflux ratio should at least equal the number of theoretical plates for optimum operation. To a lesser extent, the **holdup** of a column and the **throughput** rate can also limit the efficiency of a distillation.

To this point, the discussion has dealt only with ideal liquids. No real liquid systems, however, display ideal behavior, although some may come very close. The greatest deviations from ideal behavior are displayed by liquid mixtures that form high- or low-boiling azeotropes. Such liquids have boiling temperature–composition graphs similar to the one illustrated in Figure 5. From the graph, it can readily be shown that distilling an azeotropic mixture cannot possibly yield *both* components in pure form. In the case of a minimum-boiling azeotrope, distillation of a mixture having the composition X will yield a distillate of the azeotropic composition (indicated by the minimum in the

Relative volatility

$$\alpha = \frac{v_C}{v_D} = \frac{Y_C}{Y_D}\frac{X_D}{X_C} \quad (1)$$

The Fenske equation

$$\alpha^n = \frac{Z_C X_D}{Z_D X_C} \quad (2)$$

(A more convenient form can be derived by taking the logarithm of both sides.)

The enrichment factor α usually varies with the composition of a mixture.

The lower the HETP, the more efficient the column.

If a column having the required number of plates is not available, the distillation must be repeated one or more times, collecting and combining fractions according to a systematic plan.

The reflux ratio:

$$R = \frac{\text{liquid vol. returning to pot}}{\text{liquid vol. distilled}}$$

(measured over the same time interval)

Reflux ratios of 5–10 are common for routine separations.

holdup: the amount of liquid adhering to the surface of the column and its packing.

throughput: the distillation rate, usually expressed in ml of distillate per minute.

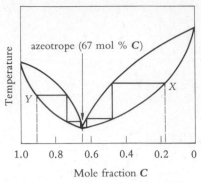

Figure 5.   Temperature-composition graph for azeotrope

Figure 6.   Vigreux column
Stainless-steel sponge pads, obtained from household goods stores, can be stretched and cut into 6–8 inch lengths for use in a typical column. The packing should be pulled into the column using a rod with a hook at one end.

graph). Since the azeotrope contains more of component **C** than does the mixture in the pot, the pot mixture will become richer in **D** until it is eventually pure **D** and the distillate is pure azeotrope. Starting at point **Y** gives the same kind of result, except that the pot liquid then becomes pure **C** instead of **D**. It can be shown by similar reasoning that distilling a maximum-boiling azeotrope will yield **C** or **D** in the distillate and azeotrope in the pot.

Azeotrope formation can cause difficulties in purifying liquids. For example, it is impossible to obtain pure ethyl alcohol by distilling 50% aqueous alcohol, since no distillate containing more alcohol than the azeotrope (95% ethanol) can be obtained. In this case, the problem is solved by "fighting fire with fire." Benzene and water form an even lower boiling azeotrope with ethanol than does water alone (b.p. 64.9°, compared to 78.1° for 95% ethanol); so by boiling 95% ethanol with some benzene, the benzene-water-ethanol azeotrope can be distilled off until all the water is removed. Other azeotropes can sometimes be purified by chemical means or by such separation methods as chromatography.

Experimental Considerations

Columns.   The column is the most important part of any fractional distillation apparatus; many kinds are available. One of the simplest is the Vigreux column, which has a series of indentations or "drip points" in a more-or-less spiral arrangement down the length of the column. The drip points provide some surface for condensation-vaporization cycles, but the total surface area is comparatively small. This kind of column is not very efficient, having HETP values in the 8–12-cm range. The most commonly used column for laboratory fractional distillation is simply a straight tube (usually jacketed) with indentations at the bottom, filled with a suitable packing material.

Column Packing.   Most columns are provided with some kind of packing, which can come in the form of rings, saddles, helices, beads, or metal turnings. Some packing materials such as Heli-Pak can provide HETP ratings of 1 cm or lower, but are quite expensive. Glass beads (or rings) and stainless steel "sponge" are more practical for undergraduate laboratories. The metal packing can be corroded by halogen compounds, or it may catalyze decomposition reactions of sensitive liquids. Packed columns are more efficient than Vigreux columns, but they suffer from higher holdup and lower optimum throughput rates. As a result, distillations must be conducted more slowly and material losses due to adsorption on the column must be allowed for.

Insulation. If either of the components to be separated boils at a temperature over about 100°C, the column and still head must be insulated to prevent heat losses that can reduce efficiency or even prevent distillation entirely. Several layers of asbestos tape or crumpled aluminum foil can be wrapped around the column and the still head. Alternatively, a layer of glass wool can be sandwiched between two pieces of aluminum foil that are crimped together at the edges, then wrapped around the column and still head.

Heat Sources. For good results, the heat source must provide constant, uniform heating. A burner is not very satisfactory unless it is provided with a **chimney** and an asbestos-centered wire gauze. Oil baths are preferred for many fractional distillations, but heating mantles and heat lamps may also be suitable.

Flooding. One problem often encountered in a fractional distillation is *flooding,* in which the column becomes partly or entirely filled with liquid. Flooding can be prevented by the proper insulation and heating rate. If it occurs, the heat source must be removed until all of the excess liquid has returned to the pot before the distillation is resumed. Flooding can greatly decrease the efficiency of a separation, since it reduces the area of contact between the liquid and the packing.

Equipment and Supplies: Boiling flask, distilling column, column packing, insulating material (optional), still head (3-way connecting tube), thermometer, thermometer adapter, condenser, vacuum adapter, receivers, rubber band or spring clamp(s), boiling chips, condenser tubing, heat source, 2 utility clamps, condenser clamp, 2 ringstands, supports for receiver and heat source.

Operational Procedure

Fill the column with packing to within a centimeter of the ground joint. Loose packing can often be introduced by laying the column on its side, adding about ¼ of the desired amount of packing, and quickly (with a hand held over the bottom) turning the column upright so that the packing is jammed together at the constriction and does not fall out; then the rest of the packing is poured in. If this method

Losses of 3–5 ml can be expected from fractional distillation with the apparatus illustrated in Figure 7. Therefore, it is not practical to distill less than about 25 ml from such a setup.

Since it is necessary to view the column during the initial heating period, the insulation should be secured in such a way that it can be easily opened for inspection.

*A **chimney** is a device that fits on top of the burner to keep out drafts.*

Flooding may also be caused by unsuitable packing or packing supports. For example, a glass-wool plug at the base of the column can hold up enough liquid to cause flooding.

The receivers can be flasks, bottles, large vials, test tubes, or graduated cylinders. They should be numbered and (if the fractions are to be weighed) tared.

Glass helices should be dropped into the column one at a time.

The vacuum-adapter drip tube should extend into the receiver to prevent losses by evaporation. For very volatile distillates, a ground-joint flask can be used as a receiver.

doesn't work, insert a support (such as a small plug of stainless-steel sponge) at the constriction. Assemble the apparatus illustrated in Figure 7. Make sure that all joints are tight, the column is perpendicular to the bench top, the boiling flask and receiver are supported properly, and none of the joints are under excessive strain. If the distilland boils much above 100°, insulate the apparatus from the bottom of the column to the top of the distilling head. Fill the distilling flask $\frac{1}{3}$ to $\frac{1}{2}$ full and add a few boiling chips.

Raise the heat source into place and begin heating at a fairly rapid rate to get the mixture to the boiling point. When it boils, adjust the heat so that the reflux ring of condensing vapor passes up the column at a *slow,* even rate (it should take 5–10 minutes to reach the top of the column). Open the insulation at intervals to observe the passage of the reflux ring, and watch for evidence of flooding. When the vapors rise above the column packing, lower the heat enough to keep the reflux ring between the packing and the sidearm for a few minutes, so that the column has time to equilibrate and all of the packing becomes moistened with liquid. Adjust the heat to initiate distillation, and record the temperature when it stabilizes. Distill at a rate of about one drop every 1–3 seconds, or at a rate that gives the desired reflux ratio. Estimate the ratio by counting the drops returning to the flask and those distilling over a short period of time. If the initial distillate is cloudy (due to dissolved water), change receivers when it becomes clear. Continue the distillation until the temperature at the head drops (indicating that all of the more volatile component has distilled) or until a predetermined temperature is reached. If another fraction is to be collected, change receivers and increase the pot temperature (if necessary) until the next fraction starts to distill.

With stainless-steel packing, the column is sometimes deliberately flooded by strong heating to wet the packing. Then the heat is reduced to drain the column before distillation is begun.

1–3 ml/min is a good rate for most packed columns. A Vigreux column may operate efficiently at 5–10 ml/min.

The temperature drops when the pot temperature is not high enough to bring the vapors of the less volatile component up to the still head.

If the separation is not sharp, the temperature may rise only gradually throughout the distillation; in that case, it is best to collect fractions continuously at regular temperature intervals and redistill them. Otherwise, continue heating until the head temperature stabilizes at a higher value, then record the temperature, change receivers, and collect the next component. Repeat this process as necessary until the pot is nearly dry (or until all the desired components have been collected), then lower the heat source and let the column drain. The intermediate fractions collected while the temperature was rising rapidly are impure; unless you wish to redistill them, they can be discarded. Disassemble the apparatus and clean it promptly.

Summary

1. Pack column, assemble and (if necessary) insulate apparatus.

2. Add liquid and boiling chips, position heat source.

3. Adjust heat so that reflux ring passes slowly up the column; hold it at the top for a few minutes.

4. Distill slowly until head temperature drops *or* until predetermined temperature (or fraction volume) is reached *or* until pot is nearly empty.
IF another fraction is to be collected,
GO TO 5

IF distillation is complete,
GO TO 7.

5. Change receivers.
IF fractions are to be collected continuously at predetermined intervals,
GO TO 4
IF not,
GO TO 6.

6. Increase heating rate until distillation resumes, change receivers when temperature stabilizes,
GO TO 4.

7. Remove heat, drain column, disassemble and clean apparatus.

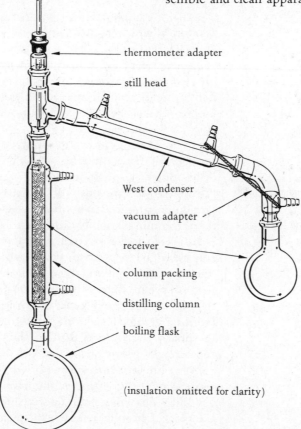

thermometer adapter

still head

West condenser

vacuum adapter

receiver

column packing

distilling column

boiling flask

(insulation omitted for clarity)

Figure 7.   Apparatus for fractional distillation

Measurement
of Physical Constants

Melting Point

Principles and Applications

The melting point of a pure substance is defined as the temperature at which the solid and liquid phases of the substance are in equilibrium at a pressure of one atmosphere. At a temperature slightly lower than the melting point, a mixture of the two phases solidifies; at a temperature slightly above the melting point, the mixture liquefies. Melting points can be used to identify organic compounds and to assess their purity.

The melting point of a pure compound is a unique property of that compound, which is essentially independent of its source and method of purification. This is not to say that no two compounds will have the same melting point. If two pure samples have *different* melting points, however, they are almost certainly different compounds. If a substance is suspected to be a certain compound, its identity can be confirmed beyond reasonable doubt by mixing it with a sample of the known compound and obtaining a melting point of the mixture. If the two substances are not the same, the melting point will generally be lowered and the melting point range broadened; otherwise, the melting point will be essentially the same as that for the two samples measured separately.

A pure substance usually melts within a range of no more than 1–2°C—that is, the transition from a crystalline solid to a clear, fluid liquid occurs within a degree or two if the rate of heating is sufficiently slow and the sample is properly prepared. The presence of impurities in a com-

Allotropic forms of an element or different crystalline forms of a compound may differ in melting point.

pound lowers its melting temperature and broadens its melting range. Since the melting point decreases in a nearly linear
fashion as the degree of impurity increases (up to a point, at
least), the difference between the observed and expected
values can give an approximate indication of the compound's purity.

Melting phenomena can be better explained by referring to a *phase diagram* such as the one for phenol (**P**) and
diphenylamine (**D**) in Figure 1. Pure phenol melts at 41°C

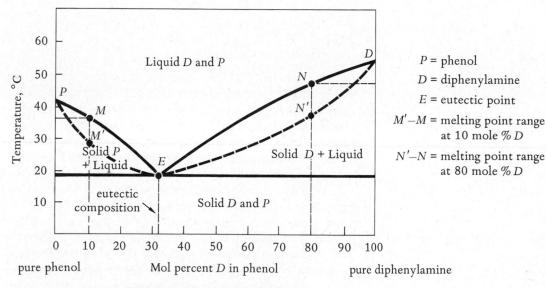

P = phenol

D = diphenylamine

E = eutectic point

M′–M = melting point range
at 10 mole % D

N′–N = melting point range
at 80 mole % D

Figure 1.   Phase diagram for the
phenol–diphenylamine system

and pure diphenylamine at 53°C. If a sample of phenol contains a small amount of diphenylamine as an impurity, its
melting point will decrease approximately in proportion to
the mole percent of diphenylamine present; likewise, the
melting point of diphenylamine will decrease upon addition
of phenol. If more than 32 mole percent of diphenylamine is
added to phenol, the melting point will begin to rise from its
minimum value as the *eutectic point* **E** is passed. Since the organic compounds you are likely to encounter contain much

*For example, the melting point of a mix-
ture containing 10 mol% **D** in phenol is
given by the point **M**; that of 20 mol%
P in diphenylamine is given by point **N**.*

smaller amounts of impurities than this, the rule that "impurities lower the melting point" is useful when not taken too literally.

Pure phenol and diphenylamine should both have sharp melting points; on the other hand, mixtures of the two (except the eutectic mixture) will ordinarily exhibit broad melting-point ranges, depending on their composition. The observable melting-point range for a mixture is given roughly by the distance between the broken line connecting **P**–**E** or **E**–**D** and the solid lines connecting the same points. For instance, the melting-point range of a 10 mol% diphenylamine mixture is given in the figure by the distance between **M'** and **M**, and for an 80 mol% mixture by the distance between **N'** and **N**.

The situation is more complicated when complex formation occurs between two compounds. In some cases, the melting point of a complex may be higher than the melting points of one or both components. For this and other reasons, melting points should be interpreted with caution.

Experimental Considerations

Melting Behavior.   The *melting-point range* of a substance is reported as the range between the temperature at which the crystals first begin to liquefy and that at which they are completely liquid. In many cases, the crystals will soften and shrink before they begin to liquefy, but the melting range does not begin until the first free liquid is visible.

This process begins at the eutectic temperature.

Occasionally traces of solvent will be present in the crystals, due either to insufficient drying or to chemical interactions with a solvent. This may be indicated by "sweating" of solvent by the crystals or by the presence of bubbles in the melt, which may resolidify when all of the solvent is driven off. If the sample is obviously wet, it should be dried and the melting point remeasured. In any case, sweating or the dissolution of crystals in the solvent should not be mistaken for melting behavior.

Bubbling and resolidification may also result from decomposition.

Some compounds can be converted directly from the solid to the vapor state by heating them in an open container. This *sublimation* can be detected during a melting-point determination by a shrinking of the sample accompanied by the appearance of crystals higher up inside the melting-point tube. The melting point of a sample that sublimes at or below its melting temperature can be determined using a sealed capillary tube. An ordinary melting-point tube should be cut ⟨OP–4⟩ short enough so that it does not project above the block of a melting point apparatus, like the one in Figure 4, or so that it can be entirely immersed in a melting bath. The sample is introduced and the open end is sealed with a microburner ⟨OP–4⟩. Then the melting point is measured by one of the methods described below.

If a melting bath is used, the sealed tube should be secured to the thermometer by a fine wire (not by a rubber band!).

Compounds that *decompose* before a true melting point can be reached do not readily give reproducible "melting" points. They usually liquefy over a broad temperature range, at temperatures that vary with the rate of heating. The "melting" or decomposition point of such a compound is best measured by heating the melting-point bath (or block) to within a few degrees of the expected decomposition temperature before inserting the sample, then raising the temperature at a rate of about 3–6° per minute.

Decomposition should be suspected when a compound becomes discolored and liquefies over an unusually broad range.

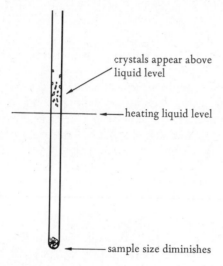

crystals appear above liquid level

heating liquid level

sample size diminishes

Figure 2.   Sublimation of a sample in a capillary tube.

Apparatus for Measuring Melting Points. Melting points can be determined with good accuracy using a special melting-point tube (Thiele or Thiele-Dennis tube) filled with a heating-bath liquid such as mineral oil. (See ⟨OP–7⟩ for information on bath liquids.) The design of such tubes promotes good circulation of the heating liquid without stirring. A capillary tube containing the solid is secured to a thermometer, which is then immersed in the bath liquid, and the solid is observed carefully for evidence of melting as the apparatus is heated. Melting points can also be measured in an ordinary beaker if the bath liquid is stirred constantly, either manually or with a magnetic stirrer ⟨OP–9⟩. Heating coils or tapes can be used, eliminating the need for an external heat source.

Commercial melting-point instruments of the "hot stage" type are available, as well as ones utilizing capillary melting-point tubes. Hot stage instruments use an electrically heated metal block containing a thermometer. The

clamp here

heat

stir with a gentle, up-and-down motion

Figure 3.   Stirred heating bath

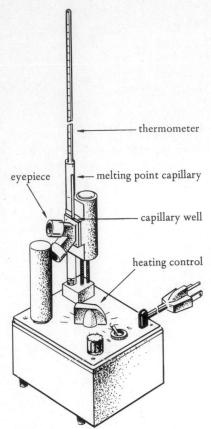

thermometer

eyepiece — melting point capillary

— capillary well

heating control

Figure 4.   Capillary melting point apparatus

Emergent Stem Correction:

Correction (to be added to t_1)
 $= 0.00017 \cdot N(t_1 - t_2)$

N = length in degrees of exposed mercury column

t_1 = observed temperature

t_2 = temperature at middle of exposed column

Melting points that have been corrected by adding a correction factor should be reported as (for example): m.p. 123–124.5° (corr.)

sample is placed between two thin glass discs, positioned on top of the block, and observed through a magnifying lens or microscope while the rate of heating is controlled with a dial. Because the sample is out in the open, it may be at a lower temperature than that measured by the thermometer bulb, which is centered in the block. Therefore, melting points obtained by this method are often too high.

Melting point devices using capillary tubes are quite accurate if operated properly and can be used to make several measurements at once. The latter feature is particularly useful when a *mixture melting point* is being determined, since the melting points of the unknown compound, the known, and the mixture can be measured at the same time and compared. The heating rate is adjusted by a knob controlling the voltage input to the heating coil. The voltage required to obtain a given heating rate at the melting point of the sample can be estimated from a heating-rate chart furnished with the instrument. Initially, the heating-rate control is often set 5–10 volts higher than this to quickly bring the temperature within 15–20° of the expected melting point. It is then reduced to maintain the desired heating rate at the melting point.

Melting Point Corrections.   The observed melting point of a compound may be inaccurate (*a*) because of defects in the thermometer used or (*b*) because of the "emergent stem error" that results when a thermometer is not immersed in the bath to its intended depth. Many thermometers are designed to be immersed to the depth indicated by an engraved line on the stem, usually 76 mm from the bottom of the bulb; slight deviations from this depth will not result in serious error. Other thermometers are designed for total immersion; if they are used under other circumstances the temperature readings will be in error. This error can be compensated for by adding an *emergent stem correction* calculated with the equation in the margin. This requires that a second thermometer be held opposite the middle of the exposed part of the mercury column.

Errors caused by thermometer defects and by an emergent stem can both be corrected by calibrating the thermometer under the conditions in which it is to be used. A thermometer used for melting-point determinations, for example, is calibrated by measuring the melting points of a series of known compounds, subtracting the observed melting point *from* the true melting point of each compound, and plotting this correction as a function of temperature. The correction at the melting point of any other compound is then read from the graph and added to its observed value. A list of pure compounds that can be used for melting-point calibrations is shown in the margin. Sets of pure calibration substances are available from chemical supply houses.

Mixture Melting Points. A mixture melting point is obtained by grinding together approximately equal quantities of two solids (a few milligrams of each) until they are thoroughly intermixed, then measuring the melting point of the mixture by the usual method. Usually one of the compounds (X) is an "unknown" that is believed to be identical to a known compound, Y. A sample of pure Y is mixed with X and the melting points of this mixture and of pure Y (and sometimes of X as well) are measured. If the melting point ranges of pure Y and of the mixture are identical (within a degree or so), then X is probably identical to Y. If the mixture melts at a lower temperature than pure Y and has a much broader melting range, then X and Y are different compounds.

Substances used for melting point calibrations

	m.p., °C
ice	0°
diphenylamine	54°
m-dinitrobenzene	90°
benzoic acid	122.5°
salicylic acid	159°
3,5-dinitrobenzoic acid	205°

b. Cut-away cork, top view

← notch in cork

← liquid level at 200°

← beginning liquid level

← capillary tube

Thiele-Dennis tube

a. Thiele-Dennis tube, assembled

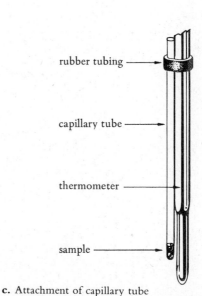

rubber tubing ⟶

capillary tube ⟶

thermometer ⟶

sample ⟶

c. Attachment of capillary tube

Figure 5. Apparatus for determining melting points

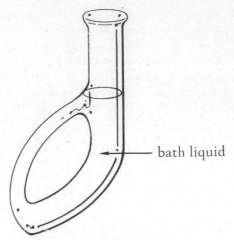

bath liquid

Figure 6.   Thiele–Dennis tube

Equipment and Supplies (Thiele Tube Method):   Thiele or Thiele–Dennis tube, clear mineral oil, thermometer, capillary melting-point tube, burner, cut-away cork, 3-mm slice of rubber tubing, watch glass, flat-bottomed stirring rod, 1-meter length of glass tubing, clamp, ringstand.

Operational Procedure (Thiele Tube Method)

The capillary melting-point tube can be constructed by sealing one end of a 10 cm length of 1 mm (I.D.) capillary tubing.

If a commercial instrument is to be used for melting point measurements, your instructor will demonstrate its operation.

Using too much sample can result in too broad a melting-point range and a high value for the melting point.

Pinch the rubber band between your fingers when inserting the m.p. tube.

The cork is bored out to accommodate the thermometer, then cut with a single-edged razor blade (or sharp knife) so that the thermometer can be snapped into place from the side rather than inserted from the top.

Clamp the Thiele tube or Thiele–Dennis tube to a ringstand and add enough mineral oil (or other bath liquid) to just cover the top of the sidearm outlet. Put a few milligrams of the dry solid on a watch glass (or a glass plate) and grind it to a fine powder with a flat-bottomed stirring rod (or a flat-bladed spatula). Press the open end of a capillary melting-point tube into a pile of the powder until enough has entered the tube to form a column 1–2 mm high. Tap the closed end of the tube gently on the bench top (or rub its sides with a file), then drop it (open end up) through a 1-meter length of small-diameter glass tubing onto a hard surface such as the bench top. Repeat the process several times to pack the solid firmly into the bottom of the tube. Secure the capillary tube at the top to a broad-range thermometer (e.g., −10°C to 360°C) by means of a 3 mm-thick rubber band (cut from $\frac{1}{4}''$ I.D. thin-walled rubber tubing) so that the sample is adjacent to the middle of the thermometer bulb (see Figure 5c for placement). Snap the thermometer into a cut-away cork (Figure 5b) at a point above the rubber band, with the capillary tube and the degree markings on the same side as the opening in the cork. Insert this assembly into the bath liquid and move the thermometer, if necessary, until its bulb is centered in the Thiele tube about 3 cm below the sidearm junction, and the temperature can be read through the opening in the cork (Figure 5**a**). The rubber band should be 2–3 cm above the liquid level so that the bath liquid, as it expands on heating, will not contact it.

If the hot oil contacts the rubber band, it may soften and allow the capillary tube to drop out.

If the approximate melting point of the compound is known, heat the bottom of the Thiele tube with a burner until the temperature is about 15°C below the melting temperature, then turn down the flame and apply it at the side arm to reduce the heating rate. The heating rate should be

A micro burner (or a Bunsen burner with the barrel removed) allows more precise heat control.

HAZARD

Mineral oil begins to smoke and discolor below 200°C and can burst into flames at higher temperatures. Oil fires can be extinguished with solid-chemical fire extinguishers or powdered sodium bicarbonate.

about 1–2°C per minute while the compound is melting, so that the sample will be in thermal equilibrium with the bath and thermometer. Record the temperature (*a*) when the first free liquid appears in the melting-point tube and (*b*) when the sample is completely liquid.

If the approximate melting point of the compound is not known, it is usually quicker to make a rough initial determination with fast heating (6°/minute or more), then measure an accurate melting point with a second sample. If two or more determinations are to be carried out in succession on the same compound, the bath should be cooled to 10–15°C below the melting point and a new sample used for each additional measurement. Cooling can be accelerated by passing an air stream over the Thiele tube. Unless the oil is too dark for further use, the Thiele tube can be stoppered and stored with the oil inside.

It is advisable to do at least two measurements on each compound.

Mineral oil can be removed from glassware by rinsing the glassware with ligroin (or petroleum ether) followed by acetone, then washing it with a detergent and water.

Summary

1. Assemble apparatus for melting-point determination.

2. Grind solid to powder, fill capillary tube(s) to 1–2 mm depth.

3. Secure capillary tube to thermometer, insert assembly in heating bath.

IF approximate m.p. is known,
GO TO 5
IF not,
GO TO 4.

4. Heat rapidly until sample melts, record approximate melting temperature, select another sample,
GO TO 3.

5. Heat rapidly to 15°C below m.p., reduce heating rate to 1–2°C per minute.

6. Record temperatures at beginning and end of melting range.
IF another determination is needed,
GO TO 7.
IF not, disassemble apparatus,
STOP.

7. Cool bath to 10–15° below m.p., select another sample,
GO TO 3.

OPERATION 29

Boiling Point

The boiling point of a liquid is defined as the temperature at which the vapor pressure of the liquid is equal to the external pressure exerted at the surface of the liquid. Its *normal boiling point* is the temperature at which the vapor pressure of the liquid equals one atmosphere (i.e., its boiling point at 760 torr). Since the atmospheric pressure at the time of a boiling-point determination is seldom exactly 760 torr, observed boiling points may differ somewhat from values reported in the literature, and should be corrected. Boiling-point values can be used to identify liquid organic compounds and assess their purity.

Experimental Considerations

Boiling points can be measured with good accuracy by distilling a small quantity of liquid or by using a micro method such as that described in ⟨OP–29a⟩. During a carefully performed distillation, it can be assumed that the vapors surrounding the thermometer bulb are in equilibrium (or nearly so) with the liquid condensing on the bulb. Since the boiling point of a liquid is (by one definition) the temperature at which its liquid and gaseous forms are in equilibrium, the temperature recorded during the distillation of a pure liquid should equal its boiling point. If the liquid is contaminated by impurities, the boiling point obtained by distillation may be either too high or too low, depending on the nature of the impurity. Therefore, if there is any doubt about the purity of a liquid, it should be distilled or otherwise purified prior to a boiling-point determination.

Volatile impurities in a liquid lower its boiling point, whereas involatile impurities raise it. In either case, the distillation boiling-point range will be broadened.

If a laboratory boiling-point determination is carried out at a location reasonably close to sea level, atmospheric pressure will rarely vary by more than 30 torr from 760. For deviations of this magnitude, a *boiling point correction*, Δt, can be estimated using Equation 1 in the margin. The value 1.0×10^{-4} is used for the constant γ if the liquid is water, an alcohol, a carboxylic acid, or another associated liquid; otherwise, γ is assigned the value 1.2×10^{-4}. For example, the boiling point of water at 730 torr is 98.9°C. Use of Equation 1 leads to a correction factor of (1.0×10^{-4}) $(760 - 730)$ $(273.1 + 98.9) = 1.1$°C, which yields the correct normal boiling point of 100.0°C.

$$\Delta t \approx \gamma(760 - P)(273.1 + t) \quad (1)$$

Δt = temperature correction, to be added to observed boiling point

P = barometric pressure (torr)

t = observed boiling point (°C)

γ = 1.0×10^{-4} for associated liquids; 1.2×10^{-4} for other liquids.

At high altitudes, the atmospheric pressure may be considerably lower than 1 atmosphere, resulting in observed boiling points substantially lower than the normal values. For example, water boils at 93°C on the campus of the University of Wyoming (elevation 7520′) and at 81°C at the top of Mount Evans in Colorado (elevation 14,264′). For major deviations from atmospheric pressure, Equation 2 can be used in conjunction with entropy of vaporization values estimated from tables in the *CRC Handbook of Chemistry and Physics* (on page D-229 in the 59th edition) or obtained from published data. For example, hexane boils at 49.6°C at 400 torr; the value of ϕ for alkanes is found (from the *CRC Handbook* tables) to be about 4.65 at that temperature. Substituting into Equation 2, we obtain a correction factor of:

$$\Delta t \approx \frac{(273 + t)}{\phi} \log \frac{760}{P} \quad (2)$$

ϕ = entropy of vaporization at normal boiling point (760 torr)

Equation 2 is a simplified form of the Hass-Newton equation found on page D-228 of the *CRC Handbook*, 59th ed.

$$\Delta t = \frac{(273 + 49.6)}{4.65} \log \frac{760}{400} = 19.3°C$$

which gives a normal boiling point of 68.9°C as a first approximation. This compares very favorably to the reported value of 68.7°C.

As in melting-point determinations, it may be necessary to correct the boiling point for thermometer error, especially for high boiling liquids. See ⟨OP–28⟩ for details.

For a second approximation, the value of ϕ at 68.9°C would be estimated from the tables and resubstituted into Equation 2. This results in a value of 68.8°C, so the difference is hardly significant.

Operational Procedure

Assemble the apparatus pictured in Figure 1, ⟨OP–25*a*⟩, using a 25-ml distilling flask and following the instructions given therein. Be sure the still head is well insulated from the heat source to prevent superheating of the vapors. Introduce 5–10 ml of the pure liquid and a boiling chip or two into the pot and distill the liquid *slowly* (1 drop a second or less), recording the temperature after the first 1–2 ml has distilled and again when only 1–2 ml remains in the pot. It may also be desirable to record a median boiling point, the temperature at which half of the liquid has distilled. If the boiling range is more than a few degrees, it may be necessary to purify the liquid by distillation (⟨OP–27⟩ or ⟨OP–25⟩) and repeat the determination.

Correct thermometer placement is essential —see ⟨OP–25⟩.

Record the barometric pressure at some time during the determination so you can make a pressure correction.

Low- or high-boiling impurities may distort the boiling range at the beginning or end of a distillation.

OPERATION 29 *a*

Micro Boiling Point

When a liquid is heated to its boiling point, the pressure exerted by its vapor becomes just equal to the external pressure at the liquid's surface. If a tube, closed at one end, is filled with a liquid and immersed open end down in a reservoir containing the same liquid, it will begin to fill with vapor as the liquid is heated to its boiling point. At the boiling point, the vapor pressure inside the tube will be balanced by the pressure exerted on the liquid surface by the surrounding atmosphere, so that the tube will be just filled with vapor. If the temperature is raised above the boiling point, vapor will begin to escape in the form of bubbles; if the tube is cooled below the boiling point, it will begin to again fill with liquid. This behavior is utilized in a micro method for determining boiling points.

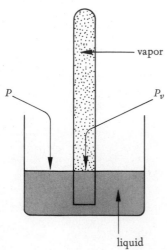

Figure 1.   Micro boiling-point principle

P_v = pressure exerted by vapor on liquid surface.

P = pressure exerted by atmosphere on liquid surface.

At boiling point, $P = P_v$

Some experimenters prefer to cut the capillary tube to a length of about 3 cm, or to use a capillary sealed near the open end.

Equipment and Supplies: Thiele tube, mineral oil, thermometer, capillary tube, boiling tube, rubber band, burner.

Operational Procedure

Add 2–4 drops of the liquid to a boiling tube constructed of a 10-cm length of 4–5 mm O.D. glass tubing sealed at one end ⟨OP–4⟩. Insert a capillary melting-point tube (sealed at one end) into the boiling tube with its open end down, and secure the assembly to a thermometer by means of a rubber band cut from thin-walled rubber tubing, as illustrated in

Figure 2. Insert this assembly into a Thiele tube as for a melting-point determination (see Figure 5, ⟨OP–28⟩) with the rubber band 2 cm or more above the liquid level.

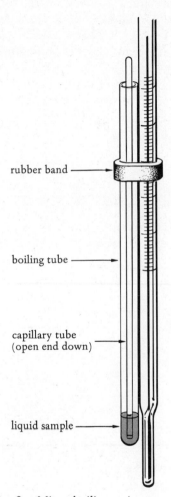

rubber band

boiling tube

capillary tube
(open end down)

liquid sample

Figure 2.   Micro boiling point assembly

Heat the Thiele tube until a *rapid,* continuous stream of bubbles (of the vaporized liquid) emerges from the capillary tube. Remove the heat, let the bath cool slowly until the bubbling stops, and record the temperature when liquid just begins to enter the capillary tube. Let the liquid partly fill the capillary, then heat very slowly until the first bubble of vapor emerges from the mouth of the capillary tube. Record the temperature at that point also. The two temperatures represent the boiling-point range; they should be within a

Avoid overheating or the liquid will boil away. It may be necessary to add more liquid if the sample is very volatile.

Sometimes the capillary tube will stick to the bottom of the boiling tube. This can be prevented by cutting a small nick in the open end of the capillary.

degree or two of each other. Cool the bath until liquid again enters the capillary tube, then repeat the determination. If repeated determinations on the same sample give appreciably different (usually higher) values for the boiling point, the sample is probably impure and should be distilled. Record the barometric pressure so that a pressure correction can be made. Clean ⟨OP–1⟩ and dry the boiling tube, saving it for future determinations.

Summary

1.   Assemble apparatus for micro boiling point, add liquid to boiling tube.

2.   Heat until continuous stream of bubbles emerges from capillary, remove heat.

3.   Record temperature when liquid enters capillary.

4.   Heat slowly until vapor bubble just emerges from capillary, record temperature.

5.   Let cool below boiling point.
IF another measurement is required,
GO TO 2
IF not,
GO TO 6

6.   Record barometric pressure, correct observed boiling point.

OPERATION 30

Refractive Index

$$n_\lambda^t = \frac{c_{vac}}{c_{liq}} = \frac{\sin \theta_{vac}}{\sin \theta_{liq}} \qquad (1)$$

n_λ^t = index of refraction at temperature t using light of wavelength λ.

c = velocity of light

Note that the index of refraction is a dimensionless ratio.

The refractive index of a substance is defined as the ratio of the velocity of light in a vacuum to its velocity in the substance in question. When a beam of light passes into a liquid at an angle, its velocity is reduced, causing it to bend downward. The index of refraction is related to the angles that the incident and refracted beams make with a line perpendicular to the liquid surface (see Figure 1) by Equation 1. The refractive index of a liquid is a unique physical property that can be measured with great accuracy (up to eight decimal places) and it is thus very useful in identifying pure organic compounds. Refractive index measurements are also used to assess the purity of known liquids and to determine the composition of solutions.

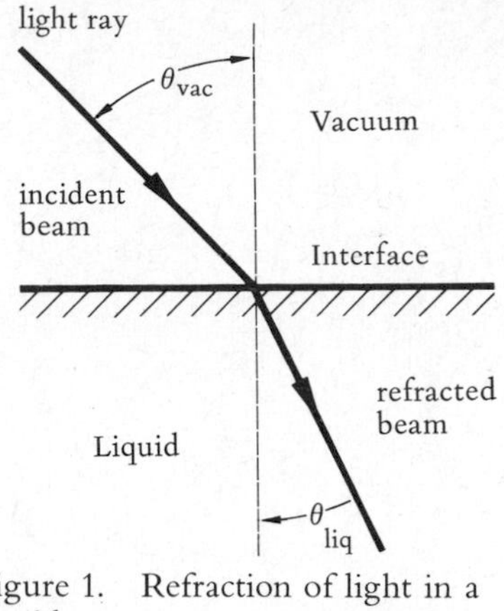

Experimental Considerations

Because of experimental difficulties associated with measurements in a vacuum, refractive index measurements are generally made in air (n_D = 1.0003) and the slight difference compensated for. Refractive index values are dependent both on the wavelength of the light used for their measurement and on the density (and therefore temperature) of the liquid. Most values are reported with reference to light from the yellow D line of the sodium emission spectrum, which has a wavelength of 589.3 nm. A refractive index in the literature may be reported as n_D^{20} = 1.3330, for example, where the superscript is the temperature in °C and D refers to the sodium D line. (Since most refractive index readings are made at this wavelength, the D is often omitted.)

The most commonly used instrument for measuring refractive indexes is probably the Abbe refractometer (Figure 3), which measures the critical angle of reflection at the boundary between the liquid and a glass prism and converts it to refractive index. The optical arrangement of the instrument is complex and will not be described here, but it utilizes a set of compensating prisms so that white light can be used to give refractive-index values corresponding to the sodium D line.

If a refractive index is measured at a temperature other than 20°C, the temperature should be read from the thermometer on the instrument and the refractive index corrected by one of the following methods:

1. A correction factor Δn is estimated from the equation $\Delta n \approx 0.00045 \times (t - 20.0)$, where t is the temperature of the measurment in °C. The correction factor (including its sign) is then added to the observed refractive index.

2. The refractive index of a *reference liquid* is measured and a correction factor is calculated as follows: $\Delta n = n^{20} - n^t$, where n^t is the measured value at temperature t for the pure reference liquid and n^{20} is its literature value at 20°C. The correction factor is then added to the observed refractive index of the sample. This method can correct for experimental errors (improper calibration of instrument, etc.) as well as temperature differences, but it is accurate only when the reference liquid is similar in structure and properties to the liquid being measured.

Figure 1.   Refraction of light in a liquid

If another light source is used, its wavelength is usually specified in nm.

This correction factor is only an approximation; the actual change in refractive index with temperature may be larger or smaller for a given liquid.

A properly calibrated B&L Abbe 3L refractometer should be reliable to 0.0002. The calibration of the instrument can be checked by measuring the refractive index of distilled water, which should be 1.3330 at 20°C and 1.3325 at 25°C. It decreases by about 1 × 10⁻⁴ for each degree increase in temperature between 20 and 30°C.

For example, the refractive index of an unknown aliphatic hydrocarbon is measured at 25.1°C and found to be 1.3874. By the first method, the corrected value would be $n^{20} = 1.3874 + 0.00045 \times (25.1 - 20.0) = 1.3897$. If the refractive index of hexane, a reference liquid, ($n^{20} = 1.3751$) is measured at 25°C and found to be 1.3726, a correction factor can be calculated as follows: $\Delta n = 1.3751 - 1.3726 = +0.0025$. Adding this to the observed value for the unknown gives a corrected refractive index of 1.3899 by the second method.

The effect of a contaminant on the refractive index of a liquid varies with the difference in the refractive indexes of the two substances.

Small amounts of impurities can cause substantial errors in the refractive index. For example, the presence of just 1% (by weight) of acetone in chloroform reduces the refractive index of chloroform by 0.0015. Such errors can be critical when refractive index values are being used for qualitative analysis, so it is essential that an unknown liquid be pure when its refractive index is measured.

Equipment and Supplies: Refractometer, dropper, washing liquid, soft tissues.

Operational Procedure (for B&L Abbe Refractometer)

Never allow the tip of a dropper or other hard object to touch the prisms—they are easily damaged.

Using an eyedropper, place about two drops of the sample between the prisms (Figure 3), either by raising the *hinged prism* and dropping the liquid on the middle of the *fixed prism* or (for a volatile and free flowing liquid) by introducing it into the channel alongside the closed prisms. With the prism assembly closed, switch on the *lamp* and move it toward the prisms to illuminate the visual field viewed through the eyepiece. Rotate the *handwheel* until two distinct fields (light and dark) are visible in the eyepiece, and reposition the lamp for the best contrast and definition at the borderline between them. Rotate the *compensating drum* on the front of the instrument until the borderline is sharp and achromatic (black and white) where it intersects the vertical reticle mark. Rotate the handwheel (some instruments have a fine-adjustment knob) to center the borderline exactly on the crosshairs (Figure 2), then depress and hold down the *display switch* (not shown, but usually on the left side of the instrument) to display an optical scale in the eyepiece. Read the refractive index from this scale (estimate the fourth decimal place) and record the temperature (if different from 20°C). Open the prism assembly and remove the sample by gently blotting it

If two fields are not observed at first, rotate the handwheel as far as it will go in either direction.

If the borderline cannot be made sharp and achromatic, the sample may have evaporated.

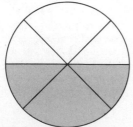

Figure 2.   Visual field for properly adjusted refractometer

with a soft tissue (do not rub!). Wash the prisms by moistening a tissue or cotton ball with a suitable solvent (acetone, ethanol, etc.) or with a non-ionic detergent solution and blotting them gently. Rinse with distilled water, if necessary, and blot dry. When any residual solvent has evaporated, close the prism assembly and turn off the instrument.

If an instrument other than the B&L Abbe Refractometer is to be used, your instructor will demonstrate its operation.

Summary

1. Start water circulating in refractometer jacket at 20°C (optional).

2. Insert sample between prisms.

3. Switch on and position lamp.

4. Rotate handwheel until two fields are visible, reposition lamp for best contrast.

5. Rotate compensating drum until borderline is sharp and achromatic.

6. Rotate handwheel until line is centered on crosshairs.

7. Depress display switch to display optical scale, estimate refractive index to 4 decimal places, record temperature.

8. Clean and dry prisms, close prism assembly, turn off refractometer.

9. Make temperature correction to observed refractive index if necessary.

A small piece of tissue paper is often placed between the prisms after use to prevent abrasion.

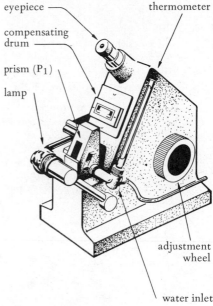

Figure 3.   Bausch and Lomb Abbe 3L Refractometer

Optical Rotation

Ordinary light can be considered a wave phenomenon with vibrations occurring in an infinite number of planes perpendicular to the direction of propagation. Plane-polarized light, which is generated when ordinary light passes through

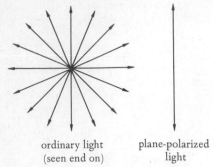

ordinary light   plane-polarized
(seen end on)   light

Figure 1.   Schematic representations of ordinary and plane-polarized light

The analyzer in precision instruments contains two or more prisms set at a small angle to each other. The reading is made when the prisms bracket the extinction point, transmitting equal light intensities. It is easier for the human eye to match intensities than to estimate the point of minimum intensity.

a polarizer (such as a Nicol prism), is light with vibrations restricted to a single plane (see Figure 1). When plane-polarized light passes through an optically active substance, molecules of the substance are capable of rotating its plane of polarization. The net rotation angle depends on the number of molecules in the light path, and therefore on the sample size and concentration. The angle of rotation is measured with an instrument called a *polarimeter* (Figure 2).

Experimental Considerations

A polarimeter consists of a polarizer, a cell to contain the sample, and an analyzer (a second polarizing prism) that can be rotated so as to compensate for the rotation induced by the sample. Usually the analyzer is rotated until the extinction point (the angle of minimum illumination) is reached. At this point the axis of the analyzer is perpendicular to the axis of the polarized light so that all the light is blocked out. By measuring the angle at which extinction occurs, the optical rotation of the sample is determined. Since the amount of

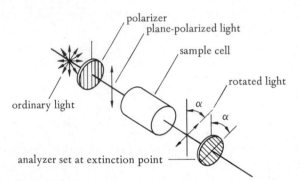

Figure 2.   Schematic diagram of a polarimeter

Specific rotation

$$[\alpha] = \frac{\alpha}{l \cdot c} \qquad (1)$$

$[\alpha]$ = specific rotation

α = observed rotation

l = length of polarimeter cell, in decimeters

c = concentration of solution, in grams of solute per milliliter solution. (For a neat liquid, substitute the density of the liquid in g/ml.)

rotation depends on factors that are not inherent properties of the sample itself (such as the length of the polarimeter tube and the concentration of the sample), this observed rotation is generally converted to specific rotation by means of Equation 1. The *specific rotation* of a pure substance under a given set of conditions is an invariant property of the substance and can be used to characterize it. Specific rotation can also be used to measure the optical purity of an enantiomer.

The optical purity of a sample can be defined as the ratio of its measured specific rotation to the specific rotation of the pure substance, multiplied by 100. Thus, an alcoholic solution of (+)-camphor ([α] = 44.3° in ethanol) having a specific rotation of 36.4° is (36.4/44.3) × 100 = 82.2% optically pure. This does *not* mean that it contains 82.2% (+)-camphor and 17.8% (−)-camphor; the "impurity" in the camphor is considered to be the racemate, (±)-camphor, of which half is (+) and half (−). Of course, optically active or inactive impurities other than the racemate can also alter the specific rotation of an optically active compound.

The optical rotation of a substance is usually measured in solution. Water and alcohol are common solvents for polar compounds; chloroform is used for less polar ones. Often a suitable solvent will be listed in the literature, along with the light source and temperature used for the measurement. The volume of solution required is dependent on the size of the polarimeter tube, but 25 ml is usually sufficient. Enough solute is weighed out (to the nearest milligram) to make up a solution having a concentration of between 4 g and 100 g per liter. (The actual concentration depends on the accuracy of the instrument and the amount of sample available.) The solution should be made up in a volumetric flask of appropriate size (a graduated cylinder is sometimes used for less precise work). If the solution contains particles of dust or other solid impurities, it should be filtered ⟨OP−11⟩.

Calculation of optical purity

$$\frac{\text{optical}}{\text{purity}} = \frac{[\alpha]\ (\text{observed}) \times 100}{[\alpha]\ (\text{pure substance})}\ \%$$

An 82.2% optically pure sample of (+)-camphor actually contains about 91.1% (82.2 + 8.9) of the (+)-camphor and 8.9% (−)-camphor, if the racemate is the only impurity present.

The CRC Handbook *entry for d-camphor reads: $[\alpha]_D^{20}$ +44.26(al), indicating that the measurement was made in an alcohol solution at 20°C using sodium D light.*

25 ml of solution thus requires between 0.1 g and 2.5 g of solute.

Operational Procedure

Remove the screw cap and glass endplate from one end of a 1- or 2-decimeter polarimeter cell (*Care:* do not get fingerprints on the glass) and rinse the cell with the solution to be analyzed. Fill the cell with the solution and slide on the glass endplate so that there is little, if any, air inside. Replace the cap and rubber washer, then shake and tilt the cell so that any air bubbles are trapped in a bulge at one end of the tube. Place the sample cell in the polarimeter, close the cover, and see that the illumination source (usually a sodium lamp, which should be allowed a half hour or so to warm up) is lined up to provide maximum illumination in the eyepiece. Rotate the analyzer scale (marked in degrees) to the left or right until two fields are clearly visible (often a vertical bar down the middle and a background field) and move the eyepiece in or out to sharpen the focus. Rotate the analyzer scale until both fields are of about equal intensity, then back off a little (toward zero) and use the fine adjustment wheel (if there is one) to rotate it, going away from zero, until the en-

This procedure applies to a Zeiss-type polarimeter with a split-field image. Experimental details may vary for different instruments.

Screw on the cap just tightly enough to make a leakproof seal. If it is too tight, the glass may be strained and cause erroneous readings.

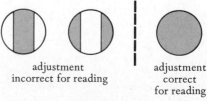

adjustment incorrect for reading | adjustment correct for reading

Figure 3. Split-field image of polarimeter

If the field is very bright and a slight turn of the micrometer knob has little effect on it, you are probably 90° off. Rotate the scale 90° back toward zero.

Because of mechanical play in the instrument, readings will differ slightly when approached from different directions. Another technique is to record readings from both directions and average them.

In reporting your specific rotation, specify the temperature, light source, solvent, and concentration.

tire visual field is as uniform as possible. Read the optical rotation from the analyzer scale and use the vernier scale (if there is one) to estimate the reading to a fraction of a degree. Note the direction of rotation (+ or −) also. Reset the analyzer scale to zero and repeat the determination several times, always approaching the final reading from the same direction (back off if you overshoot). Rinse the cell with the solvent used in preparing the solution, then fill it with that solvent and run a solvent blank by the same procedure. Rinse the cell with a volatile solvent (if necessary) and let it drain dry.

Subtract the optical rotation of the solvent blank (remember + and − signs) from that of the sample to obtain the optical rotation of the sample, and use the average value to calculate its specific rotation from Equation 1.

Summary

1. Prepare solution of compound to be analyzed (filter if necessary).

2. Rinse and fill sample cell with solution, place cell in polarimeter, adjust light source.

3. Focus polarimeter.

4. Rotate analyzer scale until field is uniform.

5. Read optical rotation, reset scale to zero.

6. Repeat steps 4 and 5 at least twice, average readings.
IF blank must be measured,
GO TO 7
IF not,
GO TO 8

7. Rinse and fill sample cell with solvent, place cell in polarimeter,
GO TO 3

8. Clean cell, drain dry.

9. Calculate optical rotation and specific rotation of sample.

Instrumental Analysis

Gas Chromatography

Gas chromatography, as generally practiced, differs from ordinary column chromatography in several respects. Instead of a liquid mobile phase and a solid stationary phase, *gas-liquid chromatography* (GLC) utilizes a gaseous mobile phase and a liquid stationary phase. The components of the sample are partitioned between the gas phase and the liquid phase; the fraction of any component in the gas phase (and therefore its rate of travel) depends on its vapor pressure and its solubility in the liquid. The separation of components can be controlled by varying the composition and temperature of the column and the flow rate of the mobile phase.

The gas chromatographic method utilizing a liquid stationary phase on a support is usually called gas-liquid chromatography (GLC). This and other terms such as vapor phase chromatography (VPC) and gas-liquid partition chromatography (GLPC) are often used interchangeably.

Experimental Considerations

Instrumentation. Although gas chromatography can be used, like column chromatography, to separate components on a preparative scale (preparative GLC), it is most often used as an analytical technique to determine the identities and quantities of components in a mixture. A typical analytical gas chromatograph is diagrammed in Figure 1 and utilizes the following components:

Injector: A port where the sample is introduced and vaporized.

Column: A long metal tube (generally coiled) packed with an inert solid *support* (such as crushed firebrick) which is coated with a high-boiling liquid used as the stationary phase.

Carrier Gas: An inert gas (the mobile phase) used to sweep the sample through the column.

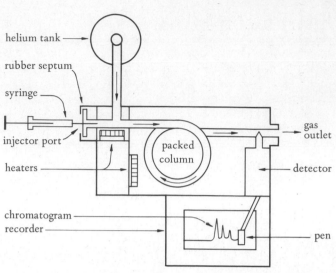

Figure 1.   Schematic diagram of gas chromatograph

Table 1.   Stationary phases for gas chromatography

Stationary Phase (Type)	Max T, °C
Apiezon-L (hydrocarbon grease)	300°
DC-550 (silicone oil)	275°
DEGS (ester)	200°
Carbowax 20M (polyglycol)	250°

The most commonly used detectors are thermal conductivity (TC) and flame ionization (FI) detectors. The latter is more sensitive, but requires hydrogen gas for the flame.

Detector: A device used to detect the presence of a component as it leaves the column.

Recorder: An instrument that records component peaks on a moving chart paper as the components pass the detector.

When a liquid mixture is vaporized and passes through the column, its components are separated between the stationary (liquid) and mobile (gas) phases according to their different partition coefficients. In general, polar compounds are best separated on polar stationary phases and nonpolar compounds on nonpolar stationary phases. Some commonly used general-purpose stationary phases, listed in approximate order of increasing polarity, are given in Table 1, along with their maximum operating temperatures. After a component passes out of the column, it encounters the detector, which sends an electrical signal to the recorder, which traces a peak on the chart paper. The resulting *chromatogram* thus consists of a series of peaks of different sizes, each corresponding to a component of the mixture.

Qualitative Analysis.   The interval between the injection of a sample and the time a component peak appears on the chart paper is called the *retention time* of the component. The retention time depends on such factors as the length of the column, the type of column packing, the temperature of the column, and the flow rate of the carrier gas. If these are kept constant, the observed retention time will be characteristic of a given component. The retention time of a component is found by measuring the horizontal distance from the starting line to the top of the component peak. (This distance can be converted to units of time if the chart speed is known.) To identify the components of a sample, their retention times can be measured and compared with those found for known compounds under identical conditions of column temperature, gas-flow rate, etc. A more reliable method is to add an authentic sample of a pure compound to the sample being analyzed and obtain a gas chromatogram of the new mixture. The peak corresponding to the added compound will grow in height and area as a result of the addition, and thus can easily be picked out from the other component peaks, whose relative areas remain the same. For example, if an esterification mixture is known to contain ethyl acetate, ethyl alcohol, and acetic acid, its gas chromatogram should show three peaks corresponding to these components (Figure 3). To find out which of the three is the ester peak, the

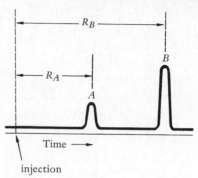

Figure 2. Retention times of two components

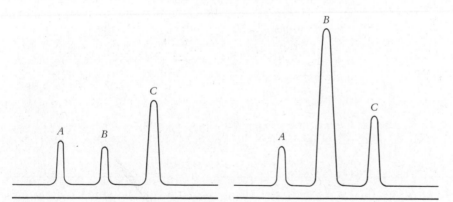

Figure 3. Use of gas chromatography for qualitative analysis. *1.* Chromatogram before adding ethyl acetate. *2.* Chromatogram after adding ethyl acetate.

mixture is "spiked" with pure ethyl acetate and another gas chromatogram is recorded. One of the three peaks will be much larger on the second chromatogram than on the first, and therefore must belong to the ethyl acetate. The process can be repeated with samples of acetic acid or ethyl alcohol to identify the other components.

Quantitative Analysis.   The area under a peak corresponding to a given component is proportional to the quantity of the

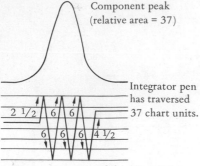

Component peak
(relative area = 37)

Integrator pen
has traversed
37 chart units.

2 1/2 | 6 | 6

6 | 6 | 6 | 4 1/2

Figure 4.   Use of the integrator

This assumes that the area ratio between peaks is equal to their weight ratio, which is not always the case. Compounds with similar molecular structures usually give reasonably good area-weight correlations.

component present. Therefore, by measuring peak areas, one can determine the percent composition of the mixture. A rough estimate of peak area can be made by penciling over the peak a triangle having an area approximately equal to that of the peak, and calculating its area from the formula $A = \frac{1}{2}bh$. A more accurate method is to carefully cut out the component peak using a sharp knife or razor blade and weigh it on an analytical balance; the weight will be proportional to its area. If the gas chromatograph is equipped with an *integrator,* the total distance (along the ordinate of the graph) travelled by the integrator pen while a component is passing the detector will be proportional to its peak area. The integrator pen will usually draw a zigzag line while a peak is being recorded. By adding up the number of chart units the pen has crossed in all of its up and down traverses, one can compute the relative area of the peak (see Figure 4). The percent composition (by weight) of a sample can be estimated by adding up all of the relative areas, dividing this value into the relative peak area for each component, and multiplying by 100. This method of determining percentage composition is suitable for semiquantitative work, but it is limited by the fact that the same mass of two different compounds will not always result in exactly the same detector response. If more precise determinations are required, it may be necessary to measure the detector response for samples of known composition and then calculate response factors for each component.

Equipment and Supplies: Gas chromatograph with recorder, microliter syringe, sample vial, syringe-drying apparatus.

WARNING   *Do not make any adjustments to the instrument. Consult the instructor if it does not seem to be working properly or if you have questions about its operation.*

Operational Procedure

It will be assumed that the instrumental parameters (column temperature, gas flow rate, etc.) have been set by the instructor and that the instrument has had time to warm up properly.

   Rinse a microliter syringe (see ⟨OP–6⟩) with the liquid sample (in a sample vial) a few times, then partially fill it as described in ⟨OP–6⟩. If there are air bubbles in the syringe, eject the sample and refill it carefully to remove the air. Expel the excess liquid until it contains the desired volume (usually 1–5 microliters). Wipe the needle dry with a tissue and pull the plunger out a few millimeters to help prevent premature vaporization of the sample. Switch on the recorder *chart drive* (and the integrator, if available), then carefully insert the needle into the *injector port* (*Warning: Hot!*) by placing its tip near the center of the rubber septum and pushing the barrel slowly but firmly until the needle is all the way in. (Be very careful not to bend the needle.) Inject the sample by carefully pushing in the plunger just as the recorder pen crosses a chart line (the starting line). Withdraw the needle without delay and mark the starting line, from which retention times will be measured. Let the recorder run until all of the anticipated component peaks have been recorded, then turn off the chart drive and tear off the chart paper. Rinse the syringe with acetone, methanol, or another volatile solvent, remove the plunger, and dry the syringe using the apparatus illustrated in Figure 5, ⟨OP–6⟩.

Sometimes air bubbles can be removed by pumping the plunger while the needle is in the sample or by holding the syringe with the needle up and tapping the barrel.

The integrator may require adjustment; the integrator pen should trace a straight line when no sample component is being recorded.

The sample should be injected soon after the needle is inserted; othewise premature vaporization may cause a "doubling" of the component peaks.

Use a straightedge or ruler for tearing the chart paper.

Summary

1. Rinse syringe with sample.

2. Fill syringe with desired volume of sample.

3. Start chart drive and integrator (if applicable).

4. Insert needle into injector port, inject sample when pen crosses a chart line.

5. Withdraw needle, mark starting line.

6. Stop chart drive when chromatogram is complete.

7. Remove chromatogram, clean and dry syringe.

8. Measure peak areas and retention times if required.

Infrared Spectrometry

Principles and Applications

The atoms of a molecule behave as if they were connected by flexible springs, rather than by rigid bonds resembling the

Figure 1.   "Ball-and-spring" model
of a chemical bond

Vibrational frequencies are usually reported in "wave numbers" ($\bar{\nu}$), the number of peak-to-peak waves per centimeter. The traditional unit of measurement is reciprocal centimeters, cm^{-1}. The relationship between wavenumber and frequency in Hertz (ν) is given by:

$$\nu = c \cdot \bar{\nu}$$

where c is the speed of light (about 3.0×10^{10} cm/sec).

The correlation chart on the back end-paper gives the approximate positions of commonly encountered infrared absorption bands. See Part III for additional information on the use of IR for qualitative analysis.

Infrared spectra cannot distinguish between enantiomers, since vibrational frequency is not a function of left- or right-handedness.

The construction and operation of such instruments is described in Bib–F1, F2, F8 and other sources.

connectors of a ball-and-stick model. Their component parts can oscillate in different *vibrational modes,* designated by such terms as rocking, scissoring, twisting, wagging, and symmetrical and asymmetrical stretching. When infrared radiation is passed through a sample of a given compound, its molecules can absorb radiation of the energy (and frequency) needed to bring about transitions between vibrational ground states and vibrational excited states. For example, a C—H bond that vibrates 90 trillion times a second must absorb infrared radiation of *just* that frequency (9.0×10^{13} Hz, 3000 cm^{-1}) to jump to its first vibrational excited state. This absorption of energy at various frequencies can be detected by an *infrared spectrophotometer,* which plots the amount of infrared radiation transmitted through the sample as a function of the frequency (or wavelength) of the radiation. Because vibrational transitions are usually accompanied by rotational transitions in the region scanned by a typical infrared spectrophotometer (about 4000–666 cm^{-1} or 2.5–15 μm), an infrared spectrum consists of comparatively broad *absorption bands* rather than sharp peaks such as those seen in NMR spectra. The bands are also usually "inverted"—strong absorption is represented by a deep valley rather than a peak.

Infrared spectrometry is extremely useful in qualitative analysis. It can be used both to detect the presence of specific functional groups and other structural features from band positions and intensities, and to establish the identity of an unknown compound with a known standard. The *fingerprint region* of the infrared spectrum (1250–670 cm^{-1}, 8–15 μm) is best for showing that two substances are identical, since the distinctive patterns found in this region are usually characteristic of the whole molecule and not of isolated groups. Infrared spectra can also be used in establishing the purity of compounds, monitoring reaction rates, measuring the concentrations of solutions, determining the structures of complex molecules, and carrying out theoretical studies of hydrogen bonding and other phenomena.

Experimental Considerations

Most of the recording infrared spectrophotometers used for routine applications are of the optical-null—double-beam type, and utilize a hot filament or rod as the source of infrared radiation. As radiation of a given frequency passes through the sample cell, some is absorbed by the sample. The remaining *transmitted* radiation activates a thermocouple, which sends an electrical signal to an attenuator placed in the *sample beam*. While attempting to balance the intensity of the sample beam with that of a *reference beam,* the attenuator moves a pen along the chart—the distance it moves being proportional to the amount of light that made its way through the sample. The frequency of the radiation is constantly varied by a rotating grating or prism, which is coupled to the chart drive. The resulting spectrum, then, is a plot of the intensity of transmitted radiation (as percent transmittance) versus the frequency (or wavelength) of the radiation.

The reference beam bypasses the sample.

Frequency Alignment. A typical infrared spectrophotometer may have the chart paper wrapped around a drum or laid flat on a recorder bed (Figure 2). In either case, the paper must be carefully aligned so that the absorption bands appear at the right place—otherwise the spectrum may be interpreted incorrectly. The chart paper is usually aligned by matching a given chart frequency (for instance, 4000 cm^{-1}) with an alignment mark on the recorder bed or drum. If precise values of the band frequencies are required (or if the instrument requires wavelength calibration) a calibration spectrum can be run using a standard thin film of polystyrene. The tips of a few polystyrene peaks (usually including those at 2850, 1603, and 906 cm^{-1}) are recorded directly over the sample spectrum and used to calculate correction factors for the absorption bands of interest.

An infrared spectrum can be linear in either wavenumber (cm^{-1}) or wavelength (μm), depending on the instrument used. These units can be interconverted by the equation:

$$\lambda \text{ (in } \mu\text{m)} = \frac{10,000}{\bar{\nu}}$$

Setting the Baseline. For maximum resolution, the "baseline" of an infrared spectrum should stay close to the top of the chart paper, but should not go off-scale (above 100% T). This can usually be accomplished by setting the attenuator control so that the pen is at 85-90% T with the sample cell in place and the recorder at its starting position (about 4000 cm^{-1}). It may occasionally be necessary to run a preliminary scan over the spectrum (manually, if the instrument allows this) and set the attenuator control so that the highest part of the baseline is just below 100%.

$\%T$ = *percent transmittance* = $100 \times T$
(transmittance)
A = *absorbance* = $\log 1/T$

The attenuator control is sometimes called the "100 percent control."

It is important not to run over 100% — this can damage the instrument.

Spectrum Intensity. An infrared spectrum should not, obviously, have absorption bands so weak that they are hard to identify or so strong that they crowd the 100% T line and cannot be resolved. Ideally, the spectrum should be recorded so that the strongest absorption band has a maximum transmittance of 5-10%. This may be difficult to accomplish with

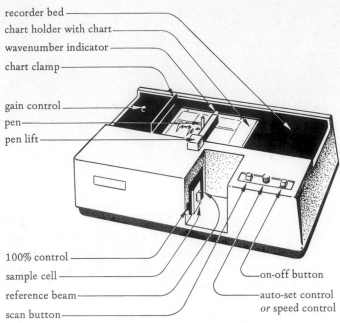

recorder bed
chart holder with chart
wavenumber indicator
chart clamp

gain control
pen
pen lift

100% control
sample cell
reference beam
scan button

on-off button
auto-set control
or speed control

Figure 2. Perkin–Elmer Model 710b
Infrared Spectrophotometer

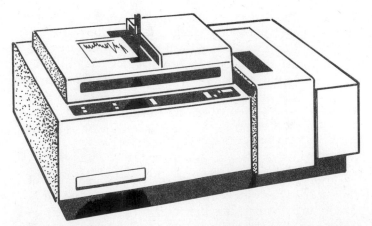

Figure 3. Perkin–Elmer Model 680 Infrared Spectrophotometer.
Note that this can be coupled with the Perkin–Elmer Data
Station to extend its problem solving capabilities.

precision, but with the right kind of cell and careful sample preparation, you should be able to obtain a good spectrum. If your first spectrum is not acceptable, try varying the following parameters, depending on the sample preparation method.

1. *Neat liquid:* Cell pathlength or film thickness.

2. *Solution:* Concentration or cell pathlength.

3. *KBr disc:* Amount of sample or thickness of disc.

4. *Mull:* Amount of sample or film thickness.

Precautions. Since most infrared spectrophotometers are comparatively sensitive instruments that do not respond well to rough handling, certain precautions should be observed in their use. *Never* move the chart holder, drum, or pen carriage while the instrument is in operation or before it has been properly reset at the end of a run—this can damage mechanical components. *Never* try to move the pen manually. Do not leave objects lying on the bed of a recorder— they might jam the chart drive. Do not unplug the instrument—many infrared spectrophotometers contain heaters to keep certain components at thermal equilibrium even when the instrument is not in operation. Do not use water or aqueous solutions near a spectrophotometer, since some of its components may be constructed of water-soluble alkali metal salts—a spill could be disastrous! Do not insert objects into the sample or reference beam openings where they can damage the attenuators and other fragile components. An infrared spectrophotometer is a precision instrument that can easily cost as much as a Lincoln Continental or a Mercedes-Benz; you must treat it with respect and follow instructions carefully to prevent damage and avoid the need for costly repairs.

Sample Preparation

Infrared spectra of liquids can be run with the liquid either neat or in solution; spectra of solids can be run using a solution of the solid, a mull, or a pressed disc. The following outline should serve as a general guide to sample preparation.

Care of Infrared Cells. Most sample cells contain metal halide windows (sodium chloride, silver chloride, etc.), which are either very fragile, water soluble, or both. All windows

Sometimes a satisfactory infrared spectrum is obtained only by trial and error.

Since sample cells and other sampling accessories vary widely in construction and application, most sampling techniques should be demonstrated by your instructor.

Even breathing on a NaCl window can cause some etching because of moisture in your breath.

should be touched only on the edges with clean, dry hands or gloves. They must be handled with great care to avoid damage; sodium chloride windows must not be exposed to moisture. Therefore, samples and solvents run in sodium chloride cells must be dry, and the cells should be kept in an oven or dessicator when not in use. When assembling demountable cells, the cinch nuts must not be over-tightened, as this can fracture the windows.

Thin films should not be used for liquids boiling below 80°—they may evaporate before the spectrum is complete.

Thin Films.   To prepare a thin film of a pure ("neat") liquid using a *demountable cell,* disassemble the cell carefully according to your instructor's directions and place 1–2 drops of the liquid on the lower window. Position the upper window by touching an edge to the corresponding edge of the lower window and carefully lowering it into place. Press the plates together so that the liquid fills the space between them, taking care to exclude air bubbles. Assemble the cell as directed and run the spectrum. When you have finished, disassemble the cell and rinse the windows with a *dry* volatile solvent (chloroform, methylene chloride, and NaCl-saturated 100% ethanol are often used). Wipe the windows dry with a soft tissue.

See J. Chem. Educ. 50, *517 (1973) for directions for preparing thin films (melts) of low-melting solids.*

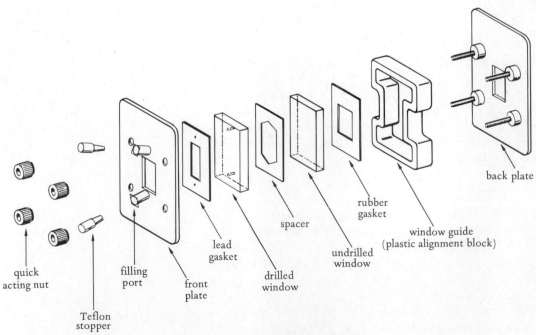

quick acting nut   filling port   front plate   lead gasket   drilled window   spacer   undrilled window   rubber gasket   window guide (plastic alignment block)   back plate

Teflon stopper

Figure 4.   Sealed-demountable cell

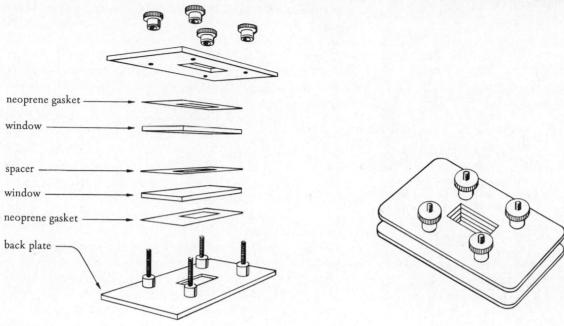

neoprene gasket

window

spacer

window

neoprene gasket

back plate

Figure 5.   Demountable cell

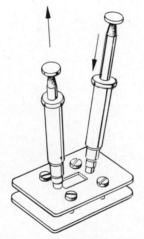

Figure 6.   Push-pull technique for flushing a sealed or sealed-demountable cell

Volatile Neat Liquids.   If a liquid is comparatively volatile, its spectrum can be run in a cell with a spacer approximately 0.015–0.030 mm thick. A demountable cell (Figure 5) is filled by disassembling the cell, placing the spacer on the lower window, adding sufficient liquid to fill the cavity in

A sealed-demountable cell with a thin spacer works well for most neat liquids, whether volatile or not.

Care: The syringe tips are fragile and break easily.

If there is much resistance to filling, use the push-pull technique described below. If there are air bubbles between the windows, they can sometimes be removed by tapping gently on the metal frame of the cell.

Twist the syringes slightly as you insert them so they will not pull out too easily.

Compressed air should not ordinarily be used to dry cells unless it is cleaned, dried ⟨OP–22⟩, and passed through the cell at low pressure.

The solvent should be relatively nonpolar and must not react with the solute. It should ordinarily dissolve enough of the solute to yield a 5–10% solution; however, more dilute solutions can be used with cells having long path lengths.

Since commercial chloroform usually contains ethanol as a stabilizer, it should be purified before use by passing it through an alumina column.

the spacer, positioning the upper window, and reassembling the cell. After the spectrum has been run, it should be disassembled and cleaned as described above.

A *sealed cell* or a *sealed-demountable cell* (Figure 4) is filled by injecting the sample into an injection port with a Luer-lock syringe body. Draw about ½ ml of the liquid into the syringe and carefully insert the tip into one of the ports with a half twist (remove the plugs first!). Holding the cell upright with the syringe port at the bottom, depress the plunger until the space between the windows is filled and a little liquid appears at the upper port. Put the cell on a flat surface, remove the syringe, insert a plug in the upper port with a slight twist, remove any excess solvent in the lower port with a piece of tissue paper or cotton, and close that port with another plug.

After the spectrum has been run, clean the cell by removing most of the liquid with a syringe and flushing it several times with a volatile solvent. A convenient way of flushing a cell is to lay it on a flat surface, insert a syringe filled with solvent in one port and an empty syringe in the other port, and slowly push on the one plunger while pulling on the other (see Figure 6). This will draw washing liquid through the cell from one syringe to the other. Do this several times, then remove the excess liquid with an empty syringe and dry the cell by passing dry, clean air or nitrogen between the windows. An ear syringe or a special cell-drying syringe can be used to force air (gently!) through the cell, or it can be dried by attaching a trap and aspirator to one port and a drying tube filled with dessicant ⟨OP–22a⟩ to the other.

Solutions.   Solutions of liquids or solids in a suitable solvent can be analyzed in a sealed or sealed-demountable cell having a spacer 0.1 mm or more in thickness. Solution spectra must always be run with a reference cell in the reference beam of the spectrophotometer, to subtract out absorption due to the solvent. Even then, strong solvent peaks will obscure certain portions of the spectrum. For example, the spectrum of a solute run in chloroform will yield no useful information in the 1250–1200 cm^{-1} and 800–650 cm^{-1} regions because the chloroform absorbs nearly all of the infrared radiation in these regions. Sometimes a sample can be run separately in two solvents to yield a complete spectrum. Carbon tetrachloride and carbon disulfide constitute such a pair; the first yields a good spectrum above 1330 cm^{-1} and the second is virtually transparent at frequencies below

1330 cm^{-1}. Prepare the sample by making up a 5–10% solution of the substance to be analyzed and filling the sample cell using a Luer-lock syringe body, as described above for neat liquids. Fill an identical cell (having a spacer of the same thickness) with pure solvent and place it in the reference beam when the spectrum is run. Clean the sample cell by removing the excess solution and flushing it (as described above) with the pure solvent used in preparing the solution. If necessary, rinse it with a less volatile solvent before drying.

Usually 0.1–1.0 ml of solution will be required. A 0.1-mm spacer can be used unless the solution is quite dilute.

It is a good idea to run a preliminary spectrum using pure solvent in both cells, to see that the cells are clean and matched.

Mulls. A solid sample can be prepared as a *mull* in a suitable oil for infrared analysis. Grind about 5–10 mg of the solid very finely in an agate or mullite mortar, add a drop or two of mulling oil (usually a mineral-oil preparation called Nujol), then grind the mixture until it has about the consistency of vaseline. Transfer most of the mixture to the lower window of a demountable cell using a "rubber policeman" or a plastic-coated spatula. Spread the mull evenly with the top window, taking care to exclude air bubbles. Assemble the cell and run the spectrum as for a neat liquid.

The particles must be ground to an average diameter of about one μm or less to avoid excessive radiation loss by scattering. The solid should cake the interior of the mortar and have a glossy appearance at this point.

Some unavoidable *scattering* of the infrared radiation may reduce the transmittance at the high-frequency end of the spectrum. For this reason, you should rapidly scan the spectrum with the pen raised to locate the frequency at which the transmittance is highest, and set the baseline just below 100%T at that point. If after this adjustment the transmittance around 4000 cm^{-1} is 50% or lower, the sample is too coarsely ground and a new mull should be prepared. The sample spectrum should be compared with a spectrum of the mulling agent so that peaks due to the oil can be identified and those regions ignored during interpretation. When Nujol is used, the aliphatic C—H stretching and bending regions (3000–2850, 1470, 1380 cm^{-1}) cannot be interpreted, but most functional groups and other structural features can be identified. If it is necessary to examine the entire spectrum, another sample can be prepared using a complementary mulling agent such as Fluorlube.

The effect of scattering can be partly compensated for by using an attenuator accessory in the reference beam.

Clean the cell windows with petroleum ether or another suitable solvent after use.

Nujol is essentially transparent at frequencies lower than 1300 cm^{-1}; Fluorlube is transparent above 1300 cm^{-1}.

KBr Discs. A potassium bromide *disc* is prepared by mixing a solid with *dry* spectral-grade KBr and using a die to press the mixture into a more or less transparent wafer. If a KBr Mini-Press is used, grind $\frac{1}{2}$–2 mg of the solid very finely in a dry agate or mullite mortar as if you were preparing a mull, then add about 100 mg of the special potassium

KBr adsorbs water readily and should be dried before use by heating it at 105–110°C for several hours. It should be

tightly capped and stored in an oven or dessicator. Even then, water may be present due to condensation of atmospheric moisture, so the O—H absorption region must be interpreted with caution.

bromide and grind it thoroughly with the sample. Screw in the bottom bolt of the Mini-Press five full turns (see Figure 7) and introduce about half of the sample mixture into the barrel. Keeping the open end of the barrel pointed up, tap it

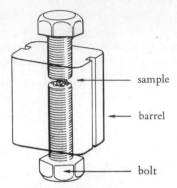

Figure 7.   KBr Mini-Press

gently against the desk top to level the mixture, brush down any material on the threads with a camel's hair brush, and screw in the top bolt. Alternately tap the bottom bolt on the bench top and screw in the top bolt with your fingers to level the sample further. When the bolt is "finger tight," clamp the bottom bolt in a vise or holder and gradually tighten the top bolt as far as you can with a heavy wrench (or to 20 ft-lbs with a torque wrench). Leave the die under pressure for a minute or two. Remove both bolts, leaving the KBr disc in the center of the barrel, and check to see that it is reasonably transparent; if not, make up a new disc. Place the block containing the disc on a special holder in the spectrophotometer sample beam, orient it for maximum transmittance, set the baseline by the method described above for a mull, and run the spectrum. Punch out the pellet with a pencil or other suitable instrument and wash the barrel and bolts with water, then rinse them with acetone or methanol and store the clean, dry press in a dessicator.

The disc may be cloudy or inhomogeneous for the following reasons: (a) the KBr or sample may be wet; (b) the two components may not be completely mixed; (c) the pellet may be too thick; (d) the sample may be low melting; (e) the sample size may be too large, or (f) the die may not have been tightened enough.

    If another kind of press is to be used, follow the manufacturer's or your instructor's directions for preparing the disc. KBr discs can also be prepared by a "wet" method in which the sample is dissolved in a volatile solvent and ground as the solvent evaporates. Since evaporative cooling causes condensation of atmospheric moisture, the sample should then be oven-dried before mixing with KBr. Consult some of the references listed in part F of the Bibliography for additional information on this and other sample preparation techniques.

See *J. Chem. Educ. 54,* 287 (1977) for additional suggestions regarding the preparation of KBr discs.

Equipment and Supplies: General: Infrared spectrophotometer, dessicator or oven, sample, cells and other accessories, chart paper.

Neat liquids: Demountable or sealed-demountable cell (with or without thin spacer), washing solvent, tissue papers.

Solutions: Two sealed or sealed-demountable cells with spacers, two Luer-lock syringes, washing solvent(s), car syringe (or other cell-drying apparatus).

Mulls: Agate or mullite mortar and pestle, mulling oil, demountable or sealed-demountable cell with no spacer, washing solvent, tissues.

KBr discs: Dried spectral-grade potassium bromide, agate or mullite mortar and pestle, KBr press assembly, washing and drying solvents.

Operational Procedure

Since the construction and operation of commercial infrared spectrophotometers vary widely, the following is meant only as a general guide to assist you in recording an infrared spectrum. Specific operating techniques must be learned from your instructor, the operating manual, or both.

Prepare a sample of the substance to be analyzed by one of the methods described above and place the cell or KBr pellet assembly in the sample beam of the instrument. If the sample is being run in a solvent, place a reference cell containing the solvent in the reference beam. An attenuator accessory can be placed in the reference beam if a mull or KBr disc is being run; otherwise that compartment remains empty. If necessary, place chart paper on the recorder bed (or wrap it around a drum), then align the paper properly and move the drum or carriage to the starting position. Set the attenuator control and any other necessary controls, lower the pen onto the chart paper, and scan the spectrum at "normal" or "fast" speed. Examine the spectrum to see that the absorption bands show satisfactory intensity and resolution. If the spectrum is not satisfactory, prepare another sample (varying concentration, path length, solvent, etc.) or consult the instructor to see if the instrument requires adjustment or repair. Then, if desired, run another spectrum at a slower speed. Remove the sample, reset the instrument (if necessary), and tear off the chart paper. Write all appropriate data on the chart paper, including the following (as applicable): identity and source of the sample, sampling method used, spacer thickness, concentration and solvent (for a solu-

Important: Make sure the instrument has been reset correctly before attempting to move the drum or carriage.

If the spectrum is to be calibrated, replace the sample by a polystyrene film, reset the instrument, move the bed or drum to the appropriate frequencies, and superimpose the polystyrene peaks over the spectrum.

tion spectrum), mulling agent (if used), scanning speed, the current date, and the name of the operator.

Nuclear Magnetic Resonance Spectrometry

Principles and Applications

The theoretical concepts underlying nuclear magnetic resonance and its application to structure determination are treated by most organic chemistry lecture texts; these concepts should be understood before any attempt is made to record or interpret an actual NMR spectrum. The following discussion stresses the practical aspects of recording an NMR spectrum and obtaining useful data from it.

In an NMR experiment, the substance being analyzed is placed between the poles of a powerful magnet generating a magnetic field H_0 and irradiated with an oscillating electromagnetic field (H_1) of frequency ν. At certain values of ν, some nuclei in the sample absorb energy from the oscillating field; this energy absorption is detected and recorded in the form of an NMR spectrum. For example, hydrogen nuclei in a magnetic field of 14,092 gauss will interact with the RF radiation at a frequency of about 60 million cycles/sec (60 MHz). Because of the shielding effect of nearby electron clouds, however, not all hydrogen nuclei will absorb energy at exactly the same value of the applied field. A proton NMR spectrum will therefore consist of a series of signals, each generated by a set of magnetically equivalent hydrogen nuclei (protons).

An NMR spectrum is recorded either by holding the magnetic field strength constant and varying the frequency of the RF field, or by holding the frequency constant and varying the field strength (*sweeping* the field). Both methods are used in practice, but the field-sweep method is more commonly encountered in the instruments accessible to undergraduate students. In a typical 60-MHz instrument used for proton NMR spectra, the field is swept over a very narrow range of about 0.25 gauss. This will encompass all of the proton signals likely to be encountered in routine anal-

NMR spectrometers operate at frequencies in the radiofrequency (RF) region of the electromagnetic spectrum, ranging from about 30 to 300 MHz.

The RF frequency needed to bring about resonance varies in proportion to H_0. For a field of 23.5 kilogauss, it is about 100 MHz.

For protons, magnetic field strength in gauss can be converted to frequency in Hz by multiplying by 4257.9

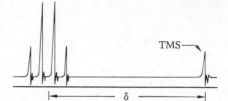

Figure 1.   Chemical shift of an NMR signal

ysis. For convenience, this sweep range is usually calibrated in frequency units and covers approximately 1000 Hz.

The parameters that can be obtained from an NMR spectrum are the chemical shift, the integrated signal area, the signal multiplicity, and the coupling constant. The *chemical shift* is the distance, measured in hertz (Hz) or parts per million (ppm), from a proton signal to some reference signal, generally that of tetramethylsilane (TMS). Since TMS protons absorb at a higher value of the applied magnetic field than nearly all other protons, the chemical shift is usually measured from right to left starting at the TMS signal. In other words, a proton that absorbs at a *lower* field strength than TMS is, by the usual convention, assigned a *positive* chemical-shift value. The chemical shift measured in parts per million is obtained by dividing the shift in hertz by the RF frequency of the instrument in megahertz. Thus, a signal that is 120 Hz downfield (to the left) of TMS on a 60-MHz instrument has a chemical shift (in ppm) of 2.00δ. On the τ scale, chemical shifts increase in the more logical left-to-right direction, with increasing field—zero on the τ scale starts 10 ppm to the left of the TMS signal, and TMS itself is assigned a τ value of 10 ppm. δ and τ values can be interconverted by the relationship $\tau = 10 - \delta$.

Since the energy absorbed by a given kind of proton depends only on the number of such protons in the sample, the *integrated signal area* is proportional to the number of protons responsible for a signal. This area is usually determined by electronic integration; it can be converted to relative proton numbers by reducing all signal areas to the lowest set of integers. If the molecular formula of the compound is known, the relative numbers can be converted to absolute values. Since electronic integrators may be in error by 5–10%, it is often necessary to round off the calculated values to whole numbers.

The *multiplicity* of a signal is simply the number of separate peaks it contains; the *coupling constant* (J) is the distance between two adjacent peaks in a signal, measured in Hz. J values of comparatively simple systems can usually be measured directly from an NMR spectrum. By comparing J values and observing the shapes of signals in an NMR spectrum, one can often identify neighboring sets of protons, since the signals from interacting protons may have the same spacing between peaks and tend to "lean" toward each other, as illustrated in Figure 2. If the interaction between two sets (*a* and *b*) of protons is first order (or nearly so), the

If an NMR signal is symmetrical, its chemical shift is measured from a point midway between two peaks of equal intensity; if not, the "center of gravity" of the signal is estimated or calculated.

$(CH_3)_4Si$

TMS

The chemical shift in Hz varies with the frequency of the instrument used, whereas the shift in ppm does not. A signal having a shift of 120 Hz on a 60-MHz instrument will absorb at 200 Hz on a 100 MHz instrument; the shift in ppm is 2.00 in both cases.

Example: For three signals of relative area 54:36:24, the lowest integral ratio is 9:6:4. This can be determined quickly by dividing through by the lowest number (24) to give the ratio $2\frac{1}{4}:1\frac{1}{2}:1$, then multiplying by the lowest integer (4) that will convert all numbers of the set to whole numbers.

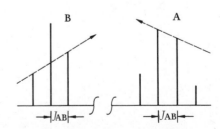

(*J* values equal, high sides facing)

Figure 2.   Signals of nearest neighbor protons

In a simple or "first order" spectrum, the chemical-shift difference for two sets of interacting protons is at least 6–7 times their coupling constant.

The NMR correlation chart on page 659 gives the chemical shift ranges for commonly encountered types of protons. For additional NMR parameters and help in interpreting spectra, consult references listed in section F of the bibliography.

CAT = computer of average transits

In a single-coil instrument, the transmitter coil is built into a bridge circuit, allowing it to detect energy absorption by the sample and thus serve as the receiver coil as well.

number of **b** protons adjacent to set **a** can be determined by subtracting 1 from the number of peaks in the signal of set **a**, and vice versa. In the Figure 2 example set **a** has three nearest neighbors (the **b** protons); set **b** has two (the **a** protons).

To determine the structure of a compound from its NMR spectrum, one must measure integrated signal areas, chemical shifts, and coupling constants for as many signals as possible, then attempt to derive the best structure that fits the data. Usually this is easier to do if one knows the origin of the substance and has additional data such as physical constants, elemental composition, and infrared or mass spectral results. Numerous auxiliary techniques have been developed to help "clean up" or simplify an NMR spectrum and aid in its interpretation. Among these are the use of chemical-shift reagents, deuterium exchange, variable temperature probes, CAT averaging of successive scans, Fourier transform NMR, and spin-spin decoupling. For information on these and other techniques of NMR analysis, refer to one or more of the references in the Bibliography.

Experimental Considerations

Instrumentation.   The schematic diagram in Figure 3 illustrates the major components of a typical cross-coil NMR spectrometer, which has a *transmitter coil* to irradiate the sample with radiofrequency energy and a *receiver coil* to detect energy absorption by the sample at resonance. The signal from the receiver coil is amplified and refined in the RF

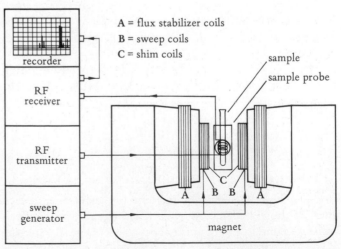

Figure 3.   Schematic diagram of a cross-coil NMR spectrometer

receiver unit, then sent on to a recorder where signal amplitude is plotted as a function of magnetic field strength.

In addition to the transmitter and receiver coils, such an instrument usually has several other coils to stabilize and sweep the magnetic field and to make it homogeneous over the sample volume.

Sample Preparation. The sample is usually analyzed in solution in a specially fabricated NMR tube, which is placed in a *sample probe* containing the transmitter and receiver coils. A typical NMR tube is 4–5 mm in outer diameter and 6–8″ long. For routine NMR analysis, samples can be made up by *(1)* dissolving 20–50 mg of a compound in 0.4–0.5 ml of a suitable solvent; *(2)* filtering the solution directly into the NMR tube through a Pasteur pipet containing a small plug of glass wool; *(3)* adding 5–15 μl of a reference standard such as TMS; *(4)* stoppering the tube carefully (NMR tubes are fragile and expensive); and *(5)* inverting the tube several times to mix the components thoroughly.

NMR Solvents. An ideal NMR solvent should have no interfering protons, a low viscosity, a high solvent strength, and no appreciable interactions with the solute. Carbon tetrachloride is the solvent of choice if the compound is sufficiently soluble in it, since it has no protons and does not interact with most solutes. Carbon disulfide, chloroform-d ($CDCl_3$), and cyclohexane are also good NMR solvents, although the former is hazardous and unpleasant to work with and the latter has a proton absorption at $\delta 1.4$. Solvents containing protons can sometimes be used if their signals do not interfere with the important solute signals. Alternatively, most common NMR solvents can be obtained in completely *deuterated* forms, such as acetone-d$_6$ (CD_3COCD_3). Since deuterium (^2H) resonates at about $6\frac{1}{2}$ times the field strength required for ^1H, deuterated solvents do not interfere significantly with proton NMR spectra; unfortunately, they are quite expensive.

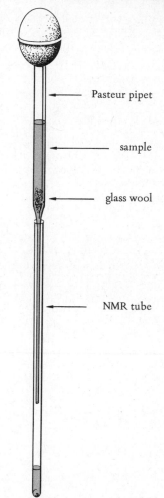

Pasteur pipet

sample

glass wool

NMR tube

Figure 4. Filtering an NMR solution

Deuterated forms of the solvents listed in Table 1 usually have ^1H peaks about $\frac{1}{2}$ –2% as strong as their nondeuterated counterparts, due to isotopic impurity.

Table 1. Properties of some NMR solvents

Solvent	^1H Chemical Shift(s) (δ, ppm)	Solvent Strength	Freedom from Solute Interactions	Viscosity
carbon tetrachloride	none	good	good	medium
carbon disulfide	none	good	good	low
cyclohexane	1.4	poor	good	medium

Table 1.   Properties of some NMR solvents (*continued*)

Solvent	¹H Chemical Shift(s) (δ, ppm)	Solvent Strength	Freedom from Solute Interactions	Viscosity
acetonitrile	2.0	good	fair	low
acetone	2.1	good	poor	low
dimethyl sulfoxide	2.5	very good	poor	high
dioxane	3.5	good	fair	medium
water	~5.2 (v)	good	poor	medium
chloroform	7.3	very good	fair	low
benzene	7.3	good	poor	low
pyridine	7.0–8.7	good	poor	medium
trifluoroacetic acid	~12.5 (v)	good	poor	medium

Note: Solvent strength refers to the ability to dissolve a broad spectrum of organic compounds. v = variable.

Commercial deuterated solvents are often prepared with 1–3% TMS added.

TMS is highly volatile, so both the liquid and the syringe should be refrigerated.

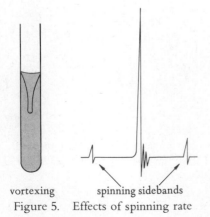

vortexing   spinning sidebands

Figure 5.   Effects of spinning rate

Important: Never touch any homogeneity control but the Y or Fine Y control without permission; any other alterations may make a lengthy retuning operation necessary.

Table 1 compares the properties of some commonly used solvents and gives the approximate positions of their proton NMR signals. The solvents containing hydrogen are ordinarily used in their deuterated forms for NMR analysis. It is often convenient to add 1–3% TMS to the bulk solvent so that it does not have to be added during sample preparation. Otherwise, the TMS should be kept in a refrigerator and added with a *cold* syringe or fine-tipped dropper.

Sample Spinning.   To average out the effect of magnetic-field variations in the plane perpendicular to the axis of the NMR tube, the sample is rotated at a rate of 30–60 revolutions per second while the NMR spectrum is being recorded. It is important to set the spinning rate correctly—excessively high rates can cause a vortex extending into the region of the receiver coil, and low rates can cause large *spinning sidebands* or signal distortion. Spinning sidebands are small peaks on either side of a main peak at a distance equal to the spinning rate from the main peak. Thus, an NMR tube spun at 30 cycles/sec gives rise to sidebands 30 Hz from each main peak. To find out whether small signals are spinning sidebands or impurity peaks, change the spinning rate and scan again to see if their positions change.

Field Homogeneity.   The most important homogeneity control is the one used to produce a uniform field along the axis of the sample tube. This is usually called the *Y control.* For routine work on a previously tuned instrument, the Y control can often be set by placing a blank sheet of paper on the instrument's platen and repeatedly scanning a strong peak in the spectrum of the sample, each time making small adjustments in the Y control until the peak is as tall and narrow as possible and shows a good "ringing" (beat) pattern. Figure 6 shows an excellent ringing pattern characterized by the high amplitude, long duration, and exponential decay of the "wiggles" following the main peaks.

Signal Amplitude.   The amplitude of NMR signals can be altered by two controls. One of them, usually called the *spectrum amplitude* control, changes the amplitude of the signals *and* the baseline "noise." It is usually set so that the strongest peak in the spectrum extends nearly to the top of the chart paper. The *RF power* control can increase the signal height without increasing baseline noise—up to the point where *saturation* sets in. Increasing RF power beyond that point eventually reduces the height and distorts the shape of a signal. Unless high sensitivity is required, the RF level is usually set at about mid-range or at some other value where there is little likelihood of saturation. Since saturation is a function of sweep time and sweep width as well as RF power, saturation at high power can sometimes be avoided by using a short sweep time and a wide sweep range.

Sweep Controls.   There are four important sweep controls on a typical NMR spectrometer, which control the reference point of the spectrum, the sweep rate, and the portion of the spectrum to be scanned. The *sweep zero* control is used to set the signal for the reference compound to the proper value (0.000 for TMS). The *sweep width* (sweep range) dial sets the total chemical-shift range to be scanned—usually 600–1000 Hz when the entire spectrum is being scanned on a 60-MHz instrument. Since most NMR spectra are scanned at a rate of about 1 Hz per second, the *sweep time* is often set so that it is numerically equal to the sweep width (e.g., 600 seconds for a 600-Hz sweep width). The *sweep offset* dial is used when only a specific portion of the spectrum is to be scanned; it sets the upfield limit of the scan. For example, if a scan between 350 and 500 Hz (about 5.8–8.3δ on a 60-MHz instrument) is desired, the sweep offset should be 350 Hz and the sweep range 150 Hz.

Other Adjustments.   A few additional controls will require adjustment in some cases. The *filter response time* (sometimes called the time constant) can be lengthened to reduce

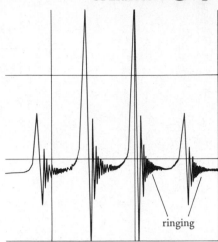

ringing

Figure 6.   Ringing pattern for the quartet of acetaldehyde

Saturation occurs when the excess proton population in the lower nuclear spin state is so depleted that increasing the RF field intensity no longer increases the number of transitions.

On a 60 MHz instrument, 1000 Hz is about 17 ppm, which encompasses virtually all protons likely to be encountered in routine work. A 600-Hz (10-ppm) range can be used if the sample is known to contain no protons absorbing above about 10δ.

The scan rate is obtained by dividing sweep width by sweep time.

noise; however, too long a time constant will distort the line shapes. According to a general rule of thumb, the magnitude of the time constant (in sec) should be no longer than the scan rate in Hz/sec. For reasonably concentrated samples, it is often set lower than this. The *phasing* control affects the shape of the baseline before and after a signal. It should be adjusted to obtain a straight baseline. Correct phasing is much more important when an NMR spectrum is being integrated than when the spectrum itself is being recorded.

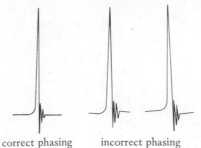

correct phasing   incorrect phasing

Figure 7.   Effects of phasing on baseline

Integration.   If you will be expected to integrate your NMR spectrum, you should consult your instructor for specific instructions; only a brief outline of the steps in the process will be given here.

1.   The RF power is optimized to provide an acceptable signal-to-noise ratio.

2.   The instrument is switched to the *integral* mode.

3.   The *integral amplitude* control is set (while scanning the spectrum rapidly) so that the integrator trace spans the vertical axis of the chart.

4.   With the sweep offset and sweep width controls set to scan a region free from NMR signals, the *balance* control is adjusted during a slow scan of that region to give a horizontal line.

Phasing should be adjusted as accurately as possible in the normal mode before starting the integration operation. It should then be readjusted in the integral mode.

When the signals on a spectrum are comparatively far apart, better integrals may be obtained with a nonspinning sample. The field homogeneity controls must be optimized on the stationary sample before the integral is recorded.

5.   The *phasing* control is adjusted while scanning over a signal to make the integrator traces before and after the signal as nearly horizontal as possible.

6.   The integral over the entire spectrum is recorded several times (preferably in both directions) using a sweep time about $\frac{1}{5}$ to $\frac{1}{10}$ that for the normal spectrum. The pen should be returned to the baseline after each scan.

7. The relative peak areas are determined by measuring the vertical distances between the integrator traces before and and after each signal, and the results for successive scans are averaged. (See Figure 8.)

Equipment and Supplies: Sample, NMR solvent, TMS, NMR tube, Pasteur pipet, glass wool, tissues, washing solvent.

Operational Procedure

Fill the NMR tube to a depth of about 4 cm with a 5–20% (w/v) solution of the compound in a suitable solvent containing 1–3% TMS. Wipe the outside of the tube carefully with a tissue paper, insert it in the sample *spinneret* using a depth gauge to adjust its position, wipe it again, and gently place the assembly into the sample probe between the magnet pole faces. Adjust the air flow to spin the sample at 30–60 Hz. Set the sweep width, sweep offset, and sweep time controls to scan the desired range at a rate of about 1 Hz/sec. Set the RF power to about midrange and the filter response time to a low value. Place a blank paper on the recorder bed, set the spectrum amplitude control to about midrange, and scan the spectrum. Readjust the spectrum amplitude to obtain good peak heights. Optimize peak shape and ringing by repeatedly scanning a strong, sharp peak (the TMS peak may be suitable) and adjusting the Y control each time. Adjust the phasing, if necessary, and set the TMS peak to zero δ with the sweep zero control (the reference peak should be very sharp and close to its expected position when the sweep offset is at zero). Record the spectrum and (if desired) the integral on a single sheet of chart paper.

If the sample is dilute, it may be necessary to optimize the RF power and filter settings to obtain a suitable signal-to-noise ratio.

Clean the sample tube immediately with an appropriate solvent. Usually, the same solvent is used for cleaning as was used in preparing the sample. (If the NMR solvent was deuterated, use the *protonated* form for cleaning.) A Pasteur pipet can be used to rinse out the tube, or a device like the one described in *J. Chem. Educ. 53,* 127 (1976) can be constructed for the purpose. Invert the tube in a suitable rack and let it drain dry. If necessary, it can be dried more rapidly with a Pasteur pipet attached to an aspirator or to a source of clean, dry air.

This is a general procedure suitable for routine analyses with many of the instruments likely to be encountered by undergraduates. Consult your instructor or the manufacturer's operating manual for more specific operating instructions.

Typical settings are:
  sweep offset: 0
    sweep width: 500–600 Hz
    sweep time: 500–600 sec
for a 60-MHz instrument.

If you obtain no spectrum at this point, recheck the settings and make sure the sample is spinning. Then, if necessary, run a standard sample such as TMS/chloroform to make sure the instrument is working properly.

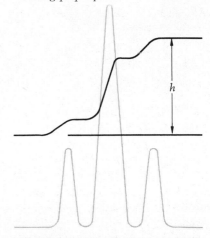

(*h* is proportional to the total area under the triplet)

Figure 8.   Measuring signal area

Visible and
Ultraviolet Spectrometry

Principles and Applications

1 nanometer (nm) = 10^{-9} m = 10 Å. The obsolete unit "millimicrons" (mμ) is sometimes used for nm.

Spectrophotometers that measure the absorption of radiation in the visible ($\sim$400–800 nm) and near ultraviolet ($\sim$200–400 nm) regions of the electromagnetic spectrum are useful in both the qualitative and quantitative analysis of certain organic compounds. Radiation in these regions induces molecular transitions in which electrons are promoted from an electronic ground state (a bonding or nonbonding orbital) to one or more excited states (antibonding orbitals). The most important of these transitions involve pi electrons in conjugated systems, so ultraviolet-visible spectrometry is used mainly to analyze aromatic compounds and aliphatic compounds that contain two or more conjugated double bonds.

The energy required for electronic transitions is much greater than that needed to induce vibrational or nuclear magnetic transitions, so the wavelength of the radiation utilized is much shorter: 0.2–0.8 μm compared to about 2.5–50 μm for IR and several meters for NMR.

Electronic transitions of single bonds and isolated double bonds generally occur at wavelengths below 200 nm, which is outside the range of most ultraviolet-visible spectrophotometers.

An electronic spectrum in either the ultraviolet or visible region is often quite featureless compared to an infrared or NMR spectrum; it may consist of only one or two broad absorption bands. The band structure is caused by rotational and vibrational transitions that accompany each electronic transition. Each different combination of vibrational and rotational transitions has a different energy; collectively, they span a broad range of wavelengths centered about the wavelength of the "pure" electronic transition. An electronic absorption band is characterized by a maximum intensity measured along the y-axis of the spectrum (usually in terms of *absorbance,* A) and a wavelength of maximum intensity, λ_{max}, measured along the x-axis. For example, the absorption band illustrated in Figure 1 has an absorbance of 0.80 and a λ_{max} of 350 nm.

Absorbance is related to transmittance by the equation:

$$A = \log(1/T) = -\log T \quad \textbf{(1)}$$

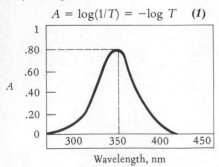

Figure 1.   Ultraviolet absorption band

Some instruments are made to scan the same range every time and thus do not have a separate control for setting the initial wavelength.

Experimental Considerations

The construction of ultraviolet-visible spectrophotometers varies widely; both single- and double-beam instruments are available, with or without recording capability. A typical recording instrument contains a closed *sample compartment* with places for sample and reference cells, a *zero adjustment* control for setting the absorbance to zero before the spectrum is recorded, a *wavelength* dial to fix the starting wavelength, a *scan* switch to start and stop the scanning of the desired wavelength range, a lever or some other control to change the *radiation source,* and a *recorder,* either as an integral part of the instrument or as a separate unit. Some chart papers, premarked with wavelength and absorbance values, are wrapped around a recording drum; others may come in the form of a roll that feeds onto a flat recorder bed during a scan, requiring that wavelength and absorbance values be written in by the user. Some instruments have a control with which the absorbance range (along the y-axis of the chart paper) can be varied. Although the slit, gain, and balance controls do not usually require frequent readjustment for routine work, they can affect the quality of a spectrum and should be adjusted periodically by the instructor or laboratory assistant. Too great a slit width can reduce the resolution of absorption peaks and decrease their intensity, thus reducing the accuracy of absorptivity and concentration measurements.

A typical absorbance range is from zero to 1 or 2 absorbance units.

Once a spectrum has been recorded, the data it contains can be presented as a tabulation of λ_{max} values giving either the absorbance, molar absorptivity (ϵ), or log ϵ at each wavelength specified. For example, the *CRC Handbook of Chemistry and Physics* reports the ultraviolet spectrum of cinnamic acid as λ^{al} 210 (4.24), 215 (4.28), 221 (4.18), 268 (4.31), where the numbers in parentheses are log ϵ values for peaks having the λ_{max} values (in nanometers) given, and "al" indicates that the spectrum was run in alcohol.

Sample Preparation

Although ultraviolet-visible spectra can be obtained using gaseous samples and KBr pellets, routine spectra are nearly always obtained in solution. The solvent must be transparent (or nearly so) in the regions to be scanned. Water, 95% ethanol, methanol, dioxane, acetonitrile, and cyclohexane are suitable down to about 210–220 nm; many other solvents can be used at higher wavelengths. 95% ethanol is a favorite solvent because it does not normally require additional purification; the other organics should be purchased as spectral grade solvents or be specially purified. The solvent must not, of course, react with the solute—for example, alcohols should not be used as solvents for aldehydes.

In making up a solution, great care must be taken to avoid contamination by impurities.

Absolute ethanol does require purification, because it contains some benzene.

The optimum *concentration* of a solution depends on the molar absorbtivity of the bands to be recorded, the cell path length (b), and the absorbance ranges available on the instrument. One can calculate from Equation 2 that the sam-

Relationship between molar concentration and absorbance (Beer's law)

$$c = \frac{A}{\epsilon \cdot b} \qquad (2)$$

ϵ_{max} is the molar absorptivity of the strongest peak in the spectrum.

If there is great variation in the intensities of different peaks, it may be necessary to analyze several solutions of differing concentrations. For example, 3-buten-2-one has a band at 210 nm with $\epsilon = 11,500$ and one at 315 nm with $\epsilon = 26$; the latter peak would be virtually undetectable at a concentration suitable for observing the former.

Fused quartz cells are used for the ultraviolet region; glass or plastic cells are suitable in the visible region.

A tungsten lamp can generally be used between 300 and 800 nm and a hydrogen lamp between 190 and 350 nm.

To find out if the sample concentration and absorbance range are satisfactory, scan the wavelength range manually (if the instrument allows this) before recording a spectrum.

ple concentration, using a 1 cm sample cell, must be less than or equal to $1/\epsilon_{max}$ to record a spectrum with a maximum absorbance of 1 or less. For example, if the maximum molar absorptivity of a sample is 10,000 (a fairly typical value), the solution concentration should be 1.0×10^{-4} M or less. To make up such dilute solutions accurately, it is usually necessary to prepare a solution that is several powers of ten too concentrated, then measure out an aliquot of this solution and dilute it. For example, to prepare a 1.0×10^{-4}M solution of cinnamic acid (M.W. = 148) one could measure 0.15 g of the solid into a 100-ml volumetric flask and make it up to volume with solvent, then pipet a 1-ml aliquot of this 0.010 M solution into another 100 ml volumetric flask and fill it to the mark with solvent. If the molar absorptivity of the sample is not known, it may be necessary to find the optimum concentration by trial and error, starting with a comparatively concentrated solution and diluting it as needed.

The most commonly used *sample cell* for ultraviolet-visible spectrometry is a transparent rectangular container with a square cross-section, having a path length of 1.00 cm and a capacity of about 3 ml. Two sides of such a cell are ground and the other two are transparent. The cell must be held by its ground sides and placed in the sample compartment so that the transparent sides are perpendicular to the light beam. Cells should *never* be touched on the transparent sides, since even a fingerprint can yield a spectrum. They must be cleaned thoroughly after use by rinsing them several times with the solvent. In some cases, it may be necessary to use detergent or a special cleaning solution to remove all of a previous sample.

Operating Procedure

Because ultraviolet-visible spectrophotometers vary widely in construction and operation, the following is intended only as a general guide and may not be applicable to all instruments. Specific operating instructions should be learned from in-class demonstrations or the operator's manual.

Be sure that the instrument's *power* and *lamp* switches have been turned on and that adequate time has been allowed for warmup. Select the appropriate radiation source for the desired wavelength range. If applicable, set the absorbance range to 0–1 (or another appropriate value) and the wavelength dial to the starting wavelength. If the instrument scans from high to low wavelength, this will be the highest wavelength of the range to be scanned. Place the reference cell and sample cell in the appropriate locations in the sample compartment, then close the compartment and set the absorbance to zero with the zero adjustment control. Fill the sample cell with the solution to be analyzed and the reference cell with the pure solvent, return them to the sample compartment, and close the compartment. Position the chart paper (if necessary) so that the scan starts on an ordinate line, label this line with the starting wavelength, lower the pen to the paper, and begin to scan the spectrum. Stop the scan when the other end of the wavelength range has been reached. If both ultraviolet and visible regions are to be scanned, change the radiation source and scan the spectrum in the other region. Raise the pen from the chart, remove the spectrum, and write down any data that may assist you in interpreting it. Rinse the sample cell several times with the solvent, then drain and dry both sample and reference cells. Measure and write down the λ_{max} and absorbance values of any peaks and shoulders of interest.

Some experimenters prefer to zero the instrument with both cells filled with the solvent.

If one of the peaks goes off scale, change the absorbance range or dilute the sample.

Cells can be cleaned and partly dried using cotton swabs, then dried with an ear syringe or a stream of dry air ⟨OP–22⟩.

Colorimetry

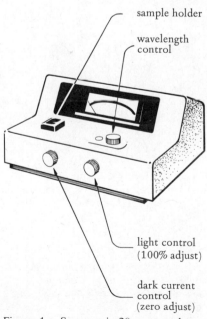

Figure 1.   Spectronic 20 spectrophotometer

Instruments operating in the near-ultraviolet and visible regions of the spectrum can be very useful for the quantitative analysis of compounds in solution, provided that the solute cither absorbs sufficient radiation in the appropriate range or can be made to absorb by combining it with a suitable reagent. Although a recording ultraviolet-visible spectrophotometer can be used for this purpose, a simple colorimeter such as the Bausch and Lomb Spectronic 20 will suffice.

Concentrations can be determined from absorbance values using either Beer's law (see Equation *1*) or a calibration curve of absorbance versus molar concentration. In kinetic studies (as in Experiment 38), absorbance values can be used to calculate rate constants directly, without first converting them to concentrations.

Operational Procedure

See that the instrument has been switched on and that adequate time has been allowed for warmup. Set the *wave-*

These directions are for operation of the B & L Spectronic 20 or a similar instrument.

Use the same solvent as you used to prepare the solution being analyzed.

Be sure the cuvettes are positioned correctly.

Calculation of concentration:

$$c = \frac{A}{\epsilon \cdot b} \qquad (1)$$

c = molar concentration
A = absorbance
b = cell path length
ϵ = molar absorptivity

length control to the desired value and set the transmittance (read from the scale) to zero with the *dark current* control. Insert a clean *cuvette* containing the pure solvent into the sample holder and adjust the *100% control* until the scale shows 100% transmittance. Replace this reference cuvette with another clean cuvette containing the solution to be analyzed and read the percent transmittance from the scale. (The absorbance can also be read from the logarithmic scale, but it is usually more accurate to convert transmittance to absorbance using Equation 1, ⟨OP–35⟩.)

If another solution containing the same solvent and solute is to be analyzed, drain the cuvette and rinse it with that solution before you make the next measurement. If another solution containing a different solute or solvent is to be analyzed, rinse the cuvette with the solvent used for preparing the first solution (clean it if necessary) and dry it before adding the next solution. After the last measurement, rinse the cuvette with pure solvent, clean it, and let it air dry.

Mass Spectrometry

Principles and Applications

When a compound is bombarded with a beam of high-energy electrons in a mass spectrometer, each of its molecules (M) can lose an electron and form a *molecular ion,* $M^{\cdot+}$. If the energy of the electron beam is high enough, the molecular ions will have enough excess vibrational and electronic energy to break apart into fragments, each pair of fragments consisting of another positive ion and a neutral molecule. Each positive ion (*daughter ion*) may in turn break down by loss of a neutral fragment to form yet another positive ion, and so on. The positive ions are accelerated through an electrostatic field into an evacuated chamber (the *mass analyzer*) where they are separated according to mass and sent to an *ion collector.* As ions of a given mass number impinge on the ion collector, it generates a signal proportional to the number of ions in the beam. The signal is then amplified and recorded on a chart. The record of all such signals, a *mass spectrum,* is essentially a graph of the relative abundance of ions in order of their mass numbers. A typical mass spectrometer is easily capable of resolving

Fragmentation of molecular ion and daughter ion

$$M^{\cdot+} \longrightarrow A^+ + X$$

$$A^+ \longrightarrow B^+ + Y, \text{ etc.}$$

(The location of the unpaired electron varies—it may be on either A^+ or X, depending on the kind of fragmentation.)

Strictly speaking, ions are recorded in order of their mass-charge ratios (m/e) rather than their masses. However, ions having charges greater than 1 are seldom encountered.

mass peaks to the nearest whole number ("unit resolution") up to a mass number of about 500.

Mass spectrometry is an extremely valuable structural tool, since it can allow one to determine the molecular weight, the molecular formula, and often the molecular structure of an unknown compound with great reliability. The method has also been used to analyze the composition of mixtures, measure bond dissociation energies, measure reaction rates, and elucidate reaction mechanisms.

Experimental Considerations

Mass spectrometers are costly, highly complex instruments ordinarily operated by skilled technicians; most undergraduate students will not have an opportunity to obtain hands-on experience with one. For that reason, and because the instruments vary so widely in construction and operation, no attempt will be made here to describe specific operating techniques; these must be learned by special instruction and by studying the manufacturer's operating manual.

Samples prepared for mass spectrometry should be very pure, since traces of impurities can make interpretation difficult. No special sample preparation is required—liquids are inserted directly into the sample inlet with a hypodermic syringe, micropipet, or break-off device, and solids are usually introduced by means of a melting-point capillary. The samples vaporize in the sample inlet system, after which their molecules flow through a controlled "molecular leak" (usually a minute hole in a piece of gold foil) into the evacuated ionization chamber.

Interpretation of Mass Spectra

One popular kind of recorder consists of five galvanometers of varying sensitivity that produce a spectrum consisting of five separate traces, increasing in amplitude from bottom to top. Since the uppermost trace is recorded by the most sensitive galvanometer, it is assigned a sensitivity factor of 1; the others are, in order, usually assigned sensitivity factors of 3, 10, 30 and 100 respectively. The height of each mass peak is measured from the baseline of the trace in which it has the highest on-scale amplitude, then multiplied by the appropriate sensitivity factor to give its relative intensity. The most intense peak in the spectrum (called the *base peak*) is assigned an intensity of 100%; the intensities of all other peaks are reported as percentages of the base-peak intensity. The

High-resolution mass spectrometers are available that can resolve peaks differing by as little as 0.001 atomic mass units.

Sample sizes for mass spectrometry range from less than a microgram to several milligrams.

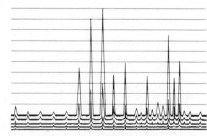

Figure 1.   Part of a mass spectrum traced by a five-element galvanometer

The molecular ion peak is sometimes, but not always, the base peak.

mass of each peak is usually determined by counting from the low-mass end of the spectrum (which can be accurately set) along the upper trace, since there will usually be at least a small "blip" at each mass unit. This data is generally tabulated, giving the normalized intensity of each significant peak in order of mass (usually from higher to lower).

To determine the molecular weight and molecular formula of the sample, the *molecular-ion peak* must be found. This is usually the last strong peak on the spectrum, since no fragment can have a higher mass than the original molecular ion. There will usually, however, be at least two discernible low-intensity peaks following the molecular ion peak. These are generated by isotopic variations of the molecular ion. Most elements occurring in organic compounds have at least one isotope of higher mass number than the common form. For example, about one carbon atom in 90 (1.11 in 100) is carbon-13 rather than the more abundant carbon-12, so the molecular ion peak of benzene (C_6H_6, mass = 78) should be followed by a peak for "heavy" benzene ($^{13}C^{12}C_5H_6$, mass = 79) having an intensity 6.66% of the parent peak intensity. Actually the mass 79 ($M + 1$) peak has a slightly higher intensity than this because ordinary benzene also contains minute quantities of benzene-d_1 (C_6H_5D). Likewise, oxygen-18 occurs naturally to the extent of about 0.20 atoms for every 100 atoms of oxygen-16, so formaldehyde shows an $M + 2$ peak (corresponding to $CH_2{}^{18}O$) having an intensity 0.20% of the parent peak.

The relative intensities of the $M + 1$ and $M + 2$ peaks can be calculated for a compound $C_wH_xN_yO_z$ by using Equations 1 and 2:

$$\%(M + 1) = 1.11w + 0.015x + 0.37y + 0.037z \tag{1}$$

$$\%(M + 2) = 0.006w(w - 1) + 0.0002wx + 0.004wy + 0.20z \tag{2}$$

For example, quinine (as the hydrate) has the molecular formula $C_{20}H_{30}N_2O_3$. From the equations, the intensity of its $M + 1$ peak should be 23.5% of the parent peak intensity and that of its $M + 2$ peak should be 3.16% of the parent peak intensity. Tabulations of these values are available (see references F8 and F21 in the Bibliography).

No two combinations of atoms having a given molecular mass is likely to yield $M + 1$ and $M + 2$ peaks of exactly the same intensity. Therefore, if the intensities of these peaks are measured from the mass spectrum of an unknown compound, its molecular formula can usually be determined

---

Except for the M + 1 and M + 2 peaks (see below), peaks having intensities of only a few percent are often not tabulated.

The molecular-ion peak is often called the parent peak. Its mass is essentially equal to the molecular mass of the sample in its most abundant isotopic variation.

Since benzene contains six carbon atoms, the chance that any one of the six will be carbon-13 is 6 × 1.11, or 6.66%.

These intensities are given as percentages of the molecular-ion (parent) peak intensity, not of the base peak intensity.

Calculation:
$\%(M + 1) = 1.11(20) + 0.015(30)$
$\qquad + 0.37(2) + 0.037(3) = 23.5.$

$\%(M + 2)$
$= 0.006(20)(19) + 0.0002(20)(30)$
$\qquad + 0.004(20)(2) + 0.20(3) = 3.16.$

from tables. The general procedure for determining molecular formulas is as follows:

1. Locate the molecular ion peak and determine its mass, M.

2. Measure the intensities of the M, $M + 1$, and $M + 2$ peaks and express the latter two as a percentage of the M peak intensity.

3. Find the formula (or formulas) listed under its M value that give $M + 1$ and $M + 2$ intensities close to the experimental values, and that make sense from a chemical standpoint.

Some formulas may be eliminated immediately because they do not correspond to stable molecules or have the expected degree of hydrogen deficiency. For compounds having the general formula $C_wH_xN_yO_z$, an *index of hydrogen deficiency* (equal to the number of rings plus the number of unsaturated bonds in each molecule) can be calculated from Equation 3. For example, the calculated I.H.D. of compound *1* is 7, which is consistent with its molecular structure (one ring and six unsaturated bonds). Another useful generalization is the *nitrogen rule*, which says that a stable compound with an even-numbered value of M can have only zero or an even number of nitrogen atoms, whereas one with an odd value of M can have only an odd number of nitrogens. Finally, some possibilities can be eliminated because compounds having those formulas would be impossible to obtain from a given reaction or source.

The use of *fragmentation patterns* to determine molecular structures is a very broad subject covered in detail elsewhere; only a few generalizations will be given here. A molecular ion (or its daughter ions) often fragments by eliminating small neutral molecules or free radicals such as CO, H_2O, HCN, C_2H_2, H·, or CH_3 ·, yielding ions with masses equal to $M - X$, where X is the mass of the neutral molecule. Certain kinds of fragment ions are encountered frequently with various classes of compounds. For example, an alkylbenzene usually undergoes bond cleavage β to the ring to form a $C_7H_7^+$ ion of mass 91, which shows up as a prominent peak in its spectrum. Tables of such ions and neutral fragments, as well as information on the fragmentation patterns of specific kinds of compounds, can be found in the mass spectrometry references listed in the Bibliography.

Formula for index of hydrogen deficiency:

$$\text{I.H.D.} = \tfrac{1}{2}(2w - x + y + 2) \qquad (3)$$

Example:

$$\text{I.H.D.} = \tfrac{1}{2}(18 - 7 + 1 + 2) = 7$$

The nitrogen rule applies only to organic compounds containing the more commonly encountered elements.

Fragmentation of an alkylbenzene

Mass = 91

Appendixes and Bibliography

Laboratory Equipment APPENDIX I

CHEMICAL GLASSWARE

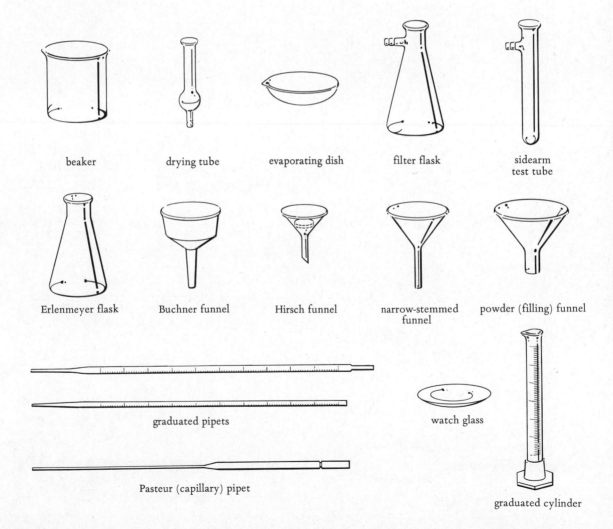

beaker   drying tube   evaporating dish   filter flask   sidearm test tube

Erlenmeyer flask   Buchner funnel   Hirsch funnel   narrow-stemmed funnel   powder (filling) funnel

graduated pipets   watch glass

Pasteur (capillary) pipet

graduated cylinder

STANDARD TAPER GLASSWARE

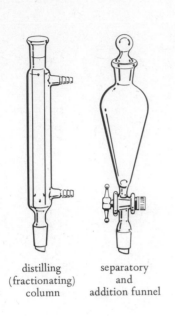

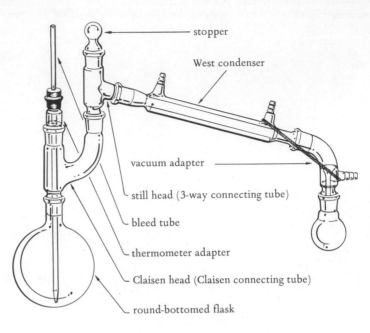

stopper

West condenser

vacuum adapter

still head (3-way connecting tube)

bleed tube

thermometer adapter

Claisen head (Claisen connecting tube)

round-bottomed flask

distilling
(fractionating)
column

separatory
and
addition funnel

HARDWARE

condenser clamp (3-finger clamp)

utility clamp

ring support

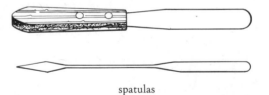

spatulas

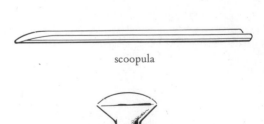

scoopula

wing top (flame spreader)

wire gauze

Writing Laboratory Reports

Each experiment will require a writeup either *(1)* on a report form provided by your instructor; *(2)* on separate sheets of paper, typewritten or handwritten in ink following a format such as that described below; or *(3)* in a bound notebook. If you are required to keep a bound laboratory notebook, refer to Appendix III. The other types of reports should include information under some or all of the following headings, as directed by your instructor:

1. Prelab: The prelab checklist, calculations, and any other prelab exercises requiring a written report.

2. Observations: Any significant observations made during the course of the experiment. This will not necessarily include a detailed description of the procedure you followed, but it should describe any important observations you made at various stages of the experiment. You should record all observations that might be of help to you (or another experimenter) if you were to repeat the experiment at a later time. These should include the quantities of solvents or drying agents used, the reaction times, the distillation ranges, and a description of any experimental difficulties you encounter. You should also make note of phenomena that might provide clues about the nature of chemical or physical transformations taking place during the experiment, such as color changes, phase separations, tar formation, and gas evolution.

3. Raw Data: All numerical data obtained directly from an experiment that may be used in further calculations. This can include quantities of reactants and products, titration volumes, kinetic data, gas chromatographic retention times and integrated peak areas, and spectrometric absorbance values.

4. Calculations: Yield and stoichiometry calculations and any other calculations based on the raw data. If a number of repetitive calculations are required, one or two sample calculations of each type may suffice.

5. Results: A list, graph, tabulation, or verbal description of the significant results of the experiment. For a preparation, this should include a physical description of the product (color, physical state, evidence of purity, etc.), the percent yield, and all physical constants, spectra, and analytical data obtained from the product. For the qualitative analysis of an unknown compound, the results of all tests and derivative preparations should be included, along with spectra and physical constants. The results of all calculations based on the raw data should be reported in this section as well.

6. *Discussion:* A discussion of the significant experimental results that attempts to explain and analyze the results. The discussion should describe possible sources of error or material losses, compare numerical results obtained with those expected from theory (if applicable), or discuss the theoretical significance of the results and attempt to draw conclusions from them. You may also use the Discussion to tell what you learned from the experiment and whether or not it fulfilled its stated objectives.

7. *Conclusions:* Any conclusions to be drawn as a result of the performance of an experiment, as suggested by the objectives for the experiment. In a qualitative analysis experiment, for example, your major conclusion would be the identity of the unknown.

8. *Topics for Report:* Your answers or reports from the assigned topics. The first two topics, which are generally mandatory, may include material covered in your Discussion or another part of your report. If so, you need not repeat your answers under this heading if you refer to the part of the report in which the answers appear.

Each report should also include (on the first page or cover) the name and number of the experiment, your name, and the date on which the experiment was turned in to the instructor. The date(s) on which the experiment was performed may also be required. The following sample laboratory report should be studied as an illustration of the kind of writeup that might be prepared by a conscientious student.

Name: Cynthia Sizer
Date: Sept. 21, 1980

EXPERIMENT NO. 5: PREPARATION OF TETRAHEDRANOL

Prelab (*checklist attached*)

Tetrahedryl benzoate needed:

$$\text{mass} = 0.075 \text{ mol} \times 172.2 \text{ g/mol} = 12.9 \text{ g}$$

$$\text{volume} = \frac{12.9 \text{ g}}{0.951 \text{ g/ml}} = 13.6 \text{ ml}$$

Aqueous sodium hydroxide (5M) required:

$$\text{volume} = \frac{0.50 \text{ mol}}{5.0 \text{ mol/l}} \times 1000 \text{ ml/l} = 100 \text{ ml}$$

Observations

The reaction was carried out under reflux for 1 hour and 10 minutes, during which time the organic layer dissolved in the aqueous sodium hydroxide and the odor of the ester disappeared. The reaction mixture was extracted with two 50-ml portions of ethyl ether and the extracts were dried over 3.0 g of anhydrous magnesium sulfate. Evaporation of the ether under aspirator vacuum (steam bath) required about 25 minutes. The residue was distilled over a 4° boiling range from 78–82°C, and solidified on cooling. None of the product was seen to solidify in the vacuum adapter during the distillation, although the procedure warned against this possibility. Its infrared spectrum was recorded using a thin film between heated silver bromide plates.

The original reaction mixture, when acidified with 88 ml of 3M sulfuric acid (~0.53 equivalents), yielded a white precipitate presumed to be benzoic acid, which was dried in a desiccator jar for 48 hours before weighing. I observed that during the precipitation of benzoic acid a white solid would appear each time a portion of sulfuric acid was added and then disappear when the mixture was stirred, right up to the last portion—then it all precipitated at once! It formed pretty white needles on cooling.

Data

Mass of tetrahedryl benzoate: 12.87 g
Mass of tetrahedranol recovered: 32.7 g
Mass of benzoic acid recovered: 8.70 g

Calculations

Theoretical yield of tetrahedranol

$$\frac{12.87 \text{ g TetOBz}}{172.1 \text{ g/mol TetOBz}} \times 68.1 \text{ g/mol TetOH} = 5.09 \text{ g TetOH}$$

Percent yield of tetrahedranol

$$\frac{3.27 \text{ g}}{5.09 \text{ g}} \times 100 = 64.2\%$$

Theoretical yield of benzoic acid

$$\frac{12.87 \text{ g Tet OBz}}{172.1 \text{ g/mol TetOBz}} \times 122.1 \text{ g/mol BzOH} = 9.13 \text{ g BzOH}$$

Percent yield of benzoic acid

$$\frac{8.70 \text{ g}}{9.13 \text{ g}} \times 100 = 95.3\%$$

(TetOBz = tetrahedryl benzoate; TetOH = tetrahedranol;
BzOH = benzoic acid)

Results

Tetrahedranol was obtained in 64.2% yield from the alkaline hydrolysis of tetrahedryl benzoate. Tetrahedranol at room temperature was a colorless, almost transparent solid having a mild "spiritous" odor. It could easily be melted to form a clear, colorless liquid. It had a distillation boiling range of 78–82°C and a micro boiling point of 81°. Its infrared spectrum displayed strong absorption bands at 3270 and 1194 cm⁻¹. Benzoic acid, obtained in 95.3% yield from the reaction, was a white solid crystallizing in the form of small needles that melted at 120–122°. The crystals were slightly discolored, but appeared dry and otherwise free from impurities.

Discussion

The infrared spectrum (attached) indicated that the product boiling at 81° was an alcohol, and a comparison of its boiling point with the literature value of 82.3° for authentic tetrahedranol provided strong evidence that this product was indeed tetrahedranol. The facts that it melted just above room temperature (reported m.p. = 28°), and that it was synthesized from tetrahedryl benzoate by a reaction designed to hydrolyze such an ester, lend support to this assumption. Further confirmation could be obtained by comparing the recorded IR spectrum with a published spectrum and by measuring other physical properties such as its refractive index.

The product melting at 120–22° would appear to be benzoic acid from its melting point (literature value = 122.4°) and its origin as a product of the hydrolysis of tetrahedryl benzoate. From the melting point, it seems to be reasonably pure. The slight discoloration could probably be removed by recrystallizing it from water, which should also narrow the melting point range.

The yield of tetrahedranol was 1.82 g less than the theoretical value. Losses could have occurred as a result of (*a*) incomplete reaction; (*b*) incomplete extraction of the alcohol from the aqueous layer; (*c*) losses in transferring the reaction mixture to the separatory funnel, the ether extract from the separatory funnel to the evaporating flask, and the residue from the evaporating flask to the distilling flask; or (*d*) losses during distillation of the product. Since the yield of benzoic acid was 95.2%, the reaction must have been at least 95% complete, so (*a*) could account for *at most* 0.25 g (5% of 5.09) of the loss. More likely the reaction was nearly 100% complete, since some benzoic acid must have been lost during the workup (see below). I tried to keep losses in transfers to a minimum by using additional solvent (ether or water) in each transfer, so (*c*) should not be a major reason for the low yield. About 0.5 ml of residue remained in the distilling flask, which can account for about 0.5 g of the loss, assuming the residue was mostly tetrahedranol. It seems likely that most of the loss (up to 1.3 g) occurred during the extraction operation, particularly since tetrahedranol is reported in the *CRC Handbook* as being "soluble" in water. This loss might have been reduced by saturating the aqueous layer with potassium carbonate to salt out the alcohol or by carrying out several more ether extractions.

The yield of benzoic acid was *very* high; it should also be reliable since the product dried to a constant weight (three weighings were performed at approximately one-hour intervals). The loss of 0.43 g can be accounted for by (**a**) losses in transfers and (**b**) incomplete precipitation from solution. Traces of benzoic acid remained on the Buchner funnel and filter paper used for the vacuum filtration. About 0.34 g might have stayed in the 188 ml of ice-cold solution from which the acid was precipitated (its solubility in water at 4° is 0.18 g/100 ml), but the high salt (Na_2SO_4) concentration in this solution probably decreased its solubility somewhat.

Conclusions

It can be reasonably concluded from the experimental results that the reaction of tetrahedryl benzoate with 5M aqueous sodium hydroxide does in fact yield tetrahedranol and benzoic acid in acceptable yield, and that the experimental objective "To prepare tetrahedranol by the alkaline hydrolysis of tetrahedryl benzoate and isolate benzoic acid as a by-product of the reaction" has been fulfilled.

(Attached: Prelab checklist, infrared spectrum of tetrahedranol, and Topics for Report)

Keeping a Laboratory Notebook   APPENDIX III

The notebook formats preferred by different instructors can vary considerably, so some of the following general suggestions may be disregarded or modified at your instructor's request.

If possible, the laboratory notebook should be bound and provided with quadrilled, numbered, perforated, duplicate pages. With this kind of notebook, a carbon is placed between identically numbered pages (usually white and yellow) so that the duplicate pages can be torn out and handed in.

Before the laboratory period, you should prepare the notebook for use by writing in some or all of the following basic information about the experiment: (**a**) experiment number, title, and date; (**b**) a clear statement of the experimental objectives; (**c**) balanced equations for all significant reactions, including possible side reactions that may lower the yield in a preparation; (**d**) a flow diagram for the separation and purification procedure to be followed; (*e*) a table of significant physical constants for the reactants, products, and other materials involved in a preparation; (*f*) calculations of the quantities of reactants required and the theoretical yield of the product; (*g*) a list of equipment, supplies, and chemicals needed for the experiment; (*h*) a sketch of the apparatus to be used. The prelab writeup should be approved by the instructor before the start of the experiment.

Observations and raw data should be recorded directly in the laboratory notebook, in ink, while the experiment is being performed. Data should be recorded in tabular form when appropriate. Your instructor may require a detailed description of the procedure as well as incidental observations.

Calculations based on raw data, a summary of experimental results, a discussion of the results, significant conclusions, and answers to the Topics for Report can be entered in the laboratory notebook after the experiment has been completed. Some instructors require equations for alternative methods of preparing a product or a discussion of alternative ways of accomplishing the experimental objectives. Refer to Appendix II for suggestions about the information to be included in each category.

APPENDIX IV	Calculations for Organic Synthesis

It is very important that students become comfortable with molar quantities so that they can appreciate the stoichiometric relationships between reactants and products and be able to recognize the limiting reactants. For this and other reasons, most reactant quantities are given in moles (mol) or millimoles (mmol) in this book; these chemical units of matter must be converted to units of mass or volume before an experiment can be performed. Interconversions of such units are carried out as follows:

(*a*) Moles are converted to mass by multiplying by molecular weight. For example, the mass of 15.0 mmol (0.0150 mol) of butyl acetate (M.W. = 116) is $\frac{15.0 \text{ mmol}}{1000 \text{ mmol/mol}} \times 166 \text{ g/mol} = 1.74 \text{ g}$.

(*b*) Moles of a liquid are converted to volume by multiplying by molecular weight and dividing by density. For example, the volume of 0.250 mol of acetic acid (M.W. = 60.1, d. = 1.049) is $0.250 \text{ mol} \times \frac{60.1 \text{ g/mol}}{1.049 \text{ g/ml}} = 14.4 \text{ ml}$. (Alternatively the mass of acetic acid in grams can be calculated, then divided by density.)

(*c*) The volume of a solution required to provide a given number of moles (or mmol) of the solute is calculated by dividing its molarity *into* the number of moles required. For example, the volume of 6.0M HCl needed to provide 0.18 moles of HCl is $0.18 \text{ mol}/6.0 \frac{\text{mol}}{\text{lit}} = 0.030$ lit or 30 ml.

In all such calculations, the validity of the result can be checked by making sure that the units cancel out to yield the correct units in the answer. The reverse calculations can be performed by reversing the operations—that is:

(*a*) grams divided by molecular weight yields moles;

(*b*) milliliters multiplied by density and divided by molecular weight yields moles;

(*c*) volume (of a solution) in liters multiplied by molarity yields moles of solute.

To calculate the *theoretical yield* of a product from a given reaction, one must know the stoichiometry of the reaction and be able to identify the *limiting reactant*. For example, suppose a procedure for preparing dimethyl adipate utilizes 58.4 g of adipic acid (M.W. = 146) and 16.0 g of methanol (M.W. = 32.0), the reaction proceeding according to the equation

$$HOOC(CH_2)_4COOH + 2CH_3OH \rightleftharpoons$$
$$CH_3OOC(CH_2)_4COOCH_3 + 2H_2O$$

This comes to 0.40 moles of adipic acid (58.4/146) and 0.50 moles of methanol (16.0/32.0); however, since the reaction stoichiometry requires only half as many moles of adipic acid as there are moles of methanol, adipic acid is in excess by 0.15 moles ($0.40 - \frac{1}{2} \cdot 0.50$). Methanol is thus the limiting reactant (the reactant that determines the theoretical yield of product). From the equation, it can be seen that the 0.50 moles of methanol will react with 0.25 moles of adipic acid to yield (in theory) 0.25 moles of dimethyl adipate, along with 0.50 moles of water. Since the molecular weight of dimethyl adipate is 174 g/mole, the theoretical yield of this product is thus $0.25 \times 174 = 43.5$ g.

The *percent yield* of a preparation is defined as:

$$\text{Percent yield} = \frac{\text{actual yield}}{\text{theoretical yield}} \times 100$$

If in the previous example a student obtains a yield of 32.4 g of dimethyl adipate, his or her percent yield is $(32.4/43.5) \times 100 = 74.5\%$.

As an exercise to test your understanding of the above concepts, calculate the theoretical yield of dimethyl adipate possible from the combination of 50.0 ml of methanol (d. = 0.79 g/ml) with 60.0 g of adipic acid. Then calculate the percent yield obtained by a student who recovers 60.0 g of dimethyl adipate from this preparation. What is the limiting reactant in this example?

APPENDIX V   Writing a Laboratory Checklist

A laboratory checklist should be prepared before the beginning of each experiment. It should include the following information:

1. Quantities of all reactants, catalysts, solvents, drying agents, and other chemicals that will be used during the experiment.

2. Equipment and supplies required for each operation (indicate the sizes of beakers, flasks, and other glassware).

3. Reaction times, distillation ranges, and other information or data that will be needed during the experiment.

4. Warnings of possible hazards.

The checklist should also include notations to help you organize your time efficiently, such as a reminder to tell you what can be done during a long reflux to prepare for subsequent operations. References to operation summaries can be made so that you can refresh your memory (as needed) prior to the performance of each operation. The checklist for Experiment 3 (Figure 1) illustrates a general format that you may wish to follow. You may have to revise this checklist to fit the equipment available in your laboratory, and you will have to include the amount of methyl salicylate calculated from the Prelab Exercise.

Figure 1.   Suggested laboratory checklist for Experiment 3

Reactants and Solvents		Step	Comments
Methyl salicylate	—*	Reaction 1	Irritant liquid
5M NaOH	50 ml	Reaction 1	*Hazard:* causes burns
2M H_2SO_4	70 ml	Reaction 2	To precipitate product
Cold water	10 ml	Separation	For washing
Water	100 ml+	Purification 1	For recrystallization
Cold water	10 ml	Purification 2	For washing

* To be calculated in Prelab Exercise.

Reaction

1. Reflux: See ⟨OP–7*a*⟩ Summary
   Time: 30 minutes
   (During reflux, prepare for addition of acid and set up filtration apparatus)

2. Addition of H_2SO_4

Separation

Vacuum filtration: See ⟨OP–12⟩ Summary
Wash liquid: 10 ml cold water (2 portions)
(While ppt is air drying, prepare for recrystallization)

Purification

1. Recrystallization: See ⟨OP–23⟩ Summary
   Solvent: water; 70 ml + 10 ml portions as needed

1a. Gravity filtration of hot solution
   (While cooling hot solution, prepare for filtration and drying)

2. Vacuum filtration
   Wash liquid: 10 ml cold water (2 portions)

3. Drying (set aside 5–10 mg for melting point)
   Oven temperature: 90–100°C
   (While drying, prepare for melting point)

Analysis

Melting-point determination: See ⟨OP–28⟩ Summary

Supplies and Equipment

100-ml r.b. flask, reflux condenser, tubing, boiling chips, wire gauze, ring, ringstand, burner

150-ml beaker, pH paper, stirring rod

Buchner funnel, filter flask, stirring rod, spatula, filter paper, wash bottle, 10-ml graduate, watch glass

2 250-ml flasks, burner, 100-ml graduate, stirring rod

powder funnel, fluted filter paper

Same as for separation.

large filter papers, watch glass, spatula

Thiele tube (with oil and cork), thermometer, rubber band, m.p. tube, ringstand, clamp, burner, watch glass, f.b. stirring rod.

Interpreted Infrared Spectra   APPENDIX VI

The following representative spectra illustrate most of the important infrared bands used in classifying organic compounds for qualitative analysis. It should be emphasized that the spectra of compounds in the same chemical family can differ widely. The spectrum of tributylamine, for instance, bears little resemblance to the spectrum of butylamine. Certain bands may be unreliable—their intensity and position may vary widely, or they may disappear entirely for certain members of a chemical family. This is particularly true of the N—H "wagging" vibration (Spectrum 9) and the C-halogen stretching vibration (Spectrum 13). Refer to Part III, pages 384–389 for a description of the four major spectral regions and a discussion of the use of infrared spectra in qualitative analysis. For simplicity, the different kinds of vibrations have been classified only as stretching (ν) or bending (δ) vibrations in most cases, rather than naming the specific type of stretching or bending vibration giving rise to an absorption band. Some bands designated as νC—O and δO—H may be combination bands involving coupled vibrations. The low-frequency bands designated as δAr—H actually include a ring deformation vibration as well as out-of-plane C—H bending. It should be noted that methyl and methylene bending vibrations appear in the spectra of nearly all compounds having alkyl chains, and give rise to two or three distinct bands near 1375 cm^{-1} and 1450 cm^{-1}.

Refer to sources in section F of the Bibliography for more information on IR spectral interpretation, and see the IR correlation chart on the back endpaper for positions of representative absorption bands.

Figure 1.   1-butanol (primary alcohol)

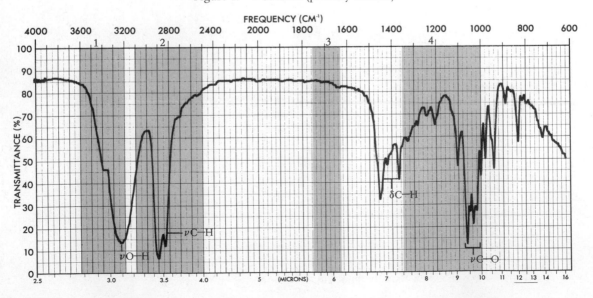

Figure 2.   *t*-butyl alcohol (tertiary alcohol)

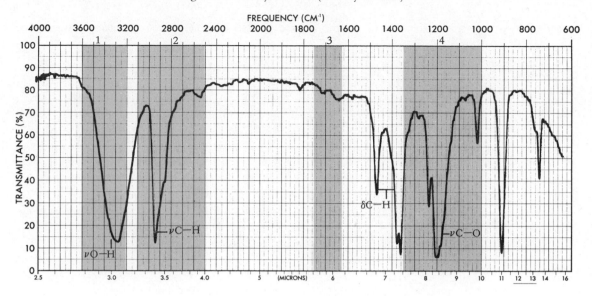

Figure 3.   benzyl alcohol (aromatic alcohol)

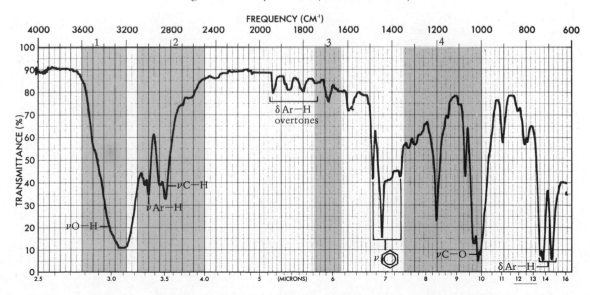

Figure 4.   *m*-cresol (phenol)

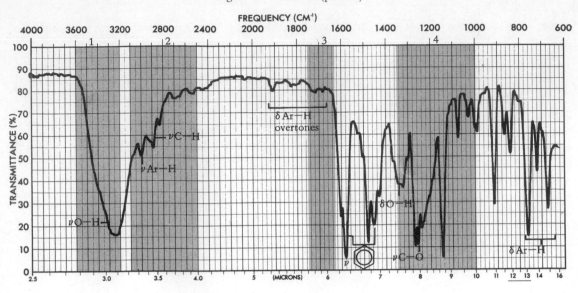

Figure 5.   butanal (aldehyde)

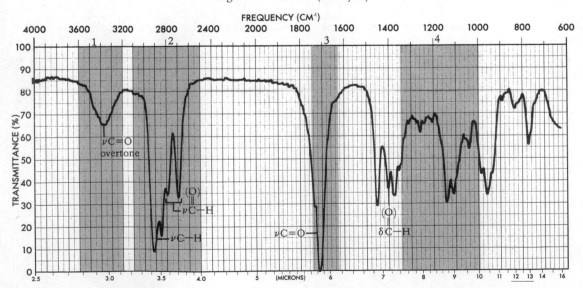

Figure 6. 2-butanone (ketone)

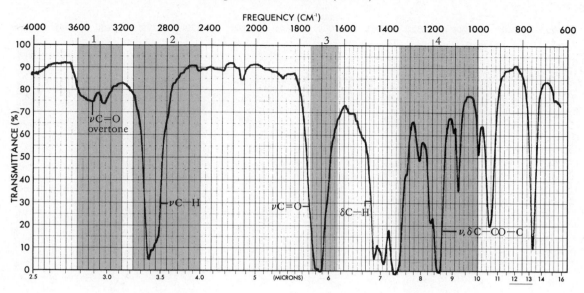

Figure 7. propanoic acid (carboxylic acid)

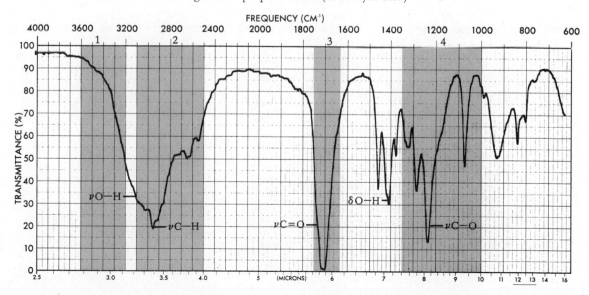

Figure 8.   ethyl butyrate (ester)

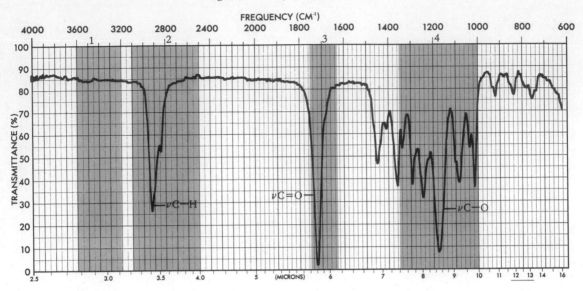

Figure 9.   *n*-butylamine (primary amine)

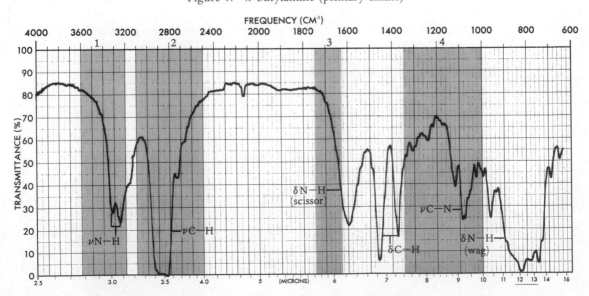

Figure 10. *o*-toluidine (primary aromatic amine)

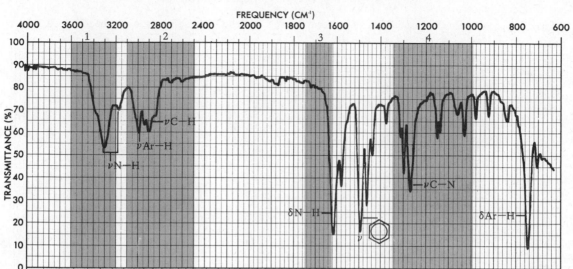

Figure 11. tributylamine (tertiary amine)

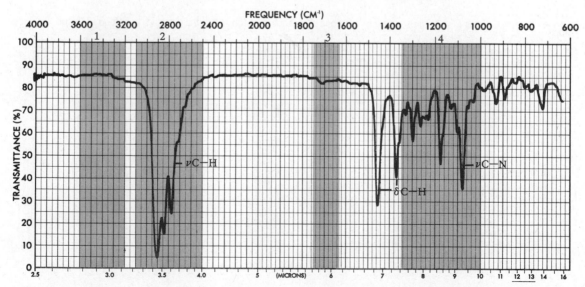

Figure 12.   ethylbenzene (aromatic hydrocarbon)

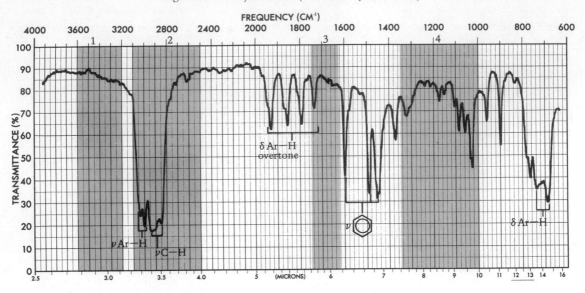

Figure 13.   1-chloropentane (alkyl chloride)

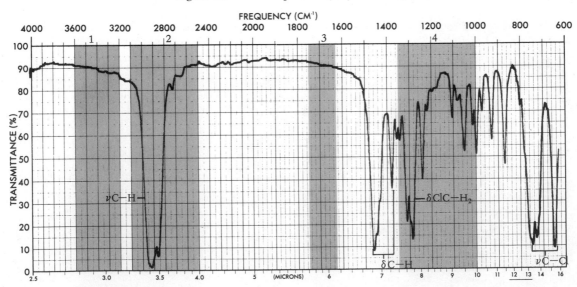

Properties of Selected Organic Compounds for Qualitative Analysis

The following tables are to be used in conjunction with the procedures described in Part III, Systematic Organic Qualitative Analysis. Compounds that melt below ordinary ambient temperatures (about 25°C) are listed in order of increasing boiling point; those that (when pure) are generally solid at room temperature are listed in order of increasing melting point. Both melting and boiling points are specified for some borderline cases. Derivative preparations are described on pages 410–423 and are referred to by number at the head of each column. Melting points in parentheses are for derivatives that exist in more than one crystalline form or for which significantly different melting points have been reported in the literature. Sometimes the recrystallization solvent will determine the form in which a derivative crystallizes, so significant deviations from the listed melting points should not be considered conclusive proof that a sample is not the expected derivative. In cases where a given compound can form more than one product (for instance, nitro and bromo derivatives), the reaction conditions for the preparation may determine the derivative isolated (a mixture of derivatives may also result). A blank (—) space in a derivative column indicates either that the derivative has not been reported in the literature or is not suitable for identification. See bibliographic references A1–A4, A7–A11, B4, G1, or G3 for physical constants and derivative melting points not listed in the tables.

Most compounds are listed by their systematic (IUPAC) names except when those names would be too lengthy.

Abbreviations used in tables

d: decomposes on melting

s: sublimes at or below melting point

m: monosubstituted derivative (such as a mononitrated aromatic hydrocarbon)

di: disubstituted derivative

t: trisubstituted derivative

tet: tetrasubstituted derivative

List of Tables

Table 1.   Alcohols

Compound	b.p., °C	m.p., °C	Deriv. m.p., °C			
			D–1 3,5-Dinitro-benzoate	D–2 1-Naphthyl-urethane	D–2 Phenyl-urethane	D–1 4-Nitro-benzoate
Methanol	65		108	124	47	96
Ethanol	78		93	79	52	57
2-Propanol	82		123	106	88	110
2-Methyl-2-propanol★	83	*26*	142	—	136	—
2-Propen-1-ol	97		49	108	70	28
1-Propanol	97		74	105	57	35
2-Butanol	99		76	97	65	26
2-Methyl-2-butanol	102		116	72	42	85
2-Methyl-1-propanol	108		87	104	86	69
3-Pentanol	116		101	95	48	17
1-Butanol	118		64	71	61	36
2-Pentanol	120		62	74	—	24
3-Methyl-3-pentanol	123		94 (62)	104	43	69
3-Methyl-1-butanol	132		61	68	57	21
4-Methyl-2-pentanol	132		65	88	143	26
1-Pentanol	138		46	68	46	11
Cyclopentanol	141		115	118	132	62
2-Ethyl-1-butanol	148		51	60	—	—
1-Hexanol	157		58	59	42	5
Cyclohexanol★	161	*25*	113	129	82	50
Furfuryl alcohol	172		80	130	45	76
1-Heptanol	177		47	62	60	10
2-Octanol	179		32	63	114	28
1-Octanol	195		61	67	74	12
1-Phenylethanol	202		95	106	92	43
Benzyl alcohol	205		113	134	77	85
2-Phenylethanol	219		108	119	78	62
1-Decanol	231		57	73	60	30
3-Phenylpropanol	236		45	—	92	47
1-Dodecanol★	259	*24*	60	80	74	45 (42)
1-Tetradecanol		39	67	82	74	51
(−)-Menthol		44	153	119	111	62

Table 1.   Alcohols (*continued*)

Compound	b.p., °C	m.p., °C	Deriv. m.p., °C			
			D–1 3,5-Dinitro-benzoate	D–2 1-Naphthyl-urethane	D–2 Phenyl-urethane	D–1 4-Nitro-benzoate
1-Hexadecanol		49	66	82	73	58
1-Octadecanol		59	77	89	79	64
Diphenylmethanol		68	141	136	139	132
Cholesterol		148	—	176	168	185
(+)-Borneol		208	154	132 (127)	138	137 (153)

* May be solid at or just below room temperature.

Table 2.   Aldehydes

Compound	b.p., °C	m.p., °C	Deriv. m.p., °C		
			D–4 Semi-carbazone	D–3 2,4-Dinitrophenyl-hydrazone	D–5 Oxime
Ethanal	21		162	168 (157)	47
Propanal	48		154	148 (155)	40
Propenal	52		171	165	—
2-Methylpropanal	64		125 (119)	187 (183)	—
Butanal	75		106	123	—
3-Methylbutanal	92		107	123	48
Pentanal	103		—	106 (98)	52
2-Butenal	104		199	190	119
2-Ethylbutanal	117		99	95 (130)	—
Hexanal	130		106	104	51
Heptanal	153		109	106	57
2-Furaldehyde	162		202	212 (230)	91
2-Ethylhexanal	163		254d	114 (120)	—
Octanal	171		101	106	60
Benzaldehyde	179		222	239	35
4-Methylbenzaldehyde	204		234 (215)	234	80
3,7-Dimethyl-6-Octenal	207		84 (91)	77	—
2-Chlorobenzaldehyde	213		229 (146)	213 (209)	76 (101)
4-Methoxybenzaldehyde	248		210	253d	133

Table 2. Aldehydes (*continued*)

Compound	b.p., °C	m.p., °C	D–4 Semi-carbazone	D–3 2,4-Dinitrophenyl-hydrazone	D–5 Oxime
			Deriv. m.p., °C		
Phenylethanal	*195*	33	153 (156)	121 (110)	99
2-Methoxybenzaldehyde		38	215	254	92
4-Chlorobenzaldehyde		48	230	265	110 (146)
3-Nitrobenzaldehyde		58	246	290	120
4-Nitrobenzaldehyde		106	221 (211)	320	133 (182)

Table 3. Ketones

Compound	b.p., °C	m.p., °C	D–4 Semi-carbazone	D–3 2,4-Dinitrophenyl-hydrazone	D–5 Oxime
			Deriv. m.p., °C		
Acetone	56		187	126	59
2-Butanone	80		146	118	—
3-Methyl-2-butanone	94		113	124	—
2-Pentanone	102		112 (106)	143	58
3-Pentanone	102		138	156	69
3,3-Dimethyl-2-butanone	106		157	125	75 (79)
4-Methyl-2-pentanone	117		132	95 (81)	58
2,4-Dimethyl-3-pentanone	124		160	95 (88)	34
2-Hexanone	128		125	110	49
4-Methyl-3-penten-2-one	130		164 (133)	205	48
Cyclopentanone	131		210 (203)	146	56
4-Heptanone	144		132	75	—
2-Heptanone	151		123	89	—
Cyclohexanone	156		166	162	91
2,6-Dimethyl-4-heptanone	168		122	92	210
2-Octanone	173		124	58	—
Cycloheptanone	181		163	148	23
2,5-Hexanedione	194		185 (m); 224 (di)	257 (di)	137 (di)
Acetophenone	202	*20*	198 (203)	238	60
Propiophenone	218	*21*	182 (174)	191	54
2-Undecanone	228		122	63	44

Table 3.   Ketones (*continued*)

Compound	b.p., °C	m.p., °C	D–4 Semi-carbazone	D–3 2,4-Dinitrophenyl-hydrazone	D–5 Oxime
			Deriv. m.p., °C		
4-Phenyl-2-butanone	235		142	127	87
4-Methylacetophenone	*226*	28	205	258	88
4-Methoxyacetophenone		38	198	228	87
4-Phenyl-3-buten-2-one		42	187	227 (223)	117
Benzophenone		48	167	238	144
2-Acetonaphthone		54	235	262d	145
3-Nitroacetophenone		80	257	228	132
9-Fluorenone		83	234	283	195
(±)-Camphor		179	237	177	118

Table 4.   Amides

Compound	b.p., °C	m.p., °C	D–6 Carboxylic acid	D–7 N-Xanthylamide
			Deriv. m.p., °C	
Formamide	195d		—	184
Propanamide		81	—	211
Ethanamide		82	17	240
Heptanamide		96	—	155
Nonanamide		99	12	148
Hexanamide		100	—	160
Hexadecanamide		106	63	142
Pentanamide		106	—	167
Octadecanamide		109	69	141
Butanamide		115	—	187
Chloroacetamide		120	61 (53)	209
4-Methylpentanamide		121	—	160
Succinimide		126	185	247
2-Methylpropanamide		129	—	211
Benzamide		130	122	223
3-Methylbutanamide		136	—	183
o-Toluamide		143	104 (108)	200

Table 4.   Amides (*continued*)

| Compound | b.p., °C | m.p., °C | Deriv. m.p., °C | |
			D–6 Carboxylic acid	D–7 N-Xanthylamide
Furamide		143	133	210
Phenylacetamide		156	77s	195
p-Toluamide		159	180s	225
4–Nitrobenzamide		201	240	233
Phthalimide		238	210d	177

Table 5.   Primary and secondary amines

| Compound | b.p., °C | m.p., °C | Deriv. m.p., °C | | | |
			D–8 Benzamide	D–9 p-Toluene-sulfonamide	D–10 Phenylthio-urea	D–12 Picrate
t-Butylamine	44		134	—	120	198
Propylamine	48		84	52	63	135
Diethylamine	56		42	60	34	155
sec-Butylamine	63		76	55	101	140
2-Methylpropylamine	69		57	78	82	150
Butylamine	77		42	—	65	151
Diisopropylamine	84	—	—	—	—	140
Pyrrolidine	89		—	123	—	112 (164)
3-Methylbutylamine	95		—	65	102	138
Pentylamine	104		—	—	69	139
Piperidine	106		48	96	101	152
Dipropylamine	109		—	—	69	75
Morpholine	128		75	147	136	146
Pyrrole	131		—	—	143	69d
Hexylamine	132		40	—	77	126
Cyclohexylamine	134		149	—	148	—
Diisobutylamine	139		—	—	113	121
N-Methylcyclohexylamine	147		86	—	—	170
Dibutylamine	159		—	—	86	59
N-Ethylbenzylamine	181		—	95	—	118
Aniline	184		163	103	154	198 (180)

Table 5.   Primary and secondary amines (*continued*)

Compound	b.p., °C	m.p., °C	D-8 Benzamide	D-9 p-Toluene-sulfonamide	D-10 Phenylthio-urea	D-12 Picrate
				Deriv. m.p., °C		
Benzylamine	185		105	116 (185)	156	199 (194)
N-Methylaniline	196		63	94	87	145
o-Toluidine	200		144	108	136	213
m-Toluidine	203		125	114	104 (92)	200
N-Ethylaniline	205		60	87	89	138
2-Chloroaniline	209		99	105 (193)	156	134
2,6-Dimethylaniline	215		168	212	204	180
2-Methoxyaniline	225		60	127	136	200
2-Ethoxyaniline	229		104	164	137	—
3-Chloroaniline	230		120	138 (210)	124 (116)	177
4-Ethoxyaniline	250		173	106	136	69
Dicyclohexylamine	255d		153 (57)	—	—	173
N-Benzylaniline		37	107	149	103	48
p-Toluidine		44	158	118	141	182d
Diphenylamine		54	180 (109)	141	152	182
4-Methoxyaniline		58	154	114	157 (171)	170
4-Bromoaniline		66	204	101	148	180
2-Nitroaniline		71	110 (98)	142	—	73
4-Chloroaniline		72	192	95 (119)	152	178
1,2-Diaminobenzene		102	301 (di)	260 (di)	—	208
3-Nitroaniline		114	155 (di)	138	160	143
1,4-Diaminobenzene		142	300 (di)	266 (di)	—	—
4-Nitroaniline		147	199 (di)	191	—	100

Note: See other Qualitative Analysis references listed in the Bibliography for the melting points of benzene-sulfonamides and 1-naphthylthioureas.

Table 6.   Tertiary amines

Compound	b.p., °C	m.p., °C	D-12 Picrate	D-11 Methiodide
			Deriv. m.p., °C	
Triethylamine	89		173	280
Pyridine	115		167	117

Table 6. Tertiary amines (*continued*)

Compound	b.p., °C	m.p., °C	Deriv. m.p., °C D–12 Picrate	D–11 Methiodide
2-Methylpyridine	129		169	230
3-Methylpyridine	143		150	92
4-Methylpyridine	143		167	152
Tripropylamine	157		116	207
2,4-Dimethylpyridine	159		183 (169)	113
N,N-Dimethylbenzylamine	183		93	179
N,N-Dimethylaniline	193		163	228d
Tributylamine	216 (211)		105	186
N,N-Diethylaniline	217		142	102
Quinoline	237		203	133 (72)
Isoquinoline	243	*26*	222	159
Tribenzylamine	—	91	190	184
Acridine	—	111	208	224

Table 7. Carboxylic acids

Compound	b.p., °C	m.p., °C	Deriv. m.p., °C D–14 *p*-Toluidide	D–14 Anilide	D–13 Amide	D–15 *p*-Nitrobenzyl Ester
Formic acid	101		53	50	43	31
Acetic acid	118		148	114	82	78
Propenoic acid	139		141	104	85	—
Propanoic acid	141		124	103	81	31
2-Methylpropanoic acid	154		107	105	128	—
Butanoic acid	164		75	95	115	35
3-Methylbutanoic acid	176		107	109	135	—
Pentanoic acid	186		74	63	106	—
2-Chloropropanoic acid	186		124	92	80	—
Dichloroacetic acid	194		153	118	98s	—
2-Methylpentanoic acid	196		80	95	79	—
Hexanoic acid	205		75	95	101	—
2-Bromopropanoic acid	205d	*24*	125	99	123	—
Octanoic acid	239		70	57	107	

Table 7.   Carboxylic acids (*continued*)

Compound	b.p., °C	m.p., °C	Deriv. m.p., °C			
			D–14 *p*-Toluidide	D–14 Anilide	D–13 Amide	D–15 *p*-Nitrobenzyl Ester
Nonanoic acid	254		84	57	99	—
Decanoic acid		32	78	70	108	—
2,2-Dimethylpropanoic acid	*164*	35	—	129 (133)	178 (154)	—
Dodecanoic acid		44	87	78	110 (99)	—
3-Phenylpropanoic acid		48	135	98	105	36
Tetradecanoic acid		54	93	84	103	—
Hexadecanoic acid		62	98	90	106	42
Chloroacetic acid		63	162	137	120	—
Octadecanoic acid		70	102	95	109	—
trans-2-Butenoic acid		72	132	118	160	67
Phenylacetic acid		77	136	118	156	65
2-Methoxybenzoic acid		101	—	131	129	113
Oxalic acid (dihydrate)		101	268 (di)	254 (246) (di)	419d (di)	204 (di)
2-Methylbenzoic acid		104	144	125	142	91
Nonanedioic acid		106	201 (di)	186 (di)	175 (di)	44
3-Methylbenzoic acid		112	118	126	94	87
Benzoic acid		122	158	163	130	89
Maleic acid		130	142 (di)	187 (di)	181 (m) 266 (di)	91
Decanedioic acid		133	201 (di)	122 (m) 200 (di)	170 (m) 210 (di)	73 (di)
Cinnamic acid		133	168	153	147	117
Propanedioic acid		135	86 (m) 253 (di)	132 (m) 230 (di)	50 (m) 170 (di)	86
2-Chlorobenzoic acid		140	131	118	140	106
3-Nitrobenzoic acid		140	162	155	143	141
Diphenylacetic acid		148	172	180	167	—
2-Bromobenzoic acid		150	—	141	155	110
Hexanedioic acid		152	238	151 (m) 241 (di)	125 (m) 224 (di)	106
4-Methylbenzoic acid		180s	160 (165)	145	160	104
4-Methoxybenzoic acid		184	186	169	167 (163)	132
Butanedioic acid		188	180 (m) 255 (di)	143 (m) 230 (di)	157 (m) 260 (di)	—

Table 7. Carboxylic acids (*continued*)

Compound	b.p., °C	m.p., °C	Deriv. m.p., °C			
			D–14 p-Toluidide	D–14 Anilide	D–13 Amide	D–15 p-Nitrobenzyl Ester
3,5-Dinitrobenzoic acid		205	—	234	183	157
Phthalic acid		210d	201 (di)	253 (di)	220 (di)	155
4-Nitrobenzoic acid		240	204	211	201	168
4-Chlorobenzoic acid		242	—	194	179	129
Terephthalic acid		>300s	—	337	—	263 (di)

Table 8. Esters

Compound	b.p., °C	m.p., °C	Deriv. m.p., °C			
			D–16 Carboxylic acid	D–16 Alcohol or phenol	D–18 3,5-Dinitro-benzoate	D–17 N-Benzyl-amide
Ethyl formate	54		8	—	93	60
Methyl acetate	57		17	—	108	61
Ethyl acetate	77		17	—	93	61
Methyl propanoate	80		—	—	108	43
Methyl acrylate	80		13	—	108	237
Isopropyl acetate	91		17	—	123	61
t-Butyl acetate	98		17	26	142	61
Ethyl propanoate	99		—	—	93	43
Methyl 2,2-dimethylpropanoate	101		35	—	108	—
Propyl acetate	102		17	—	74	61
Methyl butanoate	102		—	—	108	38
Ethyl 2-methylpropanoate	111		—	—	93	87
sec-Butyl acetate	112		17	—	76	61
Methyl 3-methylbutanoate	117		—	—	108	54
Isobutyl acetate	117		17	—	87	61
Ethyl butanoate	122		—	—	93	38
Butyl acetate	126		17	—	64	61
Methyl pentanoate	128		—	—	108	43
Ethyl 3-methylbutanoate	135		—	—	93	54
3-Methylbutyl acetate	142		17	—	61	61
Ethyl chloroacetate	145		63	—	93	—

Table 8.   Esters (*continued*)

Compound	b.p., °C	m.p., °C	Deriv. m.p., °C			
			D–16 Carboxylic acid	D–16 Alcohol or phenol	D–18 3,5–Dinitro-benzoate	D–17 N–Benzyl-amide
Pentyl acetate	149		17	—	46	61
Ethyl hexanoate	168		—	—	93	53
Hexyl acetate	172		17	—	58	61
Cyclohexyl acetate	175		17	25	113	61
Dimethyl malonate	182		135	—	108	142
Diethyl oxalate	185		101★	—	93	223
Heptyl acetate	192		17	—	47	61
Phenyl acetate	197		17	42	146	61
Methyl benzoate	199		122	—	108	105
Diethyl malonate	199		135	—	93	142
o-Tolyl acetate	208		17	31	135	61
m-Tolyl acetate	212		17	12	165	61
Ethyl benzoate	213		122	—	93	105
p-Tolyl acetate	213		17	36	189	61
Methyl o-toluate	215		104	—	108	—
Benzyl acetate	217		17	—	113	61
Diethyl succinate	218		188	—	93	206
Isopropyl benzoate	218		122	—	123	105
Methyl phenylacetate	220		77s	—	108	122
Diethyl maleate	223		137	—	93	150
Ethyl phenylacetate	228		77s	—	93	122
Propyl benzoate	230		122	—	74	105
Diethyl adipate	245		152	—	93	189
Butyl benzoate	250		122	—	64	105
Ethyl cinnamate	271		133	—	93	226
Dimethyl phthalate	284		210d	—	108	179
(+)-Bornyl acetate	*226*	27	17	208	154	61
Methyl p-toluate		33	180s	—	108	133
Methyl cinnamate		36	133	—	108	226
Benzyl cinnamate		39	133	—	113	226
1-Naphthyl acetate		49	17	94	217	61
Ethyl p-nitrobenzoate		56	240	—	93	—

Table 8.   Esters (continued)

Compound	b.p., °C	m.p., °C	Deriv. m.p., °C			
			D–16 Carboxylic acid	D–16 Alcohol or phenol	D–18 3,5-Dinitro-benzoate	D–17 N-Benzyl-amide
Phenyl benzoate		69	122	42	146	105
2-Naphthyl acetate		71	17	123	210	61
p-Tolyl benzoate		71	122	36	189	105
Methyl m-nitrobenzoate		78	140	—	108	101
Methyl p-nitrobenzoate		96	240	—	108	142

Note: Additional derivatives of the acid and alcohol portions of most esters can be found in Tables 1 and 7. * dihydrate; the anhydrous acid melts at 190°C.

Table 9.   Alkyl halides

Compound	b.p., °C	Density C–10 d_4^{20}, g/ml	Deriv. m.p., °C D–19 S-Alkylthiuronium-picrate
Bromoethane	38	1.461	188
2-Bromopropane	60	1.314	196
1-Chloro-2-methylpropane	69	0.879	167 (174)
3-Bromopropene	71	1.398	155
1-Bromopropane	71	1.354	177
Iodoethane	72	1.936	188
1-Chlorobutane	78	0.884	177
2-Iodopropane	89	1.703	196
1-Bromo-2-methylpropane	93	1.264	167 (174)
1-Chloro-3-methylbutane	100	0.875	173
1-Bromobutane	101	1.274	177
1-Iodopropane	102	1.749	177
3-Iodopropene	102	1.848	155
1-Chloropentane	108	0.882	154
1-Bromo-3-methylbutane	119	1.207	173 (179)
2-Iodobutane	119	1.595	166
1-Iodo-2-methylpropane	120	1.606	167 (174)
1-Bromopentane	129	1.218	154
1-Iodobutane	131	1.617	177
1-Chlorohexane	134	0.876	157
1-Iodo-3-methylbutane	148	1.503	173

Table 9.   Alkyl halides (*continued*)

Compound	b.p., °C	Density C–10 d_4^{20}, g/ml	Deriv. m.p., °C D–19 S-Alkylthiuronium-picrate
1-Iodopentane	155	1.516	154
1-Bromohexane	155	1.173	157
1-Iodohexane	181	1.439	157
1-Bromooctane	201	1.112	134
1-Iodooctane	225	1.330	134

Table 10.   Aryl halides

Compound	b.p., °C	m.p., °C	Density C–10 d_4^{20}, g/ml	Deriv. m.p., °C	
				D–20 Nitro derivative	D–21 Carboxylic acid
Chlorobenzene	132		1.106	52	—
Bromobenzene	156		1.495	75 (70)	—
2-Chlorotoluene	159		1.083	63	140
3-Chlorotoluene	162		1.072	91	158
4-Chlorotoluene	162		1.071	38 (m)	242
1,3-Dichlorobenzene	173		1.288	103	—
1,2-Dichlorobenzene	181		1.306	110	—
2-Bromotoluene	182		1.423	82	150
3-Bromotoluene	184		1.410	103	155
Iodobenzene	188		1.831	171 (m)	—
2,6-Dichlorotoluene	199		1.269	50 (m)	139
2,4-Dichlorotoluene	200		1.249	104	164
3-Iodotoluene	204		1.698	108	187
2-Iodotoluene	211		1.698	103 (m)	162
1-Chloronaphthalene	259		1.191	180	—
4-Bromotoluene	*184*	28	—	—	251
4-Iodotoluene		35	—	—	270
1,4-Dichlorobenzene		53	—	106 54 (m)	—
2-Chloronaphthalene		56	—	175	—
1,4-Dibromobenzene		89	—	84	—

Note: Nitro derivatives signified by (m) are mononitro compounds; all others are dinitro derivatives.

Table 11.   Aromatic hydrocarbons

| Compound | b.p., °C | m.p., °C | Deriv. m.p., °C | | |
			D–20 Nitro derivative	D–21 Carboxylic acid	D–12 Picrate
Benzene	80		89 (di)	—	84u
Toluene	111		70 (di)	122	88u
Ethylbenzene	136		37 (t)	122	96u
1,4–Xylene	138		139 (t)	300s	90u
1,3–Xylene	139		183 (t)	330s	91u
1,2–Xylene	144		118 (di)	210d	88u
Isopropylbenzene	152		109 (t)	122	—
Propylbenzene	159		—	122	103u
1,3,5–Trimethylbenzene	165		86 (di) 235 (t)	350 (t)	97u
t-Butylbenzene	169		62 (di)	122	—
4-Isopropyltoluene	177		54 (di)	300s	—
1,3-Diethylbenzene	181		62 (t)	330s	—
1,2,3,4-Tetrahydronaphthalene	206		96 (di)	210d	—
Diphenylmethane	*262*	26	172 (tet)	—	—
1,2-Diphenylethane		53	180 (di) 169 (tet)	—	—
Naphthalene		80	61 (m)	—	149
Triphenylmethane		92	206 (t)	—	—
Acenaphthene		96	101 (m)	—	161
Phenanthrene			—	—	144 (133)
Fluorene		114	199 (di) 156 (m)	—	87 (77)
Anthracene		216	—	—	138u

Note: Picrates designated u are unstable and cannot easily be purified by recrystallization.

Table 12.   Phenols

| Compound | b.p., °C | m.p., °C | Deriv. m.p., °C | | |
			D–24 1-Naphthyl-urethane	D–23 Bromo derivative	D–22 Aryloxy-acetic acid
2-Chlorophenol	176		120	49 (m); 76 (di)	145
3-Methylphenol	202		128	84 (t)	—

Table 12.   Phenols (*continued*)

| Compound | b.p., °C | m.p., °C | Deriv. m.p., °C | | |
			D−24 1-Naphthyl-urethane	D−23 Bromo derivative	D−22 Aryloxy-acetic acid
2-Methylphenol	*192*	31	142	56 (di)	152
4-Methylphenol	*232*	36	146	49 (di); 108 (tet)	135
Phenol	*182*	42	133	95 (t)	99
4-Chlorophenol		43	166	33 (m); 90 (di)	156
2,4-Dichlorophenol		43	—	68	135 (141)
2-Nitrophenol		45	113	117 (di)	158
4-Ethylphenol		47	128	—	97
5-Methyl-2-isopropylphenol		50	160	55 (m)	149
3,4-Dimethylphenol		63	142	171 (t)	163
4-Bromophenol		64	169	95 (t)	157
3,5-Dichlorophenol		68	—	189 (t)	—
2,5-Dimethylphenol		75	173	178 (t)	118
1-Naphthol		94	152	105 (di)	194
3-Nitrophenol		97	167	91 (di)	156
4-*t*-Butylphenol		100	110	50 (m)	86
1,2-Dihydroxybenzene		105	175	193 (tet)	—
1,3-Dihydroxybenzene		110	206	112 (di)	195
4-Nitrophenol		114	150	142 (di)	187
2-Naphthol		123	157	84 (m)	154
1,2,3-Trihydroxybenzene		133	—	158 (di)	198
1,4-Dihydroxybenzene		172	—	186 (di)	250

Using the Chemical Literature   **APPENDIX VIII**

The results of most chemical research are reported in scientific journals (primary sources) such as the *Journal of Organic Chemistry, Synthesis, Tetrahedron,* and the *Journal of the American Chemical Society.* Although familiarity with these sources is desirable, much of the information needed by undergraduate organic chemistry students can be obtained **(a)** from reference works that summarize or compile significant research results and **(b)** from books or articles on specific topics. Some of the important literature sources of organic chemistry are listed in the following Bibliography and referred to here by category and number.

Category A: General Reference Works.   The *CRC Handbook* (A1), *Lange's Handbook* (A2), and *The Merck Index* (A3) are condensed sources of information about specific organic compounds. The first two list physical properties and solubility data; *The Merck Index* also provides literature references to the isolation or synthesis of each compound, along with some hazard data and uses. The *CRC Atlas* (A4) includes spectral data (IR, NMR, UV, and MS) and references to published spectra (along with physical contents) for most of the compounds listed in A1. References A5 and A6 provide data and literature references for many natural compounds not covered in the preceding references. Because only the most common organic compounds are listed in A1–4, it may be necessary to consult a more comprehensive work for information about unusual or recently prepared compounds. The *Dictionary of Organic Compounds* (A7) is a useful multivolume set that includes derivatives and literature references as well as physical constants. However, it is not nearly as comprehensive as *Beilstein's Handbuch* (A8), which now covers the chemical literature up to about 1959. *Beilstein's Handbuch* is the most complete single source of information about the properties, preparation, and reactions of organic compounds, with references to literature sources for all data given. Unfortunately it is published only in German; but this is not a serious handicap to anyone with a rudimentary knowledge of German vocabulary and a good dictionary (see, for example, A19 or A20). Although there are several ways to find information in Beilstein, the simplest methods for the beginner are to locate a Beilstein reference in A1–A2 or to use the Formula Index (Volume 29) in the second supplement (*General-Formelregister, Zweites Erganzungswerk*) of Beilstein. For instance, the Beilstein reference for indigo is given in *Lange's Handbook* as XXIV-417, which means that the entry on indigo is in Volume 24, page 417 of the main series (*Hauptwerk,* **H**), which covers the literature through 1938. For more recent information on indigo, you must consult the supplements (the third and fourth are still in publication), which are keyed to the main series by the corresponding *Hauptwerk* page numbers printed at the top of each page. So to find information about indigo in any supplement (*Erganzungswerk*), you can locate the appropriate part of Volume 24 (each "volume" may contain a number of separate parts) in the supplement and then leaf through the pages having **H,** 417 printed at the top until you arrive at the entry for indigo. In using the formula index to locate indigo, you should write its formula ($C_{16}H_{10}N_2O_2$) with the elements after hydrogen listed in alphabetical order and look for the entry under that formula in Volume 29 of the 2nd supplement, which reads: Indigo **24,** 417, I 370, II 233. The first page number refers to the main series and the others to pages in Volume 24 of the first and second supplements. For further instructions on the use of *Beilstein's Handbuch,* see reference M5.

For information that has appeared in the more recent chemical literature, an up-to-date source such as *Chemical Abstracts* (A9) must be consulted. *Chemical Abstracts* summarizes articles appearing in all of the important scientific periodicals relating to chemistry, as well as listing patents, reviews, surveys, and new book announcements. The most foolproof way to find information about a given compound in CA is to use the collective indexes (appearing every five or ten years) that cover the desired time period. First locate the compound in the formula index to find the *index name* under which it appears in the corresponding subject index (the name may change from time to time). The subject index will give a brief entry and an abstract number for each article dealing with the compound. Abstract numbers are in the form **80:**12175e (from 1967 on), where the first number refers to the CA volume in which the numbered abstract appears. (The letter e is a computer check character and has no other significance.) Prior to 1967, the second number will refer to a column (there are two on each page) rather than an abstract number, and the position of the abstract in the column will be indicated by a superscript (1–9 or a–i), as in **51:**1311^g. The information you are looking for may appear in the abstract itself, or you may have to read the original article referenced in the abstract. A list of journals and their abbreviations appears in the *Chemical Abstracts Service Source Index*. To find abstracts of articles appearing after the most recent collective index, it will be necessary to consult the semiannual indexes appearing with each volume. Since 1972, the subject indexes have been divided into general subject and chemical substance indexes. Author indexes are also included; guides to the use of the indexes appear periodically.

References A10–12 compile selected properties that are useful for qualitative analysis and other purposes. References A13–17 are more-or-less comprehensive encyclopedias of chemical knowledge; of these, the Kirk-Othmer encyclopedia (A13) provides the most comprehensive and up-to-date coverage of chemical topics. *Thorpe's Dictionary* (A15) is somewhat outdated, but it is a good source of information from the earlier literature. The Hampel/Hawley *Glossary* (A18) gives concise definitions of chemical terms and the *Chemist's Companion* (A21) provides a wealth of useful data and practical information for the laboratory chemist.

Category B: Planning an Organic Synthesis.   Detailed procedures for preparing specific compounds appear in references B1–7. Of these, *Organic Syntheses* (B1) is the most reliable and comprehensive source of tested laboratory procedures. *Organic Reactions* (B2) deals with several different reaction types in each volume and includes a theoretical discussion and applications of each reaction along with some actual procedures. Shirley (B3) is a compilation of literature procedures for a number of organic compounds; Vogel (B4) includes a good description of laboratory techniques and a brief discussion of each synthetic method. Vogel also provides a large selection of classical organic preparations, most of which do not require extravagant equipment or unusual reaction conditions. *Houben-Weyl* (B8) is a monumental work that includes general laboratory techniques as well as synthetic methods. Being in German, it is not readily accessible to most undergraduates, but it provides a detailed practical discussion of each reaction type along with illustrative procedures. *Weygand/Hilgetag* (B9) and *Organicum* (B10) also provide practical descriptions of classical synthetic methods, whereas Sandler and Karo (B11) and Buehler and Pearson (B12) cover many modern synthetic reactions; all include some more-or-less detailed procedures. Theilheimer (B13) is a very comprehensive guide to synthetic procedures from the literature; each reference usually includes a brief procedural outline. House (B14) and Carruthers (B15) discuss a limited number of recent synthetic methods without giving specific procedures. References B16–19 serve primarily as guides to literature procedures; Wagner and Zook (B16) provides excellent coverage of classical preparations. (March (J8) is a good source of references to more recent synthetic literature.) *Reagents for Organic Synthesis* (B20) is an extremely useful guide to the preparation, properties, uses, and commercial availability of chemical reagents used in synthesizing organic compounds. Reference B21 tells how to purify many solvents used in organic synthesis and provides detailed information about their properties.

Category C: Laboratory Safety and Chemical Hazards.   The *MFA Guide* (C1) and *CRC Handbook of Laboratory Safety* (C4) include sections on preventing and dealing with fires and laboratory accidents as well as tables of hazard data on specific organic compounds. Muir (C2) and Sax (C3) provide detailed hazard information about many organic compounds, along with safety procedures. *Chemical Carcinogens* (C5) includes a chapter on carcinogenic hazards in teaching laboratories; it deals with a subject that should concern everyone who has frequent contact with laboratory chemicals. *The Merck Index* (A3) contains a useful section on first aid for poisoning and chemical burns, as well as a list of poison control centers in the U.S.

Category D: General Laboratory Operations.   Both the older *Technique of Organic Chemistry* (D1) series and the *Techniques of Chemistry* (D2) series are very comprehensive sources of information on organic laboratory techniques and experimental methods. Volume 1 (*Physical Methods of Organic Chemistry*) of the older series covers many of the classical laboratory techniques, some of which are left out of the corresponding volumes of D2. As mentioned above, *Vogel* (B4) also contains a good discussion of laboratory methods, and *Houben-Weyl* (B8) provides comprehensive coverage of laboratory methodology in German.

Category E: Separation and Purification Operations   These references, along with D1–4, provide more-or-less detailed discussions of the separation and purification operations used in this course (⟨OP–11⟩ through ⟨OP–27⟩). Some include extensive theoretical sections; others are geared more to practical applications. The *Handbook of Chromatography* (E11) and *The Chemist's Companion* (A21) provide much useful information and data on chromatographic methods.

Category F: Instrumental Analysis and Spectra.   Ewing (F1) and Willard-Merritt-Dean (F2) provide broad coverage of most important instrumental methods, including spectrometry. References F4–8 cover general applications of spectrometry in organic chemistry, whereas F9–20 each deal with a single spectrometric method. References F22–26 and A4 are sources of published spectra and spectral data. Of these, the *Sadtler Spectra* (F26) is by far the most comprehensive, but the Aldrich and Varian collections are more useful for observing and comparing spectral features among compounds of a given class.

Category G: Qualitative Analysis and Structure Determination.   References G1–3 are good general textbooks for organic qualitative analysis. Feigl (G4) is an occasionally useful source for organic qualitative analysis; it describes chemical tests for some specific compounds and compound types. References G5–7 rely heavily on instrumental methods for structure determination of unknown organic compounds.

Category H: Reaction Mechanisms and Theoretical Topics.   In addition to books listed under this category, *Organic Reactions* (B2) is a good source of reaction mechanisms. References H1–5 cover general reaction mechanisms, H6–9 deal with some specific reaction types, and H10–12 include other theoretical topics in addition to mechanisms.

Category J: Comprehensive and Advanced Textbooks.   The first two references are multivolume sets with comprehensive coverage of organic chemistry and biochemistry respectively. Karrer (J5) and, to a lesser extent, Noller (J6) and Fieser (J3–4), are now somewhat dated. They are, however, good sources of information from the earlier literature; Karrer and Noller provide particularly good coverage of natural and commercial products. References J7–9 are modern, up-to-date texts; March (J8) in particular is an excellent source of references to synthetic procedures.

Category K: Reports of Current Research.   These series summarize recent advances in chemical research or, in some cases, provide review articles on specific topics. Regular periodicals that review topics in organic chemistry include *Chemical Reviews, Angewandte Chemie* (English Edition), and *Quarterly Reviews*. Reference M1 lists sources of reviews on various topics.

Category L: Sources on Selected Topics.   Some of these sources will be useful as references for the Library Topics included with each experiment; others should make interesting reading for students who wish to pursue a given topic further. For additional help in finding information on topics in organic chemistry, refer to one or more of the references in Category M: Guides to the Chemical Literature.

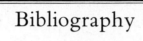

Bibliography

A. General Reference Works

1. Weast, R. C., ed., *CRC Handbook of Chemistry and Physics,* 62nd ed. (and other editions). West Palm Beach, Fla.: CRC Press, 1981.

2. Lange, N. A., *Lange's Handbook of Chemistry,* 11th ed., ed. by J. A. Dean. New York: McGraw-Hill, 1973.

3. Windholz, M., ed., *The Merck Index,* 9th ed., Rahway, N.J.: Merck and Co., 1976.

4. Grasselli, J. G. and Ritchey, W. R., eds., *CRC Atlas of Spectral Data and Physical Constants for Organic Compounds,* 2nd ed. Cleveland: CRC press, 1975.

5. Devon, T. K. and Scott, A. I., *Handbook of Naturally Occurring Compounds.* New York: Academic Press, 1972–75.

6. Glasby, J. S., *Encylcopedia of the Alkaloids.* New York: Plenum Press, 1975.

7. Pollock, J. R. A. and Stevens, R., eds., *Dictionary of Organic Compounds,* 4th ed. New York: Oxford University Press, 1965.

8. *Beilstein's Handbuch der Organischen Chemie.* Berlin: Springer, 1918 to date.

9. *Chemical Abstracts.* Easton, Pa.: American Chemical Society, 1907 to date.

10. Rappoport, Z., ed., *CRC Handbook of Tables for Organic Compound Identification,* 3rd ed. Cleveland: Chemical Rubber Co., 1967.

11. Utermark, W. and Schicke, W., *Melting Point Tables of Organic Compounds,* 2nd ed. New York: Interscience Publishers, 1963.

12. Stephen, H. and Stephen, T., eds., *Solubilities of Inorganic and Organic Compounds.* New York: Macmillan, 1963–64.

13. *Kirk-Othmer Encyclopedia of Chemical Technology,* 3rd ed., ed. by H. F. Marks, et al. New York: Wiley, 1978 to date.

14. *McGraw-Hill Encyclopedia of Science and Technology,* 4th ed. New York: McGraw-Hill, 1977.

15. Thorpe, J. F. and Whitely, M. A., *Thorpe's Dictionary of Applied Chemistry,* 4th ed. London: Longmans, Green and Co., 1937–56.

16. Hampel, C. A. and Hawley, G. G., eds., *The Encyclopedia of Chemistry,* 3rd ed. New York: Van Nostrand Reinhold Co., 1973.

17. *International Encyclopedia of Chemical Science.* Princeton, N.J.: Van Nostrand, 1964.

18. Hampel, C. A. and Hawley, G. G., *Glossary of Chemical Terms.* New York: Van Nostrand Reinhold, 1976.

19. Neville, H. H., Johnston, N. C., and Boyd, G. V., *A New German/English Dictionary for Chemists.* London: Blackie, 1964.

20. Patterson, A. M., *A German-English Dictionary for Chemists,* 3rd ed. New York: Wiley, 1950.

21. Gordon, A. J. and Ford, R. A., *The Chemist's Companion*. New York: Wiley-Interscience, 1972.

B. Planning an Organic Synthesis

1. *Organic Syntheses,* Collective Volumes 1–5 and annual volumes. New York: Wiley, 1932 to date.

2. Adams, R. et al., eds., *Organic Reactions*. New York: Wiley, 1942 to date.

3. Shirley, D. A., *Preparation of Organic Intermediates*. New York: Wiley, 1951.

4. Vogel, A. I., *Vogel's Textbook of Practical Organic Chemistry,* 4th ed., rev. by B. Furniss et al. London, New York: Longman, 1978.

5. R. Ikan, *Natural Products: A Laboratory Guide*. Jerusalem: Israel Universities Press, 1969.

6. Sandler, S. R. and Karo, W., *Polymer Syntheses*. New York: Academic Press, 1974, 1977.

7. Sorensen, W. R. and Campbell, T. W., *Preparative Methods of Polymer Chemistry,* 2nd ed. New York: Interscience Publishers, 1968.

8. Miller, E., ed., *Methoden der Organischen Chemie (Houben-Weyl),* 4th ed. Stuttgart: Georg Thieme, 1952–78.

9. Weygand, C., *Weygand/Hilgetag Preparative Organic Chemistry,* ed. by G. Hilgetag and A. Martini. New York: Wiley-Interscience, 1972.

10. Becker, H. et al., *Organicum: Practical Handbook of Organic Chemistry,* transl. by B. J. Hazzard. Reading, Mass.: Addison-Wesley Publishing Co., 1973.

11. Sandler, S. R. and Karo, W., *Organic Functional Group Preparations*. New York: Academic Press, 1968–72.

12. Buehler, C. A., and Pearson, D. E., *Survey of Organic Syntheses*. New York: Wiley-Interscience, 1970–77.

13. Theilheimer, W., ed., *Synthetic Methods of Organic Chemistry*. Basel: S. Karger, 1946 to date.

14. House, H. O., *Modern Synthetic Reactions,* 2nd ed. Menlow Park, Calif.: W. A. Benjamin, 1971.

15. Carruthers, W., *Some Modern Methods of Organic Synthesis,* 2nd ed. Cambridge: Cambridge University Press, 1978.

16. Wagner, R. B. and Zook, H. D., *Synthetic Organic Chemistry*. New York: Wiley, 1953.

17. Migridichian, V., *Organic Synthesis*. New York: Reinhold, 1957.

18. Sugasawa, S. and Nakai, S., *Reaction Index of Organic Syntheses,* rev. ed. New York: Wiley, 1967.

19. Harrison, I. T. and Harrison, S., *Compendium of Organic Synthetic Methods*. New York: Wiley-Interscience, 1971–74.

20. Fieser, L. F. and Fieser, M., *Reagents for Organic Synthesis,* Vols. 1–6. New York: Wiley-Interscience, 1967–77.

21. Riddick, J. A. and Bunger, W. M., *Organic Solvents: Physical Properties and Methods of Purification* (*Techniques of Chemistry,* Vol. II). New York: Wiley-Interscience, 1970.

C. Laboratory Safety and Chemical Hazards

1. Manufacturing Chemists Association, *Guide for Safety in the Chemical Laboratory,* 2nd ed. New York: Van Nostrand Reinhold, 1972.

2. Muir, G. D., ed., *Hazards in the Chemical Laboratory,* 2nd ed. London: The Chemical Society, 1977.

3. Sax, N. I., *Dangerous Properties of Industrial Materials,* 4th ed. New York: Van Nostrand Reinhold, 1975.

4. Steere, N. V., ed., *CRC Handbook of Laboratory Safety,* 2nd ed. Cleveland: Chemical Rubber Company, 1971.

5. Searle, C. E., ed., *Chemical Carcinogens* (ACS Monograph No. 173). Washington: American Chemical Society, 1976.

D. General Laboratory Techniques

1. Weissberger, A., ed., *Technique of Organic Chemistry.* New York: Wiley-Interscience.

2. Weissberger, A., ed., *Techniques of Chemistry.* New York: Wiley-Interscience, 1971–78.

3. Marmor, S., *Laboratory Guide for Organic Chemistry.* Boston: D.C. Heath and Co., 1964.

4. Wiberg, K. B., *Laboratory Technique in Organic Chemistry.* New York: McGraw-Hill, 1960.

E. Separation and Purification Operations

1. Perry, E. S. and Weissberger, A., eds., *Separation and Purification,* 3rd ed. (*Techniques of Chemistry,* Vol. XII). New York: Wiley-Interscience, 1978.

2. Weissberger, A., ed., *Distillation* (*Technique of Organic Chemistry,* Vol. IV), 2nd ed. New York: Interscience Publishers, 1965.

3. Berg, E. W., *Physical and Chemical Methods of Separation.* New York: McGraw-Hill, 1963.

4. Dean, J. A., *Chemical Separation Methods.* New York: Van Nostrand Reinhold Co., 1969.

5. Cassidy, H. G., *Fundamentals of Chromatography* (*Technique of Organic Chemistry,* Vol. X). New York: Interscience Publishers, 1957.

6. Bobbitt, J. M. and Schwarting, A. E., *Introduction to Chromatography.* New York: Van Nostrand Reinhold, 1968.

7. Stock, R. and Rice, C. B. F., *Chromatographic Methods,* 3rd ed., London: Chapman and Hall, 1974.

8. Heftmann, E., ed., *Chromatography: A Laboratory Handbook of Chromatographic and Electrophoretic Methods,* 3rd ed. New York: Van Nostrand Reinhold, 1975.

9. Stahl, E., ed., *Thin Layer Chromatography: A Laboratory Handbook.* Berlin: Springer, 1969.

10. Touchstone, J. F. and Dobbins, M. F., *Practice in Thin Layer Chromatography.* New York: Wiley-Interscience, 1978.

11. Zweig, G., and Sherma, J., eds., *CRC Handbook of Chromatography.* Cleveland: CRC Press, 1972.

F. Instrumental Analysis and Spectra

1. Ewing, G. W., *Instrumental Methods of Chemical Analysis,* 4th ed. New York: McGraw-Hill, 1975.

2. Willard, H. H., Merritt, L. L., and Dean, J. A., *Instrumental Methods of Analysis,* 5th ed. New York: Van Nostrand Reinhold Co., 1974.

3. Ambrose, D., *Gas Chromatography,* 2nd ed. New York: Van Nostrand Reinhold Co., 1971.

4. Brittain, E. F. H., George, W. O., and Wells, C. H. J., *Introduction to Molecular Spectroscopy: Theory and Experiment.* New York: Academic Press, 1970.

5. Dyer, J. R., *Applications of Absorption Spectroscopy of Organic Compounds.* Englewood Cliffs, N.J.: Prentice-Hall, 1965.

6. Kemp, W., *Organic Spectroscopy.* New York: Macmillan, 1975.

7. Parikh, V. M., *Absorption Spectroscopy of Organic Molecules.* Reading, Mass.: Addison-Wesley Publishing Co., 1974.

8. Silverstein, R. M., Bassler, G. C., and Morrill, T. C., *Spectrometric Identification of Organic Compounds,* 3rd ed. New York: Wiley, 1974.

9. Colthup, N. B., Daly, L. H., and Wiberley, S. E., *Introduction to Infrared and Raman Spectroscopy,* 2nd ed. New York: Academic Press, 1975.

10. Nakanishi, K. and Solomon, P. H., *Infrared Absorption Spectroscopy,* 2nd ed. San Francisco: Holden-Day, 1977.

11. Szymanski, H. A., *Interpreted Infrared Spectra.* New York: Plenum Press, 1964–67.

12. Ault, A. and Dudek, G. O., *NMR: An Introduction to Proton Nuclear Magnetic Resonance Spectroscopy.* San Francisco: Holden-Day, 1976.

13. Bovey, F. A., *Nuclear Magnetic Resonance Spectroscopy.* New York: Academic Press, 1969.

14. Haws, E. J., Hill, R. R., and Mowthorpe, D. J., *The Interpretation of Proton Magnetic Resonance Spectra: A Programmed Introduction.* New York: Heyden, 1973.

15. Jackman, L. M. and Sternhell, S., *Applications of Nuclear Magnetic Resonance Spectrometry in Organic Chemistry,* 2nd ed. Oxford: Pergamon Press, 1969.

16. McFarlane, W. and White, R. F. M., *Techniques of High Resolution Nuclear Magnetic Resonance Spectroscopy.* London: Butterworths, 1972.

17. Jaffé, H. H. and Orchin, M., *Theory and Applications of Ultraviolet Spectroscopy.* New York: Wiley, 1962.

18. Rao, C. N. R., *Ultra-Violet Spectroscopy: Chemical Applications,* 2nd ed. New York: Plenum Press, 1967.

19. Budzikiewicz, H., Djerassi, C., and Williams, D. H., *Mass Spectrometry of Organic Compounds.* San Francisco: Holden-Day, 1967.

20. Hamming, M. C. and Foster, N. G., *Interpretation of Mass Spectra of Organic Compounds.* New York: Academic Press, 1972.

21. Beynon, J. H. and Williams, A. E., *Mass and Abundance Tables for Use in Mass Spectrometry.* Amsterdam: Elsevier Publishing Co., 1963.

22. Pouchert, C. J., *The Aldrich Library of Infrared Spectra,* 2nd ed. Milwaukee, Wisc.: Aldrich Chemical Co., 1975.

23. Pouchert, C. J., and Campbell, J. R., *The Aldrich Library of NMR Spectra,* Vols. 1–11. Milwaukee, Wisc.: Aldrich Chemical Co., 1974.

24. *High Resolution NMR Spectra Catalog,* Vols. 1–2. Palo Alto, Calif.: Varian Associates, 1962–63.

25. Robinson, J. W., ed., *CRC Handbook of Spectroscopy.* Cleveland: CRC Press, 1974.

26. *Sadtler Standard Spectra,* (Collections of Infrared, Ultraviolet and NMR Spectra). Philadelphia: Sadtler Research Laboratories, Inc.

G. Qualitative Analysis and Structure Determination

1. Cheronis, N. D., Entrikin, J. B., and Hodnett, E. M., *Semimicro Organic Qualitative Analysis,* 3rd ed. New York: Interscience Publishers, 1965.

2. Pasto, D. J. and Johnson, C. R., *Organic Structure Determination.* Englewood Cliffs, N.J.: Prentice-Hall, 1969.

3. Shriner, R. L., Fuson, R. C., Curtin, D. Y., and Morrill, T. C., *The Systematic Identification of Organic Compounds: A Laboratory Manual,* 6th ed. New York: Wiley, 1980.

4. Feigl, F. and Anger, V., *Spot Tests in Organic Analysis,* 7th ed. Amsterdam: Elsevier Publishing Co., 1966.

5. Bentley, K. W. and Kirby, G. W., *Elucidation of Organic Structures by Physical and Chemical Methods (Techniques of Chemistry,* Vol. IV). New York: Wiley-Interscience, 1972.

6. Lambert, J. B. et al., *Organic Structural Analysis*. New York: Macmillan, 1976.

7. Nachod, F. C. et al., eds., *Determination of Organic Structures by Physical Methods*. New York: Academic Press, 1955–76.

H. Reaction Mechanisms and Theoretical Topics

1. Badea, F., *Reaction Mechanisms in Organic Chemistry*. Turnbridge Wells, Eng.: Abacus Press, 1977.

2. Breslow, R., *Organic Reaction Mechanisms: An Introduction,* 2nd ed. New York: W. A. Benjamin, 1969.

3. Butler, A. R., Perkins, M. J., et al., eds., *Organic Reaction Mechanisms*. New York: Interscience Publishers, 1965 to date.

4. Harris, J. M. and Wamser, C. C., *Fundamentals of Organic Reaction Mechanisms*. New York: Wiley, 1976.

5. Sykes, P., *A Guidebook to Mechanism in Organic Chemistry,* 4th ed. London: Longman, 1975.

6. Bunton, C. A., *Nucleophilic Substitution at a Saturated Carbon Atom*. Amsterdam: Elsevier Publishing Co., 1963.

7. DeMayo, P., ed., *Molecular Rearrangements*. New York: Interscience Publishers, 1963–64.

8. Miller, J., *Aromatic Nucleophilic Substitution*. Amsterdam: Elsevier Publishing Co., 1968.

9. Streitwieser, A., *Solvolytic Displacement Reactions*. New York: McGraw-Hill, 1962.

10. Hammet, L. P., *Physical Organic Chemistry: Reaction Rates, Equilibria, and Mechanisms,* 2nd ed. New York: McGraw-Hill, 1970.

11. Ingold, C. K., *Structure and Mechanism in Organic Chemistry,* 2nd ed. Ithaca, N.Y.: Cornell University Press, 1969.

12. Lowry, T. H. and Richardson, K. S., *Mechanism and Theory in Organic Chemistry*. New York: Harper & Row, 1976.

13. Hine, J. S., *Structural Effects on Equilibria in Organic Chemistry*. New York: Wiley, 1975.

14. Laidler, K. J., *Chemical Kinetics,* 2nd ed. New York: McGraw-Hill, 1965.

15. Woodward, R. B. and Hoffmann, R., *The Conservation of Orbital Symmetry*. New York: Academic Press, 1970.

J. Comprehensive and Advanced Textbooks

1. Rodd, E. H., *Rodd's Chemistry of Carbon Compounds,* ed. by S. Coffey. Amsterdam: Elsevier Publishing Co., 1964–78.

2. Florkin, M. and Stotz, E. H., eds., *Comprehensive Biochemistry*. Amsterdam: Elsevier Publishing Co., 1962–77.

3. Fieser, L. F. and Fieser, M., *Advanced Organic Chemistry*. New York: Reinhold Publishing Corp., 1961.

4. Fieser, L. F. and Fieser, M., *Topics in Organic Chemistry*. New York: Reinhold Publishing Corp., 1963.

5. Karrer, P., *Organic Chemistry,* 4th English ed. New York: Elsevier Publishing Co., 1950.

6. Noller, C. R. *Chemistry of Organic Compounds,* 3rd ed. Philadelphia: Saunders, 1965.

7. Carey, F. A. and Sundberg, R. J., *Advanced Organic Chemistry*. New York: Plenum Press, 1977.

8. March, J., *Advanced Organic Chemistry: Reactions, Mechanisms and Structure,* 2nd ed. New York: McGraw-Hill, 1977.

9. le Noble, W. J., *Highlights of Organic Chemistry: An Advanced Textbook*. New York: Marcel Dekker, 1974.

K. Reports of Current Research

1. *Advances in Organic Chemistry: Methods and Results*. New York: Interscience Publishers, 1960 to date.

2. *Annual Reports in Organic Synthesis*. New York: Academic Press, 1970 to date.

3. Chemical Society (London), *Annual Reports on the Progress of Chemistry, Section B: Organic Chemistry*. London: The Chemical Society, 1904 to date.

4. *International Review of Science: Organic Chemistry,* Series 1 and 2. Boston: Butterworths, 1973–75.

5. *Progress in Organic Chemistry*. New York: Academic Press, 1952 to date.

6. *Progress in Physical Organic Chemistry*. New York: Interscience Publishers, 1963 to date.

L. Sources on Selected Topics

1. Allen, R. L. M., *Colour Chemistry*. New York: Appleton-Century-Crofts, 1971.

2. AMA Department of Drugs, *AMA Drug Evaluations,* 2nd ed. Acton, Mass.: Publishing Sciences Group, 1973.

3. Amundsen, L. H., "Sulfanilamide and Related Chemotherapeutic Agents." *Journal of Chemical Education, 19* (1942), p. 167.

4. Baker, A. A., *Unsaturation in Organic Chemistry*. Boston: Houghton Mifflin, 1968.

5. Berkhoff, C. E., "Insect Hormones and Insect Control." *Journal of Chemical Education 48* (1971), p. 577.

6. Bethell, D. and Gold, V., *Carbonium Ions: An Introduction*. New York: Academic Press, 1967.

7. Billmeyer, F. W., *Textbook of Polymer Sciences,* 2nd ed., New York: Wiley-Interscience, 1971.

8. Blackburn, S., *Protein Sequence Determination: Methods and Techniques.* New York: Marcel Dekker, 1970.

9. Bragg, R. W. et al., "Sweet Organic Chemistry." *Journal of Chemical Education 55* (1978), p. 281.

10. Brecher, E. M. and the Editors of Consumer Reports, *Licit and Illicit Drugs.* Boston: Little, Brown, 1972.

11. Burger, A., ed., *Medicinal Chemistry,* 3rd ed. New York: Wiley-Interscience, 1970.

12. Cahn, R. S., *An Introduction to Chemical Nomenclature,* 4th ed. New York: Wiley, 1974.

13. Calvin, M., "Solar Energy by Photosynthesis." *Science 184* (1974), p. 375.

14. Carraher, C. E., Jr., "Synthesis of Caprolactam and Nylon 6." *Journal of Chemical Education 55* (1978), p. 51.

15. Coates, G. E., Green, M. L. H., and Wade, K., *Organometallic Compounds,* 3rd ed. London: Methuen, 1967–68.

16. Cook, A. G., ed., *Enamines: Synthesis, Structure and Reactions.* New York: Marcel Dekker, 1969.

17. Cremlyn, R. J. W. and Still, R. H., *Named and Miscellaneous Reactions in Practical Organic Chemistry.* London: Heinemann, 1967.

18. Eliel, E. L., *Stereochemistry of Carbon Compounds.* New York: McGraw-Hill, 1962.

19. Eliel, E. L., "Recent Advances in Stereochemical Nomenclature." *Journal of Chemical Education 48* (1971), p. 163.

20. Elmore, D. T., *Peptides and Proteins.* London: Cambridge University Press, 1968.

21. Farber, E., ed., *Great Chemists.* New York: Interscience Publishers, 1961.

22. Fieser, L. F. and Fieser, M., *Steroids.* New York: Reinhold Publishing Corp., 1959.

23. Fletcher, J. H., Dermer, O. C., and Fox, R. B., eds., *Nomenclature of Organic Compounds: Principles and Practice (Advances in Chemistry Series,* Vol. 126). Washington: American Chemical Society, 1974.

24. Forrester, A. R., Hay, J. M., and Thomson, R. H., *Organic Chemistry of Stable Free Radicals.* New York: Academic Press, 1968.

25. Fort, R. G., *Adamantane: The Chemistry of Diamond Molecules.* New York: Marcel Dekker, 1976.

26. Foye, W. O., *Principals of Medicinal Chemistry.* Philadelphia: Lea and Febiger, 1975.

27. Garratt, P. J., *Aromaticity.* New York: McGraw-Hill, 1971.

28. Geissman, T. A. and Grout, D. H. G., *Organic Chemistry of Secondary Plant Metabolism*. San Francisco: Freeman, Cooper, 1969.

29. Gilchrist, T. L. and Rees, C. W., *Carbenes, Nitrenes and Arynes*. New York: Appleton–Century–Crofts, 1969.

30. Ginsburg, D., ed., *Non-Benzenoid Aromatic Compounds*. New York: Interscience Publishers, 1959.

31. Gokel, G. W. and Weber, W. P., "Phase Transfer Catalysis." *Journal of Chemical Education 55* (1978), pp. 350, 429.

32. Goodwin, T. W., ed., *Chemistry and Biochemistry of Plant Pigments,* 2nd ed. New York: Academic Press, 1976.

33. Grant, N. and Naves, R. G., "Perfumes and the Art of Perfumery." *Journal of Chemical Education 49* (1972), p. 526.

34. Grieve, M., *A Modern Herbal*. New York: Dover Publications, 1971.

35. Guenther, E., *The Essential Oils*. New York: D. Van Nostrand Co., 1948–52.

36. Guild, W., "Theory of Sweet Taste." *Journal of Chemical Education 49* (1972), p. 171.

37. Hancock, J. E. H., "An Introduction to the Literature of Organic Chemistry." *Journal of Chemical Education 45* (1968), pp. 193, 260, 336.

38. Hansch, C., "Drug and Research and the Luck of the Draw." *Journal of Chemical Education 51* (1974), p. 360.

39. Hendrickson, J. B., *The Molecules of Nature*. New York: W. A. Benjamin, 1965.

40. Hubbard, R. and Kropf, A., "Molecular Isomers in Vision." *Scientific American 216* (June 1967), p. 64.

41. Jacobson, M. and Berzoa, M., "Insect Attractants." *Scientific American 211* (Aug. 1964), p. 20.

42. Jones, M., Jr., "Carbenes." *Scientific American 234* (Feb. 1976), p. 101.

43. Juster, N. J., "Color and Chemical Constitution." *Journal of Chemical Education 39* (1962), p. 596.

44. Kauffman, G. B., "Isoniazid—Destroyer of the White Plague." *Journal of Chemical Education 55* (1978), p. 448.

45. Kauffman, G. B., "Pittacal—The First Synthetic Dyestuff." *Journal of Chemical Education 54* (1977), p. 753.

46. Kharasch, M. S. and Reinmuth, O., *Grignard Reactions of Nonmetallic Substances*. New York: Prentice-Hall, 1954.

47. Kirmse, W., *Carbene Chemistry,* 2nd ed. New York: Academic Press, 1971.

48. Krauch, H. and Kunz, W., *Organic Name Reactions*. New York: John Wiley & Sons, 1964.

49. Lednicer, D. and Mitscher, L. A., *Organic Chemistry of Drug Synthesis.* New York: Wiley-Interscience, 1977.

50. McCullough, T., "Furfural—Ubiquitous Natural Product." *Journal of Chemical Education 49* (1972), p. 836.

51. Mead, J. F. and Fulco, A. J., *The Unsaturated and Polyunsaturated Fatty Acids in Health and Disease.* Springfield, Ill.: Thomas, 1976.

52. *Modern Drug Encyclopedia and Therapeutic Index: A Compendium,* 14th ed. New York: Yorke Medical Books, 1977.

53. Moncrieff, R. W., *The Chemical Senses,* 2nd ed. London: L. Hill, 1951.

54. Mosher, M. W. and Ansell, J. M., "Preparation and Color of Azo-Dyes." *Journal of Chemical Education 52* (1975), p. 195.

55. Nakanishi, K., ed., *Natural Products Chemistry.* New York: Academic Press, 1974–75.

56. Olah, G. A., *Carbocations and Electrophilic Reactions.* New York: Wiley, 1974.

57. Olah, G. A., *Friedel-Crafts Chemistry.* New York: Wiley-Interscience, 1973.

58. Pryor, W. A., *Free Radicals.* New York: McGraw-Hill, 1966.

59. Riegel, E. R., *Handbook of Industrial Chemistry,* 7th ed., ed. by J. A. Kent. New York: Van Nostrand Reinhold, 1974.

60. Roderick, W. R., "Current Ideas on the Chemical Basis of Olfaction." *Journal of Chemical Education 43* (1966), p. 510.

61. Sarkanen, K. V. and Ludwig, C. H., eds. *Lignins: Occurrence, Formation, Structure and Reactions.* New York: Wiley-Interscience, 1971.

62. Scientific American, *Bio-Organic Chemistry.* San Francisco: W. H. Freeman, 1968.

63. Scientific American, *The Molecular Basis of Life: An Introduction to Molecular Biology.* San Francisco: W. H. Freeman, 1968.

64. Shallenberger, R. S. and Birch, G. G., *Sugar Chemistry.* Westport, Conn.: Avi Pub. Co., 1975.

65. Shreve, R. N. and Brink, J. A., Jr., *Chemical Process Industries,* 4th ed. New York: McGraw-Hill, 1977.

66. Sidgwick, N. V., *The Organic Chemistry of Nitrogen,* 3rd ed., rev. by I. T. Millar and H. D. Springall. Oxford: Clarendon Press, 1966.

67. Simonsen, J. L. and Owen, L. N., *The Terpenes,* 2nd ed. Cambridge: University Press, 1947–57.

68. Starks, C. M. and Liotta, C., *Phase Transfer Catalysis: Principles and Techniques.* New York: Academic Press, 1978.

69. Stevens, M. P., *Polymer Chemistry: An Introduction.* Reading, Mass.: Addison-Wesley Publishing Co., 1975.

70. Sundberg, R. J., *The Chemistry of Indoles.* New York: Academic Press, 1970.

71. Swan, G. A., *An Introduction to the Alkaloids*. New York: Wiley, 1967.

72. Treptow, R. S., "Determination of Alcohol in Breath for Law Enforcement." *Journal of Chemical Education 51* (1974), p. 651.

73. Turro, N. J., *Molecular Photochemistry*. New York: W. A. Benjamin, 1967.

74. Venkataraman, K., ed., *The Chemistry of Synthetic Dyes*. New York: Academic Press, 1952–74.

75. Wallis, H. J., *Forensic Science,* 2nd ed., New York: Praeger, 1974.

76. Wassermann, A., *Diels-Alder Reactions: Organic Background and Physicochemical Aspects*. Amsterdam: Elsevier Publishing Co., 1965.

77. Weber, W. P. and Gokel, G. W., *Phase Transfer Catalysis in Organic Synthesis*. Berlin: Springer-Verlag, 1977.

78. West, J. A., "The Chemistry of Enamines." *Journal of Chemical Education 40* (1963), p. 194.

79. Wilson, E. O., "Pheromones." *Scientific American 208* (May 1963), p. 100.

M. Guides to the Chemical Literature

1. Lewis, D. A., ed., *Index of Reviews in Organic Chemistry: Cumulative Index 1971* (and annual supplements). London: The Chemical Society, 1971 to date.

2. Maizell, R. E., *How to Find Chemical Information*. New York: Wiley-Interscience, 1979.

3. Mellon, M. G., *Chemical Publications: Their Use and Nature,* 4th ed. New York: McGraw-Hill, 1965.

4. Mellon, M. G., *Searching the Chemical Literature*. Washington: American Chemical Society, 1964.

5. Weissbach, O., *The Beilstein Guide: A Manual for the Use of Beilstein's Handbuch der Organischen Chemie*. Berlin: Springer-Verlag, 1976.

6. Woodburn, H. M., *Using the Chemical Literature: A Practical Guide*. New York: Marcel Dekker, 1974.

NMR SPECTRUM-STRUCTURE CORRELATION CHART

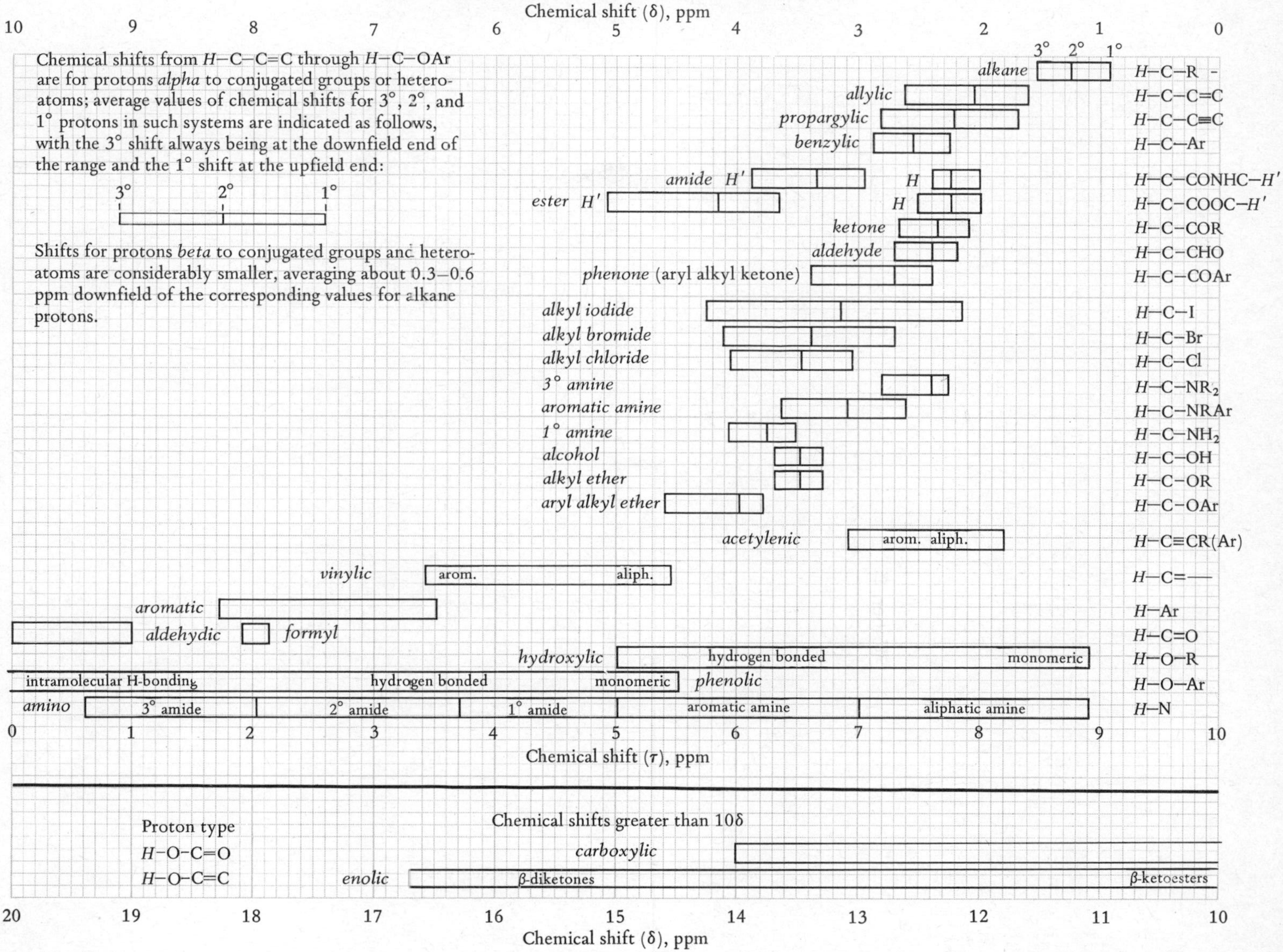

Chemical shift (δ), ppm

Chemical shifts from $H-C-C\equiv C$ through $H-C-OAr$ are for protons *alpha* to conjugated groups or heteroatoms; average values of chemical shifts for 3°, 2°, and 1° protons in such systems are indicated as follows, with the 3° shift always being at the downfield end of the range and the 1° shift at the upfield end:

Shifts for protons *beta* to conjugated groups and heteroatoms are considerably smaller, averaging about 0.3–0.6 ppm downfield of the corresponding values for alkane protons.

H–C–R –
H–C–C=C
H–C–C≡C
H–C–Ar

H–C–CONHC–H'
H–C–COOC–H'
H–C–COR
H–C–CHO
H–C–COAr

H–C–I
H–C–Br
H–C–Cl
H–C–NR₂
H–C–NRAr
H–C–NH₂
H–C–OH
H–C–OR
H–C–OAr

H–C=CR(Ar)
H–C≡—
H–Ar
H–C=O
H–O–R
H–O–Ar
H–N

alkane 1° 2° 3°

allylic
propargylic
benzylic
amide H' H H
ester H'
ketone
aldehyde
phenone (aryl alkyl ketone)

alkyl iodide
alkyl bromide
alkyl chloride
3° amine
aromatic amine
1° amine
alcohol
alkyl ether
aryl alkyl ether

acetylenic
arom.   aliph.
aliph.
arom.
hydroxylic monomeric
hydrogen bonded
phenolic monomeric hydrogen bonded
aromatic amine
aliphatic amine

vinylic
formyl
aromatic
aldehydic

intramolecular H-bonding
amino
1° amide
2° amide
3° amide

Chemical shift (τ), ppm

Chemical shifts greater than 10δ

carboxylic
enolic
β-diketones
β-ketoesters

Proton type		
H–O–C=O		
H–O–C=C		

Chemical shift (δ), ppm

INDEX

Periodic Table of the Elements

METALS

NONMETALS

TRANSITION METALS

PERIODS	IA	IIA	IIIB	IVB	VB	VIB	VIIB	VIII	VIII	VIII	IB	IIB	IIIA	IVA	VA	VIA	VIIA	O
1	1.0079 H 1																1.0079 H 1	4.00260 He 2
2	6.94 Li 3	9.01218 Be 4											10.81 B 5	12.011 C 6	14.0067 N 7	15.9994 O 8	18.9984 F 9	20.179 Ne 10
3	22.9898 Na 11	24.305 Mg 12											26.9815 Al 13	28.086 Si 14	30.9738 P 15	32.06 S 16	35.453 Cl 17	39.948 Ar 18
4	39.098 K 19	40.08 Ca 20	44.9559 Sc 21	47.90 Ti 22	50.9414 V 23	51.996 Cr 24	54.9380 Mn 25	55.847 Fe 26	58.9332 Co 27	58.71 Ni 28	63.546 Cu 29	65.38 Zn 30	69.72 Ga 31	72.59 Ge 32	74.9216 As 33	78.96 Se 34	79.904 Br 35	83.80 Kr 36
5	85.4678 Rb 37	87.62 Sr 38	88.9059 Y 39	91.22 Zr 40	92.9064 Nb 41	95.94 Mo 42	98.9062 Tc 43	101.07 Ru 44	102.9055 Rh 45	106.4 Pd 46	107.868 Ag 47	112.40 Cd 48	114.82 In 49	118.69 Sn 50	121.75 Sb 51	127.60 Te 52	126.9046 I 53	131.30 Xe 54
6	132.9054 Cs 55	137.34 Ba 56	57–71 *	178.49 Hf 72	180.9479 Ta 73	183.85 W 74	186.2 Re 75	190.2 Os 76	192.22 Ir 77	195.09 Pt 78	196.9665 Au 79	200.59 Hg 80	204.37 Tl 81	207.2 Pb 82	208.9804 Bi 83	(210) Po 84	(210) At 85	(222) Rn 86
7	(223) Fr 87	(226.0254) Ra 88	89–103 †	(260) Ku 104	(260) Ha 105													

* LANTHANIDE SERIES

138.9055 La 57	140.12 Ce 58	140.9077 Pr 59	144.24 Nd 60	(145) Pm 61	150.4 Sm 62	151.96 Eu 63	157.25 Gd 64	158.9254 Tb 65	162.50 Dy 66	164.9304 Ho 67	167.26 Er 68	168.9342 Tm 69	173.04 Yb 70	174.97 Lu 71

† ACTINIDE SERIES

(227) Ac 89	232.0381 Th 90	231.0359 Pa 91	238.029 U 92	237.0482 Np 93	(242) Pu 94	(243) Am 95	(245) Cm 96	(245) Bk 97	(248) Cf 98	(253) Es 99	(253) Fm 100	(256) Md 101	(253) No 102	(257) Lr 103